Rapid Prototyping and Engineering Applications

Rapid Prototyping and Engineering Applications

A Toolbox for Prototype Development

Second Edition

Fuewen Frank Liou

CRC Press
Taylor & Francis Group
Boca Raton London New York

CRC Press is an imprint of the
Taylor & Francis Group, an **informa** business

CRC Press
Taylor & Francis Group
6000 Broken Sound Parkway NW, Suite 300
Boca Raton, FL 33487-2742

Printed on acid-free paper

International Standard Book Number-13: 978-1-4987-9892-1 (Hardback)

Library of Congress Cataloging-in-Publication Data

Names: Liou, Frank W., author.
Title: Rapid prototyping and engineering applications : a toolbox for prototype development / authored by Fuewen Frank Liou.
Description: Second edition. | Boca Raton : Taylor & Francis, CRC Press, 2019. | Includes bibliographical references and index.
Identifiers: LCCN 2018045350| ISBN 9781498798921 (hardback : alk. paper) | ISBN 9780429029721 (e-book : alk. paper)
Subjects: LCSH: Rapid prototyping. | Prototypes, Engineering—Data processing.
Classification: LCC TS171.8 .L56 2019 | DDC 620/.0042—dc23
LC record available at https://lccn.loc.gov/2018045350

Visit the Taylor & Francis Web site at
http://www.taylorandfrancis.com

and the CRC Press Web site at
http://www.crcpress.com

Contents

Preface

Prototype development is a vital process to create a successful product, especially in such a competitive global market environment. For a company to be competitive, its products need to be adaptive to customers' changing needs, cost-effective, high quality, and quick turnaround to the market. The quick turnaround product cycle time implies that more frequent changes in products are necessary. Thus, engineers will need to use limited resources to produce a quality product in a short time. In such an environment, the successful development of one product may not be sufficient to sustain a company. A successful company will need to produce a series of good value and high-quality products with great consistency. The secret to success lies in the fine execution of the critical tasks in the product definition stage.

Many people think that product definition is just to define a product, but the actual activities involved are much more than that. Product definition should adequately define a product that will work, and this will require all critical product attributes be considered and, more importantly, the product concept be fully validated. In other words, product definition may engage the concurrent engineering of many activities in the early stage from acquiring customer requirements, problem statement, conceptual design, design for assembly, design for manufacturing, and product prototyping to validate the concept. Product prototyping is also a very critical task as it serves as the role of the integrator and evaluator of an idea or a concept. A design often needs to be validated by assessing several prototypes to yield a quality product. However, physical prototyping often is very costly and time consuming; thus, it becomes the bottleneck of the product development process.

Being an engineer is very challenging as customers want a product that is cheaper and good quality, and they want it now. These desired attributes often contradict each other, and often one factor forms a constraint to the other. Fortunately, the current technologies offer good leverage to overcome some of the issues. As prototyping activities can often be very costly in terms of time and cost, it is critical to plan them well and use state-of-the-art technologies, such as additive manufacturing (AM) and virtual prototyping tools. This book is based on this perspective and highlights the effective tools for prototype development. Depending on specific applications, some tools may be more useful than others in certain applications.

This book is geared toward senior and beginning graduate students who are interested in product design and prototyping. It can be used as a textbook for engineering senior design or capstone projects as it addresses various issues in product prototyping. It will be a good reference for engineering professionals who work in prototyping and design-related areas. This book is not intended to be the ultimate comprehensive reference, as prototyping activities are very broad, and specific references may be needed for a particular application.

This book consists of ten chapters and can be used for a two-semester class. For a one-semester course, the instructors can use selective contents, such as Chapters 2–4, 6, 7, and 9, and use the rest of the chapters as the references. Chapter 1 summarizes the background and the important role of prototyping in product definition. Chapter 2 discusses some basic tools and procedures to plan and design a prototype, including the effectiveness of using paper prototyping in the early development stage. Chapter 3 addresses the importance and examples of modeling and virtual prototyping. Chapter 4 focuses on material selection and the role of material modeling to help to prototype. Chapter 5 summarizes the modern tools of digital prototyping and digital manufacturing, and the basics of using computer-aided design/computer-aided manufacturing (CAD/CAM) technologies for prototyping. Chapter 6 gives a general overview of the current AM technologies and their applications in effective physical prototyping. Chapter 7 discusses how to use off-the-shelf components to quickly build a prototype and to solve tolerance problems on paper rather than in prototype or production. Chapter 8 emphasizes prototyping an automated system and how to use building blocks to prototype an automated system. Chapter 9 illustrates the process of using quality control tools for prototype assessment and evaluation. Chapter 10 provides an overview of optimization technologies for prototype design and improvement.

Acknowledgments

It is very important for me to thank Gail Richards for her great patience in proofreading throughout the chapters, and Sarah Kennedy, Brian Sartin, and Wenyuan Cui for their help with the outstanding artwork.

I also acknowledge the help of my colleagues, associates, and former students for their contributions in examples, and various suggestions and comments, especially Jianzhong Ruan, Todd Sparks, Lan Ren, Cory Whitaker, Sashikanth Prakash, Zhiqiang Fan, Jacquelyn Stroble, Yu Yang, Yaxin Bao, Suhash Ghosh, Krishnathejan Akumalla, Jake Strait, Joseph Dersch, David Dietrich, David Hall, Romy Francis, Rana Chanassery, Ravi Philip, Richard Reis, Tim Christensen, Jon Monroe, Shawn Woy, Pradeep Tipaji, Prashant Desai, Rodney Kestle, Sreekar Karnati, and Sriram Praneeth Isanaka.

Specific appreciation has to be given to the following companies who contributed specific thoughts and materials: 3D Systems, DeskProto, DMG/MORI, Formlabs, Optomec, Product Innovation and Engineering (PINE), Siemens PLM, and Stratasys. The images of the cover page have been provided through the courtesy of Siemens PLM, DMG/MORI, and Product Innovation and Engineering.

This book is also motivated by the research supported by the National Science Foundation Grants # DMI-9871185, IIP-0637796, and CMMI-1625736; the grant from the U.S. Air Force Research Laboratory contract # FA8650-04-C-5704; and the support from Spartan Light Metal Products, Product Innovation and Engineering, NAVAIR, and Missouri S&T Intelligent Systems Center. The help and support from Mary Kinsella of Air Force Research Laboratory, Nam Phan of NAVAIR, and Bart Moensters, Ricky Martin, Kevin Slattery, Michael W. Hayes, Hsin-Nan Chou, and James Castle of Boeing, St. Louis, Missouri, are also appreciated.

Finally, a debt of gratitude has to be tendered to my wife and family. Without the support and understanding of Min-Yu, Jonathan, and Connie, this book could not have been written.

Fuewen Frank Liou
Rolla, Missouri

Author

Dr. Fuewen Frank Liou is the Michael and Joyce Bytnar professor of mechanical engineering department, Missouri University of Science and Technology (Missouri S&T), Rolla, Missouri. He has served as the director of the Manufacturing Engineering Program at Missouri S&T since 1999. His teaching and research interests include CAD/CAM, AM, rapid prototyping, and rapid manufacturing. He has published more than 300 technical papers. Dr. Liou's research has focused on AM, including hybrid additive and subtractive processes, hybrid process path planning, multiscale multiphysics AM process modeling, and AM process monitoring and control. His research has been funded by Army Research Office, Air Force Research Laboratory, Department of Education, Department of Energy, National Aeronautics and Space Administration, National Science Foundation, Naval Air Systems Command, and companies such as Boeing, Center for Aerospace Manufacturing Technologies, DMG/MORI, GE Aviation, GKN Aerospace, Honeywell, Rolls Royce, Spartan Light Metal Products, Toyota, etc. Dr. Liou has received several teaching, research, and service awards, including several best paper awards. He is a cofounder of Product Innovation and Engineering, LLC, and a fellow of the American Society of Mechanical Engineers.

chapter one

Introduction

> Try not to become a man of success, but rather try to become a man of value.
>
> **Albert Einstein**

1.1 *Development of a successful product*

This section will help answer the following questions:

- What is world-class manufacturing?
- What is the impact of this globally competitive environment on engineers and the engineering design process?
- How would a traditional company handle developing a new product versus a world-class manufacturing company?
- Why is it required to change the way of doing things in a company in this competitive world?
- Why are some companies' products consistently better than those of the other companies?
- How does product definition impact the product development process?
- What is an effective engineering design and prototyping process?

1.1.1 *World-class manufacturing*

Due to developments in information and transportation technology, the world is getting smaller. A person can fly from one part of the world to the other in less than half a day. Sometimes, one can learn what happens in Tokyo faster than most people in Japan, since one can access the Internet and read the news online. This changing world creates many astonishing phenomena in the daily life, offers a lot of opportunities in business, and impacts the way an enterprise is run. In *The World Is Flat* [Friedman05], by Thomas L. Friedman, the changing world is described: the great reduction of trade and political barriers and current advances in the digital revolution have made it possible to do business and develop a new product or almost anything else, instantaneously with almost any other people across the world.

World-class manufacturing is achieved by those companies which are best in the field at each of the competitive priorities such as quality, price, delivery speed, delivery reliability, flexibility, and innovation. A company can no longer dominate a market with access to cheaper manpower, newer technology, convenient capital, or a particular supply chain. Now everyone can access anything from anywhere, and it impacts the entire product development process. Any company in the world now can gather the resources from anywhere, and compete for any product and any market! How would a company be able to compete in such a global environment? The answer is simple. To compete globally, one needs to have the ability to gather global resources efficiently and effectively in all of the operating processes. Whichever company is more efficient and effective becomes the winner and a world-class enterprise. Therefore, a manufacturing company which can compete globally must be able to do world-class manufacturing. Typical manufacturers' profit margins average under 1%. Profit margins for their world-class counterparts average over 4.5%. For a $500 million manufacturer, this translates into over $20 million in shareholder value annually. Typical manufacturers generate $134,000 in revenue per employee. World-class manufacturers average $189,000 or 41% more revenue per employee [MAPISC05].

Looking back in history, one can see that history has just echoed such a trend. Manufacturing automation started in 1913 when Henry Ford invented the first mass production assembly line. Everyone has benefited from this invention since products are much cheaper due to mass production [Mikell03]. This is why it is possible to buy a watch for under $10 at a supermarket. After people found such an effective tool, manufacturing companies became larger and larger to mass-produce products. Departmentalization was formed to emulate an assembly line operation. An engineering (or design) department is always the first department created, followed by the manufacturing department. When there is a problem in the manufacturing department, the quality department will follow the manufacturing department to ensure that product quality is good. In the 1970s, a slow but gradual erosion of the nation's standard of living began. People found that America's products could not quite compete with others from overseas. In the 1980s, over 700,000 companies sought bankruptcy protection—it was an expansion of global competition. Table 1.1 shows the market share of the commercial aircraft industry. In 1997, Boeing's share was about two-thirds of the world's market, whereas in 1957, it was only 2%. At Douglas Aircraft, just the opposite happened. In 1997, McDonnell Douglas became a subsidiary of Boeing. How dramatic the change has been within just 40 years! This also demonstrates how competitive the global market truly is!

Table 1.1 Commercial aircraft market share

Company	1957	1977	1997
Boeing	2%	50%	67%
McDonnell Douglas	70%	25%	2%
Airbus	—	2%	30%

Source: Bureau of Transportation Statistics, World Market Share of Large Commercial Transport Aircraft by Value of Deliveries, 2007.

Not only have the large companies experienced the impact, but medium and small companies are also under great pressure since they are part of the product supply chain. For example, automobile door manufacturing company, Company D, may serve as a supplier for the big three auto companies. It has provided various doors to these companies over the years. However, Company D may have another supplier, say Company S, which is making the door frames. In other words, Company S acts as a supplier's supplier. When the market is very competitive, each company is trying its best to optimize its process and supply chains. One or more of the big three companies may be evaluating whether to buy doors from Company D or make the doors themselves. Therefore, Company D needs to make sure that their doors will be competitive in cost, quality, and delivery time. With this in mind, Company D may need to evaluate whether to buy door frames from Company S or make the door frames themselves. This way, the competitive market impact actually propagates into all companies in the supply chain.

In such a competitive, global environment, it is a buyer's market and is very favorable to the consumers. People will be able to purchase goods with good values. For a company to have sustained growth and earnings, it needs to build customer loyalty by creating high-value products in this very dynamic global market. This means that the products need to be low in cost and high in quality. This competitive environment has changed the sequence of product development. The traditional product development starts from an idea to engineering design. On the basis of the design, product cost is estimated to determine actual product pricing, which in turn is used to assess the product market. Now due to the competitive buyers' market, a world-class manufacturer will need to reverse the sequence of how a product is developed to be competitive. In other words, the trend is that a product is launched because of market requirement, and the market determines the product pricing which drives the product development cost. The engineers' job is to make sure that the product can be designed and developed within the cost, quality, time, and function constraints. Such a global environment has placed numerous burdens on engineers.

The point here is to illustrate the impact of competitive manufacturing on the product development process, and it does not mean that the strategy of launching a product only when customers want it is the only correct option. Sometimes, customers do not know what they want, and sometimes, they cannot predict what they would want. For example, in the 1990s, a huge emphasis was made by the automakers to make sport utility vehicles (SUVs) since there was a huge market demand for these vehicles. This proved to be not quite the correct decision, when the price of fuel shot up after 9/11. Large SUV sales are now languishing because they are not fuel efficient. As a result, customers are looking for more fuel-efficient vehicles. Subsequently, the hybrid vehicle manufacturers are in a position to take advantage of this situation, and this is a big factor in their inroads into the U.S. market. Some manufacturers were also very slow to catch onto the hybrid and diesel market, and are now scrambling to license technology from European and Asian automakers. It is therefore important to watch global trends instead of focusing solely on the immediate customers.

The solutions to stay competitive in such a global market environment are still the modern technologies, and this point has been proven in the industry. While these products have various functional requirements, product cost, quality, and time-to-market are the three key factors in product development. The product development cycle time for almost all products has steadily decreased over the years, whereas the quality and cost are staying the same or getting better.

The industry has also tried to put forth their best effort in response to such a competitive environment. For example, in the 1980s, project management tools such as the program evaluation and review technique chart, quality by inspection, and statistical process control were widely adapted to improve product quality and delivery time. The customer orientation concept was also initiated, but neither concrete action nor methodology was used. People also found that inventory control was also very critical. They found that it was very easy to build up an inventory. An inventory control manager will not be popular if the shop floor workers need to wait for days or weeks to get the parts they need. It is thus human nature for the inventory control manager to order extra parts to make his or her life easier. In addition, this seems to accelerate the project progress as the workers do not need to wait. However, such practice will eventually require a much larger storage room for spare parts, and due to the dramatic increase in parts and the associated storage space, inventory management becomes an issue. When each unit's storage space becomes much larger, the distance between units is increased. The increased distance between units creates more problems, such as communication and transportation, in addition to the cost of the space and storage room.

In the 1990s, "just-in-time (JIT) manufacturing" and "lean manufacturing" were implemented. It was found that the "pull" instead of "push" strategy of moving products from one station to another in an assembly can greatly reduce the space and material handling specialties needed, thus reducing overhead cost. Also, it was found that JIT shop floor design can improve overall product quality. The Toyota JIT is one good example. However, to make this work properly, it needs to trace all the way back to product definition and design stages. Companies need to find an effective product development strategy to stay competitive, such as customer focus, system integration, and lean production systems. These are all directly or indirectly related to product definition, which will be discussed in the next section.

In addition to product cost and speed, the quality and reliability of the product are important to ensure the ability to stay functional over the designed product's life span. A good example of this idea is the B-52 bomber. While it was developed in the late 1950s, it still remains in today's market because of its operational cost, maintainability, product effectiveness, and reliability. The companies, which are willing to be world-class manufacturers, should increasingly rely on demonstrating the industry's best practices. To accomplish this objective, these companies should attempt to be the best in the field at their competitive priorities which could include quality, price, delivery speed, delivery reliability, flexibility, and innovation.

1.1.2 Product definition

What is product definition? [Crow06, Korman02, Olshavsky02]

- Product definition simply means "figuring out what to make before making it."
- A product definition is a layout of what the purpose of the product is, whom it is targeted toward, and how it will be built and manufactured. By reaching a definition for a product, not only has there been considerable customer interaction regarding the product, but many prototypes have already been completed to get the best possible design.

Table 1.2 shows the comparison of percentage effort in product development for shipbuilding over various production stages [Prasad96, Wilson90]. Since the Japanese shipbuilding industry is getting a strong market share, the data is very noteworthy and reflects a general trend in product development: substantial engineering efforts should be invested in the product definition stage to produce a competitive product. This point is further illustrated in Figure 1.1. When a product is at the definition stage, it is very easy to change, but when the product is designed or developed further, it is more difficult to make a change, since each design represents decision-making, and each decision represents a commitment to the available resources. When a company is trying to make a simple change toward production stages, the cost will be larger than the cost of change at design.

Table 1.2 Percentage effort in product development for shipbuilding over production stage

Company	Definition (%)	Design (%)	Redesign (%)
British company	17	33	50
Japanese company	66	24	10

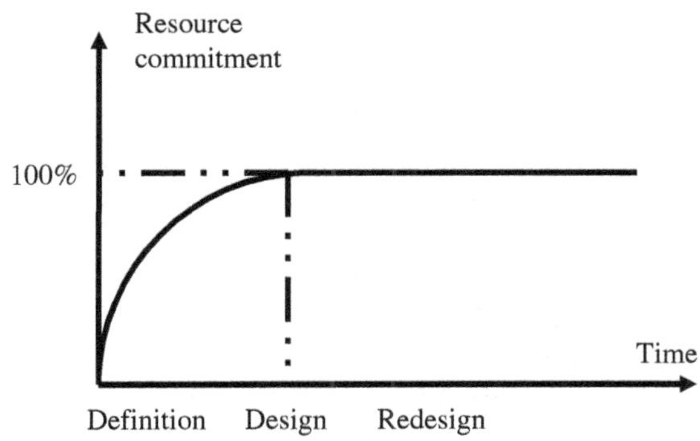

Figure 1.1 Resource commitment over various product designs and development stages.

The product designer's dream is that once it is designed, it will work well. This can possibly be achieved if the product is well designed and tested early; this early stage is called "product definition."

In other words, the solution for a successful product is to prototype early and prototype often. One needs to focus on early results and do everything one can to keep the momentum going. Promote early successes and outline the immediate short-term steps toward success by designing several levels of prototyping activities among product definition activities.

Product definition often is conceived as the process of translating customer needs into product design specifications. However, it should be more than that. It should be the concurrent engineering of all activities in the early stage from acquiring customer requirements, problem definition, conceptual design, design for assembly, design for manufacturing, and product prototyping to evaluate the concept. In other words, the result of a product definition could be a well-evaluated prototype that shows it should work! This is the best way to make sure that a product just designed will be successful. A road map with clearly defined, visible chunks of meaningful prototyping functionality can be a great help in organizing the work effort. By making each chunk of work small enough to easily build, one can ensure that there is always a constant stream of interesting results coming from the prototyping projects. One should avoid long gaps between prototypes at all cost.

Among the above product definition activities is product prototyping, especially physical prototyping, which often needs days or sometimes weeks to implement just one prototype. A good design often needs several prototypes; thus, prototyping activities become the bottleneck of the product development process. Since product definition involves concurrent engineering of many traditional engineering design activities, it is worthwhile to review the traditional engineering design process as described in the next section.

1.1.3 Engineering design process

The engineering design process can be divided into several steps: identifying customers' needs, converting these needs into product design specifications, engineering design, and product prototyping to evaluate the concepts [Wood01].

1.1.3.1 Identifying customer's needs

To develop a successful product, one needs to begin by defining all consumer-relevant product features in advance of prototype development. These attributes can be identified through careful consumer testing and a thorough evaluation of similar products on the market. The desirable attributes are then engineered into product prototypes, rather than defining product attributes after products have been designed. Therefore, the objective of engineering design is to make sure that the designed product can be successful, since customers would love to have the product.

When exploring the reason why customers would prefer a product, there is no straightforward answer, and this is due to the complex nature of the product and the differences in human beings. It is therefore very critical to be able to identify the customer's needs. Before one can develop a product, one must first know the requirements for that product. However, a product is not a blind response to customer needs and opportunities, or at least it should not be. A detailed framework and business strategy is required with such questions as follows:

- Who are your customers?
- Do you know what they want? Do they know what they want?
- Have you documented their needs?
- Does the documentation reflect their needs?
- Are you developing a product that fits with their needs?

There are many ways to find out the answers. For example, interviewing the end users, role-playing in as many ways as possible in which the product might be used, conducting a walk-through of a potential usage of the product, reversing roles with the customer during an interview, brainstorming among team members, etc. Also, it is critical to test initial prototypes on consumers to identify how closely engineered product variables match consumer needs and perceptions to receive early feedback from the customers.

1.1.3.2 *Converting needs into product design specifications*

Methodologies such as quality function deployment or house of quality can be used to convert the voice of the customer into the design specifications for the engineers. The objective here is to explore and untangle the complex relationships between customer and functional requirements to connect technology with what people need and want. After identifying the customer and functional requirements explicitly, they are captured in a formal specification that product concepts can then be evaluated against. One then screens and ranks to select the preferred concepts that are later refined. The desirable result will be that the product fits both the customer "wants" and the business "needs."

1.1.3.3 *Engineering design*

Engineering design is an iterative process and can further be classified into feasibility study or conceptual design, preliminary design, and detail design. At the conceptual design stage, one needs to validate the customers' needs, produce a number of solutions based on the needs, and evaluate the solutions mainly based on functionality and cost. At the preliminary design stage, a feasible solution is further developed. The problem requirements are further defined in greater detail; more information is gathered; critical parameters are being quantified to establish the optimal solution; and the function, strength, spatial compatibility, and financial viability of the product are evaluated. The detailed design makes a complete engineering description of a tested and producible product. At this stage, the form, dimensions, tolerances, surface properties of all individual parts, specific materials, and manufacturing processes are determined.

As mentioned in Section 1.1.2, production definition concurrently exercises all activities in product development processes in the product definition stage, and thus, other factors, such as planning for manufacturing, planning for distribution, planning for use, and planning for the retirement of the product, should also be considered. For example, in manufacturing planning, factors such as tools and fixtures, production lines or production layouts, work schedule, inventory controls, quality control system, standard working hours, labor costs for each operation, and information flow to control the manufacturing operation may need to be considered since these factors may impact product definition. Planning for distribution provides the effective production distribution to customers. At this stage, factors, such as packaging, shipping, shelf life, warehouse system, specialized sales brochures, and performance test data, should be considered. Planning for use considers consumer reaction to the product, including ease of maintenance, reliability, product safety, convenience in use (human factor engineering), aesthetic appeal, the economy of operation, and duration of service. Planning for the retirement of the product considers the disposal of the retired product that may include updated legislation and environmental protection trends, reducing raw materials, easy disassembly, and multiple levels of usages of the same material.

It is desirable that concurrent engineering can be carried out in the product definition stage so that the above product development activities can be executed simultaneously. There are many advantages that concurrent engineering can offer, such as reduction of the time needed for implementing the new product into production, reduction of the time-to-market, substantial improvement of product quality, quicker reaction to customer requirements, cost reduction, and eventually profitability improvement. Concurrent engineering requires modern technologies, as well as teamwork by various experts to make it work.

1.1.3.4 Product prototyping

As mentioned previously, among these product development activities, product prototyping is the bottleneck of the process. The focus of this book will be on how to use the virtual prototyping and additive manufacturing (AM) [AM17] technologies to accelerate this process, and thus, product prototyping will be illustrated further in the next sections.

Review problems

1. What is the impact of world-class manufacturing competitiveness? What does it mean to be a world-class manufacturer?
2. What is product definition? What type of activities should be included in product definition?
3. Why is product definition important in the product development process? Use an example to illustrate the point.
4. Describe the product design process.
5. "The world is getting smaller." What significance does this statement have in today's manufacturing scenario?
6. What are the risks of manufacturers solely following customer demands?

1.2 Product prototyping and its impact

> My great concern is not whether you have failed, but whether you are content with your failure.
>
> **Abraham Lincoln**

This section will help answer the following questions:

- Why is product innovation important?
- What are the reasons for the failure of a new product?
- What are the basic reasons for prototyping?
- Why is time-to-market so critical?
- How can bringing a product to market rapidly have an impact on the rate of return and how it affects the outcome of other competitive products?
- How can product definition impact time-to-market?

As discussed in Section 1.1.2, the world-class product development focuses effort on product definition to include all activities involved in the traditional product development process. This includes product prototyping in the product definition to evaluate various design concepts. This section further discusses how product prototyping activities can impact product development cost, quality, and time-to-market.

1.2.1 *Prototype design and innovation*

Design is to invent. As one thinks about the essential nature of an alternative design, one mentally formulates an invention. In the product invention process, many technical problems need to be resolved. Most technical problems have alternative solutions, and often there are several solutions and trade-offs associated with each solution. A viable solution is simple, easy to produce with good quality, and low cost. Product prototyping can be used as an evaluation tool in the engineering design process. Prototyping plays a key role in product innovation. Prototyping helps to quickly develop a product by providing a good tool for problem-solving and can validate a concept. Also, a prototype can play a vital role in innovation because it can be used as a visual to help communicate the product's purpose and feel. By doing this, different teams can look at the prototypes and use them as a stepping stone to further develop a new product.

Before discussion of prototyping in the product design process, one must have ideas: what is design? Is design an art or a science? Before a thing is made, an idea exists. The idea may be a clear vision or may be little more than a glimpse of possibility. If the idea is in an artist's head, he or she can make the thing directly, requiring only materials, tools, and skills needed to make the thing. The engineer must have some means of explaining his or her vision to the worker who will construct it. He or she may send the drawings to separate shops outside the city and assemble the parts in-house. Engineers have the capability of building complex machines that may need a team of engineers, and others such as pattern makers, foundrymen, machinists, and die makers. In other words, the engineer's way is to design with drawings, whereas the artist's way is to design without them. Both an engineer and an artist start with a blank page and transfer to it the vision in his or her mind's eye.

It may be fair to say that the ability to design is both a science and an art. It is a science since design can be learned through design methodologies and techniques; however, it is an art because it can only be learned by doing design to gain experience. Just like learning to cook, one may be able to learn from a recipe, but one may not be able to get it right or fast enough the first time. However, after several times, a delicious dish may be cooked, and it may even taste better than the original recipe. From another angle, a good design requires both analysis and synthesis. While analysis is the process of determining the responses of a specified system to its environment, synthesis is the process of defining the system to satisfy the needs. An engineer's goal may be reached by many different paths, and none of which is in all aspects the one best way. An engineer's decision will need to be based on intuition, a sense of fitness, and a personal preference made in the course of working through a particular design.

New product development is a complex, costly, and time-consuming process. There are many challenges. In fact, the national average success rate for start-up companies is about 10%! Therefore, to start a new company requires a lot of planning, effort, and creativity. Why did these companies fail? It is an interesting question and deserves investigation. In most companies, the failure to introduce new products does not follow from a single reason. In a study by Boston-based AMR (Advanced Market Research) research [Burkett06], 32% of respondents designated "products late-to-market/missing demand" as the top reason for new product launch failure. This was followed closely by product pricing, quality, and missing customer needs. Typical causes of product innovation failure, including market obstacles, management issues, and technology issues, are shown in Table 1.3. Note that when a company determined that the products be commercialized, the technology risks may have been considered to be minimal as most of the technical issues should have been worked out. This fact is reflected in the role of the technology issues to be minor. The market obstacles include missing demands, product pricing, and late-to-market. The management issues include poor market analysis, under-staff, and lack of capital resources. Note that some of the market issues can be indirectly related to technology. For example, the product development cost and time-to-market may be impacted by the technology used.

Table 1.3 Issues that cause product innovation failures

Rank=issues	Role	Causes	Comments
1. Market obstacles	Major	Missing demands	Need
		Product pricing	Cost
		Late-to-market	Time
2. Management issues	Medium	Poor market analysis	Need
		Understaffing	Time/cost
		Lack of capital resources	Cost
3. Technology issues	Minor	Dated technical approach	Performance Quality
		Design problems	Performance Quality
		Poor quality	Quality
4. Others	Minor		Other

Although there are many obstacles in the product innovation process, the reward may be enormous. Recall Pareto's principle—The 80–20 rule means that in anything, a few (20%) are vital and many (80%) are trivial. In Pareto's case, it meant 20% of the people owned 80% of the wealth. Similarly, 20% of the defects caused 80% of the problems. Project managers know that 20% of the work consumes 80% of one's time and resources. In product innovation, the first 20% of the companies launching the product will gain 80% of the overall product profits as they can set a higher price in the early market stage. In other words, the last 80% of the companies can only share 20% of the profit. For example, the invention of the "instant-on" computer, working like a light switch, could save a lot of time to turn on a computer and bring huge profit to the inventor. The traditional memory technologies, such as DRAM (dynamic random-access memory) and SRAM (static random-access memory), require constant electrical power to retain stored data. The instant-on computer concept does not load software from the hard drive during start-up, like traditional memory technologies. A working version of the operating system and other software is stored in memory via magnetics. Magnetic memory does not need a constant source of energy to retain the saved data. Such innovation which benefits the whole world makes all the engineering effort valuable to society.

It is interesting to observe that cost-related causes dominate the issues. However, product cost, quality, and time-to-market are all closely related, and all contribute to the success of the global market environment. As discussed previously, the result of global competition is that product life cycles have greatly been shortened. For example, the average life of Hewlett-Packard printers in the year 2000 was about 1 year and is shrinking. On the other hand, engineering changes should not be made at later stages, especially toward production, since it is too costly to make late changes. Ricoh Copier reported in 1 year that the cost of engineering orders is $35 in the design phase, whereas it is $1,777 prior to prototyping and $590,000 after the product is in production. If a product is delivered late by 6 months, then the company loses 33% of the profit, but with a cost overrun of 50% in product development, the company only loses 3.5% in profit. Product cost, quality, and time-to-market are all critical in the product innovation process, but by shortening the product development time, the cost is reduced and time is gained. Time gained can be used to improve product quality.

1.2.2 Impact on cost, quality, and time

The direct impact of a shorter product development time includes the opportunity to sell the product at premium prices early in the life cycle and enjoy longer market life cycle. In addition, the benefits include faster breakeven on development investment and lower financial risk, which leads to greater overall profits and higher return on investment (ROI). Figure 1.2 further illustrates this concept.

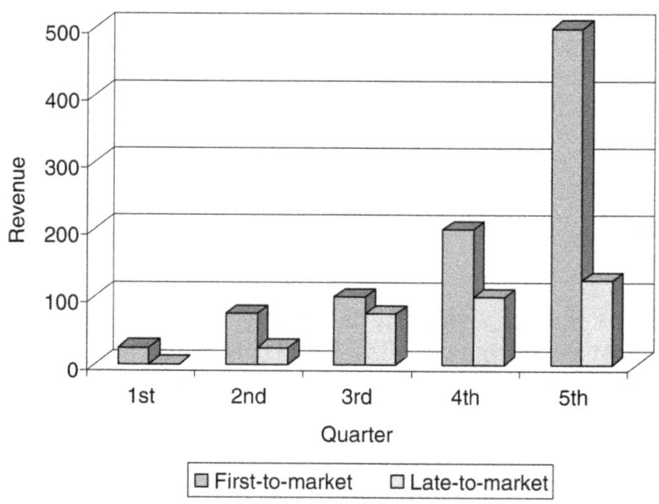

Figure 1.2 The impact of an example of timing-to-market on product revenue.

To illustrate the importance of time-to-market, here a simple model is used to observe the impact of time-to-market. The rate of ROI is an important index for the success of a product innovation project. The ROI of a product can be estimated as follows:

$$\begin{aligned}&\frac{\text{Net profit}}{\text{Product development cost}}\\&=\frac{(\text{Total profit})-(\text{Product development cost})}{(\text{Annual development cost})(\text{Time-to-market})}\\&=\frac{(\text{Total profit})-(\text{Annual development cost})(\text{Time-to-market})}{(\text{Annual development cost})(\text{Time-to-market})}\end{aligned}\tag{1.1}$$

Time-to-market normally refers to the time needed for a new product to be sufficiently debugged, so that the development personnel can be allocated to develop another product. In Equation 1.1, it is shown that the product development cost is very critical to the ROI. As time-to-market can significantly impact product development cost and can even be more critical to the ROI, it is the most important factor which needs to be addressed.

The critical issues in innovation are related to cost, quality, and time-to-market. Product definition is key to the success in the product development process, and among the activities, prototyping is the most time-consuming task. Design is an iterative process in which two or more iterations may be needed. This makes the prototyping task even longer. It is therefore very critical to be able to reduce prototyping time to shorten the entire product development cycle.

1.2.3 *Key process requirements for rapid prototyping*

Traditional design ideologies require that engineers construct a variety of physical prototypes to test and evaluate design concepts. Due to the nature of such a process, the design and analysis of new products can become very time consuming and expensive. Therefore, a traditional product design approach often yields very long product development time. Currently, new technologies involving rapid and virtual prototyping are revolutionizing the way products are designed.

Virtual prototyping, for example, integrates digital technologies such as computer-aided design (CAD), computer-aided engineering, and computer-aided manufacturing (CAM) data into a single visual environment for viewing and analysis. It provides flexibility, cost-efficient data integration, and a concurrent approach to engineering. The traditional prototyping techniques were successful in the design and analysis of products consisting of less than 100 components. However, when companies are building large assemblies, they face a bigger challenge in managing their products, and virtual prototyping technologies have proven to be very effective. For example, the Boeing 777 airplane consists of three million parts, and the fabrication involved 545 suppliers and 22,900 composites. Thousands of engineers participated in the activities, using 100% digital product definition for digital preassembly. The 777 was designed by cross-functional teams using thousands of terminals and a computer-aided, 3D interactive application system that allowed engineers to simulate the assembly for the 777 without resorting to physical prototypes. Only a nose mock-up to check critical wiring was built before Boeing workers assembled the first flight vehicle. On October 15, 1990, Boeing announced its plans to build a new twin-engine airliner to fill market demand between the smaller 767 and the massive 747. On June 12, 1994, the first 777 prototype took off on its maiden flight.

Rapid prototyping which can rapidly create physical prototypes along with virtual prototyping can be very effective in accelerating the product development process. It can increase visualization capability during the early phases of design by using rapid physical models. It can test and improve design before the manufacture of tooling.

Be first-to-market, and you will own the market. That is world-class thinking. Manufacturers that excel in time-to-market use information as a critical tool to prevent delays through lack of communications, control costs by staying advised of issues, and deliver-to-promise by collaborating with partners and customers. The key practices for world-class performers include the following:

- *Know the customers*: one should participate in the customer design process and capture the "voice of the customer"—direct discussions with the customers through interviews, surveys, focus groups, customer specifications, observations, warranty data, field reports, etc.
- *Use modern prototyping technologies*: one should use the state-of-the-art integrated business systems to save time and reduce errors.

The new technologies can help to enter new markets, for example, by lowering entry technology barriers. They can also help to increase market share, for example, by using the time saving to explore more design options to create more innovative products that customers will value. Furthermore, they can be used to change the rules of competition, for example, by offering customers individually customized products with no time or cost penalty. The new technologies can also be used to reduce errors. If mistakes can be identified before commitments are made to expensive tooling, then the costs associated with modifying such tools can be avoided. Overall, costs can be reduced in the amount of labor involved in producing models, tooling, and prototyping parts—typically very labor intensive and time consuming.

As discussed at the beginning of this section, the invention is a complex process. The new prototyping technologies can help by developing new analysis and testing procedures, manufacturing production tooling, improving communications across product divisions, and supporting customized manufacturing. The objective of this book is to introduce the readers to AM and virtual prototyping technologies and their effective applications in the product development process.

Review problems

1. What is design? Why is the ability to design both a science and an art?
2. Why is time-to-market important for the success of a new product?
3. Company A has an idea for a new product, product X. However, they need at least a 25% ROI to give the project a green light. They have compiled all the necessary data and have assigned you the task of determining if the project is a go or not. Based on the following data, judge whether product X should be launched by finding the first-year ROI.
   - Product X will be the first of its kind on the market.
   - It will sell for $30 per unit.

- It costs $18 per unit to manufacture product X (this includes all materials and manufacturing cost).
- The sales cost for product X will be $3 per unit.
- The market survey showed that there is a market for three million units of product X per year. They plan to be first-to-market (1 year after the start of development) and expect that product X can be viable for a period of at least 15 years.
- It will cost $3.5 million to develop product X.
- Company A plans on having a 33% market share (990,000 customers) for the first year.

4. Company B has an idea for a new product, product Y. They have compiled all the necessary data and have assigned you the task of determining if the project is a go or not. Based on the following data, the ROI is 28%. What is ROI if they can reduce their time-to-market by 5%?
   - It will sell for $20 per unit.
   - It costs $7 per unit to manufacture product Y (this includes all materials and manufacturing cost).
   - The sales cost for product Y will be $1 per unit.
   - Company A believes there is a market for 20,000 units of product Y annually. They plan to be first-to-market (0.5 years after the start of development) and expect that product Y can be viable for a period of at least 5 years.
   - It will cost $80,000 per developer annually to develop product Y.
   - Development staff size is two persons.
   - As this product will be new, the market share can be assumed to be 100% for the first year.
5. Suppose that for an electronic product, the time-to-market is 1 year, the annual cost per developer is $20K, and the development staff size is 25 people. The net profit of the company is $1,000K per year. Calculate the rate of ROI. Also, calculate the change in the rate of ROI for a 5% decrease in time-to-market. Discuss the significance of the result.
6. What is the significance of prototype design in product development? How is digital prototyping advantageous?
7. What is the difference between traditional product prototyping and digital prototyping?
8. Given the following scenarios, find the rate of ROI and which scenario is more favorable?

   Scenario 1:
   Development staff size = 30 people
   Annual cost per developer = $50,000 per person year
   Time-to-market = 1 year
   Profit per unit = $120,000
   Total sales = 50

   Scenario 2:
   Development staff size = 30 people
   Annual cost per developer = $50,000 per person year
   Time-to-market = 1.5 years
   Profit per unit = $120,000
   Total sales = 80

1.3 *Product prototyping and product development*

> If the only tool you have is a hammer, you tend to see every problem as a nail.
>
> **Abraham Maslow**

This section helps answer the following questions:

- What is the basic reason for prototyping?
- What is virtual prototyping? Can one totally rely on virtual prototyping for product development?
- One of the simplest methods of prototyping is a pencil and paper. Does this technique still have a place in the rapidly changing, digital workplace?
- What is the impact of the current technologies, such as CAD/CAM, AM, and virtual prototyping, on product prototyping?

1.3.1 *What is prototyping?*

As discussed in Section 1.1.1, the complexity of the global market environment not only makes the product cheaper, better, faster to the consumers, but also increases severe competition for product producers. While product complexity is increasing drastically, human resources are very limited and unstable. The trend in the industry is to use simulation technologies in all design phases from the concept to the final implementation, to minimize prototype tests on actual full-size systems by improving virtual modeling technologies, and to use automatic code generation technologies for rapid prototyping. These simulations, virtual prototyping, and AM activities are all part of the product prototyping activities.

While these prototypes play a very important role in product development as stated previously, it can be very easy to have a negative impact on the development of a product. Prototyping has huge implications on product cost, quality, and time. Obviously, prototypes are necessary for all products, and the more useful prototypes that are made, the higher the quality of the product. However, it is important when building prototypes that they are built by adding low cost to the overall product and that the final product still has the shortest time-to-market possible. Therefore, it is important that the prototype serves a purpose for the development of the product. Whether it is to study the function of the product, the appearance or "feel" of a product, to visualize improvements to a product, there is a point where prototyping can increase the cost of a product and its time-to-market. This is why material and process criteria for every prototype are important. When building a prototype, to keep cost to a minimum, it is very important to use cheap, readily available materials that will still serve the same purpose as the actual product materials. Depending on availability, function, and cost, it is also important to select a prototyping process that not only serves the prototype's purpose but also keeps cost low.

Prototyping is an approximation of the product along one or more dimensions of interest, which includes prototypes ranging from concept sketches to fully functional artifacts. It can help everyone visualize the same end result so that there is no ambiguity, and everyone is on the same page. Depending on various prototyping applications, prototyping methods can be classified into physical and analytical methods. For example, the simulation approach is an analytical method, and a clay mock-up is a physical prototype. From a different angle, prototypes can also be classified as comprehensive and focused prototypes. For example, when a prototype is used to test the "look" of a product, this prototype may be made from Styrofoam for the purpose, and thus, it is a look-focused prototype. On the contrary, a full vehicle prototype built to test its full functions would be a comprehensive prototype.

Virtually, every business uses prototyping. A wide range of businesses use prototypes, from airplane manufacturers to toy producers to computer system developers. Prototypes are one of the most useful and cost-effective quality tools businesses have. They can be a source of creativity, and they allow the user to interact with the product so the developer can receive feedback. Prototyping is not limited to product development. It can also be used as process development. Every department can use prototypes to help them excel. For example, marketing departments use prototyping to determine why consumers buy products. A nonworking mock-up of the product can be reviewed by customers prior to acceptance. Sometimes, these basic prototypes are used at trade shows. For example, the auto industry refers to them as concept cars. Rapid prototyping can be used to accelerate the design process, and it leads to high-quality, defect-free products and reduces risk. This technique has proven essential to market leaders such as Microsoft, Intel, GM, Boeing, Ford, and Cisco. In the software industry, a series of drawings that are created by the developers are used to obtain the acceptance by decision-makers. For example, sticky notes can be used when designing graphical user interfaces so that users can see the proposal.

Before a prototype is made, the goal of the prototype needs to be well defined. For example, it could be a "rough version" to answer a single or set of binary (yes/no) questions or to visualize and brainstorm possible improvements. It could also be a concept model with no working features to obtain early feedback from customers. It could be a study of the product features and models to refine difficult features, a simulated walk-through of product activities, or simply the creation of a photographic quality model to create a demonstration video for marketing and evaluating the product in use. As discussed in the previous sections, since product development is an iterative process, it usually requires building several prototypes in an iterative manner to produce a quality product. These prototypes may need to serve in various purposes and in various stages of product development. Sometimes, it is required to create as soon as possible a 3D "free-form" part for evaluation in its application context, which could include visualization, tactile feedback, function verification, and simulation of final use.

Within traditional prototyping methods, engineers created new products by using techniques, very different from today's cost-effective, faster, and quality-based processes, which demonstrate the following features:

- Traditional prototyping methods allowed the engineers to create only a static mock-up of what the interface looked like. It was not a dynamic process, and the prototypes were not alive enough to enable the designer to test the real-world usage of the product.
- Engineers had to use manual tools to create prototypes, but since these tools were manual, it is difficult to show all user requirements with a single prototype. That is why the designers, most of the time, had to develop separate prototypes for the same end product to be able to see various user requirements.
- Since the traditional prototyping methods did not allow the designer to reflect all user expectations with one prototype, the process became very costly and time consuming.
- Traditional prototyping did not give the engineer a chance to make the process iterative enough to involve all stakeholders including the users. Therefore, this situation impacted the quality of the end product.

The traditional options could include building a model from clay, carving from wood, bending wire meshing, carving from Styrofoam perhaps with surface reinforcement, and milling from a block of plastic or aluminum. However, often these methods are time consuming and sometimes lack the quality to serve its purpose such as parts fitting. Then, the better designs are selected, and physical parts or prototypes are made by various vendors, which delays the prototyping time. In order to effectively evaluate the alternative design concepts in the product definition stage, modern prototyping techniques, such as virtual prototyping and AM, are needed for world-class product development.

A virtual prototype is an analytical model of some aspect of a design. It allows the engineer to predict with some confidence the design's behavior, without building expensive, inflexible, physical prototypes.

There are many different computational approaches to address different aspects of the design process, finite element analysis (FEA), kinematics and multi-body dynamics, electronic circuit design, CAD/immersive design, and virtual reality/topological modeling. These are all analytical prototyping methods. Model-based design is a high-fidelity mathematical model that accurately predicts the behavior of a dynamic system in real time. With more improvement in simulating physical properties, there is great potential for applications such as control system design, testing and optimization, and connecting the virtual world with the real world like hardware-in-the-loop, software-in-the-loop, and operator-in-the-loop.

One example of using simulation or virtual prototyping to help in the product development process is a drivetrain prototype test bed in the auto industry. The problem is that prototype components do not arrive from suppliers at the same time and thus cause the prototype test program to be delayed until all components are in place. A flexible prototype test bed for new vehicle designs that incorporates virtual and real components for testing the whole vehicle is an ideal solution. In other words, virtual components can directly interact with the physical components. System models that were developed at the design stage by different groups or suppliers can now be incorporated into the engineering simulator. As components become available from the manufacturer, they need to be readily connected to the simulator, bypassing the component model. To make this possible, sensors, actuators, controllers, and software need to come together to deliver the complete solution, the virtual test bed.

As shown in Figure 1.3, one can mix virtual and physical components together to evaluate a prototype. In this example, the engine is not available. In order to test the other components, such as gear trains and the fuel pedal, one can create a virtual model of the

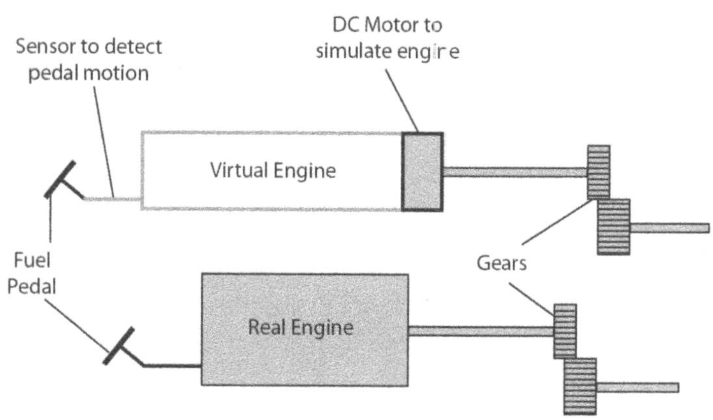

Figure 1.3 An example of mixing virtual and physical components together to evaluate a prototype.

engine. In this case, the virtual engine needs to receive input from the fuel pedal and output the rotation and torque to the gearbox. In other words, this virtual model needs to include the sensors to detect the pedal motion from the user, the engine model which describes the relationship between fuel and output rpm (revolution per minute) and torque, and a DC motor to generate the correct rpm and torque. These elements, including sensors, engine model, and DC motor, and a virtual engine model can detect the user fuel pedal level, and output rpm and torque.

The bottom line is to make the product cheaper, better, and faster. This is because engineering prototyping and testing are expensive, and there is a competitive pressure to get to market first. Since late-cycle design changes are costly to implement, GM alone spends $500 million annually on "build and test" to increase efficiency, and improve reliability and customer appeal.

1.3.2 *Rapid prototyping in product development*

Often virtual prototyping may not be able to evaluate the ultimate performance of a product. A full-scale comprehensive physical prototype is needed at least at the end. Physical prototyping enables the exploration, optimization, and validation of mechanical hardware. Physical prototyping is traditionally a very time-consuming process. Recently, AM has become a new trend to produce a physical prototype for testing. It is based on layered manufacturing, which builds a part in a layered fashion—typically from the bottom-up. It makes use of an old technology—printing. A layer of material is printed or laid down on a substrate with careful control. When various layers are stacked together, it forms a 3D object. Conceptually, it is like stacking many tailored pieces of cardboard on top of one another. Part geometry needs to be sliced, and the geometry of each slice determined. It is computer controlled and fully automated. Therefore, it is fully compatible with the CAD/CAM system for concurrent product development.

The first commercial AM system is called a stereolithographic apparatus (SLA). As shown in Figure 1.4, a UV laser traces a cross section of the part. The resin hardens under

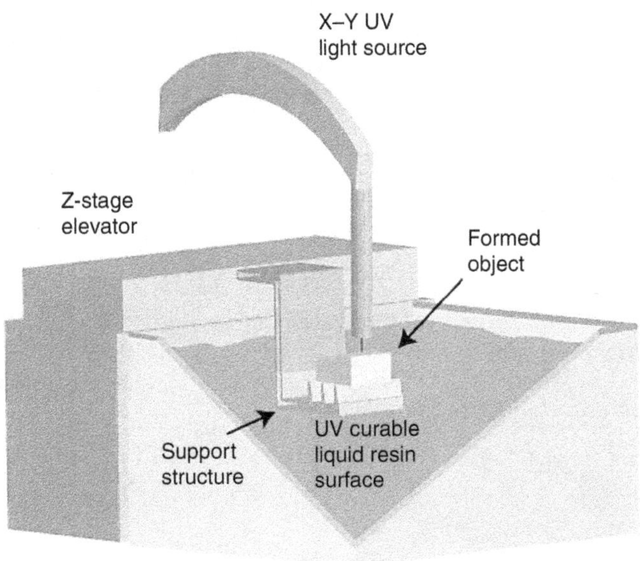

Figure 1.4 Schematics of an SLA.

the UV light. Once the top layer is selectively cured, the tray lowers, allowing a fresh layer of resin to cover the new layer. There are many other types of AM machines in the market.

Since a physical part can be quickly produced in a few minutes or a few hours, AM technology has greatly impacted the product development process. These prototypes can be used to evaluate design changes by producing free-form prototyping for complex object visualization. They can serve as a quick and simple communication of design ideas. They can be used for the verification of form and fit of a design. For example, a prototype can be used to test the fitting for product packaging, test styling, and ergonomics.

With increasing prototyping technology, it has become easier, faster, and cheaper to build a prototype, whether it be on a computer screen or a physical prototype model. Innovations in prototyping stem from the advanced technologies to form prototypes. These state-of-the-art prototyping technologies have great potential to be used to significantly shorten the product development process. The following chapter and the subsequent chapters will discuss further the prototyping approaches in detail.

Review problems

1. What is product prototyping?
2. How is virtual prototyping useful in product design?
3. How do you prototype a mechanical pencil?
4. Briefly describe layered manufacturing.
5. Why is design an iterative process and how does prototyping aid this?

References

[AM17] Additive Manufacturing, Wohlers Associates, Retrieved on August 23, 2017 from https://wohlersassociates.com/additive-manufacturing.html.

[Burkett06] Burkett, M. B., Product innovation, Retrieved on 2006 from www.mbtmag.com.

[Crow06] Crow, K., Product definition, DRM Associates, Retrieved on November 19, 2006 from www.npd-solutions.com/pdef.html.

[Friedman05] Friedman, T.L., *The World Is Flat*, Farrar, Straus, and Giroux, New York, April 5, 2005.

[Mikell03] Groover, M., *Automation, Production Systems and Computer-Integrated Manufacturing*, 2nd edition, Prentice Hall, Upper Saddle River, NJ, 2003.

[Korman02] Korman, J., Turning requirements into product definition, Cooper Newsletter, www.cooper.com, August 2002.

[MAPISC05] MAPICS, The seven keys to world-class manufacturing, Retrieved on August 26, 2005 from www.mapics.com/downloads/documents/MPWPKEY.pdf

[Olshavsky02] Olshavsky, R., Bridging the gap with requirements definition, Retrieved on November 19, 2006 from www.cooper.com/journal/2002/07/bridging_the_gap_with_requirem?

[Prasad96] Prasad, B., *Concurrent Engineering Fundamentals*, Volume I, Prentice Hall, Upper Saddle River, NJ, 1996.

[Wilson90] Wilson, P.M. and J.G. Greaves, "Forward engineering—A strategic link between design and profit," *Mechatronic Systems Engineering*, Vol. 1, pp. 53–64, 1990.

[Wood01] Otto, K. and K. Wood, *Product Design: Techniques in Reverse Engineering and New Product Development*, Prentice Hall, Upper Saddle River, NJ, 2001.

chapter two

Product prototyping

Prototyping is a quick way to learn about the characteristics of a product by incorporating direct feedback from users into a design. A prototype can be created for how it will look, how it will feel, how it will function, where to get it made, and/or how to make sure it will turn out the way one wants it. However, prototyping can also be a very time- and cost-consuming process. Therefore, unless there is a good justification, a prototype should not be conducted. It is thus critical to have a well-thought-out plan before producing a prototype. This chapter discusses when and how one should conduct a prototyping activity.

2.1 Product prototyping

This section helps answer the following questions:

- What is product prototyping?
- When is making a prototype necessary?
- How to avoid some common mistakes and issues in product prototyping?
- How to conduct prototyping activities?
- Is there a rule of thumb to decide between a functional prototype and one with limited function?
- Is there a prototyping procedure or rules of thumb that can be followed?
- What is prototyping swamp? How can this be avoided?
- What are the conditions to stop a prototyping activity?

2.1.1 When is prototyping needed?

Before a prototype is made, it is very important to define the goal of the prototype so that the level of detail of the prototype can be determined. A prototype can be used for many purposes such as the following:

- Gather initial user requirements
- Show proof of concept to senior management
- Validate system specifications
- Explore solutions to specific usability or design problems
- Deliver early proof of concept
- Resolve fuzziness in the early stages of design
- Manage change requests
- Validate evolving user requirements
- Increase constructive user participation
- Customer acceptance
- Product invisibility
- Quality assurance
- Pretrain users or to create a marketing demo

It is very important to note that a prototype does not need to be a physical complex prototype. Rather, it can be a very simple one as long as it can accomplish the learning objective(s). For example, in some cases, simple sketches or cartoons can be used to quickly demonstrate a project's architecture, functionality, and interfaces. This kind of pseudosolution prototyping proves design viability for complicated, long-term projects. They can be helpful as a go/no-go indicator for clients. A prototype can be a very complex physical prototype. In this case, the prototype is refined through various stages until it eventually evolves into the desired product. This can be a fail-safe method for rapid product development.

In addition, prototypes can be used as solid milestones to provide tangible goals, demonstrate progress, and enforce the schedule for the team. They can also be used for system integration to ensure components and subsystems work together as expected. For example, the traditional injection molding process needs a lot of trials and errors in the mold development process, especially when there are two or more mating parts. In many cases, after a mold is made, two mating parts will need to be tested for fit. The overall success rate for the molding process is about 65% since the chance of the two parts not fitting into each other due to design error is about 30% and is about 5% due to shrinkage. The additive manufacturing process, or 3D printing [3Dprinting17], can be used to make sure that the two mating parts can fit before making the mold, and thus, the overall success rate is about 95% only due to shrinkage uncertainty. In this way, the additive manufacturing process can greatly decrease the cost and time of process development. Therefore, it is very effective to build and debug the prototypes to eliminate most mistakes, to verify the robustness of a product, and to examine the production tool design.

2.1.2 *Common mistakes and issues in product prototyping*

Despite the advantage of prototyping in product development, there are some possible drawbacks to developing a prototype. For example, there is a need to decide how many prototypes and design iterations will be sufficient. Part of the purpose of prototyping is to assess customers' needs; however, there may be conflicts between developers and customers. When the customers request excessive changes, how are conflicts managed? This question leads to a related inquiry: how does one manage the schedule for a development cycle that is essentially open ended? Will the prototype produce runaway expectations from the customer? Sometimes, these issues are unavoidable, but they need to be carefully managed and planned.

How are runaway expectations prevented? The key is product performance and the expectations of the customer. There are several ways to potentially resolve the issues, such as to build the expected performance into the prototype, to develop realistic specifications, to distinguish between primary and secondary requirements, and to give customers a crash course in prototype development. Giving the customer a crash course in prototype development is also helpful, as it will allow one and one's customer to be on the same page as the project progresses. This is important since the development of realistic specifications will warrant one to distinguish between primary and secondary requirements to meet the customers' true expected needs.

How does one know when to stop prototyping? This question should be addressed by checking the prototyping goal(s). In other words, one can say that the prototyping activity is completed and should move forward the rest of the project, when (1) the prototype meets all the requirements of the final system, (2) the prototype has a limited purpose (e.g., gathering initial requirements), (3) the purpose has been achieved, or (4) developers and users jointly agree to move on to the next stage. Some people may say that it is done when one runs out of time or money. However, if a prototyping activity is terminated due to lack of time or money before the objectives are accomplished, the prototyping activity fails.

A related common mistake is beginning a product development project by constructing a full model of the final product to achieve what it will look like, how it will work, and so on. A misconception is that the closer one's prototype is to the actual thing, the better the evaluation will be. Very often, one may find that one still has so many design options to try, but the time or money is running out. In other words, overcommitment of resources to one prototype early in a process without examining other options is a poor engineering design. A prototyping swamp happens when there is a misguided prototyping effort that does not really contribute to the goals of the overall product development effort. It leads to longer product development time and more expensive product and could lead to the failure of the entire product innovation project. Table 2.1 shows some of the current prototyping activities that are influenced by two factors: prototyping cost and technological risk. For example, a low-risk and low-cost, comprehensive prototyping case could be to print a report. In this case, one comprehensive prototype, that is, a full hard copy of the report, could be used for final validation. A high-risk and low-cost case could be to develop software. In this case, many comprehensive prototypes, that is, many versions of the software, may be developed and tested. A low-risk and high-cost case could be to build a house. In this case, no comprehensive prototypes may be developed. In this case, the first prototype is also the product itself. A high-risk and high-cost case could be to build an airplane. In this case, no comprehensive prototypes will be developed. However, to reduce the risks, many analytical or focused prototypes may be developed early in the stage. In this case, the first comprehensive prototype is also the product itself.

Table 2.1 Cost and risk effects to prototyping

Technical risk	Cost of comprehensive prototype	
	Low cost	High cost
Low risk	One prototype may be built for verification, e.g., printed goods	Few or no comprehensive prototypes built, e.g., commercial buildings, ships
High risk	Many comprehensive prototypes built, e.g., software, consumer products	Analytical prototypes used extensively. Carefully planned comprehensive prototypes, e.g., airplanes, satellites, automobiles

The decision to prototype a project either physically or analytically should always be decided based on a collection of its overall cost trade-off, cycle time to complete, the accuracy of prototyping the part, material properties to be evaluated, the part size, strength, its availability, etc.

Other commonly seen mistakes in the prototyping process include (1) commitment too early to a particular design, (2) gaining a false view of how long the system will take to complete based on the time taken to prototype, (3) too many design iterations that could lead to maintenance and/or operational problems associated with previous versions, (4) the performance characteristics of prototypes misled the customer, and (5) utilizing materials or methods that do not reflect the final design, leading to erroneous performance data.

2.1.3 How to conduct prototyping?

The best reason to prototype is to save time and resources, and thus, the time and resources need to be well planned and well spent. The prototype should be built like a facade for a specific purpose—like a Hollywood set, where only the front of the building is constructed.

For a minimal investment of time and money, one can find usability and design problems and adjust one's design before one invests greater resources. This is why architects create models out of paper or cardboard, or use virtual reality tools to evaluate how the building will look before it is constructed. This is why aeronautical engineers use wind tunnels to optimize airplane profiles, why bridge builders create stress models before an actual bridge is constructed, and why software and web designers create mock-ups of how users will interact with their designs.

It is important to note that misguided prototyping efforts do not contribute to the goals of the overall product development efforts. These misguided actions can actually end up costing the precious project time, resources, and money without contributing anything to the process. Choosing prototyping for a product is a very critical issue, and it depends on many factors, such as cost trade-off, cycle time, the accuracy of the prototyped part, material property, part size, part strength, and availability. Most importantly, the selection should be based on the prototyping objective. A prototype can be a physical prototype in which artifacts are created to approximate the product (e.g., models that look and feel like the product). A physical prototype is commonly used for proof of concept to test the idea and used as experimental hardware to validate the functionality of a product. However, a prototype can also be an analytical prototype in which mathematical, analytical, simulation, or a computer model is used to analyze the product. A prototype can be a comprehensive prototype—a full-scale, fully operational version of the product. For example, a beta prototype is given to customers to identify any remaining design flaws before committing to production. A prototype can be a focused prototype to implement one or a few of the attributes of the product (work-like or look-like product). For example, a foam model is made to explore the form of a product or wire-wrapped circuit board to investigate the electronic performance of a product design.

Figure 2.1 shows the coordinate system for prototype classifications. A paper prototype which could be a sketch of a concept of the kinematics of a product is an analytical and focused prototype. A form model to show the "look" of a prototype is a physical but focused prototype. An analytical but comprehensive prototype is likely to be a full model-based virtual prototype, which could include the full geometry, kinematics, dynamics, finite element analysis (FEA) model, and many other aspects of the product. Simulation technologies are getting better and more accurate, and it is possible that an analytical model could be a comprehensive model. A physical and comprehensive model could be an alpha, beta, or production prototype. An alpha prototype is the first system construction of the subsystem that is individually proven to work while a beta prototype is the first full-scale functional prototype of a product constructed from the actual materials as the final product. A preproduction prototype is the final class of physical model used to perform a final part production and assembly assessment using the actual production tooling. In order to select the prototype of choice, there are some general principles of prototyping [Ulrich99]:

- Analytical prototypes are generally more flexible than physical prototypes.
- Analytical prototypes can be used to narrow the range of feasible parameters.
- Physical prototypes can be used to fine-tune or confirm the design.

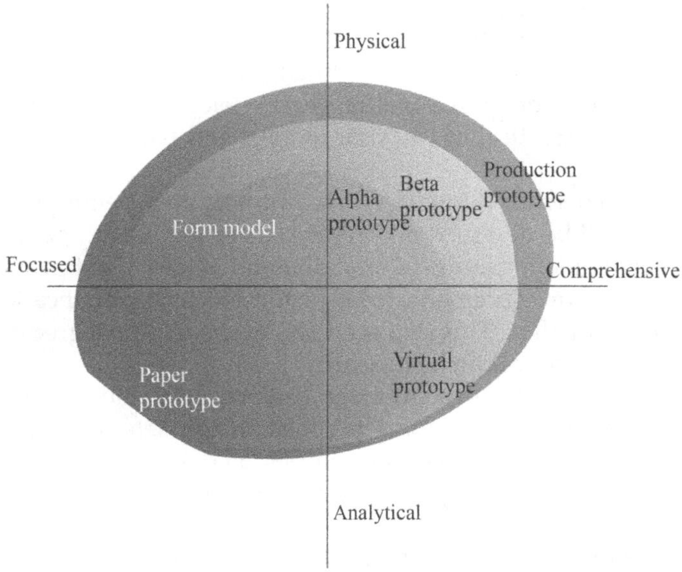

Figure 2.1 Prototype classifications.

- Physical prototypes are required to detect unanticipated phenomena.
- Analytical models can seldom reveal phenomena that are not part of the underlying analytical model.
- A physical prototype often exhibits unanticipated phenomena completely unrelated to the original objective of the prototype, for example, thermal, optical, and electrical–mechanical coupling.
- A prototype may reduce the risk of costly iterations.
- Anticipated benefits of a prototype may be weighted against time and money.
- A prototype may expedite other development steps.
- A prototype may restructure task dependencies.

Before constructing a physical prototype, the following items should be kept in mind:

1. Purpose of the prototype
2. Development time of the prototype
3. Possible forms
4. Types of tests to perform once the prototype is completed
5. Risk of constructing prototype or continuing without it

Before conducting a prototyping activity, a plan needs to be outlined as follows:

Step 1: Define the purpose of the prototype. Some examples of the purpose of the prototype include the following: (1) to learn more about customer requirements, (2) to show proof of concept to senior management, (3) to validate system specifications, (4) to resolve fuzziness in the early stages of design, (5) to explore solutions to specific usability or design problems, (6) to integrate subsystems, (7) to manage change requests, (8) to pretrain users, (9) to create a marketing demo, (10) to improve customer acceptance, (11) to assure product quality; and/or (12) to enforce milestones.

Step 2: Establish the level of approximation of the prototype. The best prototype should be the simplest prototype that will use the minimum resources to achieve the purpose of the prototyping activity. A simplified prototype is to approximate the actual product, and the level of approximation can be selected from Figure 2.1. It could be defined based on analytical or physical models and based on the level of focus or comprehension.

Step 3: Outline an experimental plan. When a prototype is being planned, the experimental plan should be designed and reviewed for the prototype. The experimental plan could include identification of experimental values, identification of test protocol, identification of measurements to be performed, and identification of the ways to analyze the resulting data. This step is critical to ensure the degree of achievement of the prototyping objectives can be measured.

Step 4: Create a schedule for procurement, construction, and testing. Prototyping is considered as a project. It is necessary to specify major dates for assembly, smoke test, and complete testing. If the prototype completion date is carried over the launching date of the next prototype or production operation, then hardware swamp occurs [Clausing94].

Step 5: Perform more detailed prototype planning that will involve extensive time and cost.

Section 2.1.4 will discuss in more detail the planning and management of a complex project.

2.1.4 *Physical prototype design procedure*

Once a physical prototype is decided upon, it generally involves more time and project cost, and thus, more detailed planning effort is needed. The concept of product definition as discussed in Chapter 1 should be actively considered concurrently here. The following sequential but overlapping prototype design tasks can be carried out. The actual implementation sequence and focus could vary from a product to the other.

2.1.4.1 *Task 1: Prototype conceptual design*

The prototype design should include customer requirements (a new product, a subassembly, or a part) for problem formulation. Problem formulation should elicit customer requirements and include a vision and market opportunity analysis. Problem formulation is a very critical step that is often overlooked. Approximately, 50% of design effort should be used to define the problem, and the other 50% should be used to find the solution. Unfortunately, most engineers jump into the solution space before an appropriate problem is defined. Problem formulation should define how a product will function, and how subassemblies or parts will function with their couplings. These alternatives in conceptual solutions should be generated, and these conceptual solutions should be evaluated analytically or by using simplified physical prototypes. The outcome of problem formulation is a problem statement and should include the following items:

1. A need statement or customer's requirements should include prototype performance (operation of the finished product), project schedule (time), and all cost before and after the design phase.
2. Prototyping goals should be set according to the need statement. Prototyping goals should present what is to be achieved rather than how it will be achieved.

3. Constraints and trade-offs should acknowledge laws of nature (gravity, etc.), interchangeability of parts and standardization, dimension limitations, legal constraints, and customer constraints.
4. Definitions of terms and conditions, such as legal terms and conditions.
5. Criteria for evaluating the design. The outcome of the problem statement should be a well-defined set of design goals, which are called product design specifications (PDSs).

2.1.4.2 *Task 2: Configuration design of prototype parts and components*

After the conceptual design, some prototype parts will result in standard components and configurations, and in this task, design involves selecting the types or classes such as a motor (DC, AC, stepper) and a spring (leaf, beam, helical). Some configuration designs are used to determine the geometric features (walls, hobs, ribs, intersections, etc.) and how these features are configured. If possible, it is strongly encouraged to use standard parts at the prototype stage to save time and cost. There is a significant cost saving in a design that comes from reusing existing parts. In addition to savings from the designers' time, such saving could be derived from the proven reliability of a particular part. For example, part reliability may require fatigue testing. Based on experience, about 20% of a product's features are new, 40% are modified from an old design, and 40% just directly adapt the previous design features. Using a standard design not only eliminates the need for new tooling in production but also significantly reduces the number of parts that must be stocked to provide service over the lifetime of the product. An effective way of design reuse is to use group technology to retrieve designs of existing parts. The fewer the number of new parts needed, the simpler are the machine and equipment requirements. In this task, design for manufacturing and automation also needs to be considered.

2.1.4.3 *Task 3: Parametric design*

This task defines major dimensions and tolerance information and specifies materials needed for the prototype.

2.1.4.4 *Task 4: Detailed design*

This task supplies remaining dimensions, tolerances, and material information for engineering drawings.

Review problems

1. What is the role of prototyping in design?
2. When performing an initial prototyping activity, which of the following may not be a good practice?
   A. Begin a design project by constructing a full model of the final product
   B. A concept model with no working features to obtain early feedback from customers
   C. A rough version to answer a single or set of binary yes or no questions
   D. Thinking that the closer your prototype is to the actual thing, the better it is
   E. A rough version to visualize and brainstorm possible improvement
3. What are some of the common mistakes to be avoided in the prototyping process?
4. What is the difference between a comprehensive prototype and a focused prototype? Give an example.
5. What is preproduction prototyping? What types of prototypes are these?
6. What is prototyping swamp? How can this be avoided?

7. Why should 50% of prototype design effort be devoted to problem formulation alone?
8. State two pros and two cons of using a physical prototype and an analytical prototype.
9. Company XYZ would like to design a new television remote for their new high-end line of TVs. What are some of the questions the customer should be asked to make sure that one is designing what the customers want and what is the reasoning for these questions?
10. Take one product and give one type of prototype for each of the four quadrants of the focused/comprehensive versus physical/analytical chart, and give a short explanation why they are placed in that quadrant and the reason for that particular prototype.
11. What are the conditions to stop prototyping activities?

2.2 *Prototype planning and management*

> The secret of getting ahead is getting started. The secret of getting started is breaking your complex overwhelming tasks into small manageable tasks, and then starting on the first one.
>
> **Mark Twain**

This section helps answer the following questions:

- Why are project definition and vision important?
- How to manage prototyping projects?
- How to define the objective of a prototyping project?
- How is project risk management conducted?

If a prototyping effort is more than a simple sketch or activity to complete, it will require proper planning and management before time and money are invested. Not only the entire product design and development effort can use project management and planning, but also each prototyping project itself may require substantial project management and planning effort. In developing a prototype, the design and fabrication phase may only take 50% of the time. It may take as much time to debug, redesign, or reconstruct as it did to create it in the first place. It is rare for a new design to work right the first time, due to errors in design, miscommunications at various activities, and mistakes in fabrication. This section provides an overview of project management and how to manage a project.

2.2.1 *Project vision in project management*

It is very easy to get caught up in the instant gratification of a prototype and include far more functionality than intended or needed. This can drag out the prototyping phase far too long and bring too little benefit to the solution. If one did not intend to demonstrate the functionality, then one should be disciplined enough not to include it. Spending time at the front end of the project defining the objective and establishing the individual tasks is a key to understanding the requirements needed to fulfill the goals of the project. Project management is a way of exercising control, a way of preventing mistakes, and a way of maximizing efficiency. It is critical for prototype development since it is directly related to the success of the prototyping project—to produce the prototype on time and within cost. There are many success factors for a project, but they normally can be related to resources, time, results, and customer satisfaction. In order to properly control and plan for the project, it is critical to (1) establish performance standards at the beginning; (2) use decision trees/matrices, quality function deployment (QFD), and/or Taguchi's methods to carry out the project; (3) assure performance standards as the project progresses; (4) verify final performance against standards; and (5) give no surprises to the stakeholders.

It is very important to establish the project vision at the beginning. There are many benefits to be gained from establishing a project vision. It helps to provide a common understanding of the project between the client and you. The project vision also helps one to know when the project will be done. The project vision can help to define the general scope of the project, get people enthusiastic about the project, get the project started, and see past obstacles. The project vision can benefit a project with postevaluation, define the expectations that need to be met, and evaluate whether to do the project at all. To set a good project vision, one needs to focus the design on the big picture and future potential.

How to define the success of a project? The general answer is "success equals met expectations," and thus, a project will never be successful unless we know when a project is successful. Whether the client is internal or external, projects are successful only if the customer is satisfied. Forgetting to keep a clear focus on what the client needs and wants will lead to disaster. Therefore, a good project vision should be SMART:

- Specific
- Measurable
- Achievable
- Relevant
- Time dimensioned

Depending on the scope of a prototyping project, a project vision may include business requirements, business opportunity, business objectives, customer/market requirements, the value provided, business benefits, and business risks. Project vision describes how the project fits into the strategic plans of the organization, and how it adds to the profitability of the company. It should also provide a vision of the solution, including a vision statement, major features, assumptions, and dependencies. A project vision should also define project scope and limitations, including the scope of the initial release, the scope of the subsequent release, and their limitations and exclusions. It should also include business contexts such as customer profiles and project priorities.

2.2.2 *How to manage prototype projects?*

Once a project vision is established, a prototype project plan should be established. Time and resources will need to be estimated and planned from top-down and from bottom-up. The release date needs to be specified, but be prepared for change. Remember Murphy's law: "anything that can go wrong will go wrong." One needs to assume that changes will be needed during project execution. Whether it is correcting design errors, fixing mistakes in construction, or reengineering the entire concept, the odds that one will not have to make any changes are very close to zero. Therefore, choose prototyping methods that allow for easy changes, leave enough space in the prototype if possible, and do not set anything in stone until it is certain that it will work.

Parkinson's law says, "work expands to fill the time available for its completion." If the boss gives 12 months for a project, the employees will fill 12 months' work to complete the job. It is critical to budget the time and resources in detail and with practical considerations. There is a balance for all tasks. Brainstorming is a good practice to follow so that

Plan B can be used if necessary. One needs to regularly monitor and assess these risks during the project's duration.

Good project management of a project team includes good communication, mentoring and brainstorming, resource allocation, tracking and measurement, and a good project manager:

- Communication: Communication is taken for granted because everyone does it so frequently, but it is a complex process. Good communication should be used with the client and within the team. This task includes sufficient planning for each meeting, project status reports, Listserv and Web site, show and tell activities, and e-mail communication. Try to use effective project management tools.
- Mentoring and brainstorming: Successful people turn everyone who can help them into mentors. This task has team members help each other and solve project problems. How does one know there is a problem? Is it important to have constant task reviews? Problem-solving techniques will be helpful.
- Resource allocation: The resource allocation decision is a plan for using available resources to achieve goals to produce the prototype. This task needs to define who should work on what task, how to avoid overload and underutilization, and what to do when things change.
- Tracking and measurement: This task requires tracking task completion, deadlines, and budgets. It also includes possible issues such as potential project risks and status.
- A good project manager: A good project manager should have good technical skills, people skills, and business skills. In the prototyping process, leader designers, analysts, fabricators, technicians, and mechanics need to work together. The project manager should be able to build a shared vision, to bring out and challenge, and to foster more systemic patterns of thinking.

Figure 2.2 outlines the key components to manage a project [Ertas96]. The first component is project planning to define prototype objectives and prototype tasks, estimate prototyping efforts, determine task interdependencies, schedule tasks, schedule resources, balance resources, and obtain plan approval. The second component is project direction to review plan objectives and tasks, assign tasks, review criteria for task computation, and provide motivation to the members. The third component is project control to review progress, control changes, report progress, replan during the change, conduct reviews, review and approve completed work, and close project. The fourth component is project administration to recruit and develop personnel; evaluate performance; develop and maintain policies, standards, guidelines, and procedures; and keep track of the budget.

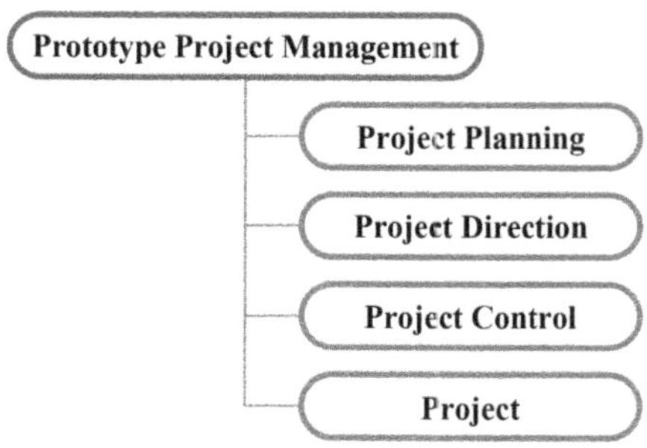

Figure 2.2 Prototype project management.

One of the most critical tasks in project planning is to define objectives to make sure that project objectives are clear and well defined. If one is not clear on why a prototype is needed, it can easily become a waste of resources. Keep it simple, make sure all intentions are clear. The project team first needs to define what the prototyping goal will be, and then state what needs to be accomplished and tested. One may think that to define an objective is easy. However, since this task is so critical, it deserves a full effort. A prototype may have one or more main objectives, and there may be alternative objectives too. If there is more than one objective, all alternative objectives of the project should be considered. If a project is too large, sub-objectives and intermediate milestones may be needed. For example, the objective "to build a vehicle prototype" is different than "to build a vehicle prototype to see if potential buyers would like the style" in many ways. Other alternative objectives may include "to build a vehicle prototype to see if the specified performances can be achieved," "if the customer will like the interior design," and "if this vehicle can be built within the budget." Since different objectives may lead to different prototyping results, it is very critical to define the objectives according to the project vision, which should be clear to all team members.

The secret to getting started on a big project is breaking the complex overwhelming tasks into small manageable tasks and then starting on the first one. It is therefore very important to plan the tasks according to project objectives. The first step is to come up with a list of tasks by choosing from the project objectives. A top-down approach as shown in Figure 2.3 can be adapted. The tasks listed in the first level (i.e., Tasks 1–3) should be the tasks needed to accomplish its upper level task (i.e., project objectives in this case). The subordinate tasks (i.e., Tasks 1.1 and 1.2) should be the tasks needed to accomplish its upper level task (i.e., Task 1 in this case). This procedure goes on until each bottom task is refined into a manageable job. This way all the necessary tasks have been defined to accomplish the project objectives. It is worth mentioning that tasks are not always in sequence. They can proceed in parallel or can be coupled, but the procedure is similar. The project team should pay special attention to designing the project tasks so that there will be as many parallel tasks as possible to dramatically shorten the overall project implementation time.

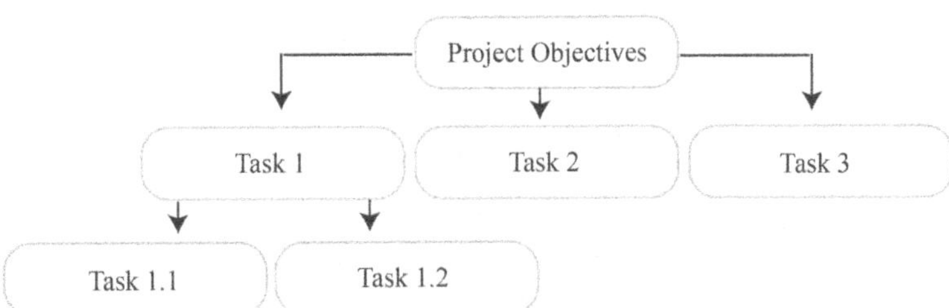

Figure 2.3 Coming up with task lists.

Once the tasks are identified, they should be planned and scheduled against available time and resources. What is the difference between planning and scheduling? In general, planning should come first to estimate, sequence, and analyze time. Scheduling should follow to extend the plan to allocate resources and to work with the details to implement the plan.

The task scheduling and sequencing are very critical. For example, often it is needed to make sure that the major parts of a prototype design will be available before basing a design on these parts. One may find an ideal part for the prototype in the catalog or website, but the part may be unavailable from distributors or may be too expensive. Thus, ensuring these parts will be available should be the first task. It is especially beneficial when the product being designed will be produced for a while. It is wise to choose these parts that are widely available.

The major elements in project planning—time, resources, and scope—are also called the project planning triangle, which is shown in Figure 2.4. "Scope" is the goals and tasks of the project and the work required to complete. Resources include people, equipment, and material used to complete tasks in a project. To determine resource needs, one needs to check the project scope, the resources available, and the tasks to be done, and decide how many resources are needed for each task. For time management, one needs to estimate the duration of each task, establish the precedence of tasks based on the procedure followed in Figure 2.3, record milestones (performance and dated), and plan start and finish dates for tasks. As shown in Figure 2.4, the purpose of planning and scheduling is to break down the entire project triangle (scope, time, and resources) into smaller project triangles (task, duration, and resources). The final results can be in the form of Gantt charts (bar charts), program evaluation review technique charts, and, in terms of resources, circle dot charts.

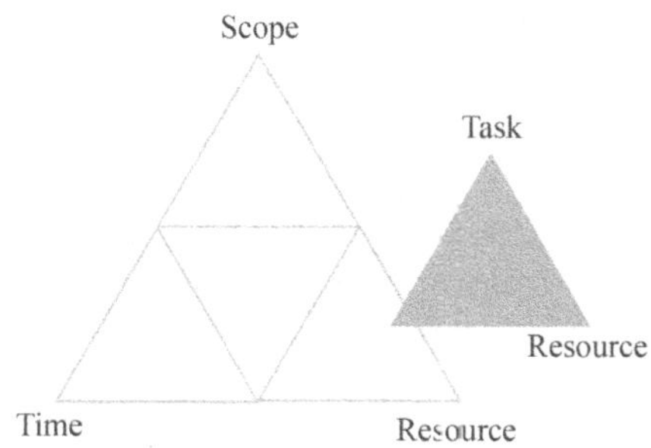

Figure 2.4 Project planning triangles.

Setting good milestones can provide critical project momentum. Due to the risks involved, not a lot of prototypes have official buy off and are generally seen as testing activities. If a prototyping project is not showing results, team members will begin to focus on other ideas or management will move team members to higher priority activities. There is almost always something higher priority than a prototype that is not going anywhere. A prototype becomes more valuable when it is showing great promise so that people will get excited about the project. They will put in the extra time if needed and it becomes more difficult for the project to have its resources taken away. A project with momentum can go through several iterations in a short period, while a project without this momentum might hit two or three iterations in the same amount of time. Both projects can have outstanding design ideas at their core, but poor execution of the prototyping phase will kill off the low momentum project before it even has a chance.

Designing each milestone with a certain level of prototype deliverables will provide momentum for the project, both for team members and for the management.

During project implementation, the following functions, such as project directing, project execution, project monitor and control, and project closing, are very important phases:

Directing: The most efficient way to complete a task is to know how to delegate the job to people who can help, and the best way to delegate the task is by defining the expectations. It is too often assumed that the coworkers or subordinates understand what are asked of them, and they do not. If the person delegated the responsibility does not understand what to expect or to what extent one expects it completed, one would always be disappointed. The best thing to do when delegating responsibility is to be sure that the expectations have been defined to a level where both parties understand what is to be expected. Review the project's objectives and determine how the project will be organized, staffed, and managed specifically to the project. Another task is to prepare the resource requirements of the project.

Execution: The project execution phase consists of the processes used to complete the work defined in the project management plan to accomplish the project's requirements, and track project execution against cost, schedule, and technical performance.

Monitor and control: The monitoring and control phase consists of those processes performed to observe project execution so that potential problems can be identified promptly and corrective action can be taken when necessary, to control the execution of the project. A task measured is a task that gets completed. Continue pursuing the project objectives such as cost, schedule, and technical performance.

Closing: Closing includes the processes used to formally terminate all activities of a project or a project phase, hand off the completed product to others, or close a canceled project.

2.2.3 *Project risk management*

To assess the alternatives and status of a project, one sometimes uses a decision tree for risk management. It will help to make decisions in succession into the future. As shown in Figure 2.5, a decision tree is started with a decision that one needs to make. This example decides what type of prototyping method to use. Draw a small square to represent a decision point that can be determined by the project team. From this box, draw outlines toward the right for each possible solution, and write that solution along the line. At the end of each line, consider the results. If the result of taking that decision will depend on event probabilities, draw a small circle. If the result is another decision that one needs to make, one can draw another square. Squares represent decisions, and circles represent uncertain outcomes. Write the decision or factor above the square or circle. If the solution has been completed at the end of the line, just leave it blank.

Starting from the new decision squares on the diagram, draw outlines representing the options that one could select. From the circles, draw lines representing possible outcomes. Again make a brief note on the line explaining what it means. Repeat this function until as many of the possible outcomes and decisions leading on from the original decisions have been drawn out.

To evaluate the decision tree, as shown in Figure 2.6, start by assigning a cash value or score to each possible outcome. Estimate how much it would be worth if that outcome came about. In this particular example, the expected values of $5,000, $2,000, and $0 are assigned to good, moderate, and poor performance of the prototype in terms of meeting the design requirements. The expected cash values (or scores) should be evaluated based

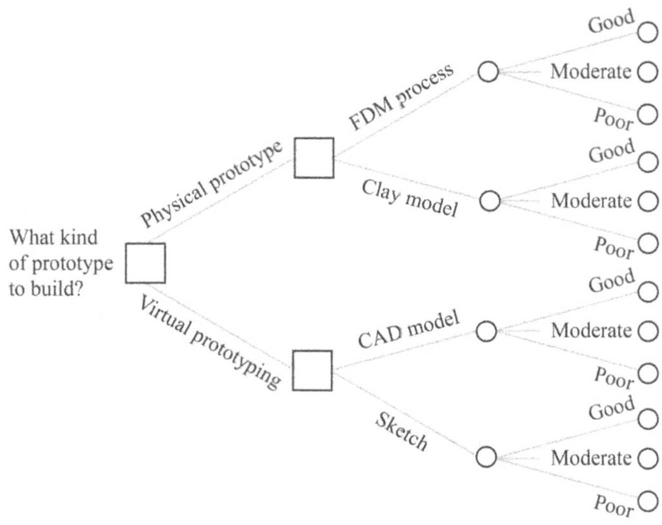

Figure 2.5 Decision tree.

on the prototyping objectives. Next, take a look at each circle (representing an uncertainty point) and estimate the probability of each outcome. If percentages are used, the total must total 100% at each circle. If fractions are used, these must add up to 1. If one has data on past events, one may be able to make rigorous estimates of the probabilities. Otherwise, write down the best guess.

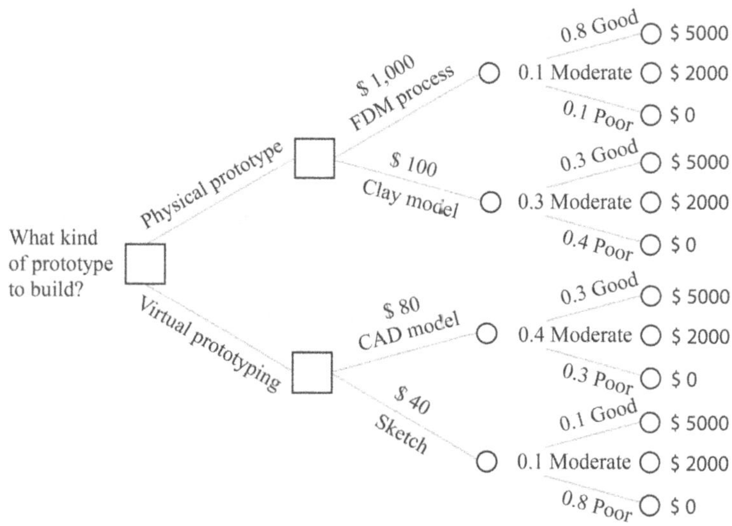

Figure 2.6 Further developed decision tree.

To calculate tree values, start on the right-hand side of the decision tree, and work back toward the left. As one completes a set of calculations on a node (uncertainty circle in the figure), all one needs to do is to record the result. All the calculations that lead to that result can be ignored from then on.

To calculate the value of uncertain outcomes (circles on the diagram), multiply the value of the outcomes by their probability. The total for that node of the tree is the total of these values. When evaluating a decision node, write down the cost of each option along each decision line. Then, subtract the cost from the outcome value already calculated. This will give a value that represents the benefit of that decision. In the example in Figure 2.6, the value for the physical prototype, fused deposition modeling (FDM) process is given as follows:

0.8 (probability good outcome) × $5,000 (value) = $4,000
0.1 (probability moderate outcome) × $2,000 (value) = $200
0.1 (probability poor outcome) × $0 (value) = $0
The total expected value = +$4,200 − $1,000 (FDM process cost) = $3,200

Once these decision benefits are calculated, choose the option that has the largest benefit, and take that as the decision made. This is the value of that decision node. In this example, as shown in Figure 2.7, expected cash values are listed on the right side of the tree. Since the FDM process can end up with the largest cash value, it should be chosen as the process for this case. Therefore, the recommended prototyping method should be a physical prototype—FDM process.

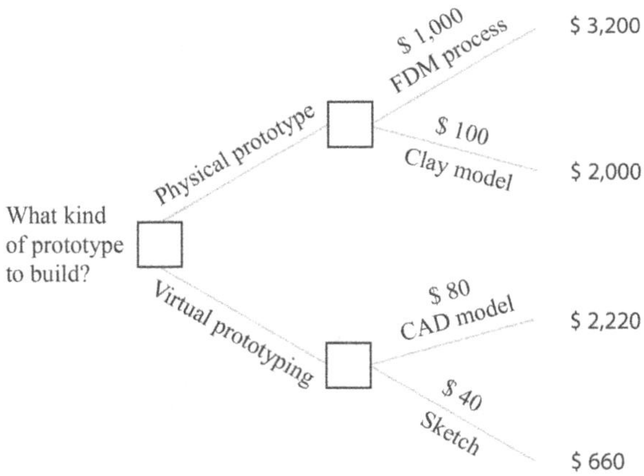

Figure 2.7 Making decision based on expected cash values.

This section has illustrated a method for project risk management. It will help the project team to thought out a project with complex project decisions. Some sensitivity analysis may be exercised for some uncertain values, such as the cost or probability. More references such as [Chapman03 and Kendrick03] are available.

Review problems

1. What does the acronym SMART stand for and what does it relate to?
2. Why is defining project objectives critical to project management?
3. What is the most important task of project management and why?
4. Identify tasks and subtasks to build a remote control airplane. This should consist of purchasing components, building, and testing tasks.
5. Determine the total expected values for each process for the following additive manufacturing tooling applications shown in Figure 2.8.

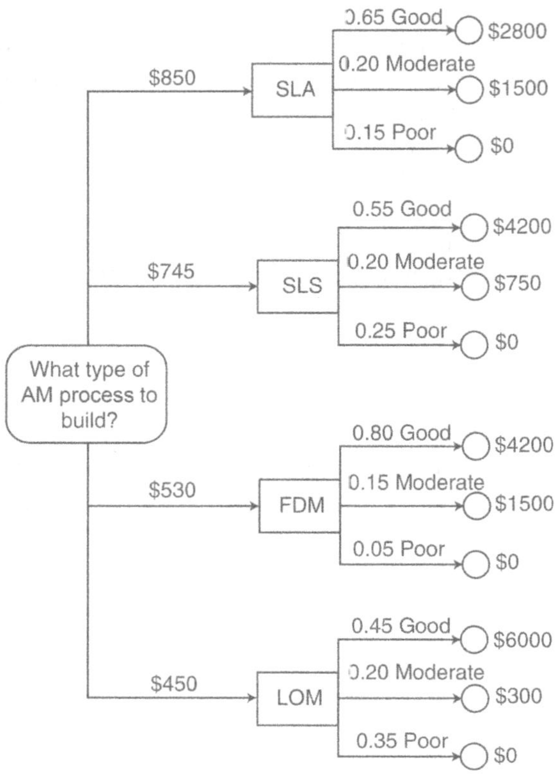

Figure 2.8 Additive manufacturing tooling applications.

6. What is resource allocation?
7. What are the advantages of establishing project vision at the beginning?
8. Are planning and scheduling the same? Why or why not?
9. You are the project manager of a design group and have to decide how to proceed with the construction of a prototype for your current product development project. Given the data below, construct a decision tree to help with the process.

		Probability of success		
Type of prototype	Cost ($)	Good	Moderate	Poor
Physical—SLS prototype	1,500	0.85	0.10	0.05
Physical—wood mock-up	800	0.50	0.30	0.20
Virtual—CAD model	200 (2 h at $100 per hour)	0.30	0.50	0.20
Virtual—sketch	100 (1 h at $100 per hour)	0.10	0.20	0.70

The expected values are $4,000 for good results, $2,000 for moderate results, and $0 for poor results.

2.3 *Product and prototype cost estimation*

Quality in a product or service is not what the supplier puts in. It is what the customer gets out and is willing to pay for. A product is

> not quality because it is hard to make and costs a lot of money, as manufacturers typically believe. This is incompetence. Customers pay only for what is of use to them and gives them value. Nothing else constitutes quality.
>
> **Peter F. Drucker**

This section helps answer the following questions:

- How to conduct cost estimation?
- What are the methods that can be used to estimate a prototype cost?
- Why define variable costs and fixed costs?
- To work on an entirely new project, how can one estimate cost if one does not even have the design of the product?

The most difficult tasks to control in project management are time and cost. As discussed in the previous section, the time required for a project can be estimated from each of the breakdown tasks. If one can estimate the time required for each manageable subtask, one can estimate the overall time required for a prototyping project. Cost estimation is without a doubt one of the most difficult and important aspects of any engineering project. A good estimation of cost can help to make each decision and even determine whether or not a particular endeavor is even worth pursuing. One of the most challenging tasks when estimating cost is doing so for the production of a prototype. The thing that makes prototype cost estimation, so difficult is that it must be done at the beginning, or even before the design stages of a project. This means that there is a minimum of historical data to use as a guide. Without the aid of historical data, there are some tried and true methods that can help an estimator get a good idea of what a particular product or process is going to cost. When planning the product definition, it is important to think ahead in all aspects especially cost. If one is working on an entirely new project, how does one estimate the cost if one does not even have the design of the product yet? As cost is a key driving factor in product development and prototyping, the estimation of the cost will help in design decision-making. It will certainly help in product risk management too. This section describes how to estimate costs at the beginning of a product design project.

2.3.1 Fundamental cost concepts

When designing a product or a prototype, it is very critical for a design engineer to learn about the cost factors—all costs that can contribute to the final cost of the project. Many opportunities for product cost optimization are embedded in the process of defining the project. Often it is too late to make a change in the later part of the product development stage. A cost estimate establishes the baseline of the project cost at different stages of the product development. Cost engineering is defined as that area of engineering practice where engineering judgment and experience are utilized in the application of scientific principles and techniques to the problem of cost estimation, cost control, and profitability. Cost estimation is also an art since sometimes cost estimates may be influenced by other factors, such as those influenced in a political way. For example, 12 months were given to do the job so that the project will be budgeted for 12 months. One wants to show the product at the trade show after 7 months, so the project needs to be budgeted for 7 months to prepare for the trade show. One knows that the competitor put in a bid of $1.5 million, so one will put $1.4 million to win the bid. The project needs 8 months, but the boss will not accept more than 6 months, so the project is budgeted for 6 months. This section will not discuss these factors, but one may need to consider these in many industrial projects.

What is a good cost-estimate procedure? Good cost-estimate practices should include the following: (1) uses the best available data; (2) adjusts data for definitions, accounting changes, time, and the unit number of production; (3) applies the appropriate estimating techniques; (4) addresses risk and uncertainty; and (5) is well documented. The sources of data for cost estimation include market research, data from prior programs, data from the current program, and some historical files. However, data should be compared apples to apples. In other words, data should be adjusted for price level changes. Then-year dollars should be adjusted according to indices to compare with the current dollars. The comparison should also be adjusted for quantity, including cost–volume–profit and learning curve.

Learning curve recognizes that repetition of the same operation results in less time or effort expended on that operation. It can be used to obtain cost estimates based on the repetitive production of parts or products. The learning curve effect reduces the labor time per unit that occurs in the early stages of producing a new product [Goldberg03]. As the number of units produced increases, the total labor cost will increase but at a decreasing rate. Labor costs are unlikely to have a linear relationship with total cost. Learning curves are used to formulate low-cost production strategies in increasing market share or to address decisions such as make or buy a product, add or drop a product line, cost–volume–profit decisions, and setting product prices. The learning curve relationship is commonly modeled with a power function described as the log-linear or constant percentage model. The following log-linear model recognizes that labor hours decrease systematically by a constant percentage each time the volume of production increases geometrically (usually a doubling of units).

$$(\mathrm{A \ or \ I_n}) = \mathrm{aX^b} \tag{2.1}$$

where

A = average cumulative labor hours for X number of units
$\mathrm{I_n}$ = number of labor hours required to produce the last nth unit

The independent variables are defined as follows:

a = number of labor hours required to produce the first unit
X = cumulative number of units produced
b = learning exponent, which is always negative

For example, to calculate the number of labor hours required to produce the 1,000th widget, given a learning curve exponent calculated at −0.25. Assuming it takes 6 h of labor to produce the first unit, using Equation 2.1, the number of labor hours required to produce the 1,000th unit can be estimated as follows:

$$\mathrm{I_n} = (6)1{,}000^{-0.25} = 1.067\ \mathrm{h}$$

The point estimate is a cost estimation method that typically is the sum of the most likely costs for each cost element. For example, the likely project cost is $1.0 million. Point estimate does not give a feeling for the possible dispersion of the cost. One can perform a sensitivity analysis by varying the assumptions one at a time, how much they cause the estimated cost to change. The estimated cost is "sensitive" to those assumptions that drive relatively significant changes.

There are some important terms in cost estimation. Manufacturing costs include direct materials, direct labor, and manufacturing overhead, whereas nonmanufacturing costs include overhead, marketing, and administrative cost. Fixed costs or capacity costs include the costs of providing the basic operating capacity. These indirect costs are constant over a relevant period and remain relatively unchanged regardless of the level of output. Variable costs have a close, often direct, relationship to the level of output. In a typical manufacturing environment, direct labor and material costs are considered as variable costs. Gasoline and tire replacement costs are good examples of variable transportation costs. Semivariable costs have both fixed and variable portions. They are indirect costs that vary with the level of output but not in direct proportion to the output. Some components of power consumption are independent of operating volume, while other components vary directly with volume.

As an example, for a manufacturing corporation, its main cost includes labor cost, material cost, tooling cost, utility cost, operating cost, and overhead cost. Labor cost is the salary of people directly producing the product. It is correlated to the level of output, so it is a variable cost. Material cost is the cost of raw materials and thus is a variable cost.

Tooling cost includes the cost of cutting tools, molds, and dies that are subjective to be worn and torn by processing products, and thus, they are correlated to the level of output. Hence tooling cost is a variable cost. Utility cost includes the cost of energy, water, waste disposal, and sewer, and thus varies with the level of the output. However, as it does not directly correlate with the level, it is a semivariable cost. Operating cost is the cost of things that are consumed during the production process such as coolant, lubricant, and towels and thus is directly related to the level of output. It is a variable cost. Overhead cost may include the cost of equipment maintenance, the salary of people that work in the office, and the shipping cost.

Such a cost will vary with the output level, but it is not directly related. So this cost is semivariable cost. Capital cost includes loan interest, insurance, and land rent. These are the costs that must be paid regardless of whether some products are produced, and thus, they are the fixed cost.

Fixed and variable costs are critical when studying other similar cases to estimate the current prototype cost. Fixed costs will not vary with respect to output quantity, whereas variable costs should be adjusted.

2.3.2 Prototype cost estimation methods

Cost estimation is the process of determining cost behavior of a particular cost item, and cost behavior is the relationship between cost and activity (often cost driver). Cost prediction uses cost behavior knowledge to forecast future costs at a particular level of activity. A cost driver is any activity or factor that causes costs to be incurred. Volume-based cost drivers assume all costs are driven, or caused, by the volume of production (or sales), whereas non-volume-based cost drivers are costs that are not directly related to production volume. Activity-based costing classifies costs into four categories: (1) unit-level activities performed for each unit; (2) batch-level activities performed for a group of product units, such as batch or a delivery load; (3) product-level activities performed for specific products or product families; and (4) facility-level costs incurred to support the whole business. Cost drivers can be identified at various levels. Cost-estimate accuracy increases with increased cost categories, but so does cost. To select the appropriate cost drivers, the input or output measure is critical. There should be a strong correlation between cost and cost driver, and cost driver must be easy to measure. One also needs to consider a long-term or a short-term perspective.

For some processes, the cost of producing a part is a simple mathematical function of some attributes of the part-cost drivers. For example, a welding process could have two cost drives: the number of welds and the length of welds. The cost drivers for programming may include effort, hardware cost, software cost, travel, and training cost. Among which, the effort may be most complex, and the cost drivers can further be identified as salaries of programmers and specialists, salaries of support staff, facility cost (space, light, heating, library, etc.), health insurance, and pensions.

As another example, the cost drivers in the forging industry could include material cost, tooling cost, manufacturing cost, and quantities produced. Material cost is the cost to purchase and process enough material to ship the product. The purchased raw material must include allowances for punch-outs, flash, other discards, and machining allowances. Tooling preparation cost quoted by forgers generally may include the cost of designing and manufacturing the tools used to produce the forging. It may also include the cost of special gauges and fixtures. Tooling cost varies with a number of factors, the most important being the forging process. Manufacturing cost includes the cost of labor plus the cost of purchasing, maintaining, and operating the required machinery and material handling equipment.

There is no simple way to estimate project costs, and each cost estimation method has strengths and weaknesses. Estimation should be based on several methods. If they do not return approximately the same result, some action should be taken to find out more to make more accurate estimates. There are several cost estimation methods [Zeil17]:

1. Estimation by analogy: This method relies on comparison to a similar available system and ability to normalize cost data. It is an estimation method that compares the new product with other products that have already been developed. This method can be conducted at the component level of the product, or it can be done at the system level. Estimation by analogy is a pretty quick and accurate way to come up with a price. The main strength of this method is that the estimates are based on actual product data and past experience. An obvious drawback of this method is that it is not applicable if another similar product or product information does not exist. By looking at other similar products, one could compare the material cost of a similar product with that of others and get a fairly accurate measurement of what the cost should be. The same goes for manufacturing techniques and assembly techniques. The trick is to make sure that one is comparing apples to apples and oranges to oranges by using this method. If two products with very different characteristics are being compared, the results can become very inaccurate. For example, one can use DC-8 aircraft to estimate the larger DC-10 since they are similar types of aircraft. However, trying to use a DC-8 to estimate supersonic transport would not be appropriate since they do not have the similar base technologies. One uses aluminum, whereas the other uses titanium as the major material.
2. Bottom-up estimation: In this method, a product is broken up into lower-level components, and costs rolled up. This method adds up the sum of the cost of each component of the product. For instance, a chair is made up of four legs, four cross supports, a seat, four back spindles, and an upper support to stabilize the back of the chair. The

cost of each of these components would be added together to come up with the cost of the overall product. This method may not be applicable during the beginning stages of the product development because not all the components of the product may be known. It can be a very accurate method because each piece is estimated separately. However, it can also be very time consuming.

3. Top-down estimation: In this method, the cost is established by considering the overall functionality of the product and how that functionality is provided by interacting subfunctions. For example, in a product, the major cost components may consist of the purchase parts, the manufacture and assembly of press parts, the manufacture and assembly of barricade parts, the tooling, and the setup. The cost estimation can be based on these functionalities. This method has its advantages because it allows for the incorporation of integration, configuration management, and documentation into the cost estimation of the project. For instance, in the building of a chair, other methods might not take into account the time it takes to integrate all of the components together into the final product. Other methods may not take into consideration all of the bookkeeping and documentation from things such as testing, prototypes, and the different product development phases. This is a faster method of estimation and efficient but may not be very accurate in many cases.
4. Expert judgment: This method uses one or more experts in both system development and the application domain to predict development costs. These experts usually have a lot of previous experience with cost estimation on the related products. The main strength of this method is the assessment of representatives and interactions, and can be used in exceptional circumstances where other normal methods are very difficult to predict. The limitation is that it could be totally biased by the experts. It can be inaccurate because one does not have any control over what the consultant is using as his guiding factors. This is a relatively cheap estimation method. Estimation by this method can be accurate only if the experts have direct experience with similar systems. Expert opinion is probably the quickest estimation method. This method could be applied to almost any other machine with equal ease and efficiency. The downside is that there is no way to judge what the estimation was based on. A different expert will likely come up with a different number but that does not make his estimation any less valid. It is completely possible that even though the overall price was pretty accurate, it could have been nothing more than a lucky guess. The most important thing to remember about the expert opinion is that it can be used in any situation, even when some of the other methods may be very difficult or impossible to apply.
5. Pricing to win: This method is commonly used in cost estimation for winning the biding process. The project costs whatever the customer has to spend on it. The cost will be fixed by the customer, and the product must be manufactured within that cost. The estimated effort depends on the customer's budget and not on the product functionality. For example, when a stapler is quoted at a low price, the additional features that are not really functionally important to the customers should be eliminated to meet the cost. The customer may be willing to pay just a few dollars for a single unit. In that case, if a plastic case can help to produce the product within the cost, then accordingly the operation and design have to be changed. Therefore, in this method, costs do not accurately reflect the work required, and the risk is the low probability of meeting cost, schedule, and performance. The contract is likely won by this cost estimation method.

6. Estimating models: This method uses a model or equation to arrive at a final cost. This method establishes an estimating model so that the input to the model is parametric data, such as mass, design and manufacturing complexity, and elements of new mechanical/electrical design, whereas the output is estimated project effort and schedule, performance, cycle time, and cost as shown in Figure 2.9. Historical data and some insight knowledge should be used to calibrate and fine-tune the model. This process has its advantages because the results are easily repeatable and easily adjusted by changing the algorithms, and the factors of the estimation can be understood well because they can be interpreted through the formulas. New products may not be as accurate with this method because new algorithms would then have to be created. These algorithms also cannot account for things such as team skill levels and teamwork that may speed the process of product development and thus save time and cost.

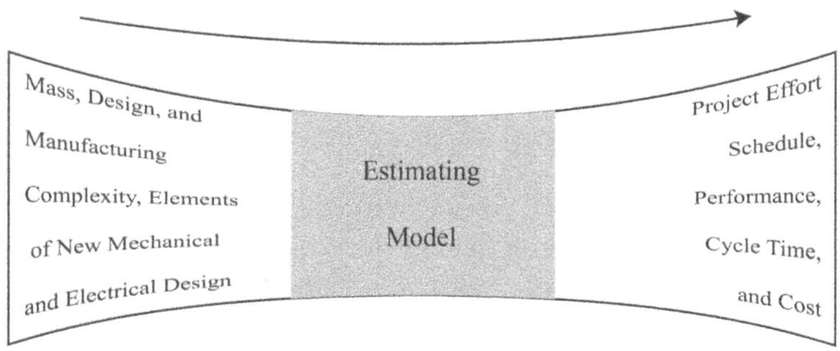

Figure 2.9 An estimating model.

As an example, the cost equation for a particular device is expressed in Equation 2.2. This equation uses three parameters: the number of A components, the number of B components, and the time allowed for completion.

$$\text{Cost} = \frac{6 \cdot 1.5(500n + 800m)}{t} = \frac{9(500n + 800m)}{t} = \frac{4{,}500n + 7{,}200m}{t} \tag{2.2}$$

where

n = number of A components
m = number of B components
t = time allowed for completion in months

The equation was devised as follows: the cost of component A is assumed to be $500, and the cost of component B is assumed to be $800. The factor of 1.5 is to allow for assembly/welding since the amount of assembly/welding should be proportional to the number of parts involved. The factor of 6 is to normalize for the typical amount of time required to complete one of these machines, 6 months. The solution to the model in Equation 2.2 given the typical parts (64 A components and 115 B components) and time (6 months) for this device is

$$\text{Cost} = \frac{4{,}500n + 7{,}200m}{t} = \frac{4{,}500(64) + 7{,}200(115)}{6} = \$186{,}000 \tag{2.3}$$

7. High–low method: In this method, a regression equation is derived by getting the two extreme limits of the observations in the graph between a dependent variable (total cost) and an independent variable (activity measures). The limitations are the reliance on two extreme observations. Since each cost element has a reasonable high and low value, often the cost estimate is a range. Add all the low values together to get the lowest cost value, and add all the high values together to get the highest cost value. The result is a very conservative range estimate as shown in Figure 2.10.

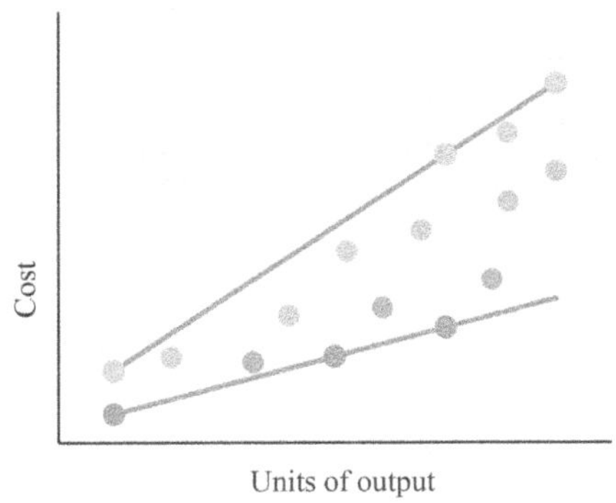

Figure 2.10 An example of a high-low method.

The high–low method may not be ideal for prototype estimation, as the historical data may not be available. However, the historical data may be from existing components. It may also be possible that there are options when building a prototype. Several different versions of prototypes may be under consideration, all with different prices. In this case, the high–low method would display the cost of the different models and show a range of prices for the prototype.

8. Graphical or scatter graph method: Past observations are plotted on a graph, and a line of best fit is shown in Figure 2.11. No attempt should be made to make the smooth curve actually pass through the plotted data points; rather, the curve should pass between the data points leaving approximately an equal number of points on

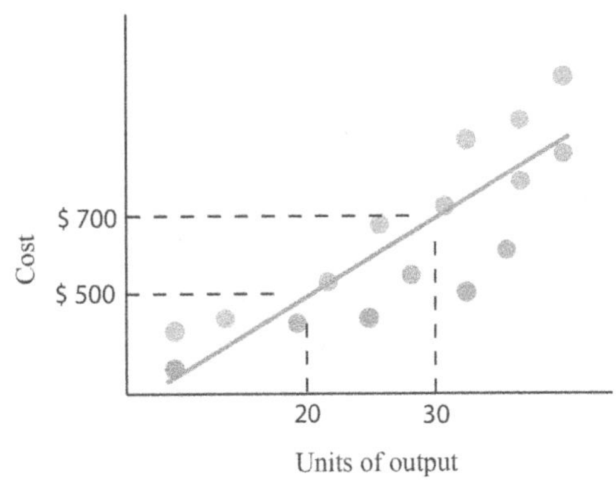

Figure 2.11 An example of a graphical method.

either side of the line. For linear data, a clear ruler or another straight edge may be used to fit the curve. For the example in Figure 2.11, the cost per unit = (\$700−\$500)/(30−20) unit = \$20 per unit, and therefore, the following equation can be used to represent the cost model:

$$\text{Cost} = \$500 + \$20 \text{ per unit}(\text{units of output} - 20) = \$100 + \$20 \text{ per unit}(\text{units of output})$$

The objective for fitting the curve is to "best-fit" the curve to the data points plotted; that is, each data point plotted is equally important, and the curve fitting must consider each and every data point. Although considered a rather outdated technique today, plotting the data is still always a good idea. By plotting the data, one gets a picture of the relationship and can easily focus on those points that may require further investigation. Hence, as a first step, one should plot the data and note any data points that may require further investigations before developing a graphical forecasting curve or mathematical equation. The advantages of graphical methods are that they are quick and easy to use and make visual sense, calculations can be done with little or no special software, and the visual testing of the model is an additional benefit. Whereas the limitations are that they are biased even with large samples, they do not provide minimum variance, and they do not give confidence intervals for the parameters.

One can use regression analysis to use statistical techniques for estimating cost behavior by minimizing the difference between the cost line and the data points as follows:

$$\text{Least squares regression: } Y = a + bX \tag{2.4}$$

where

X = independent variable

Y = dependent variable

$$a = \frac{\Sigma Y - b(\Sigma X)}{n} \tag{2.5}$$

$$b = \frac{n(\Sigma XY) - (\Sigma X)(\Sigma Y)}{n(\Sigma X^2) - (\Sigma X)^2} \tag{2.6}$$

Let us use a simple example to illustrate the procedure. The X and Y values are listed in Table 2.2.

$$b = \frac{n(\Sigma XY) - (\Sigma X)(\Sigma Y)}{n(\Sigma X^2) - (\Sigma X)^2} = \frac{10(2{,}330) - 36 \times 555}{10(162) - 36^2} = 10.25$$

$$a = \frac{\Sigma Y - b(\Sigma X)}{n} = \frac{555 - 10.25 \times 36}{10} = 18.6$$

$$Y = a + bX = 18.6 + 10.25X$$

9. Inspection of account method: The method of inspection of accounts looks at historical data and classifies all expenses as fixed, variable, or semivariable. The departmental

Table 2.2 An example of regression analysis

X	Y	X^2	XY
3	50	9	150
5	65	25	325
2	30	4	60
1	30	1	30
6	90	36	540
5	75	25	375
4	65	16	260
6	70	36	420
3	45	9	135
1	35	1	35
$\Sigma X = 36$	$\Sigma Y = 555$	$\Sigma X^2 = 162$	$\Sigma XY = 2{,}330$

manager and accountant inspect each item of expenditure within the accounts for a particular period of time and classify each item as fixed, variable, or semivariable. Cost can then be estimated based on these methods.

For example, Table 2.3 shows the breakdown costs of a production run. The data in Table 2.2 will be used to estimate the cost of an upcoming production of 50,000 units.

Table 2.3 Data for inspection of accounts estimation

Production	10,000 units
Labor	\$1,600
Material	\$100,000
Tooling	\$2,500
Overhead	\$5,000

Equation 2.7 could be used:

$$\text{Cost} = \text{FC} + (\text{n} \cdot \text{VC}) \tag{2.7}$$

where
FC = fixed cost
n = number of units produced
VC = variable cost

First, the costs must be classified as fixed or variable. In this case, all of the costs are variable except for overhead. All of the variable costs will be calculated on a per unit basis to make future estimating easier. The per-unit costs are shown in Table 2.4.

Now that this data is available, Equation 2.7 can be applied:

$$\text{Cost} = \text{FC} + (\text{n} \cdot \text{VC}) = 5{,}000 + 50{,}000(0.16 + 10 + 0.25) = \$525{,}500$$

Table 2.4 Variable costs per unit

Labor	$0.16
Material	$10
Tooling	$0.25

This estimation was found simply by studying the costs associated with a past project. The inspection of the account method requires good record keeping.

10. Engineering methods: Engineering analysis is based on direct observations of physical quantities required for an activity and then converted into cost estimates. Engineering studies are performed with a focus on the relationships that should exist between inputs and outputs. Time and motion studies (task analysis or work measurement) are the observation of the steps required and time taken by employees to perform particular activities. These methods are useful for estimating the costs of repetitive processes where input–output relationships are clearly defined, such as for estimating the costs associated with direct labor, materials, and machine time.
11. Prototype cost estimation: To estimate the prototyping cost, one can use the previous methods to assess the major elements, such as staff time, hardware and software in the system, consumables and office expenses, space costs for the development team, and travel costs. From a different angle, one can also use the black-box approach as shown in Figure 2.12. The total prototyping cost is the expenditures of all inputs and the disposal of the scraps and wastes. Figure 2.13 shows another way to break down cost estimation. The component cost is referred to as the standard parts purchased from the suppliers, such as motors, sensors, and screws. In this case, a bill of materials can be a good way to estimate the cost. To estimate a new component, one can compare it to a similar part the firm is already producing or purchasing in a comparable volume. One can also solicit price quotes of the major components from vendors

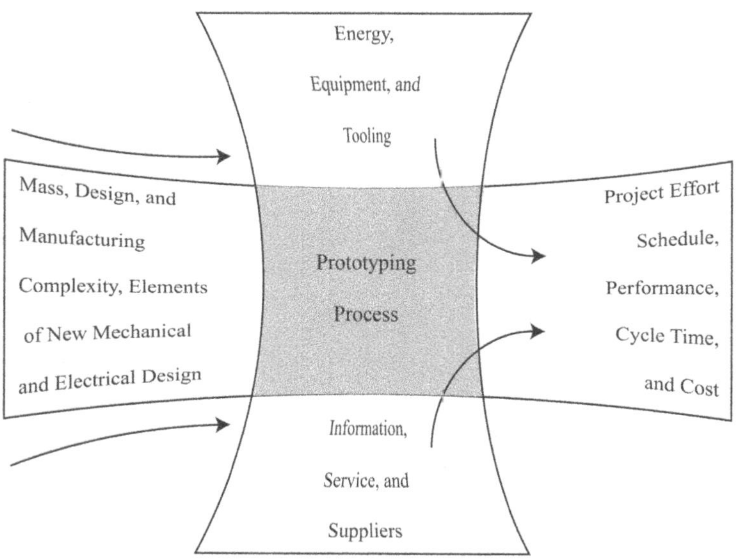

Figure 2.12 An estimating model for prototyping.

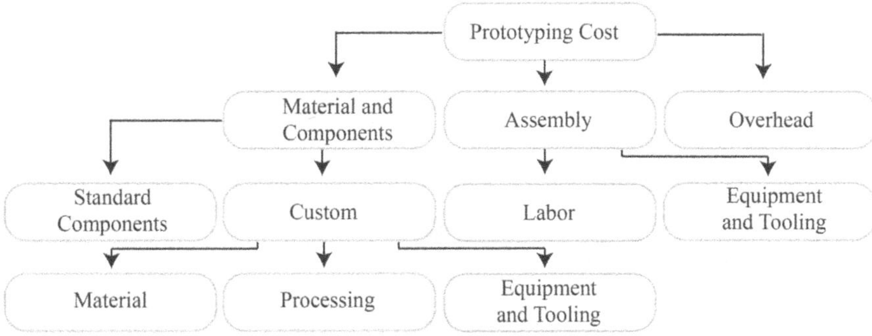

Figure 2.13 Prototyping cost breakdowns.

or suppliers, and the costs of minor components, such as bolts and springs, which are based on experience. The vendor list can be obtained from the Thomas Register of American Manufacturers (www.thomasregister.com). To estimate the custom component cost for a single part, raw material, operator cost, and equipment and tool costs are the major cost items. For example, the raw material needed is the part mass plus 5%–50% of scrap for an injection molding process, and raw material is part mass plus 25%–100% scrap for sheet metal. The operator cost can range from $25 to $90 per hour depending on location and time. In addition, equipment and tool costs should be considered.

2.3.3 *The cost complexities*

There are many complexities in cost estimation. For example, cost structures in modern manufacturing may change with time. Recently, fewer costs vary with production volume since automation replaces labor with machinery, and there is a relatively stable workforce that cannot be varied in the short run. A lot of major equipment is quoted free on board—a trade term requiring the seller to deliver goods on board a vessel designated by the buyer. The sellers fulfill their obligations to deliver when the goods have passed over the ship's rail. Costs, such as equipment installation, are not included, and thus, it may impact cost estimation. For example, in the year 1970, a 10 hp electric motor costs $1,400 to purchase the motor, but it costs $5,000 to install the motor in the plant.

Also remember that as discussed in Chapter 1, when a product is at the definition stage, it is very easy to change. However, when the product is designed or developed further, it would be more and more difficult to make a change, since each design represents a decision–making, and each decision represents a commitment to the available resources. These add additional complexity in estimating the product cost.

Example 2.1

The size of a prototype phone base is 6 in. by 8 in. by 1.5 in. The desired prototype material is ABS plastics. What are the methods that can be used to estimate the prototype?

Method 1: Estimate by analogy

Compare with a similar prototype that was previously made. This method is fast and easy; however, it can only be applied when there is a similar part that has already had a cost estimation.

Method 2: Engineering method prototype cost includes the following:

- Plastics materials: \$50
- Prototype processing fee: \$100 per hour for 5 h = \$500
- Labor fee: \$30 per hour for 5 h = \$150
- Overhead: 50%
- Total estimated cost = \$1,050

This method can calculate the cost for the prototype. It can be precise if all cost factors can be estimated exactly.

Method 3: Expert estimate

One or a group of experts can give their estimate according to their experiences. This method can be fast and convenient when one cannot estimate by analogy or directly calculate. However, this will work only when a true expert is available.

Method 4: From the business quote

For example, a prototype part with a volume of less than 5 in.^{3} (81,935 mm^3) is \$25.00; for prototype parts with volumes over 5 in.^{3} (16,387 mm^3), there is an additional charge of \$5.00 per in.^{3}. The part has about 72 in.^{3}, so it will cost \$360 if the prototype is conducted in a prototyping shop.

Example 2.2

Estimate the cost of a double-ear headset with microphone for a computer. Three different cost estimation methods are used:

Method 1: Bottom-up approach

Assume that the headset is made up of headphones and a microphone packaged together. The two main subsystems are functionally independent except for the fact that physically placing the two systems close together is extremely convenient. Therefore, estimating what it would cost to purchase each product independently would give a good idea of how much the whole system should cost.

It was found that a set of headphones, with reasonable quality, cost \$16.99 and a microphone costs \$14.99. Using this method, the headset should cost around \$32.

Method 2: Cost estimation by analogy

This is based on the fact that there are similar headsets for hands-free phones. Most of those are a single-earpiece design, but the one being considered is a double-earpiece design, so one will have to factor into the estimation that the prices for the phone headset lack one earpiece. In addition, one will have to keep in mind that the phone headset is mainly meant only to play voices not the full depth of music, so the speaker quality of the phone headset may be skewed downward because it will likely lack range and power. It was found the phone headset costs \$19.99. One would then expect the price to go up about \$5 for a second earpiece and \$5 for the quality of the headphones for a total price of about \$30.

Method 3: High–low-cost estimation method

Looking at the costs of the different headsets, one can find the following prices on headphones, microphones, and headsets as shown in Figure 2.14. Note that some very high-end products are offered, and one needs to make one's own judgment when one thinks that a product was out of the scope.

The highest price for a headset is \$52. The highest price for a headphone/microphone combination is \$66. The lowest price for a headset is \$15. The lowest price for a headset/microphone combination is \$27. The average price of the headsets is about \$29. The average price of the headphones/microphones combination is about \$43. Therefore, the cost for the headphones/microphones combination is about \$27 to \$66.

A	B	C	
Headphones ($)	Microphones ($)	headsets ($)	
17	10	15	
30	10	15	
100	10	20	
12	15	17	
13	19	20	
15	28	20	
15		20	
19		26	
19		26	
25		30	
25		42	
25		50	
38		50	
38		52	
[redacted]			
27.93	15.33	28.79	Average
100	28	52	High
12	10	15	Low

Figure 2.14 Costs of the different headsets.

Example 2.3

A computer mouse needs to be mass-produced (10,000 units). It consists of a rolling ball, a left button, a right button, a rolling wheel, a mouse body, a top cover, and a brand panel. The prototype should reflect the geometry of the real mouse and feeling, which one can have when using the real mouse. Use two methods to estimate the cost of the prototype:

Method 1: Bottom-up estimation

The product is broken up into lower level parts, and then, an estimation of each part's cost is shown in Table 2.5. Therefore, the estimated cost for mass production will be about $10.5 per unit.

Table 2.5 Bottom-up estimation

Number	Part	Cost estimate ($)
1	Rolling ball	2
2	Left button	1
3	Right button	1
4	Rolling wheel	1
5	Mouse body	3
6	Top cover	2
7	Brand panel	0.5
	Total	10.5

Method 2: Top-down estimation

Consider the overall functionality of the product and how the functionality is provided by the interacting subfunctions as shown in Table 2.6. Therefore, the estimated cost for mass production will be about $19 per unit.

Table 2.6 Top-down estimation

Number	Function	Cost estimation ($)
1	Support	3
2	Cover	1
3	Rolling	2
4	Moving	2
5	Display	1
	Total	9

Review problems

1. Looking through the newspaper, you saw an advertisement for a 2005 Ford Escape. It did not list a price for it. Approximate the price by comparing it to other vehicles in the paper (i.e., estimation by analogy).
2. Place a tick in the appropriate columns.

Name of cost	Direct material	Direct labor	Manufacturing overhead	Nonmanufacturing cost
Cost of wood in a wooden table				
Delivery cost (out)				
Wages of carpenter who builds tables				
Advertising costs				
Cost of glue m table				
Rent cost of factory equipment				
Assembly line workers salary				
Depreciation of factory equipment				
Depreciation of admin. office equipment				

3. Using the data given in the table, find the total cost of production of pens when the labor works for 9 weeks. Compare the results using linear regression and graphical method of the line of best fit and comment on the results.

Labor time (weeks) with 8 h per shift, double shifts per day [x]	Total cost of production of pens ($) [y]
2	250
4	300
5	200
6	500
7	550
8	400

4. For the following set of data, use the least squares method to find the least squares regression equation.

Assembly time of part A (h)	Cost ($)
3	500
2.75	430
2	405
2.25	415
1.75	345

5. Name five cost drivers for machining a part and describe why they are cost drivers.
6. Come up with five items in a typical manufacturing company for each fixed cost, variable cost, and semifixed cost, and briefly describe why you placed them in that category.
7. Find a cost estimation example by estimation by analogy.
8. Find a cost estimation example by top-down estimation.
9. Figure 2.15 shows an example of a cost graph. In the table, variable X is units, and variable Y is cost. Use regression analysis to obtain the best fit line that will minimize the difference between the line and the given data points.

X (number of units)	Y ($)
50,000	500,000
10,000	71,000
100,000	500,000
1,000,000	4,000,000
15,000	135,000
500,000	3,000,000
350,000	1,050,000
800,000	2,400,000

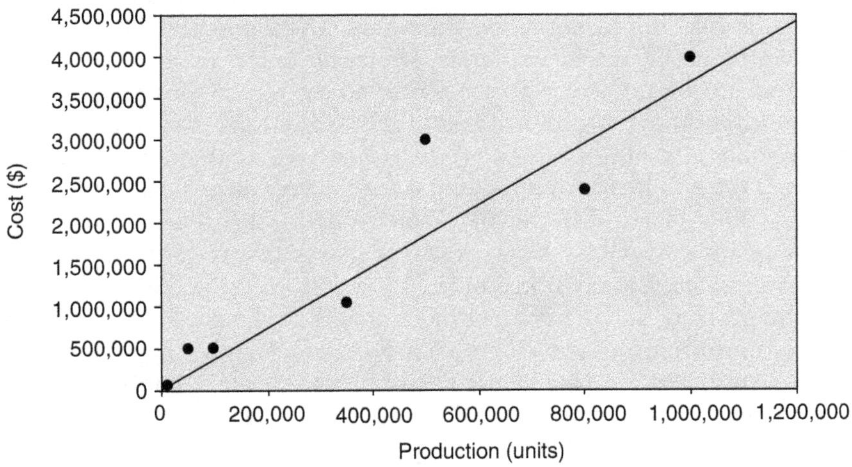

Figure 2.15 An example of a cost graph.

2.4 Prototype design methods

> Design is the method of putting form and content together. Design, just as art, has multiple definitions; there is no single definition. Design can be art. Design can be aesthetics.
>
> Design is so simple, that's why it is so complicated.
>
> **Paul Rand**

This section helps answer the following questions:

- What is engineering problem-solving?
- What is a heuristic method?
- Are there problem-solving techniques that can be used in product prototyping?
- How can problem-solving techniques be used in prototype design and development?
- What makes a good design?
- Is there a systematic way to acquire customers' requirements and incorporate these requirements into product design and prototyping?
- What is house of quality? How would this help in designing and prototyping a product?
- What are PDSs?

As discussed in Sections 2.1–2.3, prototyping a product needs to involve many factors, such as cost trade-off, cycle time, the accuracy of the prototyped part, material property, part size, and part strength. It is a very complex process and thus needs systematic approaches to help decision-making and design. This section discusses engineering problem-solving, design principles, house of quality, and PDS that can be used in the process.

2.4.1 Engineering problem-solving

As with anything, there will always be problems. So the same should hold true when talking about prototyping activities. Regardless of the number of hours of planning or amount of research done, there will be some glitch that no one foresaw. In today's market, speed to market is very important to the financiers. Therefore, the ability to quickly and effectively solve the problem that will arise is integral to the success of the product. For this reason, problem-solving techniques are needed during the design and development of a prototype. Prototyping is problem-solving. It is a culture and a language. One can prototype just about anything—a new product, a service, or a special promotion. What counts is moving the ball forward, achieving some part of the goal, and not wasting time. Prototyping is a state of mind [Kelley01]. Prototyping and problem-solving are mainly aimed at the same idea, which is displaying an idea or issue in a form that allows its issue to be displayed, simulated, and worked out.

Engineering problem-solving is a technique that is applicable for general engineering problems, including prototype development. The characteristics of general engineering problems include frequent changes, available resources, optimal solution, and uncertainty. Engineering change causes a transition from state A to state B, and the optimal solution must be consistent with the available resources. In general, engineering problem-solving is challenging since the knowledge about the system before, during, and after the transition is incomplete or inconsistent. This section discusses a general approach to solve such engineering problems.

One common practice in engineering problem-solving is to use heuristics to cause the best change in a poorly understood situation within available resources. A heuristic is anything that provides a plausible aid or direction in solution. A heuristic does not guarantee a solution but reduces the search time for a solution. Furthermore, it may contradict other heuristics, and its acceptance depends on the immediate context instead of an absolute standard. Some examples of engineering heuristics are listed below:

- At some point of the project, freeze the design.
- A project should be continued when confidence is high enough to permit further allocation of resources for the next phase.
- Allocate sufficient resources to the weak link.
- The yield strength of a material is equal to a 0.02% offset on the stress–strain curve.
- Air has an ambient temperature of 20°C and a composition of 80% nitrogen and 20% of oxygen (rule of thumb, not exact).
- The cost of building construction scales the same as the price of meat.
- Solve problems by successive approximation.

In other words, many heuristics are based on previous experience to help find an optimal solution. For example, uncertainty in the calculated value is always present, and factors of safety (FSs) are to compensate for this uncertainty, for example,

- Use FS = 1.2 for leaf springs.
- Use FS = 1.5 for commercial airplanes.
- Use FS = 2.0 for the bolts in the elevated walkway.
- Use FS = 10–13 for cast iron flywheel.

Some heuristics can be used for attitude determination because engineers like to quantify or express all variables in numbers. How much is a human life worth? This can be critical in designing some objects. For example, to optimize a product that has potential danger (e.g., flights and vehicles) to human life, how much consideration should be taken into account? EPA's (United States Environmental Protection Agency) or EPS's, number is $7.4 million ($2,006) [EPA17].

Some heuristics can be used for risk control. For example,

- Make small changes in the state of the art or SOTA.
- Always give yourself a chance of retreat.
- Use feedback to stabilize engineering design (study past successes and failures).

Some heuristics can be helpful in product prototyping. If heuristics are based on experience and previous knowledge, does it mean that heuristics can be a function of time? In other words, when time changes, will some heuristics become invalid? The answer is yes. Since science and technology are changing as time goes by, heuristics can also change. A set of heuristics can be judged to represent the best engineering practice at the present time and is referred as "SOTA".

2.4.2 Prototype design principles

Among the heuristics, some design principles are important and general enough to call "design principles." For example, the general guidelines for product design are given as follows:

1. Simplicity: Keep it simple for success (KISS) principle. This is because cost correlates with size/mass and complexity/number of parts.
2. Clarity: Is the function of the machine, product, or component clear for all to see?
3. Safety: Is the machine inherently safe? A product should be designed to be safe for all people who will encounter it during its life. Is it reliable?

Some design principles can be used to evaluate whether a design is good or not, or why it is better than the others. The critical parameters in designing a product are called design parameters (DPs), and the critical functions a product needs to satisfy are called functional requirements (FRs). One critical question, which is critical for design judgment, is how many DPs will a product need to satisfy FRs? An independent axiom or functional decoupling axiom stated by Nam Suh [Suh01] is that an optimal design always maintains the independence of FRs. In an acceptable design, the DPs and FRs are related in such a way that specific DP can be adjusted to satisfy its corresponding FR without affecting other FRs. As an example, when designing a refrigerator door, assume that there are two FRs to be considered: providing an insulated enclosure to minimize energy loss and providing access to the food in the refrigerator. Is a vertically hinged door, just like most of the current refrigerators used at home, a good design? When the door is opened, cold air will be lost, and thus, the two FRs are coupled. Therefore, it is not a good design. On the contrary, a horizontally hinged and vertically opening door, as seen in some stores, is a good design. Why are most of the house refrigerators vertically hung? Are not they bad designs? Yes, they are bad designs if there are only two FRs. However, if one considers other factors, such as convenience in the design, a vertically hung door may still be a popular choice. Therefore, it is very critical to define the true FRs to end up with a design that fits the customers' needs.

Another axiom is called the information or physical coupling axiom. It states that the best design is a functionally uncoupled design that has minimum information (design and manufacturing) content. For example, to design a mouse for a laptop, two FRs are considered for the mouse: portability and maneuverability. The design is the touch pad. Portability, which is the most important FR, is very satisfying because the touch pad is embedded in the laptop and can be used anywhere. Its operation is acceptable, and some people even prefer to use the touch pad rather than the traditional mouse. Not only has it successfully fulfilled the portability issue without sacrificing the maneuverability, but it has also been set up in the proper place without interfering with other parts of the laptop. Since both FRs are embedded in the same physical device rather than two separate components, minimum information content (design and manufacturing) is achieved. Thus, according to the functional decoupling axiom and physical coupling axiom, the touch pad is a good design.

2.4.3 House of quality

QFD or house of quality is a term used to describe a strategy for focusing engineering design attention on quality issues as perceived by customers [Dixon95, Otto01]. It is getting the voice of the customer into technical design specifications. The customer's requirement must be translated into measurable design targets. One cannot design a car door that is "easy to open" when one does not know the meaning of the word "easy." QFD is a systematic technology that translates customer's requirements to engineering specifications that are measurable and thus can be designed using a scientific approach.

As shown in Figure 2.16, customer needs or requirements are stated on the left side of the matrix. Customer requirements, or "WHAT" in the matrix, may be expressed qualitatively, that is, customers may want something to be easy, smooth, or fast. If the number of needs or requirements is too excessive to manage, decompose the matrix into smaller modules or subsystems to reduce the number of requirements in a matrix. For each need or requirement, state the customer priorities using a 5- or 10-point scale rating. Use ranking techniques and paired comparisons to develop priorities. The rankings are listed under "Importance rating." The "HOW" in the matrix of Figure 2.16 is a list of engineering characteristics or FRs. These characteristics are design variables that designers can control to meet the FRs of the customer, and thus, they should be meaningful, measurable, and global. They should also be stated in a way to avoid implying a particular technical solution so as not to constrain designers' options.

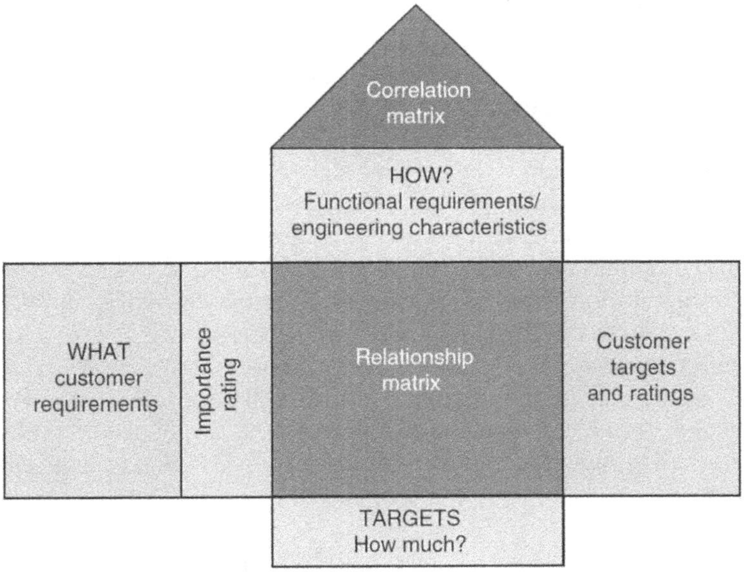

Figure 2.16 House of quality.

The "relationship matrix" in Figure 2.16 is the relationship between customer needs and the company's ability to meet those needs. One can use symbols for strong, medium, and weak relationships. Associated scores can be used to represent these symbols. For example, scores 1, 3, and 9 can be assigned to weak, medium, and strong relationships, respectively. A blank represents 0 or no relationship. This is a critical step that makes sure that all customer needs or requirements have been addressed and are related to the engineering characteristics. If any product requirements or technical characteristics stated are not related to customer needs, they are not needed and thus can be eliminated. The idea is that all engineering characteristics or FRs should satisfy the customer's requirements.

Technical competitive assessment or customer targets compare a firm's product to the products of the best competitors to establish performance metrics for the firm. One can use surveys, customer meetings, or focus groups/clinics to obtain feedback. Several items will need to be identified to develop a product strategy. Price points and market segments for products under evaluation, warranty, service, reliability, and customer complaint problems to identify areas of improvement are all useful types of feedback needed for product strategy. One needs to consider the current strengths and weaknesses relative to the competition, and how these strengths and weaknesses compare to customer priorities. The opportunities for breakthroughs to exceed competitors' capabilities, areas for improvement to equal competitors' capabilities, and areas where no improvement will be made all need to be identified.

The roof of the quality house or correlation matrix shows the degree of coupling between engineering characteristics. The correlation matrix is potentially positive and negative interactions between FRs or engineering characteristics using symbols for strong or medium, positive or negative relationships. Too many positive interactions suggest potential redundancy in "the critical few" product requirements or technical characteristics. One should focus on negative interactions that indicate trade-offs between characteristics in which product concepts or technology need to overcome these potential trade-offs.

Target value in the matrix represents the target number and units of each engineering characteristic (or FR). Often below "TARGET," one can calculate "importance ratings" by multiplying the customer importance rating by the weighting factor in each box of the matrix (e.g., 9-3-1) and adding the resulting products in each column. The importance ratings represent the importance of each engineering characteristic to the customer's requirements. Later, these ratings will be useful in design trade-offs.

Figure 2.17 shows an example of the house of quality to design a compliant microrobot. The arm of a compliant microrobot needs to be able to bend in two axes with the use of a piezoelectric actuator moment in only one axis. The moment will be applied by a rectangular piezoelectric strip that will expand and contract along its length at a high frequency. The arm will also have to support a load of one milligram at room temperature. Excessive vibrations will also be present during use. The arm will have to withstand approximately one million cycles at very high frequencies before failure. Target product cost is about $500 per robot. The result shows that "Small/simple shape" (34.7%) is most critical based on customers' input.

2.4.4 *Product design specifications*

PDS is a detailed listing of the requirements to be met to produce a successful product or process. It is the formal means of communication between the buyer and the seller. It represents the commonly used product specifications and should be used as a checklist for product design. When the design goals are developed and matured, they are a list of PDS. Since PDS has been used so often, it can also be used as a checklist in specifying a new product.

- Performance: the basic need or expectation of the user
- Environment: range of temperature, pressure, humidity, dirt, dust, corrosive environments, shock loading, vibration, etc.

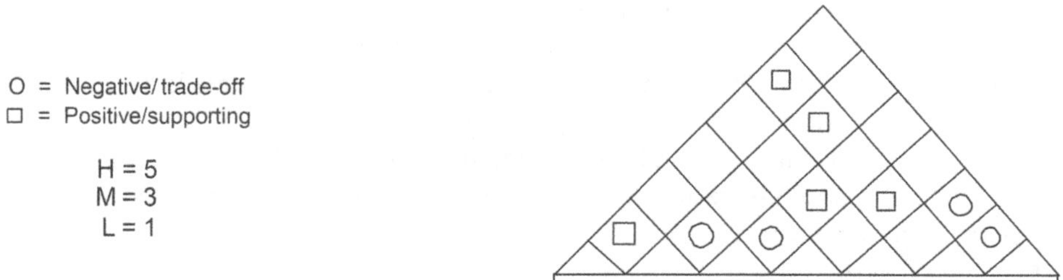

Compliant microrobot arm			Engineering characteristics: Durability		Provides forward movement			Manufacturing	
Customer attributes		Importance (60 pts total)	Vibration resistant	Long lasting	High speed	Precise movement	Low no. of piezostrips	Small/simple shape	Large-scale etching
Handling	Easily placed	7		L				M	
Functionality	Withstand up to 1 mg	12		M				M	
Functionality	Low bulk	10			M	L	H	H	
Functionality	Requires less piezostrips	15			L		H	H	
Functionality	Durable	5	M	H				L	
Miscellaneous	Cheap	11					H	L	H
Abs. importance			15	68	45	10	180	198	55
Rel. importance (%)			2.6	11.9	7.9	1.8	31.5	34.7	9.6

Figure 2.17 The house of quality of designing a compliant microrobot.

- Service life: expected service life and duty cycle
- Maintenance and logistics: specify ease of access to components, spares, tools and test equipment, operator and maintenance manuals, and training
- Target product cost: experience shows that almost all target costs are unrealistically low
- Competition: future and existing
- Shipping: size of boxcar doors, cargo hatches, weight, limits on truck trailers, shipping costs
- Packing: storage, corrosion, or shock loading
- Quantity: cost per unit, the methods of production, and the manufacturing cost
- Manufacturing facility: whether the product is to be produced in an existing plant or whether a new plant will be built
- Aesthetics: visual appearance of a product, color, shape, form, and texture of finish

- Materials: performance and manufacturing
- Product life span: whether the product is likely to remain marketable for 3 or 30 years
- Standards and specifications: SI or U.S. units and existing standards and specifications
- Ergonomics: human–machine interface requirements
- Quality and reliability: high-risk areas should be identified
- Shelf life in storage: batteries or bake goods, construction site, and spare parts
- Company constraints
- Market constraints
- Testing and inspection: acceptance testing and quality requirements
- Safety: critical parts should be identified. Warning labels should be devised and operating manuals should clearly spell out the abusive use of the product
- Patents: prevent a costly patent infringement suit
- Social and political factors: EPA, FDA (US Food and Drug Administration), or NRC (US Nuclear Regulatory Commission)

Example 2.4: Conceptual design of a vehicle lifting device

PROBLEM STATEMENT

Farmers, ranchers, and others who frequently work or travel in remote areas often find themselves the victim of a tire failure or a stuck vehicle. They could make use of a device that would aid in repairing or recovering the vehicle. Often help is far away and could be very expensive. A wide range of vehicles could be involved—from light trucks weighing 4,000 lbs to tractors weighing 20,000 lbs. The device or system should be able to lift a portion of the vehicle weight to allow tire changing or repair, and also be capable of pulling the vehicle. The device should be relatively low in cost approximately $200 or less.

After the problem is investigated, the following PDS, house of quality, and conceptual designs can be obtained.

PRODUCT DESIGN SPECIFICATIONS

1.0 Performance
1.1 The conceptual designs will include focused analytical prototypes for each alternative.
1.2 The conceptual designs will provide lifting and pulling capability.
  1.2.1 of 10,000 lbs
  1.2.2 of 60 in.
1.3 The device or its individual components will weigh less than 50 lbs each.
1.4 If the device is human powered, the input force required to operate it at maximum load will not exceed 50 lbs.
1.5 The device will be assembled by the user without tools.
2.0 Environment
2.1 The device will be expected to operate in an environment including frequent contamination with dirt and water.
2.2 The device will be expected to operate in temperatures ranging from −40°F to +110°F.
3.0 Service life/duty cycle
3.1 The device will be expected to operate on continuous duty for 8 h.
4.0 Maintenance
4.1 The device shall require minimal maintenance including lubrication only.
4.2 All electrical components will be sealed and maintenance free.

5.0 Target product cost
5.1 Cost is not to exceed $200.
6.0 Competition
6.1 Competing products will be evaluated using the same criteria.
7.0 Packaging and shipping
7.1 Packaging and shipping are not considered at this time.
8.0 Quantity
8.1 Production volume is not considered at this time.
9.0 Manufacturing facility
9.1 No exotic manufacturing processes should be required.
10.0 Aesthetics
10.1 Aesthetics are not considered
11.0 Materials
11.1 Materials and components will be selected for low cost, ease of manufacture, and durability.
12.0 Product life span
12.1 The product is expected to have a useful life of 5 years with usage once per month.
13.0 Standards and specifications
13.1 There are no constraints on the measurement system used in the conceptual design.
14.0 Ergonomics
14.1 Ergonomics is not considered at this time.
15.0 Quality and reliability
15.1 Quality and reliability are not considered at this time.
16.0 Shelf life and storage conditions
16.1 Shelf life and storage conditions are not considered at this time.
17.0 Testing and inspection
17.1 Testing and inspection are not considered at this time.
18.0 Safety
18.1 Conceptual designs incorporating automatic safety devices and overload protection will be favored
19.0 Constraints
19.1 Conceptual designs are to be submitted by September 30, 2020

Focus areas for the design activities are prioritized by the results from the house of quality as shown in Figure 2.18. There is an opportunity to provide a product with increased capacity if this can be done within the weight and cost limitations. The competing product is VLift jack. Following is a description of its features and specifications:

JACK FEATURES

- Every VLift jack comes complete with adjustable top clamp clevis for use in clamping and winching.
- Safety bolt is designed to shear at 7,000 lbs (3,175 kg, rated capacity in jack mode).
- For speedy disengaging, the VLift lifting unit automatically drops away when the load is removed.
- 41/2 in. long lifting nose for positive contact with load.
- Low pickup of 41/2 in. (11 cm).
- 28 in.^{2} base plate.
- 48 or 60 in. models available.

Roof: "+" = Synergy, "–" = Compromise

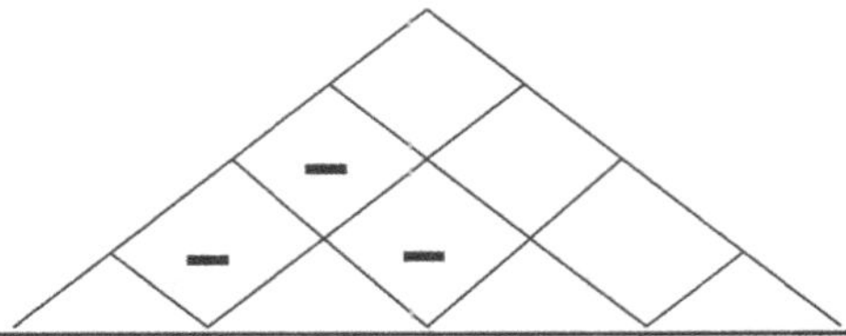

Customer Requirements	Importance rating 1–5	FR's and EC's	Capacity Target	Stroke Target	Weight Minimize	Safety devices Maximize	Competitive assessment '+ = exceeds, 0 = meets '– = does not met
Lift and pull 10,000 lb	3		Strong-9				-
Lift and pull 60 in.	2			Strong-9			0
Weight <50 lb	4		Strong-9	Weak-1	Strong-9		+
Safe	5				Weak-1	Strong-9	-
Low cost <$200	4		Strong-9			Strong-9	+
Importance of FR's and EC's			99	22	41	81	
Percent importance of FR's and EC's			40.7%	9.1%	16.9%	33.3%	
Benchmarking	Vlift Jack		4660	60	35	1	
	Our product		10,000	60	<50	?	

Figure 2.18 House of quality for the vehicle lift device.

JACK SPECIFICATIONS

- Approximate weight 30 lbs (14 kg).
- Pulling capacity of 4,660 lbs (2,113.74 kg, rated capacity in winch mode).
- Climbing pins of specially processed steel with 125,000 PSI tensile strength and 100,000 PSI yield.
- VLift Steel bar is manufactured of specially rolled extra high carbon steel with 80,000 lbs minimum tensile strength and carbon content of 0.69%–0.82%.
- VLift Steel handle of 14 gauge high-yield structural tubing with a minimum yield of 55,000 PSI. 15⁄16 in. diameter × 30 in. long.
- Cost for VLift is $55.

CONCEPTUAL DESIGNS

Design no. 1: Worm gear system with gearbox (Figure 2.19)

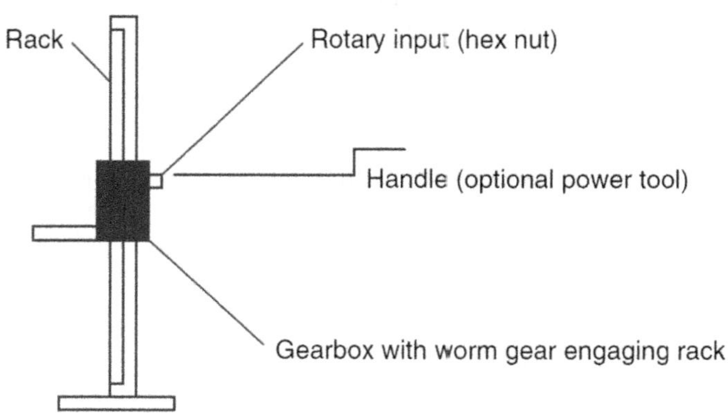

Figure 2.19 Conceptual design number 1: a worm gear system with a gearbox.

The device would provide both lifting and pulling force using an arrangement of gears. The final stage of the gear mechanism would consist of a worm gear engaged with a rack, integrated with the main column of the device. The worm gear would provide an automatic safety device to prevent the load from being released. In addition, the gearbox could be constructed to provide the necessary mechanical advantage to reduce the input force required. Automatic overload protection could be provided using a shear pin in the input shaft. Input power would be rotary. The input could be either a hand-operated crank or an optional rotary power tool.

Design no. 2: Hydraulic system with a hand pump (Figure 2.20).

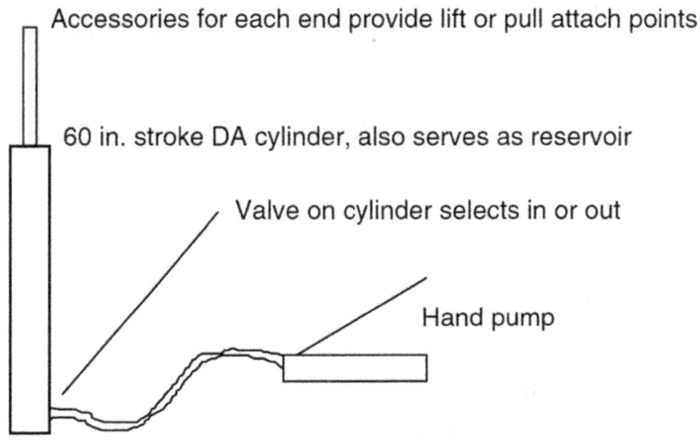

Figure 2.20 Conceptual design number 2: a hydraulic system with a hand pump.

The device would operate using a double-acting cylinder with selector valve to both lift and pull a load using a hand-operated hydraulic pump. Load control safety could be provided by a release valve with a safety lock. Overload protection could be provided by an internal pressure relief circuit.

EVALUATION OF ALTERNATIVES AND SELECTION

The two conceptual designs will be compared to the existing competing design based on the same criteria used in the house of quality. A comparison matrix is constructed as shown in Table 2.7. Estimates of cost and weight are rough but are believed to be close enough to analyze the alternatives.

The hydraulic concept failed due to cost and weight. The cost of the cylinder alone will be approximately $250. The cylinder drives the design over the weight limit as well, weighing approximately 80 lbs alone. An additional drawback to this design, which was not built into this analysis, is the speed of operation. This device would probably be extremely slow to use.

The concept selected based on this analysis is the gearbox concept. The gearbox concept scored the best overall and appeared to be good enough to result in a competitive product in the marketplace. The trade-off between performance and cost appears to be appropriate. The change to the rotary input also resolves a safety problem with the VLift jack. The user must be cautious to make a complete downward stroke on the handle of this jack. If the user fails to do this and releases the handle, the handle will spring back violently and can injure the user.

Table 2.7 A comparison matrix for various designs

Customer requirements	Importance rating 1–5	VLift DATUM	Gearbox concept	Hydraulic concept	Comments = Notes
Lift and pull 10,000 lbs	3	0.47	1	1.1	Ratio of capacity to specifications
Lift and pull 60 in.	2	1	1	1	Assume all will meet specifications
Weight <50 lbs	4	35 lbs = 1	50 lbs = 0.7	100 lbs = 0.35	Ratio of estimated weight to datum
Safe	5	1	1.5	1.5	Datum has known safety concern
Low cost <$200	4	$55 = 1	$100 = 0.55	$400 = 0.14	Ratio of estimated cost to datum
Weighted score		16.41	17.5	14.76	

Review problems

1. What is engineering problem-solving?
2. What is a heuristic method? Give examples of heuristics in product design.
3. What is PDS?
4. Describe QFD and its main functions.
5. Create a correlation matrix using five FRs, weight, package size, battery life, the cost for life of the product, and storage capacity, for an MP3 player and give a brief reflection on the results.
6. Complete an absolute importance matrix for an MP3 player and briefly discuss your findings.

		Technical requirements							
		Rating: High (H) = 9; Medium (M) = 6; Low (L) = 3							
Customer requirements	Weights	Battery		Housing		Screen		Interface	
Weight	7	H	9	M	6	L	3	L	3
Package size	10	H	9	H	9	M	6	L	3
Life	6	H	9	H	9	H	9	H	9
Cost of for life of product	4	L	3	M	6	L	3	L	3
Storage capacity	5	M	6	L	3	L	3	L	3
Ease of use	10	L	3	L	3	H	9	H	9
Durability	8	H	9	H	9	H	9	H	9
Absolute importance									
Ranking									

7. Write a simple PDS sample for a USB MP3 player that can play music and store files. The battery should last long (8 h a day) and be used outdoors for at least 3 years. It should be impact resistant. One million pieces will be produced, and the target cost to consumers should not exceed $300.

8. What are the characteristics of an engineering problem?
9. According to the functional decoupling axiom and the information axiom, which in a manufacturing context is a physical coupling, give one example of a good design and a bad design and state why.
10. Knowing the importance of the "voice of the customer" in developing a successful product, describe some tools used to capture this information.

2.5 *Prototype design tools*

> A good scientist is a person with original ideas. A good engineer is a person who makes a design that works with as few original ideas as possible. There are no prima donnas in engineering.
>
> **Freeman Dyson (1923)**
> *English physicist, educator*

This section helps answer the following questions:

- Why is distributing weighting among various design factors important?
- How to determine the weighting between various decision factors?
- What techniques are available that will stimulate creativity?
- What is an objective tree?
- What are the idea generation methods that can be used during prototyping and design activities?
- What is the functional efficiency technique?

Design is a decision-making process. As discussed in Section 2.4, often a design may have multiple objectives, and, depending on the emphasis, the design results may be completely different. Therefore, how to generate useful ideas and how to evaluate these ideas are critical to the success of a product or prototype. This section discusses these two issues.

2.5.1 *Evaluating alternatives*

The amount of time and effort that one dedicates to specific factors in the design process, that is, cost, speed, convenience, size, and quality, will vary significantly from product to product. Some factors will always be present, but in some instances, there could be 10–12 factors that the customer considers key to the product. The question is, "How does one weight each factor?" A useful technique for this endeavor is a priority matrix.

Let us assume that a printer is being designed, and the considerations may include factors such as cost, speed, convenience, size, and printing quality. How could one determine the factor that is more important than others and which factor or factors should one include in the decision-making? The answer is to assign a different weight to each factor and combine all factors into a performance index of a design:

$$P = W_A A + W_B B + W_C C + W_D D + W_E E \tag{2.8}$$

where
W_i = the weight for factor i, where i = A, B, C, D, E

2.5.1.1 First approach

The next question is how to determine these weights that are very critical to the design outcomes. Let us use an example to illustrate this point. To design a product that satisfies five product design objectives (say, objectives A, B, C, D, and E) and determines the best alternative among three designs, one needs to first assign weight to each design objective. To determine the weights, it is easier to compare two factors at a time instead of trying to determine all the weights at the same time. As shown in Table 2.8, each design objective is listed as compared to other objectives, two at a time. For example, when objective B on the left-hand side of the table is considered to be more important than objective C, it is scored one point and thus "1" is recorded under row B and column C. When objective C on the left-hand side of the table is considered to be more important than objective A, it is scored one point and thus "1" is recorded under row C and column A. The overall scores for factors in each row are added and recorded into the "Total" column. Since objectives B and D have the highest scores, they are the most important factors in the design consideration.

Table 2.8 Comparison between factors for importance

	A	B	C	D	E	Total
A	—	1	0	0	1	2
B	0	—	1	1	1	3
C	1	0	—	0	0	1
D	1	0	1	—	1	3
E	0	0	1	0	—	1
					Total	10

The next step is how to assign the weight to each objective. One approach is to directly use the grand total T, each subtotal t_i (i = A, B, C, D, E), and the weight for an individual factor

$$W_i = \frac{t_i}{T} \text{ and } i = A, B, C, D, E \tag{2.9}$$

where

T = grand total (Σt_i)
t_i = subtotal
W_i = weight for individual factor

Computing each weight, one can find the weights: $W_A = 0.2$, $W_B = 0.3$, $W_C = 0.1$, $W_D = 0.3$, and $W_E = 0.1$.

Table 2.9 shows another example. One can calculate the weights: $W_A = 0.20$, $W_B = 0.27$, $W_C = 0.33$, $W_D = 0.13$, $W_E = 0.07$, and $W_F = 0.00$.

2.5.1.2 Second approach

Another approach is to place the weights on a 0–10 scale as shown in Table 2.10. The basis of this approach is because directly using the binary numbers in Table 2.8 may not be

Table 2.9 Another example to show weight calculation

	Durability	Accuracy	Repeatability	Base of control	Base of manufacture	Cost	Total
Durability	—	0	0	1	1	1	3
Accuracy	1	—	0	1	1	1	4
Repeatability	1	1	—	1	1	1	5
Base of control	0	0	0	—	1	1	2
Base of manufacture	0	0	0	0	—	1	1
Cost	0	0	0	0	0	—	0

the best way to assign the weights. One may like to have the flexibility to do some fine-tuning of the weights by using a completely new scale. In other words, although both objectives B and D receive a score of 3, one can use intuition and engineering judgment to decide B to be slightly more important than D, and thus, object B will be assigned a weight of 10 and D with 9 (i.e., $W_B = 10$ and $W_D = 9$). Similarly, although E and C received a score of 1 from Table 2.8, one can assign $W_E = 2$ and $W_C = 1$. Since object A is ranked between B, D, and E, C, one can assign $W_A = 7$. As more engineering judgment is required to work on the problem, this approach is used when an engineer has a good grasp of the problem.

Table 2.10 Assigning weights based on intuitions and engineering judgment

10	9	8	7	6	5	4	3	2	1	0
B	D		A					E	C	

2.5.1.3 *Third approach*

One can also make use of the weights from the first approach and distribute 100 points among the design objectives. For example, W_B was 0.3 based on Table 2.8, and based on 100-point distribution, it should be assigned 30 points. However, one can base on intuition and assign 35 points, and thus, W_B is 35 instead of 30. Similarly, one can assign $W_D = 30$, $W_A = 18$, $W_C = 9$, and $W_E = 8$ as shown in Table 2.11.

Table 2.11 Weights fine-tuned on a 100-point scale

	B	D	A	C	E	Total
Weight from Table 2.10	0.3	0.3	0.2	0.1	0.1	1.0
Fine-tuned weight	35	30	18	9	8	100

Example 2.5

Establish a criterion to purchase a laptop using the digital logic approach. The decision factors and their symbols are shown in Table 2.12.

Table 2.12 Decision factors

Factor	Processor (CPU)	Total memory	Hard drive	Display	Graphics	Battery
Symbol	A	B	C	D	E	F
Factor	Exterior	Cost	Weight	Volume	Reliability	Guarantee
Symbol	G	H	I	J	K	L

Solution:

Step 1: Each factor is listed as compared to other factors, two at a time.

	A	B	C	D	E	F	G	H	I	J	K	L	Total
A	—	1	1	1	1	1	1	1	1	1	1	1	11
B	0	—	1	1	1	1	1	1	1	1	1	1	10
C	0	0	—	1	1	1	1	1	0	1	1	1	8
D	0	0	0	—	1	1	1	0	0	1	0	1	5
E	0	0	0	0	—	1	1	0	0	1	0	1	4
F	0	0	0	0	0	—	1	0	0	0	0	0	1
G	0	0	0	0	0	0	—	0	0	1	0	0	1
H	0	0	0	1	1	1	1	—	1	1	1	1	8
I	0	0	1	1	1	1	1	0	—	1	1	1	8
J	0	0	0	0	0	1	0	0	0	—	0	0	1
K	0	0	0	1	1	1	1	0	0	1	—	1	6
L	0	0	0	0	0	1	1	0	0	1	0	—	3

Step 2: Weight of each factor is calculated.

$$N = I_A + I_B + I_C + I_D + I_E + I_F + I_G + I_H + I_I + I_J + I_K + I_L$$

$$= 11 + 10 + 8 + 5 + 4 + 1 + 1 + 8 + 8 + 1 + 6 + 3 = 66.$$

$W_A = 0.167$, $W_B = 0.152$, $W_C = 0.121$, $W_D = 0.076$, $W_E = 0.061$, $W_F = 0.015$, $W_G = 0.015$,

$W_H = 0.121$, $W_I = 0.121$, $W_J = 0.015$, $W_K = 0.091$, $W_L = 0.045$.

Step 3: Distribute 100 points among the factors.

A	B	C	D	E	F	G	H	I	J	K	L
17	15	12	8	6	1.5	1.5	12	12	1.5	9	4.5

Results:

$W_A = 17$, $W_B = 15$, $W_C = 12$, $W_D = 8$, $W_E = 6$, $W_F = 1.5$, $W_G = 1.5$, $W_H = 12$, $W_I = 12$,

$W_J = 1.5$, $W_K = 9$, $W_L = 4.5$.

Sometimes when there are too many factors, and if some of the factors are not in the same category, it may not be very easy to compare them. For example, one would like to establish a performance index to design an aircraft, and the factors to be considered are initial cost, lift, interior, exterior, options, maintenance, ease of handling, comfort, taking off/landing, reliability, etc. It may not be easy to consider "maintenance" with "comfort" since these may not be in the same category. One way to resolve such a problem is to use the objective tree as shown in Figure 2.21. First, the objectives should be classified into various levels, and the tree will start from the highest level into a global weighting of 1.0. At each lower level, the objectives are given weights relative to each other but also total 1.0 locally. The second level includes body, cost, reliability, and performance of the aircraft, and thus, it is easier to compare factors between these four categories. One can use the previous methods to assign weights among these four items. In this case, $W_{body} = 1/8$, cost = 3/8, $W_{reliability} = 1/4$, and $W_{performance} = 1/4$. Note that the total should be equal to 1.0 locally. On the second level, local weight is also equal to the global weight since global weight is equal to the local weight times the global weight assigned in the above level, which is always 1.0 for the first level.

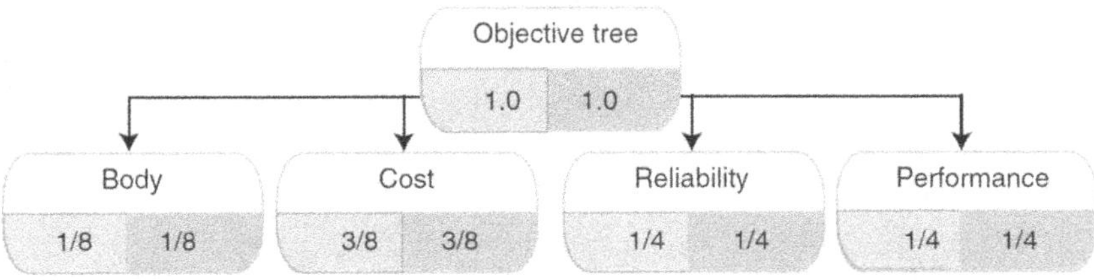

Figure 2.21 First and second levels for the objective tree for designing an aircraft.

The elements in the third level have been classified into four categories: body, cost, reliability, and performance. For example, "body" includes interior, exterior, and options as shown in Figure 2.22. Following the same weight assignment method, 1.0 point can be assigned among interior, exterior, and options, and their local weights are

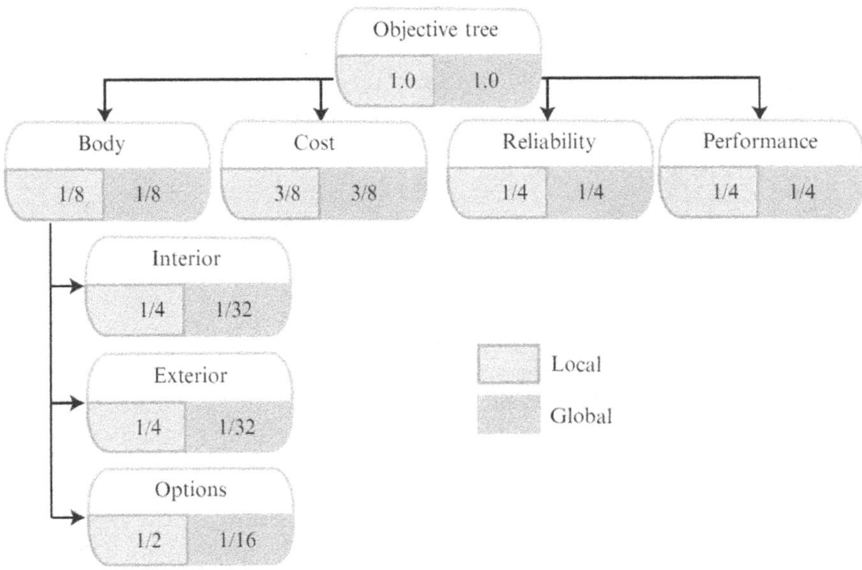

Figure 2.22 Third level is added for body.

assigned to be 1/4, 1/4, and 1/2, respectively. Their global weights can be computed by multiplying each of the local weights with the global weight of body, which is 1/8. The global weights for these are 1/32, 1/32, and 1/16, respectively.

As shown in Figure 2.23, the other third-level components can be branched out from cost, reliability, and performance. Cost includes initial cost and maintenance cost; performance includes ease of handling, ease of taking off/landing, comfort, and lift; and reliability has no subelement. This process continues until the lowest level is completed, and the global weight of each element in the lowest level represents the final weight of each basic element. For example, $W_{options} = 1/16$ and $W_{reliability} = 1/4$. This is a very effective method to handle multiple and complex objectives.

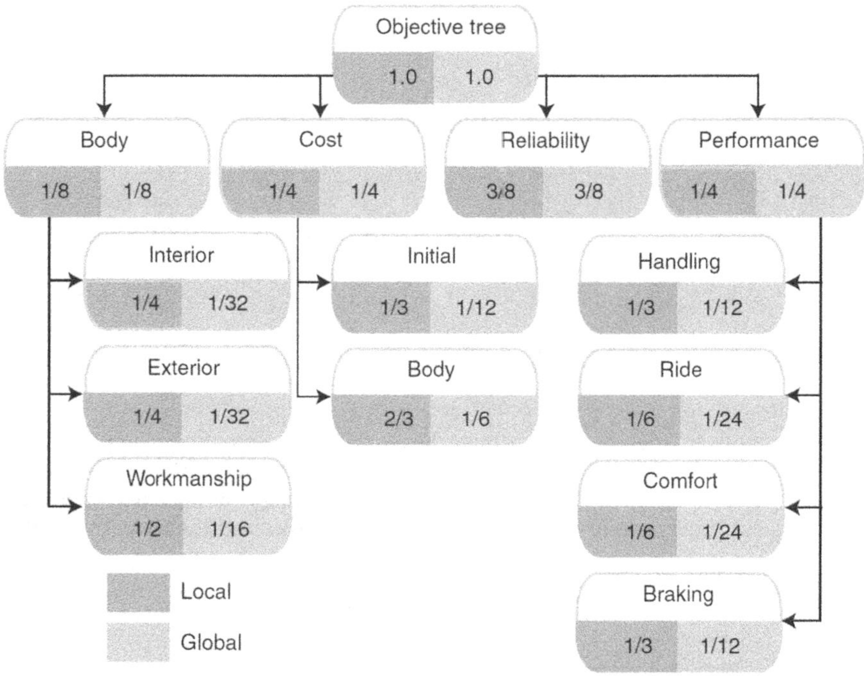

Figure 2.23 Objective tree for designing an aircraft.

Now from Equation 2.8, the weights are decided as shown below:

$$P = 0.2A + 0.3B + 0.1C + 0.3D + 0.1E \quad (2.10)$$

The formulation may not be complete since the objectives A, B, C, D, and E may not have the same unit. For example, objective A is to minimize the cost, and thus, its unit may be dollars ($); objective B is to minimize the weight, and thus, its unit may be pounds (lb); C may be in size (in.); and D may be to feel comfortable (unit is unknown). Equation 2.10 may not be very useful in decision-making since one cannot conveniently combine them together. To resolve this issue, one can convert values obtained for various design objectives into a consistent scale of values using point scales such as 5- or 11-point scales. A 5-point scale is shown in Table 2.13. It can convert the values of various objective functions into the same scale so that these objective functions can be combined for evaluation.

Table 2.13 Using 5-point scale to convert the values of various objective functions into the same scale

Points	Cost ($)	Weight (lbs)	Size (in.)
4	<1,000	<10	<2
3	1,000–3,000	10–15	2–3.5
2	3,000–4,000	15–20	3.5–5
1	4,000–9,000	20–30	5–10
0	Over 9,000	Over 30	Over 10

2.5.2 Useful idea generation methods

To design or prototype a product, creativity and new ideas are the key for product innovation. Can a person's creativity be improved? Or is it even possible? Are there systematic idea generation methods that can be used when developing a product?

In the *Engineering and the Mind's Eye* [Ferguson94], several very creative scientists and engineers were investigated about why they were so creative. A common approach was found to be the use of visual thinking to design, as revealed by a number of creative engineers.

"The machine was in my mind's eye long before I saw it in action," "Build up in the mind mechanical structures and set them to work in imagination, and observe beforehand the various details performing the perspective functions ..."

In other words, creativity is embedded in everybody's mind, but it is deep inside. One can make creativity work by thinking about the problem day and night. When the time is right, ideas may just appear. As a result, a major factor that could assist in improving creativity is having adequate time to spend on producing a creative solution for a given problem. Creativity can be squashed if sufficient time is not allocated.

Ideas can be everywhere, but often it depends on one to search for them. This is why Mark Twain said "Name the greatest of all inventors: 'Accident.' Intelligence is not to make no mistakes, but quickly to see how to make them good." In 1970, Spencer Silver at 3M was to find a strong adhesive. An adhesive was developed but was much weaker than the existing products. It stuck to objects but could easily be lifted off. No one found it useful at that time. Four years later, Arthur Fry used the adhesive on markers to keep his hymnal in place and found that it could be lifted off without damaging the pages. In 1980, 3M began to market Post-it Notes nationwide.

In 1907, one of the paper suppliers sent a shipment of paper that was very wrinkled and heavy. It was not suitable as a tissue. Arthur Scott perforated the tissue into individual sheets and sold them in railroad stations, hotels, etc., and it became a very popular product that is now called paper towels.

In 1903, Edouard Benedictus, a French chemist, accidentally knocked a glass flask to the floor. The glass shattered, but the broken pieces still hung together in the shape of the flask. He tried to sell this to auto manufacturers, but no one was interested because of its cost. In 1914, during World War I, manufacturers used their safety glass for gas mask lenses, which were later developed into today's safety glasses.

The underlying goal of idea or concept generation is to develop as many ideas as possible. There are several useful idea generation methods, such as morphological analysis, functional efficiency technique, brainstorming, bionics, and inversion.

2.5.2.1 *Morphological analysis*

Similar to combinative ideas, morphological analysis can be used to uncover combinations of factors to create a new design. For example, to develop a new design for a robotic mechanism, one can identify two or three major functions that must be performed. The next step is to construct a matrix series in which a major function is an axis and each of the methods as shown in Figure 2.24. Each coordinate in the matrix represents an idea. For example, the chosen coordinate represents the idea that the power source is electricity, the input type is linear, and the output type is also linear. Once these ideas are identified, one can use the concept selection to find the best solution. If needed, one can repeat the process for the cells that appear most promising. More in-depth ideas are focused based on the electricity–steam–vertical oscillation higher-level idea.

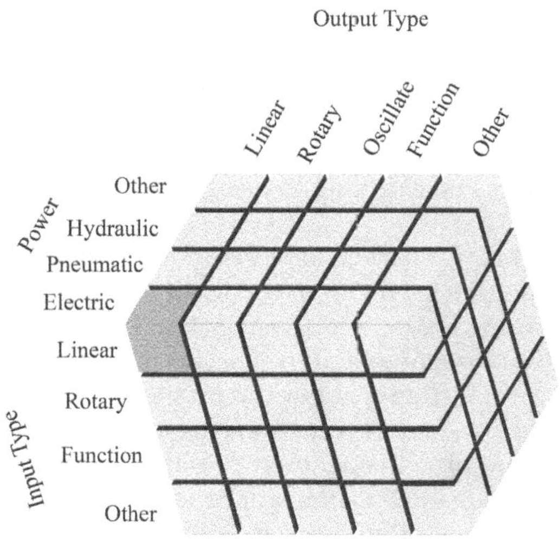

Figure 2.24 An example of morphological analysis.

2.5.2.2 *Functional efficiency technique*

The functional efficiency technique is a top-down approach to define and understand product function and how functions relate to one another. To do this, one needs to display functions in a logical sequence, prioritize them, and test their dependency. The first step is to start with a list of the derived functions for a new product with a verb and a noun. One needs to select a quantifiable noun so that its cost can be evaluated. For example, a wire that may conduct current should be described in this manner instead of conduct electricity since electricity is hard to quantify, but current can be quantified. Later on, this technique can be used to effectively estimate the cost of a product, since all the functions of the product are measurable. Similarly, the verb should be specific and precise. For example, to describe the one who turns the handle of a faucet, "open valve" is not as good as "increase flow rate" or "decrease pressure difference" since the first term is not specific and not quantifiable. Table 2.14 shows some commonly used verbs and nouns.

The second step of the functional efficiency technique is to find the relationships among the functions by arranging the functions into a logical, orderly fashion and find the subfunctions and related units. Two functions will be related to each other if (1) Function A

Table 2.14 Commonly used verbs and nouns in functional efficiency technique

Verbs	Accelerate, decelerate, stop, prevent, protect, exclude, prohibit, transmit, conduct, insulate, create, emit, modulate, resist, distribute	Nouns	Mass, pressure, velocity, acceleration, voltage, current, diameter, intensity, density, work, energy, friction, torque, force, weight, movement, oscillation, damage, overspeed, undervoltage

is a logical answer to the question "why Function B is executed?" and (2) Function B is the logical answer to the question "how is Function A executed?" For example, as shown in Figure 2.25, Function A: project transparencies is a logical answer to the question "why Function B: 'illuminate transparency' is executed?" and Function B: illuminate transparency is the logical answer to the question "how is Function A: 'project transparencies' is executed?" Figure 2.26 shows an example of using a functional efficiency technique to find the bottom design components. When the physical object can be replaced by a function, the process can be stopped, and thus, the bottom components are the physical objects needed for the design.

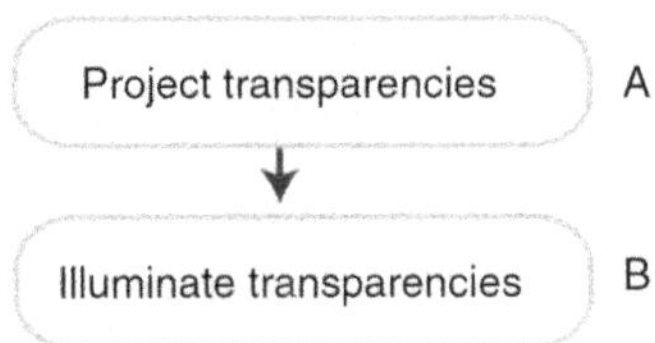

Figure 2.25 An example of how two functions are related.

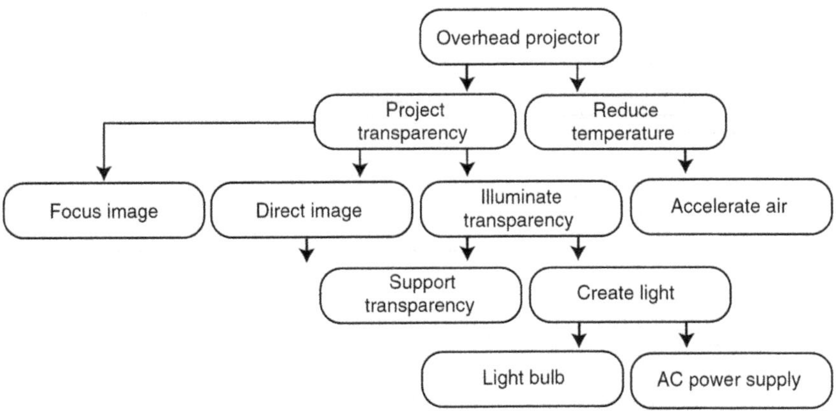

Figure 2.26 An example of using functional efficiency technique to find the bottom design components.

Review problems

1. A motor corporation wants to improve their design by using rapid prototyping technology and assign the weighting factors for its relationship matrix. Distribute 100 points total weight among the design factors.

	Cost	Accuracy	Material selection	Strength	Product size	Lead time	Heat resistance	Total
Cost	×	0	0	1	0	0	0	1
Accuracy	1	×	1	1	1	1	1	6
Material selection	1	0	×	0	0	0	1	2
Strength	0	0	1	×	0	0	1	2
Product size	1	0	1	1	×	0	0	3
Lead time	1	0	1	1	1	×	1	5
Heat resistance	1	0	0	0	1	0	×	2

2. Establish the objective tree to make a decision based on factors needed to purchase your vehicle. The decision factors include interior, exterior, fuel, ride, breaking, workmanship, initial cost, vehicle handling, and comfort of the vehicle.
3. A company has been asked to design a steering wheel for a new General Motors truck.

   The wheel is to be used on three different trucks and allows for varying electronics on the steering wheel. In the space provided below, list five things that should be considered when designing the steering wheel. Also, provide a brief explanation for why each design criterion was chosen.
   a. Use five design criteria to evaluate them using the digital logic approach.
   b. Use digital logic approach to determine the weight given for each design criteria.
4. If a high-school student wants to choose between different colleges, when considering college ranking, location, specialty, and expense, try to find the relative importance of the following criteria: college ranking, location, specialty, and expense. Complete the following table:

	College ranking	Location	Specialty	Expense	Total
College ranking					
Location	0				
Specialty	1	1			
Expense	0	1	0		

5. Continuing the last question, if another student wants to select one of these four colleges, but the four factors in sequence are college ranking, specialty, location, and expense, help him to make the selection.

	College ranking	Location (mile)	Specialty	Expense ($) (per year)
A	3	2,200	Satisfying	55,000
B	35	1,100	Very good	34,000
C	48	Local	Good	18,000
D	120	600	Very good	8,000

2.6 Paper prototyping

Everything should be made as simple as possible, but not simpler.

Albert Einstein

This section helps answer the following questions:

- What is paper prototyping?
- How can various levels of prototyping activities be categorized?
- When is a low-fidelity prototyping technique used?
- Is a high-fidelity prototyping technique always more effective?
- What types of paper prototyping are available?
- How to proceed with paper prototyping?

Prototypes are products that are used exclusively for eliciting feedback. Therefore, prototypes should be developed only to the point of beginning to be able to solicit the desired feedback. Viewing prototypes as deliverable products often leads to too much expenditure of developer effort when prototyping. There are several resources needed for prototyping including an appropriate prototyping tool; skill and knowledge necessary to use the tool effectively and efficiently; a valid means of collecting reliable feedback about the prototypes, including the time, skill, and motivation to make changes to the product; and willingness to collect feedback on the changed product. This section discusses prototype fidelity, paper prototyping that is quick and requires less effort, and user test issues for prototyping.

2.6.1 Selecting a prototype

Since there are several levels of prototyping activities, the objectives of the prototyping should first be reviewed so that proper prototype fidelity can be selected. Proper questions such as the following should be asked:

- Which prototype will take the shortest time to learn?
- Which prototype will elicit the fastest performance?
- Which prototype will users most prefer to use?

These questions are asked to decide on features, that is, what will the prototype do? One needs to choose an interaction paradigm, define possible constraints, and decide size and shape of the prototype, user interface, and environment. Goals should be realistic because 100% is never realistic. The testing goals should be detailed, and the result is not just a simple good or bad. The selected prototyping tool should enable the easy creation and modification of prototypes, be easy to learn and remember, and allow team development where many members can develop and modify prototypes.

2.6.1.1 Prototype fidelity

Prototype fidelity is the degree to which the prototype accurately represents the final product. Because often nobody knows exactly what the features of the final product will be, it is difficult to determine the fidelity. Fidelity then focuses on what level the prototype can represent the current understanding of the product. Low-fidelity prototypes are an efficient way to search the design space, are predictive of preferences in the actual product, enhance user participation in the design process, enable visualization of possible design solutions, and provoke innovation. Low-fidelity prototypes are easy to create, inexpensive to change, and are good for providing a basic "high-level view" of the overall system structure. Examples would be a jumbled mass of sticky notes on an office wall: multicolored, cryptic drawings on a whiteboard, or even as a set of "back of the napkin" scribbles. The limitations of low-fidelity prototypes include that certain aspects of a design cannot be adequately simulated and they are less effective in detecting problems. Low-fidelity prototypes are useful in the early stages of design, and the advantages are listed below:

- Paper prototypes (sketches) are very quick. In the early stages, there is no need to worry about the details. It is relatively easy to make changes and for other people to make their comments and suggestions.
- It is easy to give a simple sketch to a nontechnical person and receive feedback.
- It is time efficient and cost-effective. One probably spends only a few minutes sketching a concept. As one has nothing invested in the earlier designs time, it is easy to make the changes.

If properly performed, medium-fidelity prototypes are a best-of-both-worlds implementation that allows for a combination of both high-level and detailed views; rapid, iterative changes; and the ability to conduct meaningful user tests to evaluate complex functionality and to help determine system specifics.

High-fidelity prototypes, on the other hand, are often deployed with an almost-full set of features, functionality, and tie-ins. They are very detailed systems that are expensive to create and difficult to change quickly, and usually require more rigid development practices in order to complete. High-fidelity prototypes are effective in demonstrating products to clients, effective as an adjunct to user interface specifications, and effective as a means of communication with documentation and training specialists.

Prototypes can be classified by breadth and depth. Some prototypes have great breadth, but little depth. They cover a broad range of issues, while not providing much information on any one issue. A software prototyping example is a user interface that shows access to all product requirements, but none of the actual functionality works. Another software example is prototyping all windows in a product and accessing each new window by pressing the spacebar. Some prototypes focus on great depth, but little breadth. They cover a narrow range of issues, but in great detail. An example in software development includes prototyping only the menu bar and all pull-down menus, where each menu item really works.

Figure 2.27 shows various levels of prototypes used at various stages of the design process. In general, low-fidelity prototypes are utilized in the early design stage such as the conceptual design stage, medium-fidelity prototypes are used in the preliminary design stage, whereas high-fidelity prototypes are used during usability testing.

2.6.2 *Paper prototyping*

The most commonly used low-fidelity prototypes are paper prototypes that are a paper mock-up of the object, interface, look, feel, and functionality. They are quick and cheap to prepare and modify. Paper prototypes help ensure that early design decisions are based on the requirements of the customer's perspective, rather than from the viewpoint of technical departments. Like the general features of low-fidelity prototypes, a paper prototype

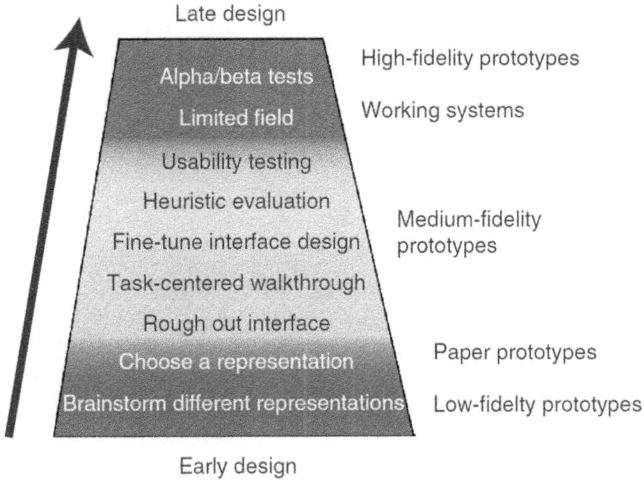

Figure 2.27 Various levels of prototypes used at various stages of the design process.

provides many benefits. It is inexpensive and quick. In particular, it supports the participatory design, in which a variety of stakeholders get together to mock-up designs that are designed to meet user requirements. Paper prototypes can be used to gain feedback from within the business and can also be used to validate design decisions by having customers walk through the design. This approach to design is also an excellent way to identify and address issues that may otherwise not surface until after implementation when they are extremely costly to remedy.

The purpose of paper prototyping is often used to brainstorm competing representations, elicit user reactions, and elicit user modifications/suggestions. A surprising finding is that in many circumstances, sketches work better than higher quality prototypes for user evaluation, since users suggest more major changes, and users focus on a high level rather than color, labels, or graphical details. Many user interface designers prefer to sketch early prototyping ideas on paper. It is the simplest form of rapid prototyping. Developers or users can use a pen or pencil to sketch windows, menus, widgets, etc., resolve flowchart navigation issues, and develop storyboards. Favored media for sketching include paper, transparencies, and whiteboards.

The main advantage of paper prototyping is the capability to prototype a concept in short turnaround time. It requires skills that most people have. It is easy to add written annotations and verbal descriptions, and useful for team prototyping and sharing of ideas. Paper prototyping is effective to keep participants focused on requirements, and design issues, as opposed to issues such as "how will the computer do this?", "are these the best background and foreground colors?", and "how does this widget work?". The disadvantages of paper prototyping include the inability to adequately portray some types of user interactions, and its visual orientation makes it difficult to prototype some other issues. Figure 2.28 shows an example of a simple sketch for paper prototyping for a remote control device for the curtains and windows. When the transmitter points to the window or curtain, it can achieve three functions:

1. Window open/close (push once) or incremental feed (press and hold)
2. Curtain open/close (push once) or incremental feed (press and hold)
3. Light dark/bright (turn the knob clockwise, more light will come in)

Figure 2.28 An example of the simple sketch for paper prototyping for a remote control device for the curtain and window control.

Although the description of the above problem may not be complete, questions and suggestions are entertained because of the simple sketch.

Another type of paper prototyping is storyboarding, which is a series of keyframes to tell the prototyping story. The idea was originally from film and was used to get the idea of a scene. It can be snapshots of the prototype interface at particular points in the interaction so that users can quickly evaluate the direction the interface is heading. Figure 2.29 shows an example of a storyboard that shows a roof mechanism for the solar house. The plates of silicon set up on the roof can rotate with the position of the sun. The house can absorb the sunlight as much as possible. Moreover, the rotation needs to be adjusted monthly because the direction of sunlight will change.

Figure 2.29 An example of storyboard to show a roof mechanism for the solar house.

Another paper prototyping method is a plastic interface for collaborative technology initiatives through video exploration (PICTIVE) approach, which was originally a plastic interface for collaborative technology initiatives through video exploration. The design can be multiple layers of sticky notes and plastic overlays where differently sized stickies represent icons, menus, windows, etc. Interaction is demonstrated by manipulating the notes. The contents were changed quickly by a user/designer with pen and note repositioning. Often a session is videotaped for later analysis. This process usually ends up with a mess of paper and plastic. Figure 2.30 shows an example of premake paper interface components. The purpose of this project is to prototype a cost-effective gas closing and warning system in a laboratory environment for argon gas tanks. In the early design stage, instead of making a physical prototype, the actual equipment is photographed, and the components of the gas closing and warning system and wires are arranged in their respective positions to show the design as illustrated in Figure 2.31.

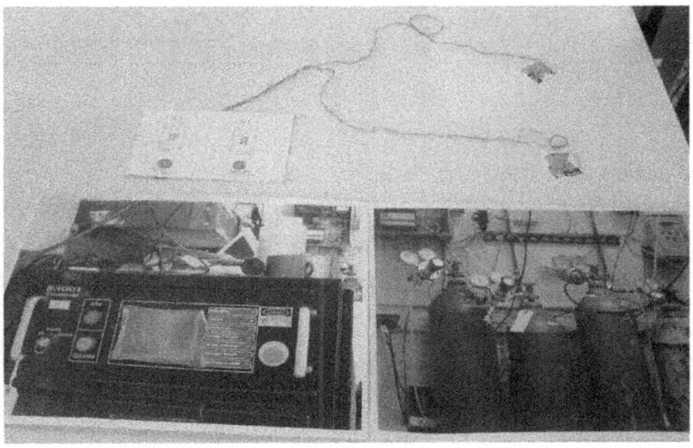

Figure 2.30 PICTIVE prototype: the actual equipment is in pictures, and the components of the gas closing and warning system and wires are ready to be arranged in their respective positions.

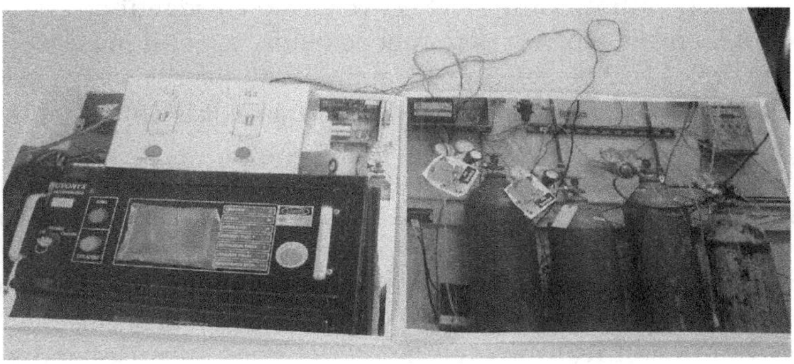

Figure 2.31 PICTIVE prototype: the actual equipment is in pictures, and the components of the gas closing and warning system and wires are arranged in their respective positions to show the design.

Medium-fidelity prototypes for software engineering can be used to prototype with a computer. This type of prototype can be used to simulate some but not all features of the interface. The purpose is to provide a sophisticated but limited scenario for the user to try and test more subtle design issues. The dangers include the following: user's reaction is often "in the small"; users are reluctant to challenge designer; users are reluctant to touch the design; and management may think it is real. Medium-fidelity prototypes also include drawing each storyboard scene on the computer as shown in Figure 2.32. Medium-fidelity prototypes could also include scripted simulations to create a storyboard with media tools. The user is given a very tight script/task to follow. It appears to behave like a real system, and thus, any script deviations may blow the simulation.

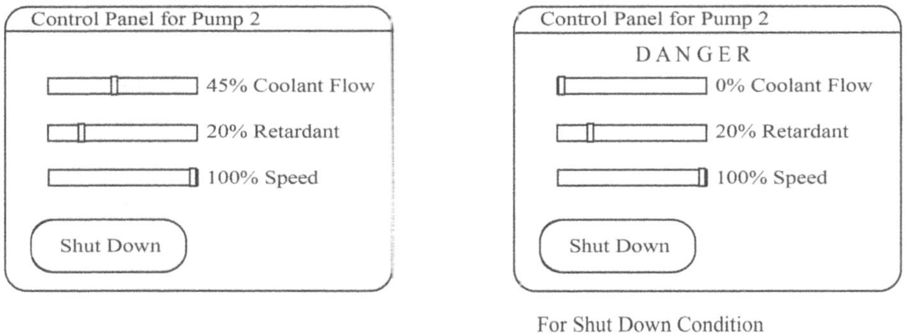

Figure 2.32 Storyboard scene as a medium-fidelity prototype.

Wizard of Oz is a method for testing a nonexisting system. In the system, the human "wizard" simulates system responses and interacts with the user via a simulated software user interface. It is useful for adding complex functionality, such as speech and gesture recognition, and language translation.

2.6.3 *User tests*

Usability testing helps in finding out if the product meets the user expectations and needs. It can assist in delivering a product that is more intuitive to use. Also, remember that effective usability testing is not a one-time effort, but continuous throughout the system development cycle. Usability testing is also an effective method of identifying design error at an early stage in development, when system functionalities are still in basic form or not yet implemented. As errors will be determined, appropriate solutions have to be developed. At this stage, changes will be easier and cheaper to make. It is important in ensuring the success of the system at every stage of the design process.

Usability testing is an essential part of the design process. It allows the introduction of new ideas, clarifies user needs, and overcomes designer bias. The general approach to set up a usability test is first to develop a prototype, choose a task for the user to perform, ask the user to complete the task, take notes, revise and modify the prototype, and repeat the process. The reason to involve users is evident: expectation management and ownership. Involving users can project realistic expectations, offer no surprises and no disappointments, and provide timely training. It can also make the users active stakeholders that are more likely to forgive or accept problems, which can make a significant difference in acceptance and success of the product. The principles for product definition are to keep an open mind but never forget the users, the user tasks, and environments, discuss ideas with all the stakeholders as much as possible, use (low-fi) prototyping to get rapid but well-informed feedback from stakeholders, and iterate until the prototype is satisfactory.

How can one turn a good idea into a successful system? It is not the laws of physics, but the experience of working with the system! Experience will speed up the design process by an order of magnitude and facilitate the creation of excellent systems. It is therefore important to select the proper users for the task. The following are the aspects to be considered:

- Representative subjects of specified user community: Should the users have a background in computing? Should the users have experience with the task? Should the users have a motivation? Should the users have a certain level of education?
- Physical concerns: Should the users have good eyesight? Should the users have certain age limitation? Should the users be left/right handed? Should the users be of a certain gender?
- Experimental conditions: Should the test be run during a certain time of day or a certain day of the week? Should the test be in physical surroundings? Should the test be with a certain level of distractions, noise, or room temperature?

During user tests, specific tasks to be executed using the product should be described in terms of the task. For example, "Send an e-mail to your friend," "Find a website about Abe Lincoln," and "Play your 'Hotel California' MP3." During the test, be aware of the following:

- What confuses the user?
- Does the user try to do something that is not possible?
- Is there anything hard to find?
- Does the user lose focus during the task?
- Is more explanation needed? Is less?

For a more complex prototype, testing guidelines should be provided early in the design process. It is basically a summary of a set of working guidelines, for example, Apple Human Interface Guidelines. The content could include screen layout issues, I/O devices, action sequences, and training. Be sure to have the users review and test the guidelines before the test because one may like the users to focus attention on interface early, the controversial issues can be discussed early, and fewer design changes will be made during the implementation phase.

Review problems

1. Define low-, medium-, and high-fidelity prototypes.
2. What is paper prototyping?
3. What is the importance of usability testing?
4. There are many types of prototypes. Match the following prototypes to their key features:

1. Paper prototypes	a. Can premake paper interface components
2. Storyboarding	b. Users can evaluate quickly the direction the interface is heading
3. PICTIVE	c. The simplest form of rapid prototyping
4. Painting/drawing packages	d. User given a very tight script/task to follow appears to behave like a real system
5. Scripted simulations	e. Drawing each storyboard scene on computer docs not capture the interaction "feel"

5. What is the "PICTIVE" prototyping method?
6. Using a sketch, give an example of a prototype for a personal webpage.
7. Conduct an interface paper prototype for a student's schedule.
8. What is the uniqueness of paper prototyping? Briefly, explain the applications of paper prototyping.

2.7 Learning from nature

Nature is a proven concept, a system that works.

William McDonough

This section helps answer the following questions:

- What is the relationship between product/prototype development and nature?
- Are there ideas and prototyping examples of learning from nature that have made large changes from the way we used to do things and may have revolutionized an industry?
- How does one transform ideas from nature into creative product innovation?
- What is biomimetics?
- What is the synectics technique?

One great way to find designs and product ideas is to learn from nature. "We need a new design. We have to recognize that every event and manifestation of nature is design." William McDonough said, "Nature is a proven concept, a system that works." [Frenay17]. It is full of the most advanced working prototypes that last. Therefore, when looking for design or prototyping ideas or when in doubt about whether a concept or prototype will work or when looking for the direction of a product, nature is the best resource.

2.7.1 What can we learn from nature?

Nature's blueprints are logical principles of how the Earth creates and maintains life, principals that could give new shape and meaning to human endeavor. We can learn to live within laws of nature, to accept the interdependence with forces larger than ourselves. Biomimetics is the abstraction of designs from nature. Nature works for the greatest achievement at minimum effort. Biomimicry (from bios, meaning life, and mimesis, meaning to imitate) [Dollens17] is a new science that studies nature's best ideas and then imitates these design and process insights to solve human problems. For instance, one can study a leaf to invent a better solar cell. Nature has already solved many problems. Animals, plants, and microbes are the consummate engineers. They have found what actually works, what is appropriate, and most importantly, what lasts here on Earth. After 3.8 billion years of development, failures are fossils, and what surrounds us is the secret to survival. The conscious emulation of life's genius is a survival strategy for us, the human race, a path to a sustainable future.

There are many cases that the current technologies are far from what nature can offer [Benyus97, Benyus17]. For example, one uses extreme high temperature, force, and heat treatment to make materials to make them strong and durable like Kevlar, a premier, high-tech material. To make such material, petroleum-derived molecules are poured into a pressurized vat of concentrated acid and boiled at several hundred degrees. It is then subjected to high pressures to force the fibers into alignment. Nature takes an entirely different approach. An organism makes materials like bone or silk right in its own body, no need to process it in an extreme environment. A spider produces waterproof silk that is tougher and more elastic than Kevlar. Ounce for ounce, it is five times stronger than steel! The spider manufactures it in water, at room temperature, without using chemicals or pressure. There is very little waste of material, energy, and pollution to the environment!

One can study nature's models and then imitate from these designs and processes to solve human problems. One can use nature as a mentor to judge the "rightness" of the innovations. After 3.8 billion years of evolution, nature has learned: what works, what is appropriate, and what lasts. One can use nature as a measure to view and value nature. It is not on what one can extract from nature but on what one can learn from it. For instance, one can learn a lot from abalone mussel nacre, which is the mother of pearl coating that is extremely lightweight but fracture resistant, as shown in Figure 2.33. It is a crystalline coating that self-assembles in perfect precision atop protein templates. In the abalone, it is a 3D masterpiece, tougher than anything one can manufacture! The applications include hard coatings for windshields, bodies of solar cars, and airplanes. Antlers, teeth, bones, and shells are the blueprints of living organisms. The layer-by-layer assembly or layered manufacturing was inspired by natural biomineralization. Dolphin and shark skins have a submarine-hull-like material that deforms slightly to shrug off water pressure. An example of an application of bionics is orb-weaver spider silk made from a new fiber manufacturing technique—a way to manufacture fiber without using extreme heat, pressure, or toxic chemicals. The fiber is stronger and more resilient than anything one now has and could be utilized in parachute wires, suspension bridge cables, sutures, and protective clothing. Rhinoceros horn is a self-healing material that is both compressively and laterally strong. Blue mussel adhesive is an underwater adhesive. Unlike human-made glues, the blue mussel adhesive sets underwater and does not need a primer, an initiator, or a catalyst to work. It could revolutionize paints and coatings, and enable surgeons to operate without sutures.

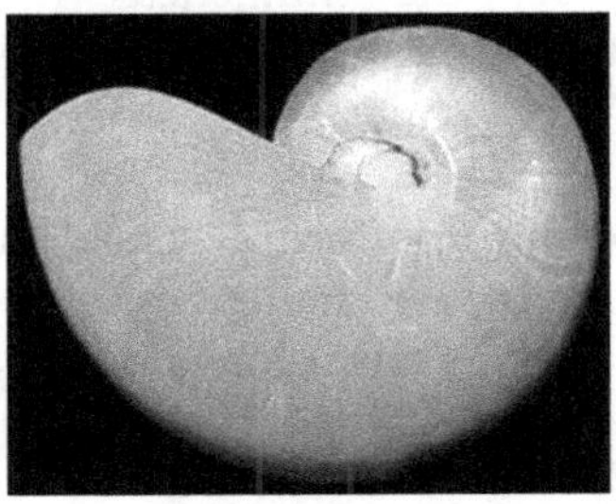

Figure 2.33 Nautilus whose shell was buffed off to reveal the beautiful mother of pearl.

Biomimetics, in design, is a process that instead of imitating the final product of biological systems, it imitates the process by which they are designed. Many examples in nature provide hints for future manufacturing methods. Teeth and seashells might be excellent strong building materials, but their growth is typically too slow to be attractive for human manufacturing. However, could one learn from the process so that one can manufacture using a similar but faster technology? Other examples are also very stimulating, such as the fact that a ruby-throated hummingbird can make a 600-mile journey on less than 0.1 oz (3 g) of fuel; crocodile and alligator hides can deflect arrows and even bullets; the nautilus has special chambers that enable it to regulate its buoyancy; and the squid uses a form of jet propulsion. Could one design a car or airplane that consumes minimum energy? Could one invent and manufacture an armor that is as strong as or even stronger than that of the crocodile? Could an underwater vehicle that takes full advantage of regulating its buoyancy be invented? Has the squid's jet technology been applied to the current technology?

One can also learn sensor fusion technology from animals and apply this to the task of flying a drone to target using onboard video. Without sophisticated eyes, flies can land accurately, bees find flowers, and bats catch evading insects in flight. What do they "know" that humans do not? One possibility is that they use variations of optic flow and represent sensory image field by a motion vector field. This way a fly can land simply by maintaining constant optic flow, and a dog can track by maintaining constant sensory flow and following the gradient using sniffing as a form of "chopper amplifier."

2.7.2 *Synectics*

To learn from these excellent prototypes around us, synectics is a good approach. Synectics is a method of problem-solving, usually performed by groups, which seeks to illuminate and utilize the factors involved in creative thinking. Creative thinking is impaired in two potential ways: (1) the problem is beyond one's everyday experience that one cannot imagine how it could be solved; and (2) the situation is so familiar that one cannot think of a better way to resolve the problem. Synectics aims to promote creative thinking using two principal techniques: making the strange familiar and making the familiar strange.

- Making the strange familiar

  The mind tends to analyze a new situation by forcing the problem to fit existing preconceptions. This is also a reflection of the fact that human thought tends to be conservative. In this approach, the strangeness is compared with data previously known to eliminate as much of the strangeness as possible, and this requires a paradigm shift.
- Making the familiar strange

  This process is very tough to perform because strangeness and uncertainty are uncomfortable. To overcome these challenges, synectics makes use of analogies, which include personal analogy, direct analogy, symbolic analogy, and fantasy analogy.

2.7.2.1 *Analogy*

The analogy is an inference that if two or more things agree with one another in some respects, they will probably agree in others. This is similar to some particulars between things otherwise unlike. The following analogies are commonly used.

2.7.2.1.1 Direct analogy The direct analogy makes links between the present issue and similar issues that have already been solved. For example, Sun Tzu's "The Art of War" [Cantrell04] is now used for business strategy. Lego is analogous to real building bricks. To miniaturize an electronic device, one can see how smartphones are miniaturized.

To make a lightweight, strong laptop, look to see how other lightweight and strong objects are made (e.g., planes).

2.7.2.1.2 Fantasy analogy Fantasy here also means "beyond belief." Many of today's conventional technologies were imagined by earlier science fiction/fantasy writers. For instance, escalator moving staircases were imagined by Arthur C. Clarke [Adam17], the laws of robotics were first mentioned by Isaac Asimov in 1942 [Wikipedia17], and submarines were described by Jules Verne [Verne17]. Fantasy analogy can be used to remove a block in the design process: "imagine the solution to this exists, and let's carry on." It can be used to approach a practical solution from the reverse as discussed by Albert Einstein, "When I examine myself and my methods of thought, I conclude that the gift of fantasy has meant more to me than any talent for abstract, positive thinking."

2.7.2.1.3 Symbolic analogy Symbolic analogy sums up the objective in a way not technically accurate but captures the essence of the situation. For example, one wants a car that moves like "greased lightning," a seal that is tighter than a "clam shell," a solution that is "outside the box," and a basketball shoe that sticks to the floor "like glue." Many of these subconscious similes can suggest ways in which the problem can actually be solved. This can also involve "pictorial" thinking. For example, one can imagine electrons in an atom that orbit the nucleus like planets around a sun or electrons in a semiconductor that act like balls on a hill.

2.7.2.1.4 Personal analogy In personal analogy, the designer imagines being part of the system. When teaching how transistors work, one can encourage students to think like an electron. Electrons move in response to voltage gradients like a ball may do to physical hills and valleys (a symbolic analogy!), so one can imagine how one would respond to the electrical environment if one were the electron. This requires expertise and familiarity with the situation.

2.7.2.1.5 Biological analogies or bionics Biological analogies are directly learned from nature. In many situations, particularly in mechanical and civil engineering, nature has already solved the problem. For example, animal backbones are similar to bridges, and artificial neural networks are based on models of the brain, etc.

Asking the appropriate questions can often hasten a solution, and creating a checklist is one approach to study how to adapt nature's footprint. Questions such as the following can be investigated:

- What is similar to it?
- Why is it necessary?
- What can be eliminated?
- How can its assembly be improved?
- What could new materials be used?
- In what way is it costly?
- Are there any other applications?
- In what way is it inefficient?
- Can it be improved ergonomically?
- What is wrong with it?
- What does it not do?

2.7.3 Better products—back to nature

There are nine design elements in nature [Althen17] that may be learned when a new product or prototype is developed:

- Runs on sunlight.
- Uses only the energy it needs.
- Fits form to function.
- Recycles everything.
- Rewards cooperation.
- Banks on diversity.
- Demands local expertise.
- Curbs excess from within.
- Taps the power of limits.

For example, nature has produced products that are not polluting the Earth and save energy. There are many ideas to make better products. For example, high-volume, low-pressure spray paint guns can decrease the release of toxic chemicals, provide greater worker protection, and decrease the amount of paint needed for greater cost savings. For conventional uses, it can save up to $13,000 on paint costs every year. How about the solar-powered jacket, as seen on the secret service and the president; with integrated solar panels to charge electrical devices, within its pockets?

Another example is in car design [Butler17]. Daimler developed a new high fuel efficiency concept vehicle based on the body shape of a boxfish, a common cube-shaped fish found in marine habitats. The bionic car offered 20% lower fuel consumption and up to 80% lower nitrogen oxide emissions.

Examples are available everywhere such as composite structures. It is learned that the strength-to-weight ratio of a tree is provided with longitudinal fibers embedded in a resinous solution. The study and adaptation of the multiple wing tips of sailing birds leads to multiple winglets on a real sailplane. This solution allows the decomposition of the wing tip vortex in multiple lower intensity vortexes. One can develop light-emitting organic materials that will achieve energy efficiencies similar to those found in the insect world. This idea comes from the fact that insects such as fireflies can emit light while staying cool. This means that the insect's body is very efficient. Scientists are studying these insects and attempting to make new materials that will make the world more energy efficient. Bone, in general, is very tough, and bird bone is very stiff, yet birds still manage to fly. Furthermore, the ceramic content of bird bone is extremely high. Research in greater detail may happen to consider the nature of the stiffness of bone, and whether it can be mimicked using other components. The results could be applied to suggest ideas for tougher helmets and general impact protection.

Golden ratio is a geometrical proportion with pleasing, harmonious qualities. The golden ratio, also known as divine proportion, golden mean, or golden section, is an irrational number with several curious properties. It can be defined as a number that is equal to its own reciprocal plus one (or $x = 1/x + 1$). Solving the so-formed quadratic equation gives a value of 1.618033989 or −0.618033989.

The golden ratio seems to get its name from the golden rectangle, a rectangle whose sides are in the proportion of the golden ratio. The golden rectangle is an aesthetic one. The ratio is considered to be aesthetically pleasing and so found spontaneously in a great deal of art.

Nature is full of working prototypes that may be far more advanced than the current technologies. Should not the invaluable knowledge nature provide be fully utilized to improve the world?

Review problems

1. What is biomimetics? Why is it needed? Give some examples.
2. Consider a leaf for design exploration, in terms of function, structure, circulation, waterproof properties, etc.
3. Use fireflies as an example to explain what can be learned from nature to solve human energy problems.
4. Below are some examples of what we can learn from nature. Match the following characteristics on the left to the lessons on the right:

1. Antlers, teeth, bones, shells	a. Used to make decorative objects (mother of pearl coating)
2. Dolphin and shark skin	b. It can be blueprints of living organisms because of the layer-by-layer assembly of layered manufacturing
3. Orb-weaver spider silk	c. Deforms slightly to shrug off water pressure
4. Abalone mussel nacre	d. A way to manufacture fiber without using high heat, high pressure, or toxic chemicals

5. There are many manufacturing characteristics that we can learn from animals. According to their characteristic, match the following:

1. Crocodile and alligator	a. Has special chambers that enable it to regulate its buoyancy
2. Squid	b. Uses a form of jet propulsion
3. Bees	c. Applies sensor fusion to flying a drone to target
4. Nautilus	d. Can deflect spears, arrows, and even bullets

6. Define the term "synectics."
7. Research the industrial application of hexagonal wax cells built by honeybees. What are the benefits of using it?
8. What is the "golden ratio"? Can you find any connections between nature and human-made structures?

References

[3Dprinting17] Wikipedia, 3D printing, Retrieved on August 28, 2017 from https://en.wikipedia.org/wiki/3D_printing.

[Adam17] Adam, D., The cheap way to the stars—by escalator, 2003, Retrieved on August 24, 2017 from www.guardian.co.uk/spacedocumentary/story/0,2763,1041360,00.html.

[Althen17] Althen, A., The nature of LEED: how natural laws inspire and inform green building, Retrieved on August 24, 2017 from www.usgbc.org/articles/nature-leed-how-natural-laws-inspire-and-inform-green-building.

[Benyus97] Benyus, J., *Biomimicry: Innovation Inspired by Nature*, Perennial, New York , 1997.

[Benyus17] Benyus, J., Biomimicry, Retrieved on August 24, 2017 from https://biomimicry.net.

[Butler17] Butler, R., Biomimetics, technology that mimics nature, August 24, 2017, Retrieved on September 24, 2006 from https://news.mongabay.com/2005/07/biomimetics-technology-that-mimcs-nature.

[Cantrell04] Cantrell, R.L., *Understanding Sun Tzu on the Art of War*, Center for Advantage, Arlington, VA, July 15, 2004.

[Chapman03] Chapman, C., *Project Risk Management: Processes, Techniques and Insights*, John Wiley & Sons, Hoboken, NJ, December 2003.

[Clausing94] Clausing, D.P., *Total Quality Development: A Step-By-Step Guide to World Class Concurrent Engineering*, American Society of Mechanical Engineers, New York, NY, 1994.

[Dixon95] Dixon, J. and C. Poli, *Engineering Design and Design for Manufacturing*, Field Stone Publisher, Norwich, VT, 1995.

[Dollens17] Dollens, D., Design biomimetics: an inquiry and proposal for architecture and industrial design, Retrieved on August 24, 2017 from https://arl.human.cornell.edu/linked%20docs/DennisDollens_DesignBiomimetics.pdf.

[EPA17] Mortality Risk Valuation, retrieved on August 24, 2017 from www.epa.gov/environmental-economics/mortality-risk-valuation.

[Ertas96] Ertas, A. and J.C. Jones, *The Engineering Design Process*, 2nd edition, John Wiley & Sons, Hoboken, NJ, 1996.

[Ferguson94] Ferguson, E.S., *Engineering and the Mind's Eye*, Reprint edition, The MIT Press, Cambridge, MA, 1994.

[Frenay17] Frenay, R., Biorealism: reading nature's blueprints, 1995, Retrieved on August 24, 2017 from www.ibiblio.org/london/permaculture/mailarchives/permaculture-WA/msg00024.html.

[Goldberg03] Goldberg, M.S. and A.E. Touw, *Statistical Methods for Learning Curves and Cost Analysis*, CNA Corporation, Alexandria, VA, 2003.

[Kelley01] Kelley, T. and J. Littman, *The Art of Innovation: Lessons in Creativity from IDEO*, Broadway Business, New York, 2001.

[Kendrick03] Kendrick, T., *Identifying and Managing Project Risk: Essential Tools for Failure-Proofing Your Project*, American Management Association, New York, April 2003.

[Otto01] Otto, K. and K. Wood, *Product Design*, 1st edition, Prentice Hall, Upper Saddle River, NJ, 2001.

[Suh01] Suh, N.P., *Axiomatic Design: Advances and Applications*, The Oxford Series on Advanced Manufacturing, Oxford University Press, UK, 2001.

[Ulrich99] Ulrich, K. and S. Eppinger, *Product Design and Development*, 2nd edition, McGraw-Hill/Irwin, New York, 1999.

[Verne17] Jules Verne invention that came true: electric submarines, (National Geographic) Retrieved on August 24, 2017 from http://oceanleadership.org/jules-verne-invention-that-came-true-electric-submarines.

[Wikipedia17] Wikipedia, Three laws of robotics, Retrieved on August 24, 2017 from http://en.wikipedia.org/wiki/Three_Laws_of_Robotics.

[Zeil17] Zeil, S.J., Lecture notes & slides, Retrieved on August 24, 2017 from http://www.cs.odu.edu/~price/cs451/Lectures/04mgmt/costest/costest_ht.html.

chapter three

Modeling and virtual prototyping

Many people built a prototype of the wrong thing. For instance, they invested a lot of time and resources making a system prototype when only a subsystem prototype was necessary or built a prototype when a model or simulation would have resulted in the same conclusion. Of course, in prototyping activities, there are always risks involved. However, spending 90% of the resources to address a 10% risk area is not a good strategy. Modeling and virtual prototyping are powerful tools in prototyping, especially in the early stage. A model may not be able to come up with the exact solution to a problem. However, it may be able to provide answers to the following questions:

- Which approaches are not likely to work?
- What parameters may be important?
- What might the answers be?
- How sensitive is the answer to assumptions?

This chapter provides a notion about the usefulness of using modeling and virtual modeling in developing a prototype. Some examples are also used to demonstrate the idea.

3.1 Mathematical modeling

This section helps answer the following questions:

- Is there a way to quickly find out whether a prototype will work even before a physical prototype is built?
- How can a model be used to help in prototyping activities?
- What is considered a mathematical model?
- How to apply the derived formulae, or laws of physics, in design and prototyping? What are the examples?

As discussed in the previous sections, modeling and virtual prototyping are very efficient in the product development process, especially in the early design phase. This section discusses the relationship and importance between mathematics and physics and provides some examples of mathematical models and physics in product design and prototyping. You may need to read further references to come up with a good model in a specific area of application.

A mathematical model can be analytical, numerical, or a fuzzy model, and it can also be a static, quasi-static, or dynamic model. A mathematical model can be governed by fluid mechanics, heat transfer, thermodynamics, solid mechanics, solid model, or electromechanics, and could at the same time be a lumped mass or finite element model. Just like prototyping fidelity, model descriptiveness, model detail, or model fidelity determines the level of a model. Selecting model fidelity is a trade-off decision between model descriptiveness versus model construction and resolution determination time.

3.1.1 Relationship between mathematics and physics: an example

Consider the law of gravitation, how is the phenomenon described? Or, how would one model the phenomenon? It is known that the following Newton's equation works:

$$F = ma \text{ or Force} = (\text{mass}) \text{ times } (\text{acceleration}) \tag{3.1}$$

What are the conditions associated with this equation? It is also known that this equation is from the law of gravitation:

$$F = GmM/r^2 \tag{3.2}$$

Two bodies, with masses m and M, exert a force upon each other, which varies inversely as the square of the distance between them and varies directly as the product of their masses. Mathematics is a useful tool in complex situations to figure out what in given circumstances is a good move. Much of fundamental physics can be described in elementary mathematical forms. The abovementioned equations look simple, but it actually took several hundred years to come to this conclusion. This section summarizes the findings of this gravity model as an example of trying to find a model. It represents an unknown state just like early scientists' knowledge trying to understand gravity [Feynman70].

Early scientists had many questions about the universe, and even the Sun and planets:

- Exactly how do planets go around the Sun?
- Do they go with the Sun as the center of the circle?
- How fast do they go with the Sun?
- What does the planet do?
- Does a planet look at the Sun, see how far away it is, calculate the inverse of the square of the distance, and decide how much to move?

Figures 3.1–3.3 show Kepler's (1571–1630) model based on his observations about how the Earth moves around the Sun. From these figures, we can see that an important step in

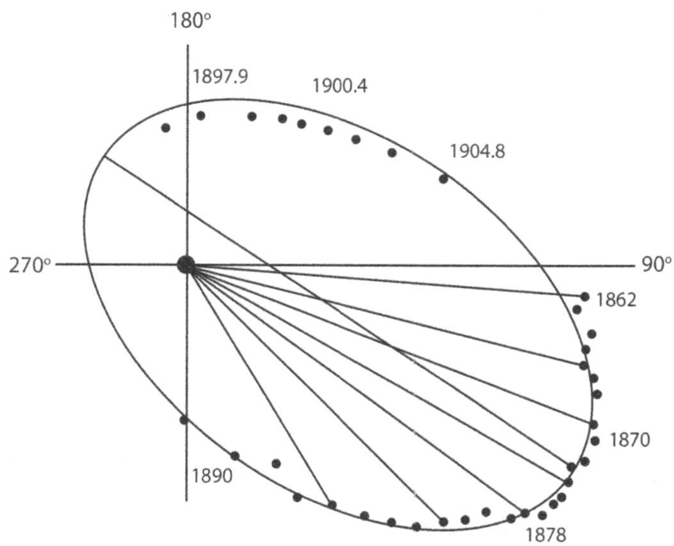

Figure 3.1 Kepler's observation about how the Earth moves around the Sun.

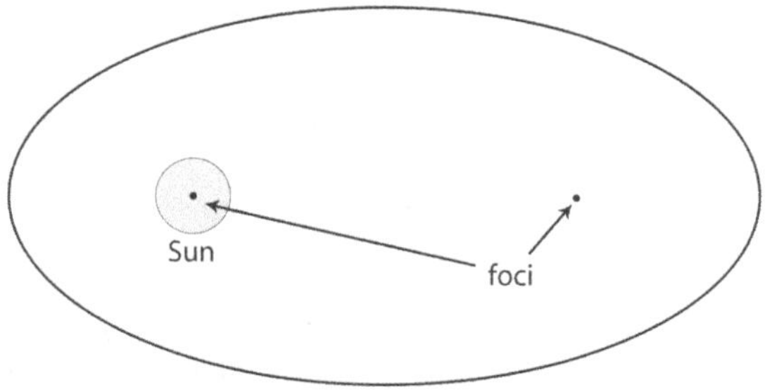

Figure 3.2 Kepler's observation about the Sun located at one of the loci of the ellipse.

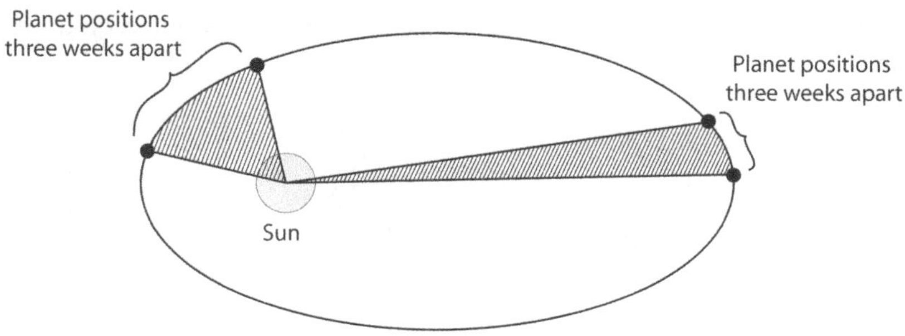

Figure 3.3 Equal areas are swept in equal times.

Kepler's model is the careful experimentation and observation about the physical phenomena. This concept applies to establishing a model or virtual prototype. In Figure 3.2, the orbit of the planet around the Sun is an ellipse with the Sun at one focus. Kepler further found that equal areas are swept in equal times as shown in Figure 3.3. In other words, the planet has to go faster when it is closer to the Sun.

During Kepler's time, there was no concept of gravity, and thus, scientists were greatly confused about this phenomenon. If Newton's model is correct, then if there were no force at all, the planet would just go straight. Now that the planet is not moving straight, there must be a huge force acting on the planet as shown in Figure 3.4. There was an old saying, "The angels have to do is to beat their wings in toward the Sun all the time." As shown in Figure 3.5, Kepler found that if there is no external force acting on the planet, the planet should move in a straight line. At the same time, it would move exactly the same distance from points 1, 2, and 3 in the figure, and thus, equal areas are swept out in equal times. It is therefore concluded that the external force be a normal force which changes just the planet's path along the normal direction but not the magnitude so that in equal times the planet moves exactly the same distance, and thus, equal areas are swept out. In other words, there must be an external force between the Sun and the planet and the force is

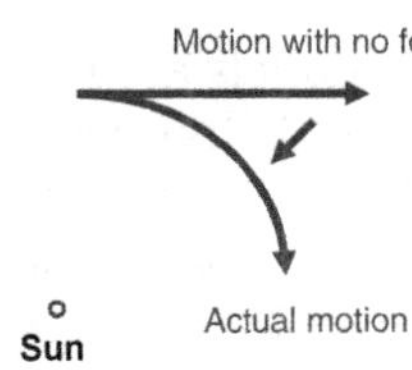

Figure 3.4 Since the planet is not moving straight, there must be a huge force acting on the planet.

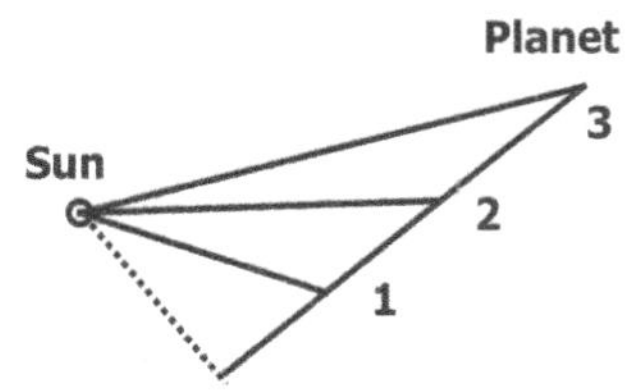

Figure 3.5 If no external force is offered, the planet should keep right on going in a straight line, move exactly the same distance, and thus, equal areas are swept out in equal times.

always pushing the planet toward the Sun. The similar force between the Earth and an object is known as the gravity force on the Earth.

In mathematical form, as shown in Figure 3.6, the area change and change rate can be expressed as

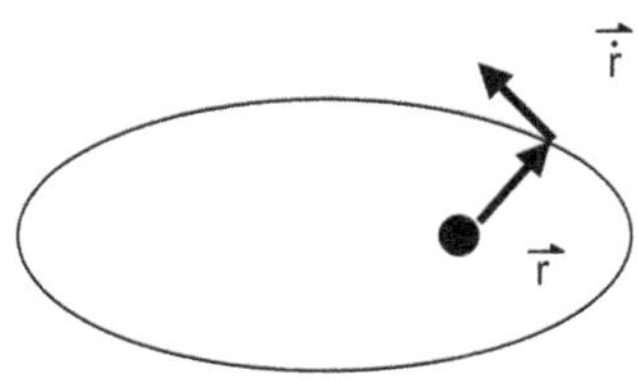

Figure 3.6 A mathematical form of the system.

$$\vec{\dot{A}} = \vec{r} \times \vec{\dot{r}} \tag{3.3}$$

$$\vec{\ddot{A}} = \vec{\dot{r}} \times \vec{\dot{r}} + \vec{r} \times \vec{\ddot{r}} = \vec{r} \times \vec{\ddot{r}} = \vec{r} \times \overrightarrow{F/m} = 0 \tag{3.4}$$

Since the area change from observation is a constant, its derivative is zero, and thus, the force has the same direction as the radius. This force is what is known as gravitational force between two planets (or objects). Do planets actually go in ellipses? The answer is—not quite. Since there are many other planets in addition to the Sun, they interfere with each other. Based on calculations, Jupiter and Saturn went according to the calculation but not Uranus, as shown in Figure 3.7. After Uranus was discovered, it was found that its orbit was not as it should be in accordance with Newton's laws. It was then predicted that another planet must be perturbing Uranus' orbit. In 1846, Adams and Leverrier proposed that the motion of Uranus was due to another unseen planet—Neptune. Adams told the other scientists, "Turn your telescope and look there and you will find a planet," and their answers were: "How absurd!" "Some guy sitting with pieces of paper and pencils can tell us where to look to find some new planet." Although Adams' calculation was not quite correct, Neptune was found. This is a good example of using a model to guide experimentation.

Currently, it is known that several models are related to gravity. In Newton's model, forces depend on something at a finite distance away as described previously using $F = GmM/r^2$. There is also a field model and a minimum potential model. In a local field model, there is a number at every point in space, and the numbers change when going from place to place. If an object is placed in space, the force on it is in the direction in which the numbers change most rapidly (potential changes). The force is proportional to how fast the potential changes as the object moves. In the minimum potential principle,

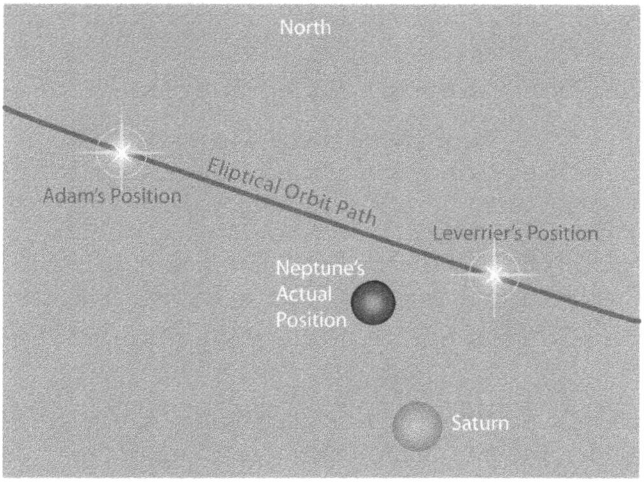

Figure 3.7 The solar system: in 1846, Neptune was not yet known.

say when the particle wants to go from x to y in 1 min and one wants to know what route it should use, one can calculate the energy (kinetic and potential energy) on each quantity and choose the path with the minimum energy (see Figure 3.8). The question is, which model is correct? Why are there several models to describe basically the same things? Although they are very different psychologically, they are equivalent scientifically. Models may look different when trying to guess a new law or use them in design.

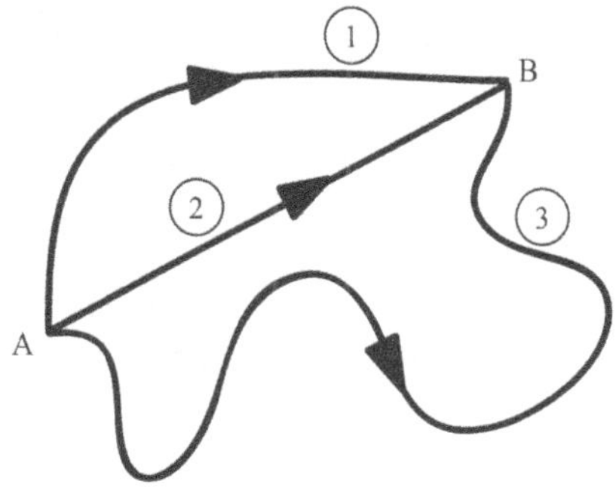

Figure 3.8 Example of choosing the route with the minimum energy.

Among so many models or equations, is there a place to begin to deduce the whole work? Is there some particular pattern or order by which one can understand that one set of statements is more fundamental and consequential?

There are two schools of thought in mathematics. One is the Babylonian tradition method to learn something by doing a large number of examples until one caught on to the general rule. He or she would know a large amount of geometry, many properties of circles, and formulae for the areas of cubes and triangles, and the tables of numerical quantities were available to solve elaborate equations. Another school of thought is the Greek tradition in which all the theorems of geometry could be ordered from a set of the axioms.

This is also the basis of modern mathematics and is not so suitable for physics or design. The Babylonian approach seems to be most appropriate for physics and design.

For example, the volume of a simple cylinder can be calculated as

$$\text{Volume} = \pi(\text{Radius})^2(\text{Height}) \tag{3.5}$$

This is a simple but well-known mathematical model. It could represent the volume of a hollow container, but could also be used to estimate the machining volume. Along with other well-known models, such as the cylinder surface, it could also be used to represent many other physical objects, such as shaft and motor volume and cross sections, bearing contact area, rocket volume, and surface, wheel contact area, gears, and motors.

3.1.2 *Using models for product and prototype design and evaluation*

Some of the commonly used basic models that can be used in product development are summarized in the following paragraphs [Basmadjian02].

3.1.2.1 *Conservation of mass*

The law of conservation of mass states that the mass of a system of substances is constant, regardless of the processes acting inside the system. In other words, the mass that enters a system must either leave the system or accumulate within the system:

$$(\text{Rate of mass in}) = (\text{Rate of mass out}) + (\text{Rate of change of mass content}) \tag{3.6}$$

For example, if the "rate of mass in" and "rate of mass out" are combined into mass outflow, the increment of mass inside the control volume must equal the negative of the outflow mass:

$$d\left(\iiint_V \rho dv\right) = -\left(\iint_S \rho V \cdot nS\right)dt \tag{3.7}$$

Example 3.1: Estimate the volume flow rate to select a pump for a prototype hydraulic system

Select a pump for a prototype hydraulic system as shown in Figure 3.9. The system has a cylindrical tank of radius R = 2m, which is being filled with liquid by a pump. The tank

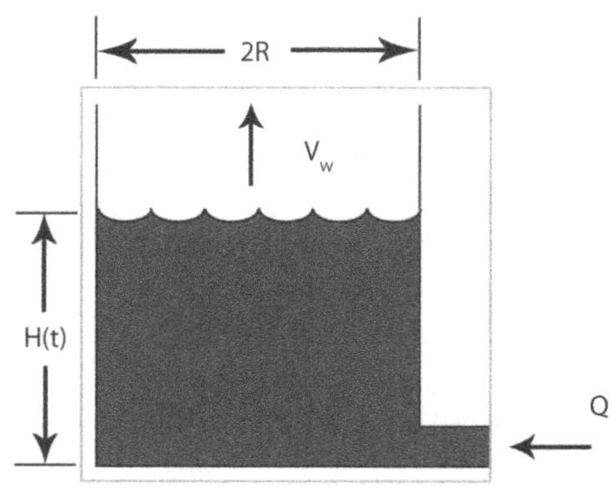

Figure 3.9 An example of prototype pump selection.

axis is vertical as shown in the figure, and the level of the liquid in the tank needs to be rising at a rate of V = 1 mm/s. Estimate the volume flow rate Q of the liquid needed for the pump.

Solution:

The mass conservation can be expressed as

$$\frac{d}{dt}\left(\iiint_V \rho\, dv\right) = -\left(\iint_S \rho V \cdot nS\right) = \dot{m}_{in} - \dot{m}_{out} = 0$$

$$\dot{m}_{out} = \frac{d}{dt}\left[\rho\left(\pi R^2 H(t)\right)\right] = \rho\left(\pi R^2\right)\frac{d}{dt}\left[H(t)\right] = \dot{m}_{in} = \rho Q$$

$$Q = \left(\pi R^2\right)\frac{d}{dt}\left[H(t)\right] = \pi R^2 V = \pi 2^2 (0.001) = 0.01256\ m^3/s$$

Therefore, a pump with the volume flow rate of at least 0.01256 m^3/s is needed.

3.1.2.2 *Conservation of momentum*

Conservation of momentum states that in the absence of external forces, a system will have constant total momentum. It is commonly used in collision models. A physics problem often requires the use of this fact in the collision of two elastic objects as shown in Figure 3.10. Due to momentum conservation, the sum of the momentum before the collision must equal the sum of the momentum after the collision:

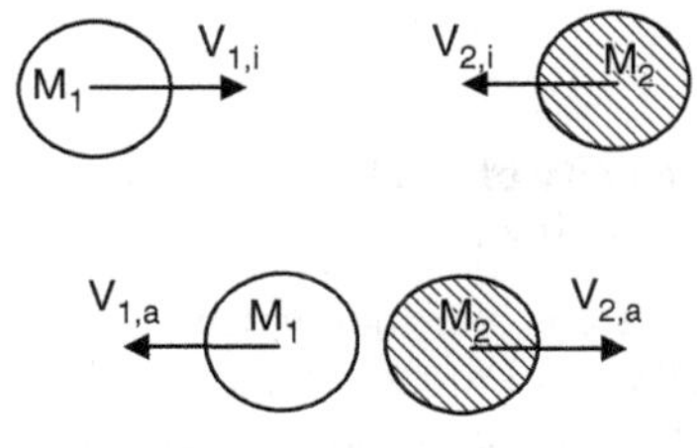

Figure 3.10 An example of two elastic objects in collision.

$$M_1 V_{1,i} + M_2 V_{2,i} = M_1 V_{1,a} + M_2 V_{2,a} \tag{3.8}$$

where
the subscript "i" signifies initial (before the collision)
"a" signifies after the collision

An inelastic case is when two objects have an impact, and the impulse, I, is introduced as

$$M_1 V_{1,i} - M_1 V_{1,a} = -M_2 V_{2,i} + M_2 V_{2,a} = M\Delta V = I = F_{average}\Delta t \tag{3.9}$$

where
$F_{average}$ = average impact force
Δt = impact duration

Example 3.2: Estimate the bird-striking force to design a prototype protecting structure for an airplane

In order to design a bird-strike protecting structure, one needs to estimate the amount of force acting on an airplane traveling at 900 km/h when it hits a bird of 1 kg. This will be important information when designing the protecting structure for the aircraft. Assume that the initial bird speed is negligible, using Equation 3.9, the following results can be obtained:

$$F_{average} = M\Delta V / \Delta t \tag{3.10}$$

where $\Delta V = 900{,}000\ m/3{,}600\ s = 250\ m/s$.

Assume the impact distance is 1 m, then the impact time $\Delta t = 1/250\ s = 0.004\ s$.

$$F_{average} = M\Delta V / \Delta t = 1\ kg(250\ m/s)/(0.004\ s) = 62{,}500\ N(7\ \text{tons of force!})$$

This is an estimation based on the 1 m impact distance, but it can provide some idea about its magnitude. The actual impact distance could be smaller than 1 m, and thus, the impact force will be even larger. If more accurate data is needed, a more in-depth model or some type of focused physical experiments may then be needed.

3.1.2.3 Conservation of angular momentum

The angular momentum of a rigid object is the product of the moment of inertia, I, and the angular velocity, ω. Without applying torque to the object, the angular momentum is constant. The angular momentum is a measure of the amount of torque that has been applied to the object over time.

$$I_1\omega_1 = I_2\omega_2 \tag{3.11}$$

Example 3.3: Estimate the velocity of a satellite for a prototype space camera aiming device

One use of the conservation of angular momentum is to model and find the velocity of a satellite. Note that gravitational force is an internal force resulting in the planet's centripetal acceleration. A subscript of 1 represents behavior at position 1, a subscript of 2 represents behavior at position 2, m represents the mass of the satellite, and R represents the radius of the satellite. The following equations can be true:

$$mR_1{}^2\omega_1 = mR_2{}^2\omega_2 \tag{3.12}$$

$$R_1v_1 = R_2v_2 \tag{3.13}$$

or

$$R_1/R_2 = v_1/v_2 \tag{3.14}$$

Thus, the satellite's speed is inversely proportional to its average distance from the Earth. This is similar to the velocity relationship between the Earth and the Sun as discussed earlier in this section. This model can save a lot of measurements and experiments in designing the aiming device.

3.1.2.4 Conservation of energy

Conservation of energy states that the total inflow of energy into a system must equal the total outflow of energy from the system, plus the change in the energy contained within the system. Thus, energy can be converted from one form to another, but it cannot be created or destroyed. Conservation of energy is used in the first law of thermodynamics and in the conservation of kinetic and potential energy:

1. First law of thermodynamics: the increase in the internal energy of a system is equal to the energy added to the system by heating, minus the amount lost in the form of work done to the system on its surroundings.

$$dU = \delta Q - \delta W \tag{3.15}$$

where
 dU = small increase in the internal energy of the system
 δQ = small amount of heat added to the system
 δW = small amount of work done by the system

Example 3.4: To estimate the work done by a prototype engine

One example of this law is adding heat δQ to a volume of gas and using the expansion of that gas to do work δW to push down a piston in an internal combustion engine. Before the gas is used to do the work, the energy is stored in the gas dU. Therefore, by calculating the energy stored in the gas dU, and the heat δQ added to the system, one can estimate the maximum work to be output by the prototype engine.

Another similar example is in refrigerator design in which work is done on a refrigerant substance to collect energy from a cold region and exhaust it in a higher temperature region, thereby, further cooling the cold region. Refrigerators have made use of fluorinated hydrocarbons, such as Freon-12 and Freon-22, which can be forced to evaporate and condense by successively lowering and raising the pressure. They can, therefore, pump energy from a cold region to a hotter region by extracting the heat of vaporization from the cold area and dumping it in the hotter area outside the refrigerator. These conditions about refrigerators apply to air conditioners and heat pumps.

2. Conservation of kinetic and potential energy (or together called conservation of mechanical energy):

$$PE + KE = \text{Constant} \tag{3.16}$$

with the absence of outside forces such as friction.

KE or kinetic energy is often defined informally as the energy of motion. It is defined as the work it would take to get an object moving with linear velocity v and angular velocity ω:

$$KE = \tfrac{1}{2} mv^2 + \tfrac{1}{2} I\omega^2 \tag{3.17}$$

PE or potential energy is defined as the path integral:

$$U = -\int_C \mathbf{F} \cdot ds \tag{3.18}$$

Example 3.5: To measure the mechanical equivalent of heat

Figure 3.11 shows James Joule's famous Joule experiments [Wikipedia17]. In this experiment, the amount of work done on the paddle was done by lowering a weight, so that the work done = mgz. The energy involved in increasing the temperature of the bath was shown to be equal to that supplied by the lowered weight. This design can also be used to measure the associated liquid thermal properties. In other words, Equation 3.18

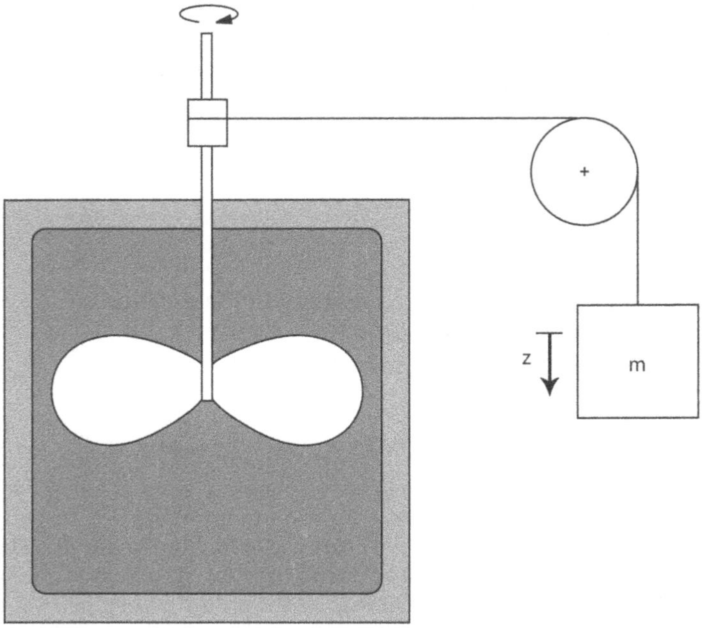

Figure 3.11 Setup for Joule experiments.

can be used to find the work, and by measuring the temperature change due to work, the liquid properties can then be obtained.

Example 3.6: Designing pneumatic toggle clamp general stress models

In mechanical design, often when designing the shape of a component, some basic stress patterns are useful in calculating or estimating the stress level of the components. To illustrate these basic stress patterns, a pneumatic toggle clamp as shown in Figure 3.12 is used as an example. When the clamp is disengaged, the spindle arm opens a minimum of 888 for easy workpiece removal. The spindle moves easily to any position along the U-shaped, open spindle arm.

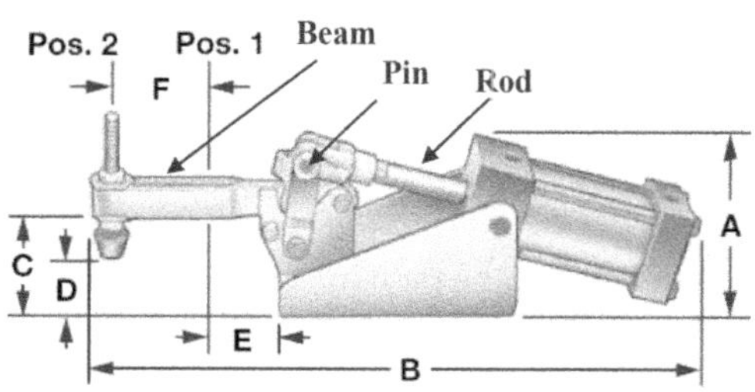

Figure 3.12 A pneumatic toggle clamp [Mcmaster07].

1. Tension or compression stress
   Tensile or compressive stress is the applied load divided by the cross-sectional area:

$$\sigma = F/A \tag{3.19}$$

As shown in Figure 3.12, the rod of the piston is subject to the tension and compression loads from the pneumatic pressure, and thus, the model can be used to help determine the rod material and size.

2. Bending
When a beam is subjected to bending, the stress can be calculated as

$$\sigma = My/I \tag{3.20}$$

where (as shown in Figure 3.13)
M = maximum bending moment of the beam
y = location in which stress σ occurs
I = moment of inertia of the beam. The bending moment is positive if it tends to bend the beam section concave facing upward

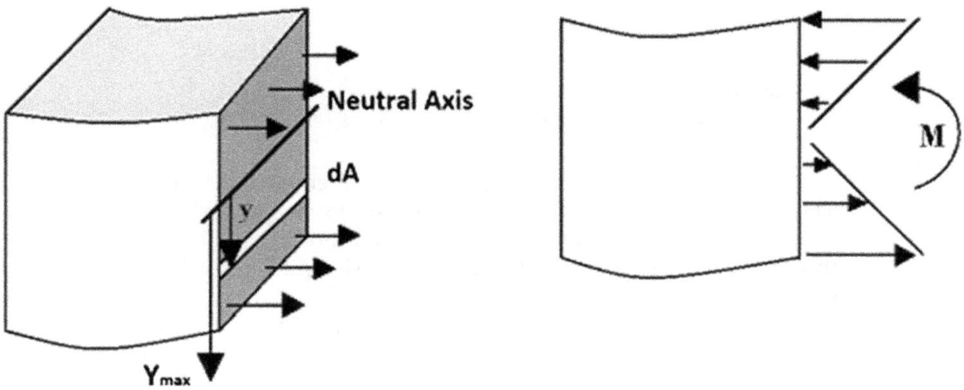

Figure 3.13 Cross sections of a beam under bending.

As shown in Figure 3.13, the beam is subject to a bending moment when the clamp on the left side is in contact with a workpiece.

3. Transverse shear
For a beam in bending, the transverse shear force will be the same along the entire length of the beam, and the shear stress will be maximum at the neutral axis and zero at the outer fiber. The shear force is defined as positive if it tends to rotate the beam section clockwise with respect to a point inside the beam section as shown in Figure 3.14.

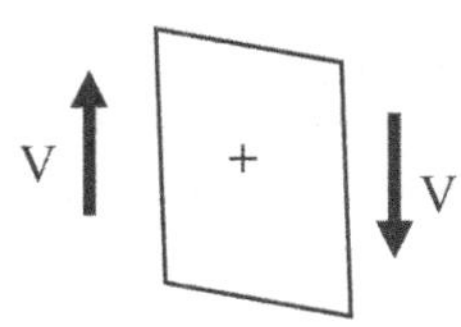

Figure 3.14 Sign conventions for shear force V.

As shown in Figure 3.12, when the clamp on the left side is in contact with a workpiece, the beam is subject to a bending moment as well as the shear force; thus the models can be used to help determine the beam material and size.

4. Torsion
When designing the stress level of a beam or bar subjected to twisting couples or torques, the torsion model as shown in Figure 3.15 can be used.

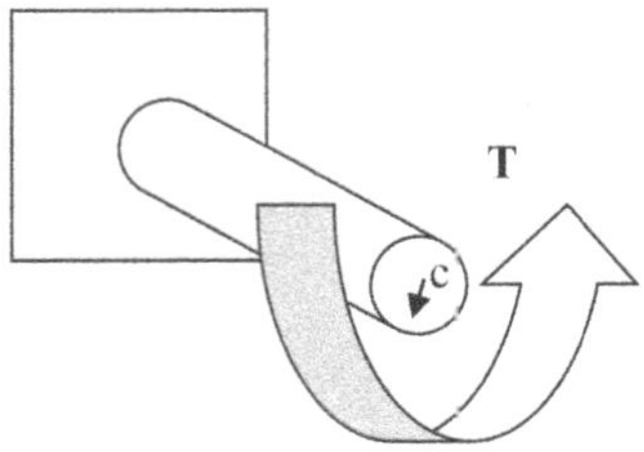

Figure 3.15 A simple cylindrical bar loaded with torsion.

The shear stress

$$\tau = Tc/J \tag{3.21}$$

where
J = polar moment of inertia
c = radius of the point of interest

Therefore, the maximum shear occurs when c is equal to the radius of the bar.

The applications of this model can be used to design various mechanical components, such as connection shafts between motors, gearboxes, couplings, and flywheels. It can be seen that there are no torsion pins in Figure 3.12.

5. Direct shear
An example of direct shear, V, is shown in Figure 3.16. The direct shear stress is

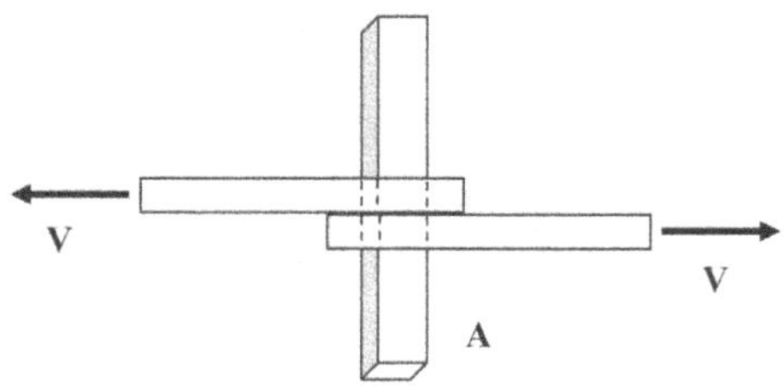

Figure 3.16 A beam loaded with direct shear.

$$\tau = V/A \tag{3.22}$$

where A is the cross-sectional area of the beam. As shown in Figure 3.12, when the clamp on the left side is in contact with a workpiece, the pin is subject to a direct shear force, and thus, the model can be used to help determine the pin material and size.

6. Contact stress
When two objects are in contact, the contact stress, σ, is

$$\sigma = P/A \tag{3.23}$$

where P is the uniform concentrated load on area A. This is a simplified model. Depending on contact geometry, more complex models are available. As shown in Figure 3.12, when the clamp on the left side is in contact with a workpiece, the contact point of dimension D is subject to contact stress, and thus, the model can be used to help determine the contact tip material and geometry.

3.1.2.5 Linear models

Some of the designs can be modeled or approximated with first-order relations. For example, in control theory, linear relations are often used to establish the system model, and good control can be achieved through iterations. Some of the commonly used models are explained in the following sections.

3.1.2.5.1 Modeling of mechanical systems To model a mechanical system, besides Newton's law (Equation 3.1), in Hook's law, an object is elastic if it behaves like a spring. If it is stretched or compressed, it will try to return to its original shape. It produces a restoring force, F:

$$F = kx \tag{3.24}$$

where
- F = force exerted by the spring
- k = spring constant
- x = displacement from equilibrium position

A similar relation is to model the damping with a linear model:

$$F = bv \tag{3.25}$$

where
- F = force exerted by the damper
- b = damper coefficient
- v = velocity

A simple mass–spring–damper equation (Fa, M, B, and K are the actuating force, mass, damper, and spring constant, respectively)

$$F_a(t) = M\ddot{X} + B\dot{X} + KX \tag{3.26}$$

can be used to model the suspension of a car, an airplane, or a bicycle. With this model, one can find the position, velocity, and acceleration of the vehicle to design the associated devices.

3.1.2.5.2 Modeling of electrical systems Electrical systems are electrical circuits that are often treated as lumped elements. Example applications include a sensor, a motor, and a controller. Ohm's law formulated the relationships among voltage, current, and resistance as "the current in a circuit is directly proportional to the applied voltage and inversely proportional to the resistance of the circuit":

$$V = IR \tag{3.27}$$

where
- V = potential difference between two points with a resistance R
- I = current flowing through the resistance

A similar relation is to model the inductor with a linear model:

$$V = L(dI/dt) \tag{3.28}$$

where

V = potential difference between two points with inductor L
I = current flowing through the inductor

Also, the relation to model the capacitor is

$$I = C(dV/dt) \tag{3.29}$$

where

V = potential difference between two points which include a capacitor C
I = current flowing through the capacitor.

3.1.2.5.3 Modeling of thermal systems Thermal systems are systems in which the storage and flow of heat are involved. Example applications include a refrigerator system, engine cooling system, and an oven. The modeling of a thermal system is very similar to modeling the electrical systems. For example, the thermal resistance R of a wire is

$$R = d/A\alpha \tag{3.30}$$

where

A = cross-sectional area
d = length of the wire
α = thermal conductivity

The flow of heat Q by conduction from a body at temperature T_1 to a body with a temperature T_2 obeys the relationship

$$Q(t) = (T_1(t) - T_2(t))/R \tag{3.31}$$

when R is the thermal resistance.

The relation to model the thermal capacitor C is

$$dT/dt = (Q_{in}(t) - Q_{out}(t))/C \tag{3.32}$$

where

T = temperature of a body with a thermal capacitor C
$(Q_{in}(t) - Q_{out}(t))$ = net heat flow rate into the body

3.1.2.5.4 Modeling of hydraulic systems Hydraulic systems involve the flow and accumulation of liquid. They can commonly be seen in hydraulic control units or chemical systems. Most systems involve fluids.

Modeling a hydraulic system is also very similar to the electrical or thermal systems. For example, the hydraulic resistance R of a pipe, a valve, or an orifice is

$$R = \frac{2\sqrt{\Delta p}}{k} \tag{3.33}$$

where
- Δp = pressure difference
- k = constant depending on the characteristics of the pipe, valve, or orifice

The flow rate w of the fluid can be expressed as

$$w = Rk^2/2 \tag{3.34}$$

The relation to model the hydraulic capacitor C is

$$C = dv/dp = A(h)/\rho g \tag{3.35}$$

where
- v = volume of the vessel
- p = pressure
- A(h) = cross-sectional area of the vessel at height h above the bottom
- g = gravitational constant
- ρ = density of the liquid

3.1.2.5.5 Newton's viscosity law Industrial processes, such as refinery and chemical reactors, require the transport of fluids such as liquids, gases, and steam. They involve the motion of fluid in a conduit, and Newton's viscosity law states that the shear stress, t, in a lubricating film is given as

$$\tau = \mu \frac{du}{dy} \tag{3.36}$$

where
- μ = lubricant viscosity coefficient
- du/dy = velocity gradient in the film normal to the surface

If one looks closer at the liquid flow as shown in Figure 3.17, the laminar layer obeys Newton's viscosity law.

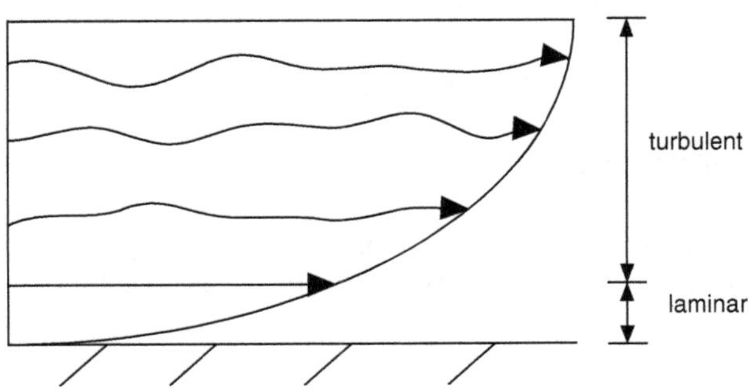

Figure 3.17 The laminar flow and the turbulent flow.

3.1.2.5.6 Fick's law Random motions of molecules or random motions in fluids (turbulence) have a tendency to transport materials down gradients of matter or energy. Fick's law is related to mass flux, Nx, with concentration:

$$N_x = AD\frac{\Delta C}{\Delta x} \tag{3.37}$$

where

N_x = mass flux (mass × time^{-1})
A = area of plane of diffusion (m^2)
D = diffusion coefficient (molecular or turbulent [m^2/min])
ΔC = concentration difference (mg/m^3)
Δx = distance over which concentration gradient exists (length)
ΔC/Δx = concentration gradient

3.1.2.5.7 Fourier's law Heat transfer without mass transfer is by heat conduction. Fourier's law is an empirical law based on observation. As shown in Figure 3.18, it states that the rate of heat flow, dQ/dt, through a homogeneous solid is proportional to the area, A, of the section at right angles to the heat flow direction, and to the temperature difference along the path of heat flow, dT/dx:

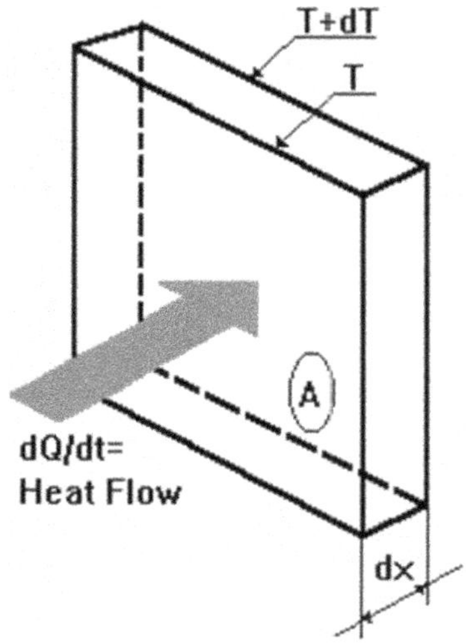

Figure 3.18 Fourier's law to model conduction heat transfer.

$$dQ/dt = -\alpha A(dT/dx) \tag{3.38}$$

where α is the thermal conductivity of the material.

3.1.2.5.8 Newton's law of cooling Newton's law of cooling says that the rate of change of the temperature of an object is proportional to the difference between its own temperature and the ambient temperature:

$$\frac{dT}{dt} = -k(T - T_a) \tag{3.39}$$

where
T_a = ambient temperature
k = positive constant

This is a simple but useful equation. The change of temperature in a laser deposition process is also called cooling rate, which is a major factor in controlling the microstructure of the deposited part. If the ambient temperature, T_a, is considered to be the same, one can control the molten pool temperature to control the microstructure.

As there are many possible models that can be used in product or prototype design, this chapter does not intend to and it is almost impossible to list all the models. A model does not necessarily have to be very complex to be useful in design. The simple models were particularly useful in judging a design decision. For example, a simple "stress = force/area" can be used to do a few short calculations to determine if certain components will be able to carry out their desired duties. By knowing the load that a simple part must carry the designers will be able to determine just how large a part they can build. This can also be used to help identify the type of materials that can be utilized in a product.

Review problems

1. As shown in Figure 3.19, a simply supported 20 ft beam with a load of 10,000 lbs is acting downward at the center of the beam. The beam used is a rectangular 2 in. by 4 in. steel beam. Determine the maximum bending (axial) stress that develops in the beam due to the loading. Disregard the intrinsic weight of the beam.

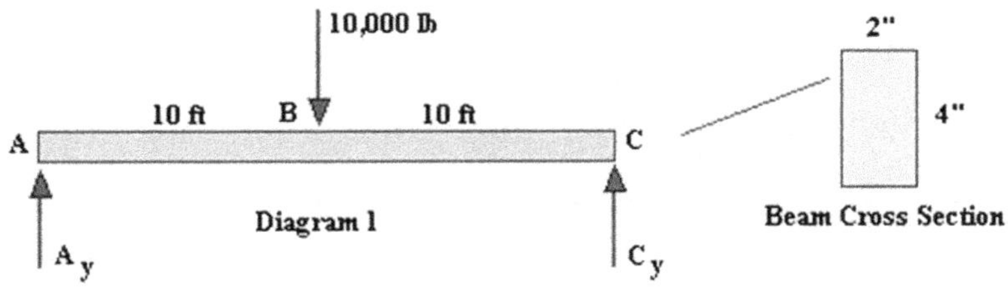

Figure 3.19 To determine the maximum bending (axial) stress which develops in the beam due to the loading.

2. Construct a model to help design a motor speed control system, which requires the use of a DC motor to control the output angular speed by adjusting the input voltage. The simple model for this system is shown in Figure 3.20 below.

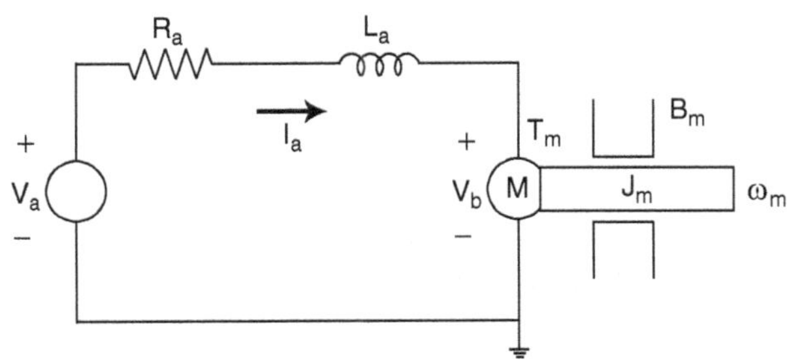

Figure 3.20 To model the motor speed control system.

In Figure 3.20, V_a represents the control voltage, R_a represents the electrical resistance, L_a represents inductance, I_a represents the electrical current, V_b represents the voltage applied to the DC motor, J_m represents the moment of inertia of the output axle, B_m represents the damper, v_m represents the output angle speed, and T_m represents the torque output of the motor.

3. Develop a model for a common system such as the automotive braking system as shown in Figure 3.21. Brakes use the following principle for operation: When the brake pedal is depressed, the force from your foot is transmitted to the brakes through a fluid. Since the actual brakes require a much greater force than can be applied with a human leg, the braking system must multiply the force of your foot. It does this in two ways: leverage and hydraulics. The brakes can transmit the force to the tires using friction, and the tires transmit that force to the road using friction also. Explain the function of three subsystems (leverage, hydraulics, and friction) of the model.

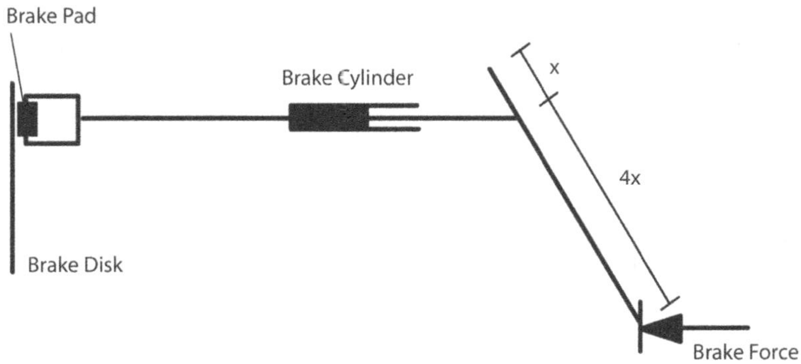

Figure 3.21 To develop a model for a car's braking system.

4. Establish a mathematical model for designing an electronic weighing machine. For the typical machine, the person or object whose weight has to be calculated is mounted on a platform. The platform rests on a spring. The platform has a light source and facing the light source on the side is a strip with a large number of very closely spaced holes. The other side of the hole is a sensor that detects light, and corresponding to each hole, each value is displayed.
5. A safety inspector wishes to check the speed of a train along a straight piece of track.

   A sketch of the model is shown in the Figure 3.22. The inspector stands 10 m to the side of the track and uses a radar gun. If the reading on the gun is to be within 5% of the true speed of the train, how far away from the approaching train should the reading be taken?

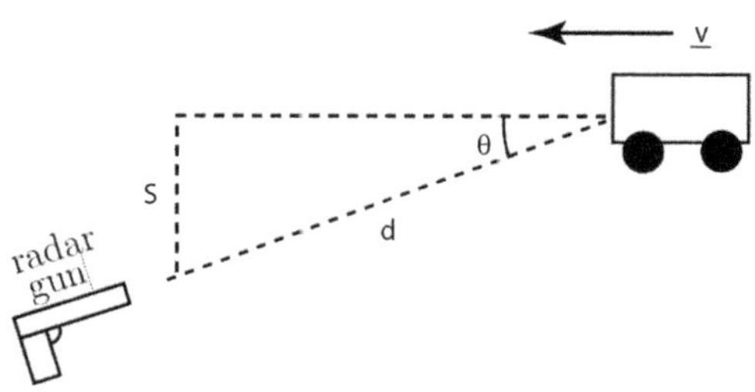

Figure 3.22 To check the speed of a train along a straight piece of track.

6. Develop a model for the tank–pipe system shown in the Figure 3.23. The input is the pressure, p_1, and the output is the pressure, p, in the tank. Assume that laminar flow is in the pipe.

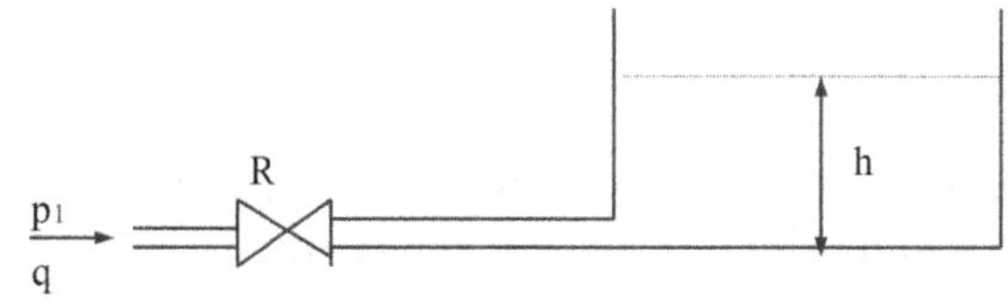

Figure 3.23 To model the tank–pipe system.

7. Suppose we have a 3 m long hall connecting an underground parking garage polluted with CO at 24 mg/m³. The hall is 1 m wide and 2 m tall. At the end of the hall is an office where ventilation maintains the CO level at 10 mg/m³. If the diffusion coefficient is 0.5 m²/min what is the CO maximum flux toward the office?
8. As part of his summer job at a restaurant, Jim learned to cook up a big pot of soup at night, just before closing time, so that there would be plenty of soup to serve customers the next day. He also found out that, while refrigeration was essential to preserve the soup overnight, the soup was too hot to be put into the fridge when it was ready. (The soup had just boiled at 100°C, and the fridge was not powerful enough to accommodate a big pot of soup if it was warmer than 20°C.) Jim discovered that by cooling the pot in a sink full of cold water, (kept running, so that its temperature was roughly at 5°C) and stirring occasionally, he could bring the temperature of the soup to 60°C in 10 min. How long before closing time should the soup be ready so that Jim could put it in the fridge and leave on time?
9. A 12 in. cube of steel weighing 320 lbs is being moved on a horizontal conveyor belt at a speed of 6 miles/h. (a) What is the kinetic energy of the cube? (b) If the conveyor in the above example is brought to a sudden stop, how long would it take for the steel block to come to a stop and how far along the belt would it slide before stopping if the coefficient of friction μ between the block and the conveyor belt is 0.2 and the block slides without tipping over?
10. Figure 3.24 shows a piece of material subjected to the opposite and equal 1,000 lbf forces in tension. Measurements indicate that the sample elongates a total of 0.0025 in. under force. Find the stress and strain in the sample.

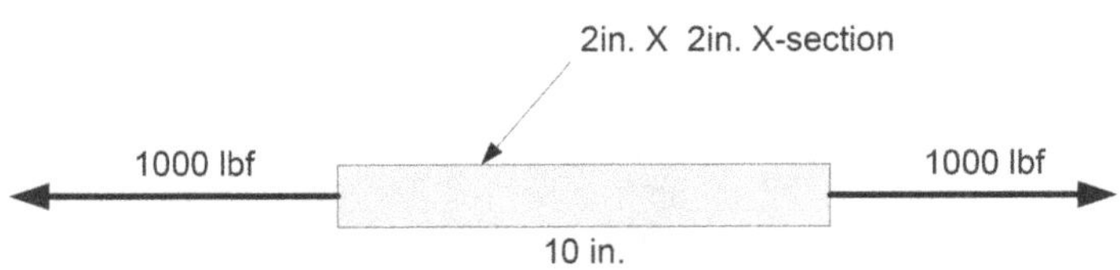

Figure 3.24 To find the stress and strain of the rod.

11. You have been asked to design a cylindrical pressure vessel. The ends will be bolted on with 3/8 in. bolts, if the diameter of the vessel is 48 in. The pressure of the fluid inside is 50 psi. How many bolts will be required to safely hold the ends in place with a factor of safety of 2? Create a mathematical model. Assume infinite stiffness of bolted materials. In other words, ignore the effects of preloading.

3.2 *Modeling of physical systems*

Let our advance worrying become advance thinking and planning.

Winston Churchill

This section helps answer the following questions:

- How are mathematical models used to represent products and ideas?
- Can nonmathematical models be used in modeling?
- If there are other mathematical models that are used, what are their applications?
- When should mathematical models be applied in the prototyping process?
- Are mathematical models limited to the actual design of a product or are there other applications for them?
- Are there practical examples on how to use a model to help product design and prototyping?

An engineer can build a descriptive model of a system as a hypothesis of how the system could work or how an unforeseeable event could affect the system. A mathematical model is the description of a system by means of variables. The values of the variables can be practically anything: real or integer numbers, Boolean values, or strings. The variables represent some properties of the system. The model is a set of functions that describe the relationships between the different variables.

3.2.1 *Types of modeling*

As shown in Figure 3.25, a model can be viewed as a black box, and the input and output are the variables or parameters that one can manipulate. There are several different types of models, such as mathematical, stochastic, neural, and fuzzy. The properties of the model can be one or a combination of the following:

- Continuous or discrete time
- (Non) linear
- (Non) stationary
- (Non) causal
- (Non) deterministic
- (Un) stable
- Bias or systematic error
- Variance or random error

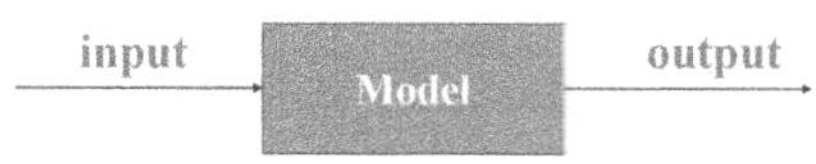

Figure 3.25 A black box model.

Most of the real-world problems are nonlinear, but often a linear function is used to approximate a nonlinear system, such as Taylor series expansion. For example, Figure 3.26 shows a piece-wise linearization of a function. The function or system can be approximated as linear within a certain range of the function, and the error can be acceptable. Linearization can often lead to a good approximation in the environment, thus, the

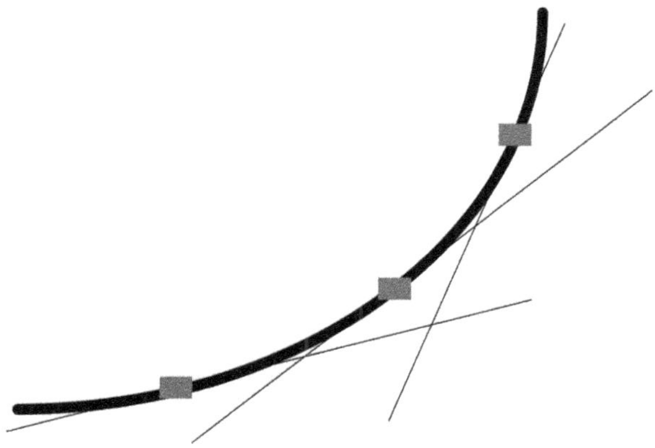

Figure 3.26 Piece-wise linearization of a function.

function is easy to calculate. Since there is a well-developed linear theory, there is known error propagation.

How does one know if a mathematical model describes the system well? In general, if the model was built well, the model will adequately show the relationship between system variables for the measurements at hand. How does one know that the measurement data is a representative set of possible values? This can be further broken down into two questions: Does the model describe well the properties of the system between the measurement data (interpolation)? Does the model describe well the events outside the measurement data (extrapolation)? These two questions may be addressed by splitting the measured data into two parts: training data and verification data. Training data is utilized to train the model to approximate the model parameters, whereas verification data is used to evaluate model performance. If the model describes the verification data well, then the model describes the real system well.

How well does this model describe events outside the measured data? The answer is open depending on various situations. Consider the Newtonian classical mechanics model. Newton made his measurements without advanced equipment, so he could not measure properties of particles traveling at speeds close to the speed of light. Therefore, his classical mechanics model would not be able to describe events outside the measured data. However, the classical mechanics model is still widely used in the world. This does not mean that one trusts all the models. When using a mathematical model, careful attention must be given to the uncertainties in the model, in particular for the products related to safety.

3.2.2 *Examples of physical modeling*

In this section, case studies will be used to illustrate the idea of using a model to solve engineering problems.

Example 3.7: A simple model to estimate air flow rate

A model does not need to be complex. This example is to find the air flow volume of a chamber with a ventilation dock as shown in Figure 3.27. It is necessary to quantify the flow in order to determine the fan required to pump the air out of the chamber.

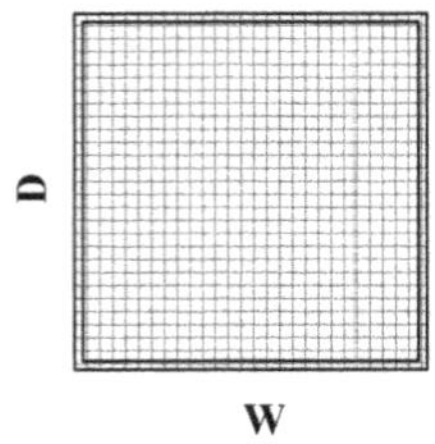

Figure 3.27 The ventilation dock of a chamber.

In order to find out the air flow, a simple model as shown in Figure 3.28 is used. The dimensions D and W represent the dimensions of the ventilation dock. Using this model, and assuming that the air is a rectangular pattern, one can estimate the air flow of the chamber. What is needed is an air flow meter as shown in Figure 3.29 to find the flow velocity, H, per unit time or dH/dt. Therefore, the airflow is equal to DW(dH/dt). This is an example of using the elementary concept to solve an engineering problem.

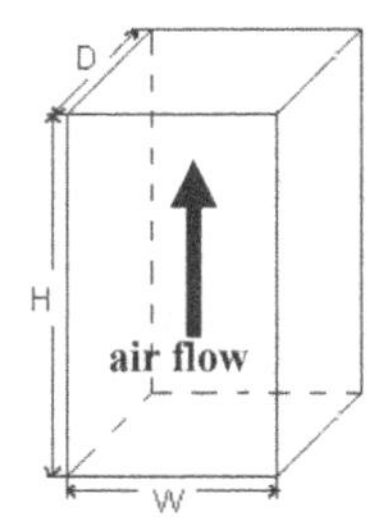

Figure 3.28 The air flow model: volume (V) = H × W × D per unit time.

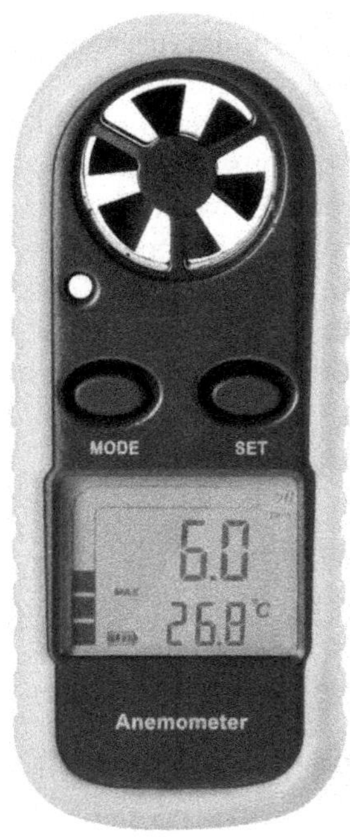

Figure 3.29 An air flow meter.

Example 3.8: Design and prototype a powder feeder

This modeling example is to design a screw-type powder feeder as shown in Figure 3.30 to deliver the metal powder uniformly at a certain flow rate. Before a physical prototype is made, it would be helpful if a model of the powder flow rate can be obtained.

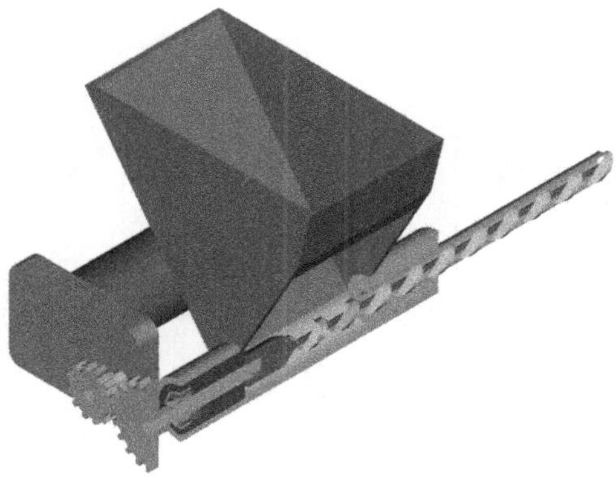

Figure 3.30 To design a screw-type powder feeder. A gravity feed powder feeder uses a screw to advance powder.

The model for powder mass flow rate (g/s) can be estimated to be

$$\dot{M} = 2\rho A_d r V_m \tag{3.40}$$

where
- r = feed screw helix lead (mm)
- ρ = powder density (g/mm^3)
- A_d = the cross-sectional area (mm^2)
- V_m = motor speed (rev/s)

If a low volume powder feeder is of interest, from the model in Equation 3.40, one can minimize r, A_d, or V_m to reduce the flow rate. From the model, the major factors which may contribute to the design objective can be quickly obtained. If the screw lead and cross-sectional area are held constant, the only parameter that can be used to vary the flow rate is the motor speed, and thus, it is a critical parameter used for the powder flow control.

Example 3.9: Use a laser beam model to redesign a nozzle

In one instance, a laser nozzle as shown in Figure 3.31 was investigated. Right after a new laser fiber is installed into an existing laser deposition system, the deposition nozzle was overheated during laser deposition. It was suspected that the laser beam was colliding with the nozzle assembly. It was necessary to find out this suspicion was true, and if so, where the collision was located and what caused the collision. This information was useful in correcting the issue or would even be useful in redesigning a new nozzle.

The beam model for a spot diameter was used:

$$d_{spot} \approx f\theta = \frac{2.4\lambda f}{D} \tag{3.41}$$

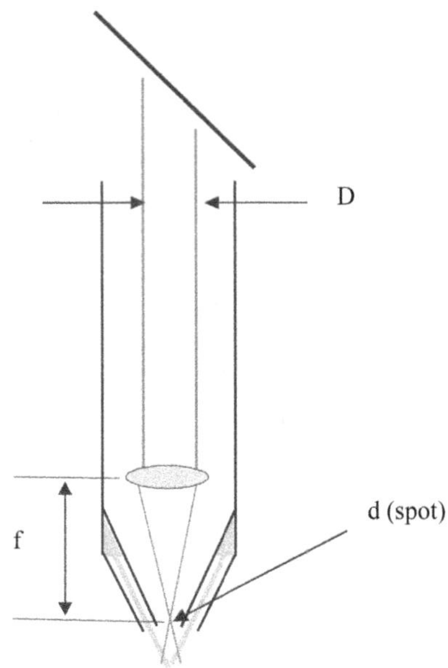

Figure 3.31 A schematic diagram of a nozzle assembly for a laser deposition system.

For the laser beam shown in Figure 3.31, f is the focal length, D is the laser beam input diameter, and λ is the laser wavelength. Since the laser source and the lens are the same before and after the laser fiber is changed, the only parameter that may modify the laser profile is the input laser beam diameter. After some experiments, it was found that somehow the new laser fiber indeed has a different input laser beam diameter. This is a good example of using a model for problem-solving and fault detection.

From this model, one can also conclude that, in order to design a system for micro application, a fine spot diameter will be needed. Therefore, one should choose a laser with a smaller wavelength, smaller focal length, and larger laser input diameter.

Example 3.10: Use a table to model laser–material interaction

This example illustrates that sometimes, a table, instead of an equation, can be a good model. As shown in Table 3.1, the heating times are summarized. One can see that at 5 W/mm^2, it took 23 s to heat the nickel close to the melting temperature at about 1,456°C. At 50 W/mm^2, it will only take 0.236 s. One can see that at 2,000 W/mm^2, the sample is overheated and could damage the sample. This table can be viewed as a model in which the input and output parameters are obvious, and one can interpolate the data from within. For additive manufacturing purposes, 0.236 s to melt nickel is still too slow, and 2.8E−3 s is about right except that the sample is heated much above the melting temperature. It can be concluded that at around (or less than) 500 W/mm^2, it is suitable for laser deposition of nickel.

Table 3.1 Summary of experimental results for a laser deposition process

Heating times as a function of power for a 0.025 mm melt depth in nickel				
Power density (W/mm^2)	5	50	500	2,000
Surface temp (°C)	1,456	1,469	1,590	1,960
Melting time (s)	23.1	0.236	2.8E−3	2.6E−4

Example 3.11: Use a graph to model laser–material interaction

This example illustrates that a graph can be a good model. As shown in Figure 3.32, the graph is obtained based on experimental data. Dilution occurs when the substrate is overheated and subjected to excessive laser power (relative to the amount of powder). Porosity may occur when there is not enough power or when the excessive powder is dumped into the melt pool. From the previous example, it is known that laser power should be above a certain threshold so that the material can be melted quickly enough for additive manufacturing purposes, and thus, the dashed line in Figure 3.32 represents the threshold. The input–output can be related well and the desired conditions to obtain good laser cladding conditions can be obtained from this model.

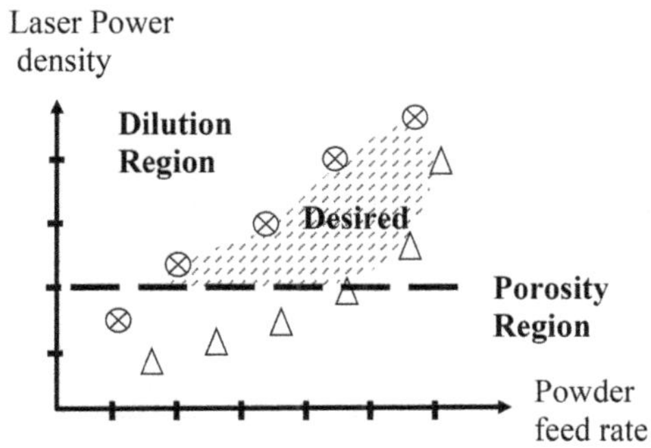

Figure 3.32 Desired parameter graph for laser–material interaction.

Example 3.12: Control of deposited bead in laser cladding

The production of parts with desired physical characteristics by selecting appropriate parameter values before processing requires an understanding of process mechanics. A model is needed to describe the laser deposited bead. As shown in Figure 3.33, a model is used to describe the geometry of a bead. To relate the geometry with the process parameters, a model can be established at constant laser power:

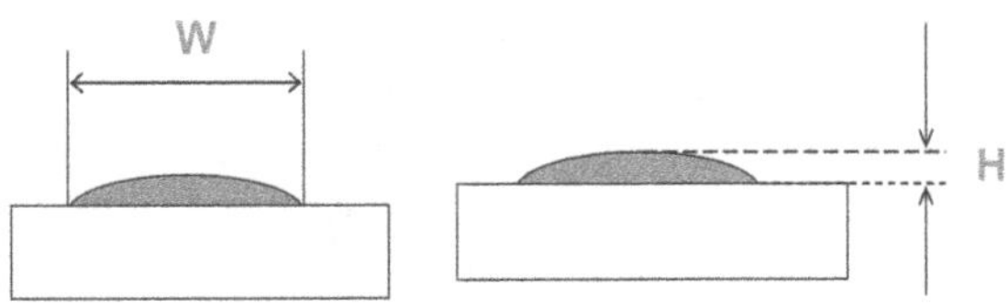

Figure 3.33 Model of a laser deposited bead.

$$W = D'(1 - bV) \tag{3.42}$$

where

W = bead width
V = traverse speed (mm/s)
b and D′ = parameters depending on other factors

D′ is approximately equal to the beam spot size D. For a particular setup and materials' applications, D′ and b can be found experimentally.

The following statistical relationship can be used to relate bead height as a function of powder flow rate and workpiece traverse speed (V) as

$$H = a(M/V) \tag{3.43}$$

where
- H = bead height (mm)
- M = powder flow rate (g/s)
- a = parameter depending on laser power, mode structure, surface condition, and other factors

These two models are established based on experimental observations. They relate well and explain how the key parameters play in the system. For example, from Equation 3.43, it is known that in order to increase the height, one needs to either increase powder flow rate or decrease laser travel speed. These may be intuitive, but they are instrumental in system design and control.

Figure 3.34 shows laser dilution in which diluted substances in the substrate occur when the laser melts the powder as well as the substrate. Dilution, Di, can be modeled as

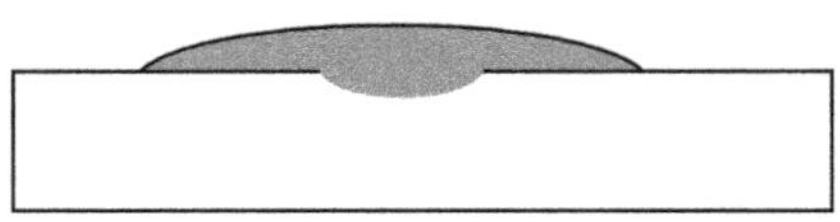

Figure 3.34 Dilution in laser cladding.

$$Di = cP - d_m M \tag{3.44}$$

where
- P = laser power
- c and d_m = decided through experiments

This model echoes the model in Figure 3.32 where the powder flow rate and the laser power are the key parameters for dilution.

The three equations (3.42–3.44) can be used to establish a control model for laser cladding or deposition to control the bead geometry as shown in Figure 3.35. The desired control parameters are the bead width, bead height, and degree of dilution, and these parameters will need to be monitored directly or indirectly by a set of sensors. In order to close the loop, it is necessary to find the difference between the desired and measured parameters:

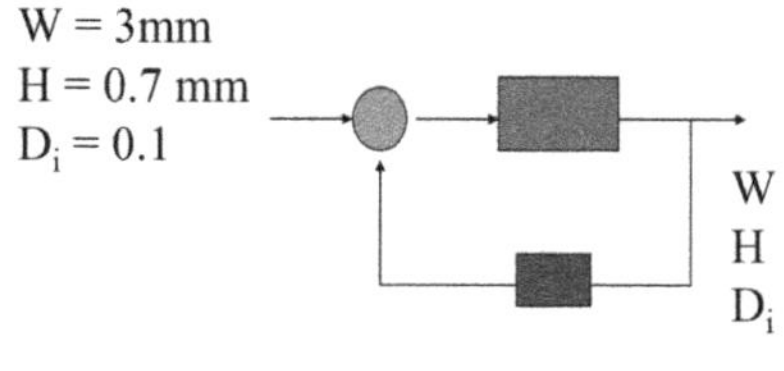

Figure 3.35 Feedback control in laser cladding.

$$dW = W_r - W[k] \tag{3.45}$$

$$dH = H_r - H[k] \tag{3.46}$$

$$dD = Di_r - Di[k] \tag{3.47}$$

where

W_r = desired bead width
H_r = desired bead height
Di_r = desired dilution
W[k], H[k], and Di[k] = measured data at the kth iteration

All parameters are known in the control process. By taking the derivatives of Equations 3.42–3.44, the following can be obtained:

$$dV = -K_v \delta W \tag{3.48}$$

$$dM = K_m V \delta H \tag{3.49}$$

$$dP = K_p \delta Di \tag{3.50}$$

where K_v, K_m, and K_p are constants and can be determined from experiments.

From Equations 3.48–3.50, the following equations can further be defined:

$$V[k+1] = V[k] - K_v \delta W \tag{3.51}$$

$$M[k+1] = M[k] + K_m V[k] \delta H \tag{3.52}$$

$$P[k+1] = P[k] + K_p \delta Di[k] \tag{3.53}$$

where V[k], M[k], and P[k] are the table velocity, mass flow rate, and laser power in the kth iteration, respectively.

Note that these parameters can be controllable parameters. V[k + 1], M[k + 1], and P[k + 1] are the table velocity, mass flow rate, and laser power, respectively, in the (k + 1) th iteration and represent the command value to the next step. Equations 3.51–3.53 relate nicely to the control values of the next time step with the known parameters (on the right-hand side).

Example 3.13: Energy model of laser–material interaction

The understanding of interactions between the laser and the metallic material is critical in metal-forming processes. Laser–metal interaction depends on a lot of parameters. Some are related to the laser beam itself, such as wavelength (from ultraviolet to infrared), irradiation duration (a few nanoseconds for pulse laser or few seconds for continuous laser), and fluence or energy intensity (J/cm^2). Some interactions depend on material characteristics and properties such as reflectivity, thermal diffusivity, and melting temperature.

Figure 3.36 shows the interaction of a laser beam and material. It can be seen that not all the energy, E, from the laser is absorbed and used to heat the material. Only the absorbed energy, AE, is utilized to heat the material.

Some fractions of energy are reflected and transmitted away from the material resulting in lost energy. The fraction of energy lost is substantial for most metals, which gives a low percent energy utilization, approximately 10%–20%. The fractions of energy during interaction with a metal can be expressed as [Laeng00]

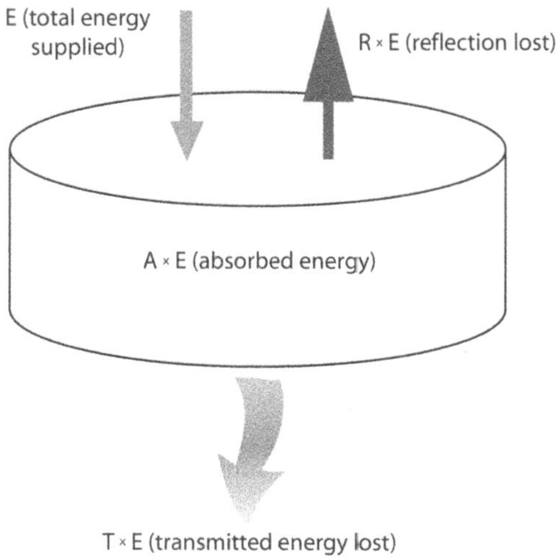

Figure 3.36 Fractions of laser beam energy during interaction with solid material.

$$R + A + T = 1 \tag{3.54}$$

where
R = reflectivity
A = absorptivity
T = transmissivity

The total energy, E, during interaction with a material, can be expressed as

$$E = AE + RE + TE \tag{3.55}$$

At larger wavelengths, such as that of the CO_2 laser (10.6 μm), metallic materials generally have a large reflectivity and, consequently, laser treatment of metals is relatively difficult.

AE enters the material and is progressively absorbed. At a given depth, z, the portion of available energy after being gradually consumed or used to heat the material as laser penetrates through a material is given by Beer–Lambert's law as

$$EZ = AE_0\left(\exp^{(-\mu z)}\right) \tag{3.56}$$

where
EZ = available energy at a given depth z
AE_0 = initial AE on the surface (z = 0)
μ = absorption coefficient
z = depth

At the surface of the material (z = 0), all the AE is available for melting the material, and the AE at depth, z, equals to the initial AE on the surface. Thus, the equation above becomes

$$E_Z = E_0 \tag{3.57}$$

This model can be used to help estimate and design the system parameters needed to melt the metal powder properly.

In order to model the melt pool formation during the laser deposition process, a thermal model is necessary:

$$P = Q_b + Q_v + Q_p \tag{3.58}$$

where

P = heat input (laser power) (W)
Q_b = heat conducted away by the boundary of control volume (W)
Q_v = enthalpy change (W)
Q_p = heat absorbed by the powder particles (W)

Equation 3.58 shows where the laser power energy is distributed. From this equation, one can find that the resulting heat in the deposition melt pool will depend on powder flow rate, substrate temperature, substrate material, and laser power. Since the melt pool temperature is directly related to the resulting heat in the melt pool, and the microstructure is directly linked to the melt pool temperature and cooling rate, one can further develop this model to predict the microstructure formation.

Example 3.14: Contact model

Suppose one is interested in finding out the stiffness of a contact point between a fixture and a workpiece (Figure 3.37) as a critical parameter for proper fixturing, monitoring, and detection. From the Hertz theory, the relation between deformation δ resulting from the applied compression P assumes the form of a power function, namely:

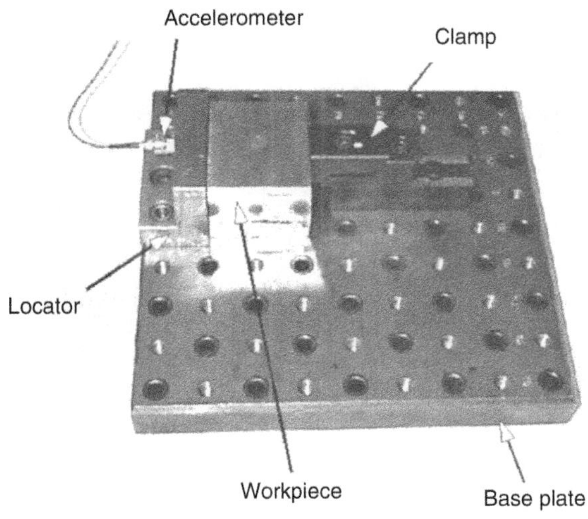

Figure 3.37 An example to detect whether clamping fixture workpiece is secured [Yeh00].

$$\delta = c \cdot P^n \tag{3.59}$$

in which the coefficient c depends on workpiece surface finish and hardness, as well as the area of contact between the workpiece and fixture.

Consideration of surface hardness and the contact area is more practical when dealing with the contact condition of the fixture elements. The relationship between the force exerted on the fixture and the resulting deformation has been experimentally studied [Shawki66]. The equation to express two plane contact surface deformation, d, and applied pressure, p, is

$$\delta = (a_1 - a_2 H_B + b_1 A_c + b_2 R_h) p^n \quad (3.60)$$

where

δ = deformation per contact (mm)
H_B = Brinell hardness value of the workpiece material (kgf/mm^2)
A_c = contact area (cm^2)
R_h = surface finish of the workpiece (μm)
p = pressure (kgf/cm^2)
a_1, a_2, b_1, b_2 = coefficients which depend on the material properties, such as Young's modulus
n = index which depends on the object material

Thus, Equation 3.60 can also be represented as

$$\delta = C_1 p^n \quad (3.61)$$

Thus, the contact stiffness between the workpiece and the fixture can be obtained from

$$K = \frac{\Delta P}{\Delta \delta} = \frac{P_1 - P_2}{C_1 (p_1{}^n - p_2{}^n)} \quad (3.62)$$

where

P_1, P_2 = total loadings subjected to the different pressures p_1 and p_2
δ_1, δ_2 = contact deformations subjected to the different pressures p_1 and p_2

Equation 3.62, thus, can be used to model the stiffness for the deformation taking place at the contact surfaces between the workpiece and the fixture locating elements under the action of steady loads.

Review problems

1. Find the maximum stress model of the beam as shown in Figure 3.38.

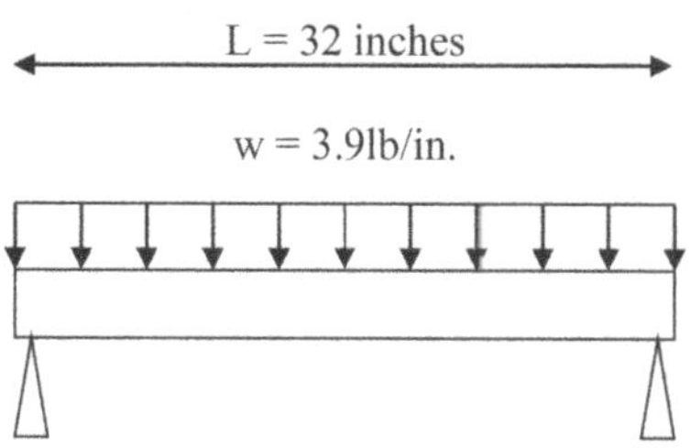

Figure 3.38 To find the maximum stress model of the beam.

L = length of frame
w = distributed load (1,000 lbs over eight tubes = 125 lbs/tube)
I = 0.015625 in.^{4} = moment of inertia of the beam
c = 0.25 in. = distance from the bending plane to the outside of the tube

2. Conduction is heat transfer by means of molecular agitation within a material without any motion of the material as a whole. If one end of a metal rod is at a higher temperature, then energy will be transferred down the rod toward the colder end because the higher speed particles will collide with the slower ones with a net transfer of energy to the slower ones. For heat transfer between two plane surfaces, such as heat loss through the wall of a house, the rate of conduction heat transfer is shown in Figure 3.39. Give a simple model for this phenomenon.

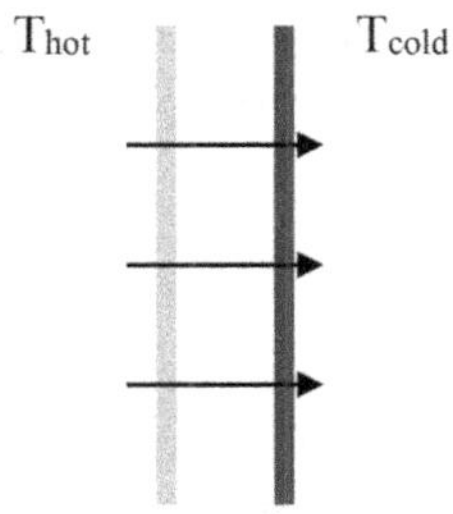

Figure 3.39 To model heat transfer between two plane surfaces.

Q = heat transferred in time t
k = thermal conductivity of the barrier
A = area
t = temperature
d = thickness of the barrier

3. Model the functional relation of a computer mouse in block diagram form or function tree form.
4. Model the momentum conservation of a gun firing a bullet. The outgoing bullet velocity is v, and the resulting gun velocity is V. The mass of the gun is M and the bullet mass is m.
5. Model the effects of cutting speed on a mill, using a graph model.
6. It was found that the fuel causes a certain vibration pattern in an aircraft, and thus, an investigation was conducted to understand this issue. We can model the aircraft wing as a transverse beam and fuel as tip mass. Since the mass of the wing is constant, it is ignored; model and see how the frequency of the system changes as the fuel is changed from full to used up. Based on the model, we can find the stiffness of the wing as well as the frequency of the lumped mass as

$$K = \frac{3EI}{l^3} \text{ and } \omega_n = \sqrt{\frac{3EI}{ml^3}}$$

Assuming the mass of the rod is 10 kg when empty and 1,000 kg when full, $I = 5.2 \times 10^{-5} m^4$, $E = 6.9 \times 10^9 N/m$, and $l = 2 m$. Find the natural frequencies when fuel is empty and when full.
7. A lot of real-world problems are nonlinear, but often linear models are used instead. Many models are approximations within an error that is acceptable. Find examples of a common model that is an approximation.

3.3 Product modeling

In all planning, you make a list and you set priorities.

Alan Lakein

This section helps answer the following questions:

- Is there a systematic way to model a product or prototype?
- How to organize all aspects of the model design phase such that all the intended functionality is captured, researched, and developed?
- How to choose which model type to use and are there times when both formal and informal models should be used?

3.3.1 Product model

A product model is a model of a product metric that is a representation, simplification, or estimation of a product's realization to aid in making product decisions. Modeling of product metrics separates engineering from other professions. To avoid trial and error methods, product model can be based on applied mathematics and science (or on physical prototypes). In order to establish a model, a wealth of data needs to be collected to develop analytical models, including customer needs, PDS (product design specifications), the house of quality, and chosen product concepts. The model will permit decisions to be made with an understanding of their impact on performance.

To prepare a product model, the following steps can be followed:

Step 1: Model preparation to relate the customer need/weights to the product functions
Step 2: Model prioritization to identify the functions that relate most strongly to the customer needs
Step 3: Model quantification to choose the metrics (engineering characteristics) that may be used to quantify the material, energy, or signal flows for these functions, and to identify target values for these metrics based on benchmarking results

It is also important to know not all aspects of a product may be modeled mathematically, thus, Step 1 is to relate the customer needs and importance to the product functions. Most of the customer requirements and attributes are modeled informally. Only a few of the requirements and attributes are critical and may require excessive computational resources or take time to develop accurately. Step 2 simplifies the modeling process by focusing on the primary function carriers, and Step 3 chooses a means of measuring the performance of the product and places bounds on the metrics.

In the product and prototype development phase, one often does not analyze a real system. The abstract of the system is analyzed through informal and formal models. An informal model is when a typical design concept is developed and first conceived in informal terms, PDS, and rough sketches. For example, the vehicle must have a "comfortable" ride; the battery must recharge "quickly," and the door can be "conveniently" opened. An informal model is a designer's interpretation of a description of the customer's needs, engineering characteristics, manufacturing requirements, etc. A formal model constructs a computable or analytical model of the design problem. It is not possible to compute the universe of all possible designs that satisfy the same requirements. Most of them can only be considered informally.

Questions such as the following should be examined:

- What is the problem really about?
- What implicit expectations and desires are involved?
- Are the customer's needs, requirements, and constraints truly appropriate?
- What characteristics or properties must the product have?
- What characteristics or properties must the product not have?
- What are the technical conflicts inherent in the design task?
- What aspects of the design task should be quantified now?
- Are there any other features of the product?

Example 3.15: Product modeling for retractable earphones

Suppose a designer is going to define a set of retractable earphones that one would use for a portable music device such as a pocket radio or MP3 player. The main attribute to look for in this particular device is that it comfortably takes up the slack in the headphone cord when in use and then winds the cord up, so there is no slack when not wearing them. Therefore, it is necessary to interpret the customer's comments to try and quantify what the product should do in some measurable ways.

- What is the problem really about?

  The real reason this product is needed is that earphones are often removed when users are using the product, and they should not get caught on something. When the earphones are removed and stuffed into a pocket, the cord should not get tangled.
- What implicit expectations and desires are involved?

  It is expected that the usability, comfort, and portability of the device will not be affected by using this product. Assume the circular part of the device must be less than 1.5 in. and it must be less than 0.75 in. thick.

  Another implicit requirement is that the earphones themselves be of decent quality because the best retracting device in the world will not mean a thing if the music coming out of them sounds awful.

  One of the most implicit requirements is that the length of cord that the device lets out must be of sufficient length to allow the user to put the audio device at least as far down as the bottom of a pocket in their pants.
- Are the customer's needs, requirements, and constraints truly appropriate?

  Yes. Generally, the reason MP3 players and pocket radios are chosen in the first place is that they are small and lightweight, which makes them comfortable and easy to accommodate on the go.
- What characteristics/properties must the product have?

  The product must be comfortable and easily operated.
- What characteristics/properties must the product not have?

  The product must not be bulky and awkward for the user. The retractable device must not make it feel like someone is trying to pull the earphones out of the user's ears while he or she is using them.
- What are the technical conflicts inherent in the design task?

  The technical conflicts are that one will probably use a coil spring to provide the rotational reaction to wind the cord. So one will probably want the coil to only accommodate as few rotations as possible in winding and unwinding, but to do this, the diameter of the spool that the cord is wound around needs to be large. Making the spool large means that the package size must increase as does the force the spring needs to exert increases because the moment arm it is now working against from the center of rotation to where the cord meets the spindle is longer.

- What aspects of the design task should be quantified now?
  The total length of the cord, the envelope for the retracting device, and the acceptable tension on the cord the retracting device can exert while the earphones are in use.

Example 3.16: Product modeling for a binder clip

The format of an informal model is not necessarily the same as the previous example. This example illustrates another way to establish an informal model.

1. Interview (see Table 3.2)

Table 3.2 Interview results and interpretation

Questions	Answer	Interpretation
1. What is the need?	Binder clip	Clip together papers that other paperclips cannot handle
2. For official use or personal use? "For official use" needs to be formal and not	Official	
3. For use where people can see or general use?	General	Need not be exclusive
4. For holding how many pieces of paper?	50–100	Should be rigid and probably hold 150 pieces of paper
5. Ease of use?	Easy	Should have good leverage
6. Compactness?	No concern	Can give enough length for leverage and strong material for rigidity
7. Cost?	Average	Metal should not be too costly and should not have a complicated design procedure or manufacturing technology

2. Present cost scenario
   There are not many binder clips on the market that can hold 150 pieces of paper. Even the ones available are a bit costly, around $3 each. The clips are not worth that amount because the material used is plastic with an aluminum spring. The plastic tends to break easily in a short period of use. The present market only sells 10,000 clips bimonthly. It is expected to improve with proper marketing and good design.
3. Cost breakups are summarized in Table 3.3.

Table 3.3 Cost breakups of the current and proposed design

	Current cost scenario	Proposed cost scenario
Assembly (per piece)	$0.9	$0.75
Transportation (per piece)	$0.1	$0.1
Design cost (per piece)	$0.05	$0.045
Material cost (per piece)	$0.12	$0.2
Packaging (per piece)	$0.08	$0.09
Advertising (per piece)	$0.01	$0.15
Total cost (per piece)	$1.26	$1.335
Profit%	142.85%	31.08%
Profit (monthly)	$8,700	$10,300

It is expected that the sale of the proposed design will be increased (to 20,000) monthly, and thus, more profit will be obtained.

4. Technical questioning
   - What is the basic use?
   - To hold paper using a clip arrangement
   - Customer needs
   - Good enough to hold 50–100 pieces of paper
   - Company's aim
   - With enough room for flexible use and make big enough to hold 150 pieces of paper
   - What are the customer's needs regarding usage?
   - It should be easy to operate, which means that more leverage should be provided to increase the ease of use. The tension of the clip should not be very high, but there should be a balance between leverage and tension
5. Mission statement

   Binder clip will hold 150 sheets of paper; easy to use and not too stylish but more of formal use intending office use to replace the existing product in the market delivered by the competitor. Avenues for design are that one can use full metal instead of plastic, thus increasing the product life cycle and selling more units.
6. Technical evaluation

   A sheet of paper is 0.1 mm thick. Thus, 150 sheets mean 15 mm. The mouth of the binder clip should open 15 mm wide. Using a material with a stiffness of spring constant equivalent of 130 N/mm, it would need a force of 1.95 N. With the handle protruding at an angle of 308 and a length of 30 mm, the torque for compression of 15 mm is 1.95 [cos(30/1,000)].
7. Material selection

   Since longevity is the most important issue, steel with powder coating was selected.
8. Manufacturing technique used

   After procuring a steel sheet of the required thickness, it is pressed into shape with the help of an automated forge and the handle is bent into shape using another strip of wire. The clip part is powder coated and the handle part is sleeved with a plastic resin strip. Then both are assembled and packed.

3.3.2 *Formal model*

The analytical and numerical modeling of a product should represent a physical understanding of the product. The goal is to find the optimal design variables to satisfy design requirements. The solutions can be analytical or numerical. The following sections illustrate the formal modeling approach and the examples. The basic modeling steps are summarized as follows [Otto01]:

1. Identify a flow: Identify a material, energy, or information flow associated with each effect of the product concept. In other words, the flow is the possible interactions of the product with its usage environment. When the most critical flow of the object is picked, and if there are multiple flows of equal criticality, some trade-offs are needed in deciding the most critical flow for the system.
2. Identify a balanced relationship: Identify a balanced relationship for the flow to find the equilibrium equation to model the system. In other words, similar equations and models as discussed in the previous sections can be used to find the balance relationship of the system.

3. Identify a boundary for the balance relationship: Recognize the boundary conditions of the product concept. How is it loaded, and how does it interact as a system with its environment (including the human interface)? What are the inputs and outputs across the boundaries, and what are the limits or ranges of these inputs and outputs?
4. Formulate an equation for the balanced relationship: Convert the balanced relationship to a mathematical form by assigning geometric variables, material property constants, and so forth to formulate a set of equations. Applied mathematics and science are required. A number of assumptions and simplifications are needed.
5. Use the model to explore design configuration options.

Example 3.17: Product modeling for retractable earphones (Continued from Example 3.15)

As discussed in Example 3.15, the critical flow explored is the force that describes the tension in the cord compared to the unwinding of the apparatus. The cord in the previous design appears to wind twice per revolution of the spindle. The description of the product indicates that the cord has a length of 32 in. The length to the wind (c) is then 16 in. Assume that the device provides the winding action by using a spiral torsional spring, otherwise known as a clock spring, which is made out of a flat strip of steel. This is the type of spring that is used in tape measures to retract the tape automatically. The number of wraps is given by the following equation:

$$x = \frac{c}{\pi D_s} \tag{3.63}$$

where

x = number of times the cord is wound around the spindle
c = wound length of the cord
D_s = diameter of the spindle

Deflection of the spindle in revolutions is given by the following equation (from the *Machinery's Handbook*):

$$U = \frac{Sl}{3Et} \tag{3.64}$$

where

U = deflection of the spindle in revolutions
S = safe tensile strength of the material in psi
E = modulus of elasticity (30,000,000 psi for steel)
t = thickness of the material
l = length of the spindle in inches

The force to unwind the spiral torsional spring at its maximum length of cord let out is given by the following equation (from the *Machinery's Handbook*):

$$W = \frac{Sbt^2}{6R} \tag{3.65}$$

where

W = force tangent on the spindle
b = width of the spring material
t = thickness of the spring material
R = distance the force is acting (assume $R = D_s/2$)

Variables x and U from Equations 3.63 and 3.64, respectively, are really the same, and thus, the right-hand sides of each equation can be set equal to one another:

$$\frac{c}{\pi D_s} = \frac{Sl}{3Et} \tag{3.66}$$

Solve Equation 3.66 for S to obtain the equation

$$S = \frac{3cEt}{\pi l D_s} \tag{3.67}$$

Substitute Equation 3.67 into Equation 3.65 and the fact that $R = D_s/2$ to obtain

$$W = \frac{bct^3E}{D_s^{\ 2}l} \tag{3.68}$$

The preceding equation is a single equation with four variables, so some boundaries are needed for values of these variables to solve this equation. Variable b, the width of the spring material, must be greater than 0.001 in. because there must be some minimum value and it must be less than 1.5 in. because the width of the apparatus must be 0.75 in. at most to constrain the width of the spring element to something less than the total package width. The variable t, the thickness of the spring material, must be greater than 0.001 in. because there must be some minimum value and it must be less than 0.030 in. because this would be a practical limit to the thickness of the spring material. The variable l, the length of the spring material, must be greater than 4.712 in. because there must be some minimum value and it must be less than 100 in. because there must be some upper limit to the length allowed. The variable D_s, the diameter of the spindle, must be greater than 0.25 in. because there must be some minimum value and to manufacture anything smaller than this would be very difficult. D_s must be less than 1.5 in. because this would be the maximum allowable diameter of the spindle for an acceptable package. Using Excel Solver as discussed in Chapter 10 by inputting Equation 3.68 with the limits listed above and finding a solution to the problem, the following results shown in Table 3.4 can be obtained.

Table 3.4 Design results of the retractable earphones

Description	Variable	Value	Argument	Limit	Units
Cord length to be wound	c	16	=	16	in.
Diameter of spindle	D_s	0.7551755	>=	0.25	in.
			<=	1.5	in.
Length of spring material	l	4.712389	>=	4.712389	in.
			<=	100	in.
Width of spring material	b	0.4928122	>=	0.001	in.
			<=	0.5	in.
Thickness of spring material	t	0.0047486486	>=	0.001	in.
			<=	0.03	in.
Modulus of elasticity	E	30,000,000	=	30,000,000	psi
Tension on cord	W	2.9999996	=	3	lbs
	Change cell				
	Constraint				
	Target cell				

The value for the tension on the cord, W, will have to be experimentally defined. This is how much tension the user can comfortably tolerate while using the earphones. It can be seen that the design was able to stay within the original package size of a diameter of 1.5 and 0.75 in. thickness. This may be reduced if it is found through testing that it is possible to get acceptable performance out of a smaller spring.

Example 3.18: Product model for jar opener

Often jar tops are difficult to open with the bare hand, particularly for older people and people with arthritis or physical impediments. Such a tool to open jar tops has a market. With this example, it is known that in the stage of product modeling, one can follow the procedure of basic modeling approach to

- Understand customer needs
- Construct the model that can represent a physical understanding of the product
- Find the optimal design variables to satisfy the customer needs and improve the product with such a model

Step 1: Identify a flow (as shown in Table 3.5).

Table 3.5 Identify a flow for the jar opener

Customer need	Flow
Cost	
Easy to store	Kitchen
Easy to hold	Handle
Receive the torque	Force
Increase the torque	Force
Grasp the jar tops	Jar tops

Step 2: Identify a balanced relationship. A part of the force applied by human hands will be transmitted to friction between the jar opener and the jar top. And the moments about the fulcrum must sum to zero. In other words, the moment generated by the friction will equal to the moment generated by force applied by hands. Figures 3.40 and 3.41 show the jar top and jar opener, respectively, whereas Figure 3.42 shows the force analysis of the jar opener.

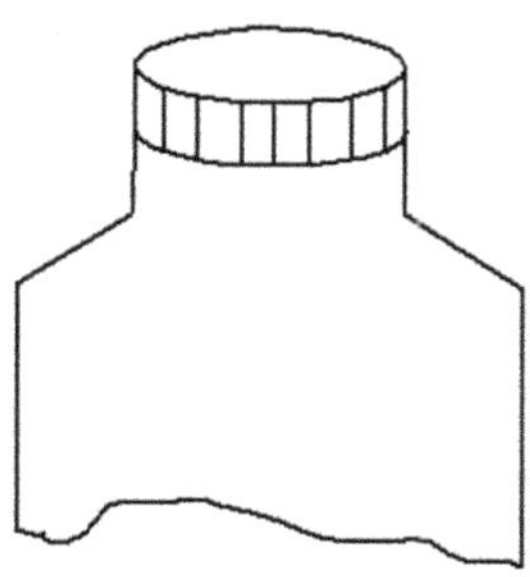

Figure 3.40 Jar top.

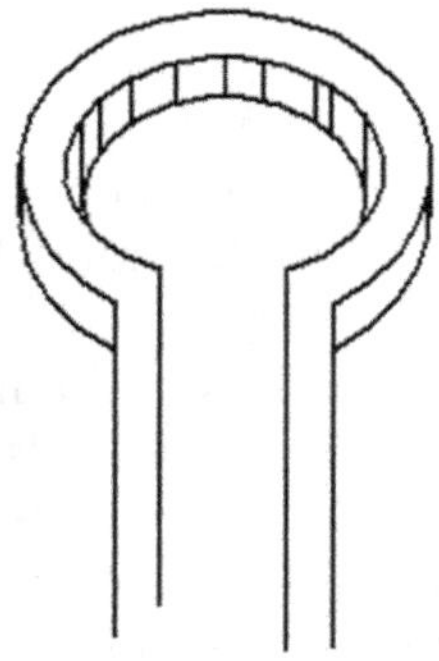

Figure 3.41 Jar opener.

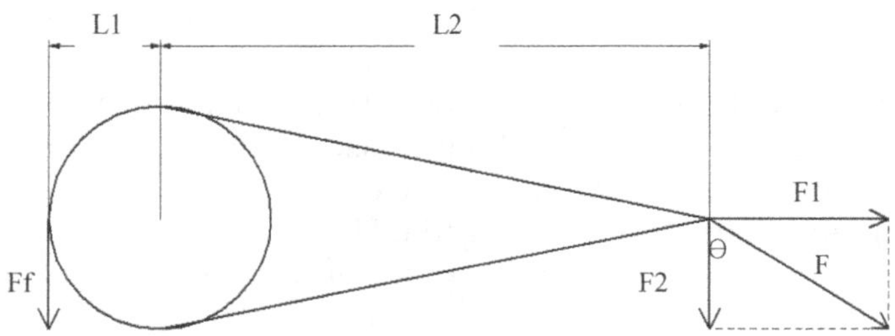

Figure 3.42 Jar opener force analysis.

Step 3: Identify the boundaries. As shown in Figure 3.42, the force applied by hands will be divided into two perpendicular parts. One will be transmitted to friction and another will generate the moment to open the jar top.

Step 4: Derive equations. As the figure showed, the balanced relationship to mathematical equations can be converted into the following relations.

$$F_1 = F\sin\theta \quad (3.69)$$

$$F_2 = F\cos\theta \quad (3.70)$$

$$F_f = F_1\mu = F\sin\theta\mu \quad (3.71)$$

and μ represents the coefficient of friction.

$$F_f \cdot L_1 = F_2 \cdot L_2 \quad (3.72)$$

Step 5: Explore design configuration options. With this model, the working principle of this kind of opener is known, and this product can be improved. The length of L_2 needs to be optimized in order to get the best effect of increasing the torque and the facility to be operated.

Moreover, one can change the rigid opener to some flexible strap that is equipped with enough friction with jar tops. In this way, the length of the strap can be adjusted to adapt to different sizes of various jar tops.

Review problems

1. There are some unique differences between an informal and a formal model. Match all the choices from the right side with the choices on the left side.
2. What are the characteristics of the product model?
3. What is a product model?
4. The problem: A few years ago, I was looking for new stereo speakers. I visited stores, read articles, books, and magazines. I talked to friends and salespeople. In the end, I was disappointed that the really "good" speakers were too expensive, and the affordable speakers were ... well, "cheap." I wanted great sound without a mortgage.

   The bright idea: Design a high-end speaker that can be made without paying a fortune for it. Use informal product modeling to model the product idea.
5. Conduct product modeling of a new key ring. It should hold several keys together. The key business goals are to obtain a 50% profit margin and the initial 2% market share. The primary market is for adults of all ages, focusing on key ring users, and the secondary market is for key ring collectors. The ring should be small, of compact storage volume, and have a long life (over 10 years). The avenues for creative design include ergonomic shape, compact storage, ease of putting keys into the ring, reasonable tightness of key inlet, and long life cycle.
6. A new computer mouse will be designed. Use this mouse as an example to write down technical questions (What is the problem really about? What implicit expectations and desires are involved? Are customers' needs, requirements, and constraints truly appropriate? What avenues are open for creative design? What characteristics or properties must the product have? What avenues are limited for creative design? What characteristics or properties must the product not have? What aspects of the design task should be quantified now?) and mission statement.
7. Give the correct order to the following steps with respect to the basic modeling approach:
   - Use the model to explore design configuration options.
   - Identify a balanced relationship.
   - Formulate an equation for the balance relationship.
   - Identify a boundary for the balanced relationship.
   - Identify a flow.
8. Hunters often have to travel a long distance away from camp to find games. After they kill their game, they face a daunting task of transporting the dead animal back to camp. Some of the kills can be upward of a few hundred pounds, and hence, to move the animal back to camp in one trip is virtually impossible even for two hunters working together. Create a formal model to design a frame capable of transporting an elk back to the campsite.
9. Develop a formal model for a coffee thermos. Focus the energy flow on the energy transmitted through the walls of the thermos through conduction and then transmitted into the ambient surroundings through conduction, radiation, and convection.
10. Develop a product model (informal and formal) for a disposable camera to be used in an underwater environment. The formal model should be focused on underwater pressure.
11. Develop a formal model for a manual inflator. Tires on bicycles, motorcycles, and cars always need to be inflated. The manual inflator should be small and convenient, and can be carried and used by hand. Develop a formal model to relate force input and the pressure output from this inflator.
12. Develop a product model for a mechanical pencil.

3.4 Using commercial software for virtual prototyping

> Planning is bringing the future into the present so that you can do something about it now.
>
> **Alan Lakein**

This section helps answer the following questions:

- What can commercial software do in a product design and prototyping process?
- Can the current software replace a physical prototype? If yes, to what level?
- What types of software are available for virtual prototyping?
- How can the dynamic behavior of a product be simulated in a computer-aided design (CAD) model?
- How to evaluate the response of a part subjected to an external load?

Many types of commercial software are excellent virtual prototyping tools, and often they start with a CAD model to define the product geometry. CAD models could be a wireframe, surface model, or solid model. Solid modeling stores many details of the model and hence needs more space to store the data. However, today's computer systems are fast and with ample memory; thus, solid modeling is being extensively used. The benefits of solid modeling include better visualization with greater detail, the ability to perform engineering analysis and computer simulation, and efficient transfer of data files between different software packages. Figure 3.43 shows an example of mold design details. Similar to this example, often a picture is worth a thousand words and expresses the concept effectively.

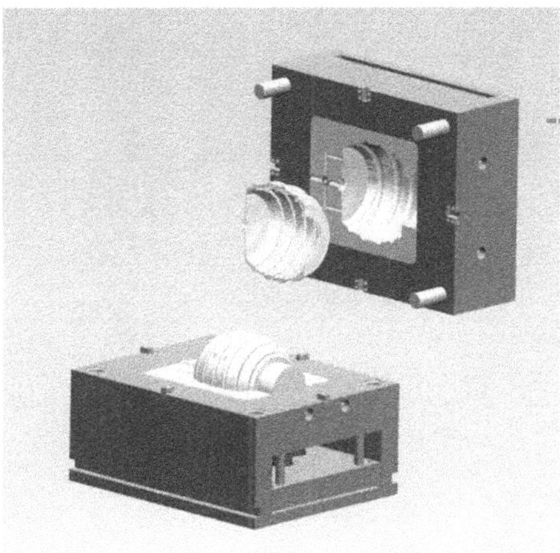

Figure 3.43 A solid model to show a mold design.

In addition, the same CAD data can be used for other analyses, including finite element analysis (FEA), additive manufacturing, and computer-aided manufacturing (CAM). The availability of the computer as a tool for analysis and modeling has changed the trend in the industry, which includes features such as preassembly in the CAD system. A lot of steps are reduced and condensed into the product-process definition, which saves a lot of time and in turn money. Also, this has contributed to concurrent engineering technology.

For example, using FEA technology, one can perform moments of inertia calculations, mass properties, linear and nonlinear static and dynamic analysis, and fatigue analysis for a given structure before it is fabricated. Other analyses, such as mold flow and heat transfer, can be conducted. Using a solid model for computer simulation has several industrial applications, such as architectural design, aircraft design, manufacturing cell operations, and virtual reality. The solid model can be applied in the CAM area so that CNC (computer numerical control) codes can be generated to drive the manufacturing equipment, such as machining centers, robots, and material handling devices, to make the part. Figure 3.44 shows an example of the CAD/CAM simulation for part manufacturing.

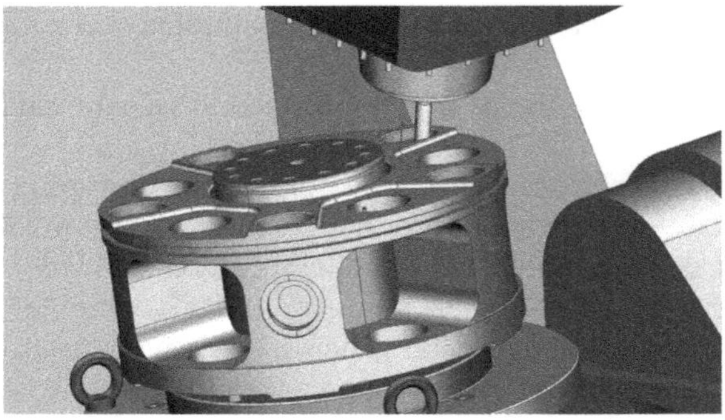

Figure 3.44 Five-axis machining simulation in a CAD model. (Courtesy of Siemens.)

The traditional model of industrial operation starts from the following sequence:

1. Sales and marketing
2. Product development
3. Design
4. Analysis
5. Release the design
6. Planning
7. Tooling
8. NC program
9. Product control
10. Fabrication
11. Inventory

The impact of the current technology has changed the way the industry operates, and the steps have been greatly reduced and will continue to reduce:

1. Sales and marketing
2. Product definition
3. Product and process definition
4. Preassembly
5. Build inspection

This will greatly reduce time and expenditure and significantly increase the efficiency of the operation. Thus, top quality designs can be achieved.

To be effective, these CAD tools must be used with concurrent engineering at the product development stage. A good example is the use of a 3D solid model for Boeing 777. It is the single largest trial of CAD technology, and the total 3D design for airplane design was conducted using CATIA. The project involved 2,200 workstations, eight mainframes, 3 terabytes of data, and 238 teams. A solid model was used for part design and preassembly, which eliminated the need for a full-scale nonflying mock-up. On average, there were four changes on each drawing—it eliminated costly iterations. Noticeable benefits are better part fit and more first-time successes in fit. The fuel tanks did not leak, and avionics systems worked. The wing tip was off by 0.001 in. on the 777, while it was off by 4 in. on 747. Fuselage alignment was off by 0.023 in. for a total length of 200 ft.

One primary application of CAD is in concurrent engineering, a systematic approach to creating a product design that considers all elements of the product life cycle from conception through disposal. Concurrent engineering simultaneously defines the product, its manufacturing processes, and all other required life cycle processes, such as logistic support. It allows different groups to create individual parts of mechanical design at different locations and at different stages of the design process. For instance, one user might be designing an electrical component while another user is testing it for installation or maintainability, while a third user is performing training on that same component.

Concurrent engineering is not the arbitrary elimination of a phase of the existing, sequential, feed-forward engineering process but rather the codesign of all desired downstream characteristics during upstream phases to produce a more robust design that is intolerant of manufacturing and use variation, at less cost than sequential design. The benefits of concurrent engineering include reduction of the time needed for implementing a new product into production, reduction of the time-to-market, substantial improvement of product quality, quicker reaction to customer requirements, cost reduction, and profitability improvement. The commercial model tools can effectively be used for concurrent engineering activities. Some of the example applications in prototyping are summarized in the following sections.

3.4.1 Dynamic analysis for prototype motion evaluation

In addition to creating an exploded view of an assembly, CAD software can also demonstrate how parts within an assembly will move with respect to one another after the assembly is appropriately constrained.

An example of virtual prototyping software for product design is Siemens NX. The mechanism models of NX can be defined directly by information from Master Assembly, including solid geometry and inertia properties. Rigid bodies are automatically created as joints are created, and constraints are defined simply by selecting appropriate topology on assembly or subassembly instances. Joints can also be created automatically from constraints, including revolute, translations, cylindrical, universal, spherical, planar, fixed, rack and pinion, screw, and constant velocity. Primitive joints can be used together with standard joints. For example, multi-joint capabilities enable the modeling of gears, cams, and followers. Motions and functions can be applied within the joint user interface. The software can also apply force fields, such as gravity and contact loads. Figure 3.45 shows an example. The components of the vehicle and mechanism were assembled and constrained in the 3D model. After constraining the components with the desired joints, the mechanism can be manipulated with respect to the constraints. Figure 3.46 shows another example of landing gear design in a tight space constraint.

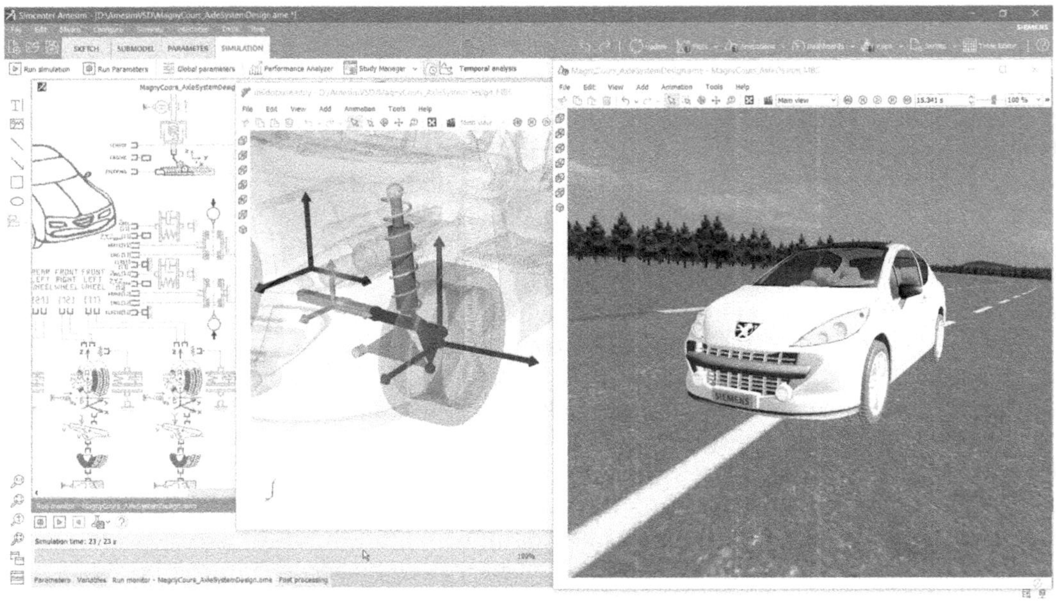

Figure 3.45 Solid-based kinematics and dynamic analysis simulate complex motions of mechanisms. (Courtesy of Siemens.)

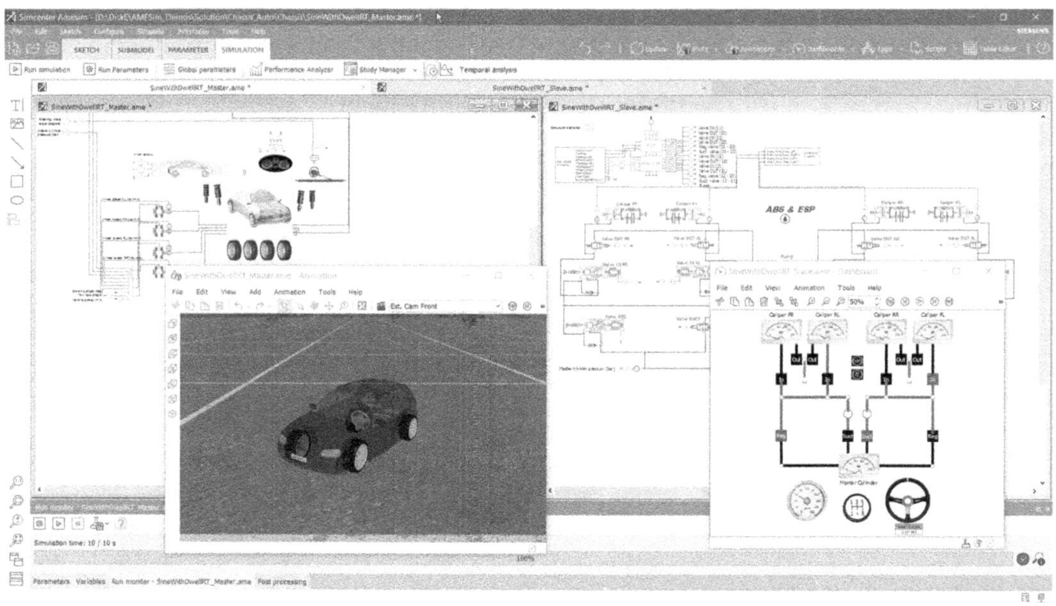

Figure 3.46 Solid-based kinematics and dynamic analysis simulate the motions of a landing gear. (Courtesy of Siemens.)

NX mechanism simulation can automatically simulate interactions among complex shapes with collision detection and response as shown in Figure 3.47 in which the dynamic response (displacement, velocity, acceleration, force, etc.) of a specified point can be monitored and displayed. This is a virtual prototyping tool.

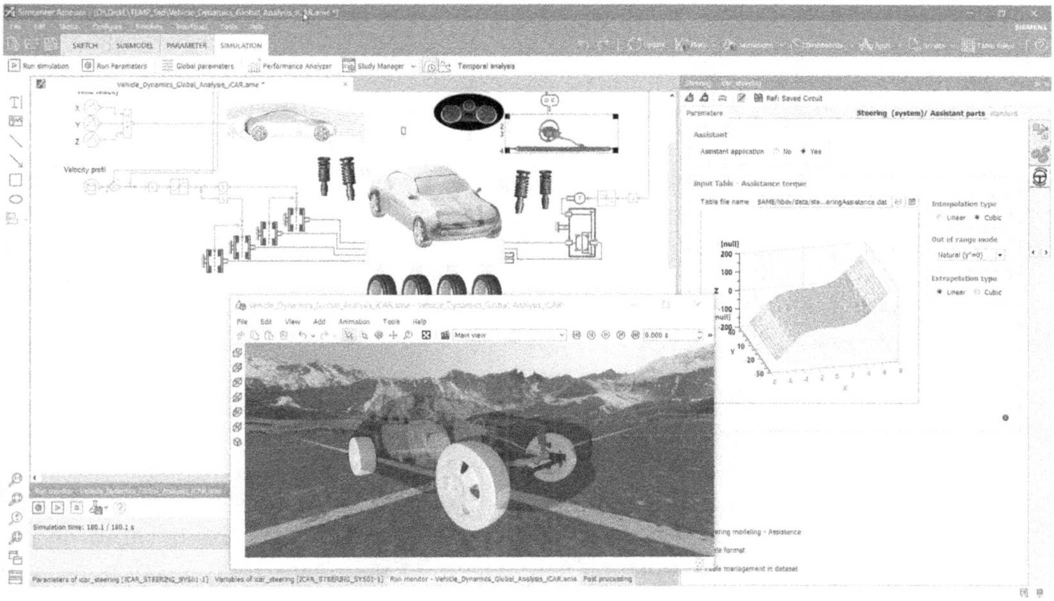

Figure 3.47 The dynamic response of created objects can be monitored in the model. (Courtesy of Siemens.)

Similar software is the Working Model in which default gravity could also be assigned; and when the software was run, the object fell out of sight when hit by the car. The simulation includes real-time collision detection and dynamic response. The car and test dummy could be created with pin joints to revolve one with respect to another to constrain the motions, and thus, it can be used to simulate the test dummy's response during a collision to understand how the human body may be impacted during a car accident.

3.4.2 FEA for prototype structure evaluation

The CAE (computer-aided engineering) method of product development is very convenient for analyzing the complicated components at the design stage itself. FEA is a computer-based numerical method for calculating the strength and behavior of engineering structures. It can be used to calculate deflection, stress, vibration, buckling behavior, and many other phenomena. It can be used to analyze either small- or large-scale deflection under loading or applied displacement. It can analyze elastic deformation or "permanently bent out of shape" plastic deformation. In the finite element method, a structure is broken down into many small, simple blocks or elements. The behavior of an element can be described by a relatively simple set of equations. Just as the set of elements are joined together to build the whole structure, the equations describing the behaviors of each element are joined into a large set of equations that describe the behavior of the entire structure. The computer can solve this set of simultaneous equations. From the solution, the computer extracts the behavior of each element. From this, it can get the stresses and deflections of all the parts of the structure. The stresses will be compared to allowable values of stress for the materials to be used, to see if the structure is weak or strong.

For example, in CAD model building, the user can start with a blank file. The working coordinate system, which is very important to pay attention to, starts off in the absolute coordinate system. For this starting point, the designer can start with primitive curves that define the outside perimeter of the product and start the solid modeling task. In this session, a simple box assembly for a computer card is used. The requirements of this example include ensuring that the interior envelope of the box provides adequate volume for the computer card and the accessories, wires, and knobs. The card itself will be mounted to integrated stands in the box. The box assembly will consist of a lower section and an upper section. These two sections will be designed to fit snugly together without any preloading and will be held together with four pan head screws. The upper section will also have exterior slots to allow for any ambient cooling air to flow.

There are many ways to create a solid model; although the final product will usually look the same, different modeling methods can make a big difference in other factors of the solid model, that is, editing. For this example, a feature in Siemens NX called "sketching" was used. In sketch mode, the user defines a 2D perimeter of the part with dimensional and physical constraints. From these sketches, it is possible to return to the modeling mode and extrude the created curves into solid bodies. The power behind the sketcher is that after the solid body is created, the designer can go back into the sketcher and easily modify a dimension or constraint for whatever reason, and NX will automatically adjust the related solid body to the new requirements. This feature allows the user to effectively sketch the part without knowing the full physical dimensional requirements of the product.

Both sections of this box assembly were created with the sketching mode. The existing computer card was reverse engineered, and by using the sketching mode, it enabled the user to create the box assembly prior to receiving the computer card to obtain the actual dimensions. One could approximate dimensions to begin the process and quickly and easily adjust the solid model once the actual part was received. Figure 3.48 shows the bottom section of the box assembly, and Figure 3.49 shows the top section.

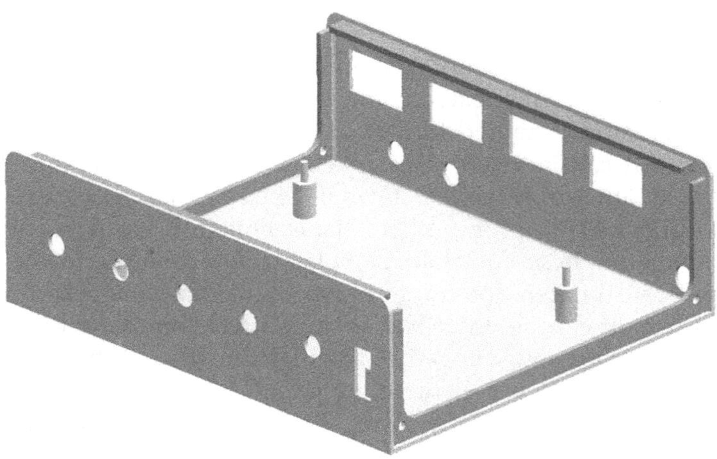

Figure 3.48 The bottom section of the box assembly.

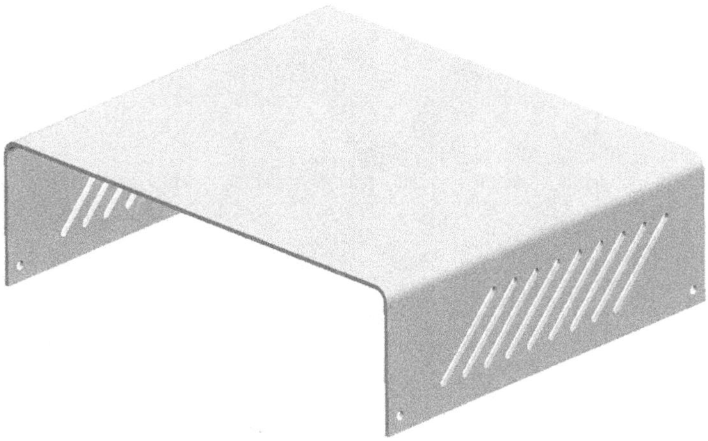

Figure 3.49 The top section of the box assembly.

The use of the sketcher also ensured that the boxes could be created in separate models, and when the two parts were assembled together, the parts could be modified to ensure a good fit. The gaps between the parts were adjusted to allow for manufacturing tolerances after the determination of what parts were going to be made from.

Again, once a solid body of a part is created, there are lists of analysis tools that become available to the user. These analysis tools have also revolutionized the industry and allowed the designer to locate "hot spots" on their design prior to fabricating a part and actually attempting destructive testing. The tools require good computing power and simplify the analyzing process to allow many people to utilize them for their parts.

For this box assembly, as stated previously, the two sections were determined to be made from an SLA (stereolithography apparatus) rapid prototyping machine. The material to be used is the RenShape SL7520 resin. With properties pulled from the supplier's website, it becomes possible to analyze the upper section of the box. One requirement for the box assembly is to allow for lightweight miscellaneous equipment to be haphazardly stored on the top of the box. The first step in using the finite element modeling tool is to set the material properties for the part. For cured SL7520, the following properties were inputted:

Mass density = 0.0426 lbm/in.^{3}; Young's modulus = 450,000 psi; Poisson's ratio = 0.33
Shear modulus = 415,000 psi; Yield strength = 9,200 psi

With these properties assigned to the upper box section, it is possible to create a finite element mesh of the part along with load and boundary conditions. For this part, a 5 lbs downward normal force on the upper surface of the box is assumed. The boundary conditions consisted of four fixed locations where the screws attach to the lower section. Figure 3.50 shows the meshed parts with the load arrow and fixed boundary conditions.

An initial run of the finite element model gave a deflection output graph as shown in Figure 3.51.

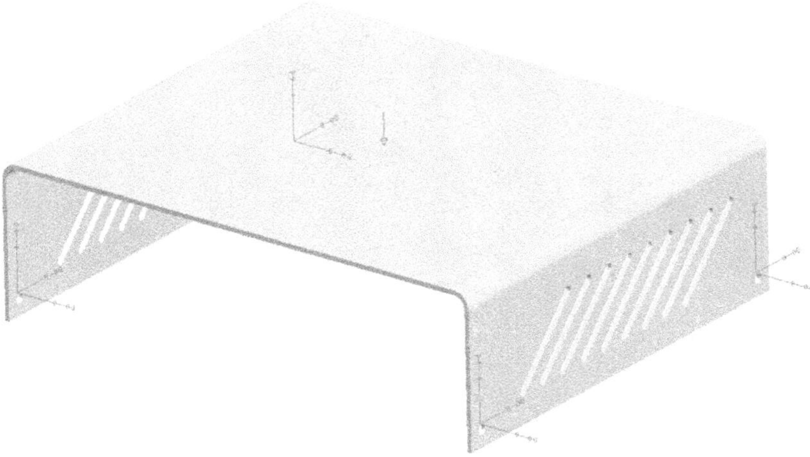

Figure 3.50 Load and boundary condition for FEA on the upper section.

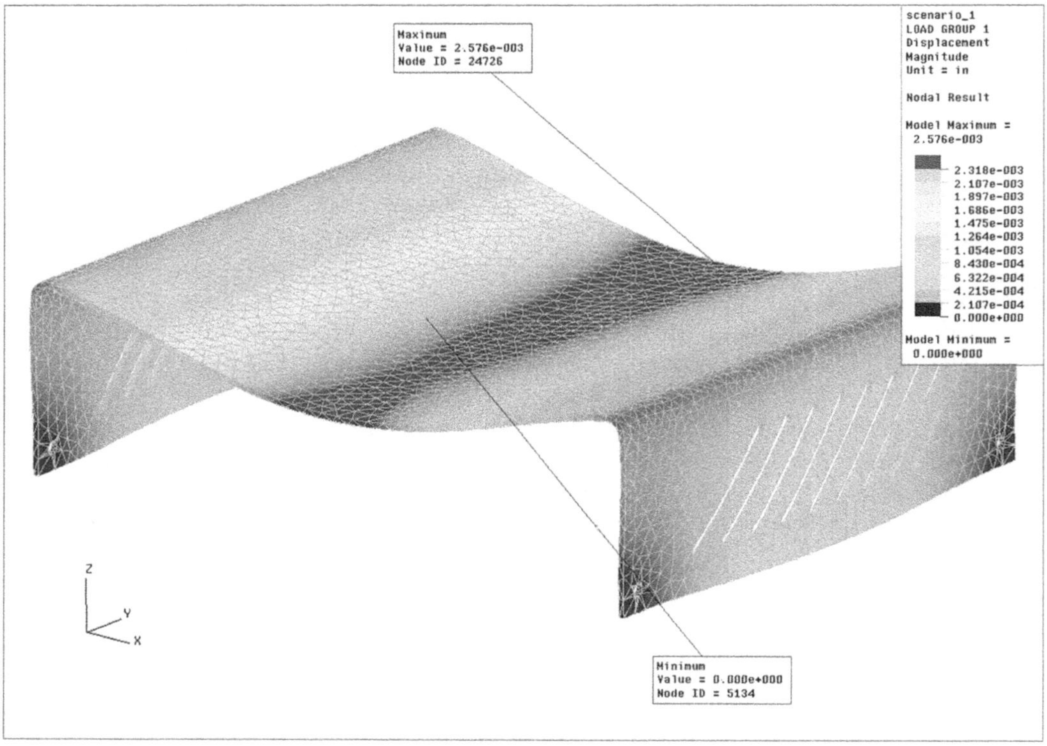

Figure 3.51 Finite element run with initial boundary conditions.

Note that the boundary condition that fixes the upper edges in the z-translation was not included since these edges will lay on edges of the lower section. Figure 3.52 shows the meshed finite element model along with the revised boundary conditions.

The updated boundary conditions that produced the deflection output are shown in Figure 3.53.

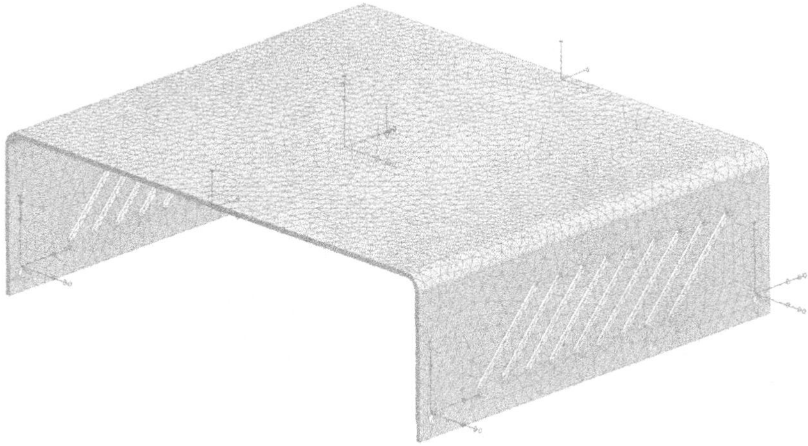

Figure 3.52 Finite element meshed model with load and boundary conditions.

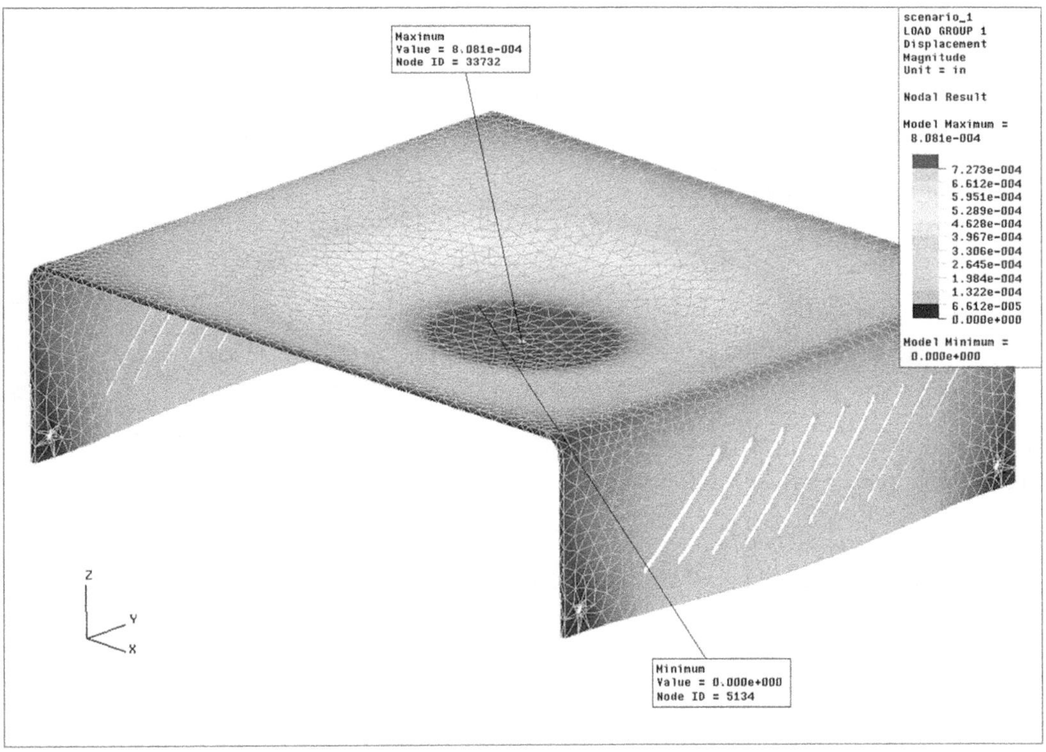

Figure 3.53 Finite element deflection model.

The model represents the deflection that the part will see with the given load. The point was input as a normal point load, but to be better represented, it should have been given an area and loaded as a pressure load. Figure 3.54 shows another example with more complex geometry for stress analysis. The finite element tools can be used to quickly identify problem areas of the solid body and allow the designer to go back and redefine the part, as required.

Figure 3.54 Finite element stress analysis of a part. (Courtesy of Siemens.)

Review problems

1. What is a solid model?
2. What is the role of computer simulation in prototyping?
3. Discuss some of the benefits of using a 3D modeling software to perform virtual prototyping.
4. What is concurrent engineering?
5. What are the benefits of solid modeling?
6. How can the dynamic behavior of a product be simulated in a CAD model?

3.5 Virtual reality and virtual prototyping

A good plan violently executed now is better than a perfect plan next week.

George S. Patton

This section helps answer the following questions:

- What is virtual reality?
- Why does one use virtual reality in product development?
- What is augmented reality (AR)?
- Why is AR used in product design or prototyping?

3.5.1 Virtual prototyping

Virtual prototyping can be used for a designer to interact with the product before it is actually created. Virtual prototyping is combining the virtual environment with engineering design to enable a designer to play an active role in the design sensitivity and optimization process. Virtual reality is a system that provides a nonreal environment but gives the feeling of real sense. There are several technological requirements of virtual prototyping, such as hardware, input devices to sense user interaction, output devices to replace user's sensory input into computer-generated output, and software to handle real-time simulation.

In terms of product development, before a product is built, virtual reality provides the ability to do human-in-the-loop in real time to test "what -if" scenarios and thus reduces development time and reduces time to deliver products to market.

Several application areas of these tools are design, training, engineering, military, education, and medical markets. The user could interact with the design and quickly "drag" and "click" applications with a facility for constant modifications. A new tool is World2world which could be used to interact with intranet and LAN (local area network) deployed systems. It supports collaborative design and visualizing, and distance learning. An example of a virtual reality application is a training program to train drivers on the safe operation of a forklift truck in a busy and complex real warehouse environment. Similarly, other applications of virtual reality, such as the Biogen demo, were created to visualize the process of how the patient's health improved when he took new drugs and how these drugs reacted with DNA and brain cells. The virtual real estate is another example to visualize different properties on a 3D map where the user can move around the virtual environment and look for desired features.

Technology requirements

- Hardware capable of rendering real-time 3D graphics and stereo sound.
- Input devices to sense user interaction and motion.
- Output devices to replace user's sensory input from the physical world with computer-generated input.
- Software to handle real-time input/output processing, rendering, simulation, and access to the world database.

Let us use an example to illustrate a virtual environment for parts handling applications in a factory. Some related technical challenges are also discussed. The applications of a virtual system can be used to prototype a parts feeding system in an assembly line, robotic parts handling, parts in a fixture, parts on a conveyor, or the components interactions in mechanical assemblies. In this way, one can prototype a manufacturing work cell to observe whether the parts are going to get jammed in a transfer line, study whether a robot can successfully pick up the parts with different orientations, and test whether a modular fixture can firmly clamp a part for machining. This way the detailed effects of clearance, friction, chamfers, etc., can be studied and the parts and components of the equipment can be checked for their handling feasibility. A design can be improved before a part or equipment is fabricated or purchased. The common physical nature of the above prototyping activities is the involvement of the interactions of objects. A virtual environment for this type of application may involve all or part of the following functions: (1) the designer can observe a visual representation of the virtual parts and equipment involved; (2) the designer can interactively manipulate a virtual part or equipment component through haptic interface devices, and the virtual part or component will react according to the force or motion the designer provides; and (3) the designer can feel the objects and hear sounds from the virtual work cell.

To implement a virtual environment system for parts handling applications, at least two major enabling technologies are needed:

1. Simulation technology to model and process the interaction between objects in the virtual world. In other words, a physically based simulator should be provided to simulate part and equipment component interaction.

2. Human–machine interface technology to effectively generate the appropriate sensory output of the simulation system to the user and receive the input command from the user.

One technical challenge, critical to developing a virtual environment in parts handling applications, is the graphical hardware and software capability to deliver adequate frame rate and at the same time present sufficient detailed information for the parts handling work cell. In general, a virtual environment should not have a frame rate of less than 10 frames per second so that the visual illusion of real time can be sustained. Actually, it is highly desirable to have a frame rate greater than 30 frames per second and the total delay time including simulation and computation to be less than 0.1 s. If the parts or equipment components to be modeled exhibit high frequencies of motion, such as impact between parts, it will require even higher frame rate and shorter time delay.

3.5.2 An AR system: an example

This section summarizes an example study for an AR system for collaborative product development using the internet and AR technology. The system allows fruitful interaction between experts at geographically dispersed locations through a flexible environment for sharing of information, which could be in the form of live video, CAD, audio, textual, or conceptual. This system will provide a new paradigm for flexible interaction between the vendors and the domain experts at different stages of the product development cycle in order to reduce turnaround time. A case study on an AR environment with 3D tracking and dynamic simulation technologies for the parts feeding systems is also summarized so that engineers can run a high-fidelity simulation to test new materials, components, and systems.

One major barrier in a collaborative product development process is the difficulty in information exchange between personnel from various departments and vendors, especially those working at different locations. The aim of the research here is to provide a tool for rapid product development using resources and expertise at different locations. These distributed resources, expertise, and products may belong to the same manufacturing companies at different locations or to independent consultants or suppliers. Advancements in communication systems, such as tele- and video-conferencing, have enhanced the information sharing capability between different engineering domains. However, mere sharing of information does not complete an ideal collaborative product development environment. An equally critical aspect is the "interaction" capability between various domains. Data sharing ability merely emphasizes the structured and syntactic aspects of information flowing between disparate domains; it is the interaction between disparate domains that captures the true spirit of integration. This interaction could be in the form of online audio–visual communication, textual information exchange, live video of the working environment or product, online manipulation of CAD, or virtual environment.

Advances in virtual reality and simulation technologies have demonstrated the capability of performing significant product trade-offs in a virtual environment. However, VR has a major limitation in effectively modeling complex environments, such as an existing machining work cell or an entire factory environment. This information is often critical in the product development process so that the design team can understand the physical constraints of the actual working environment. AR along with the advancements in web-based distributed environment provides an exciting possibility to make this interaction possible. AR is a relatively new technology, which seeks to augment a user's view of the real world with computer product data or graphics models. It is a more natural and effective means to exhibit and analyze a design in its real-world context, and it can provide new levels of interactivity and much broader knowledge bandwidth. It is hoped that soon this system will facilitate real-time "dialogue" between distributed engineering domains.

The aim of the project shown in Figure 3.55 is to integrate AR, telerobotics, and an Internet-based distributed communication environment to develop a truly collaborative product development tool. One can explore effective ways to integrate conceived 3D geometric information of the part with the actual 3D production environment at various stages of the product life cycle and to share this augmented knowledge with other team members for effective product development. The approach is based on the distributed AR that displays a design in its real-world environment through the Internet. With such a system available, engineers and personnel can sit in their offices and design, evaluate, or modify a product in a collaborative manner. Effective communication between the team members can be achieved through visualization and online modification of the part as well as its associated environment. Therefore, new levels of interactivity can be achieved, and knowledge bandwidth can be increased. Such augmented 3D knowledge can greatly impact knowledge sharing in the product development process, through enabling technologies of AR and distributed collaboration.

Figure 3.55 The conceptual representation of the collaborative product development environment.

AR is a new and emerging technology, which seeks to enhance the virtual reality environment by integrating real-world information in the form of live audio/video into the virtual environment. A purely virtual reality environment seeks to create a realistic model of the real environment and subsequently simulate real-world behavior in a virtual environment. AR, on the other hand, tries to superpose the virtual computer model onto a live video of the real world, thus, eliminating the need for precise modeling of the environment as shown in Figure 3.56. The augmentation of the virtual world with the real world is particularly important when the real world is too complicated to model and maintain or when very high accuracy is required.

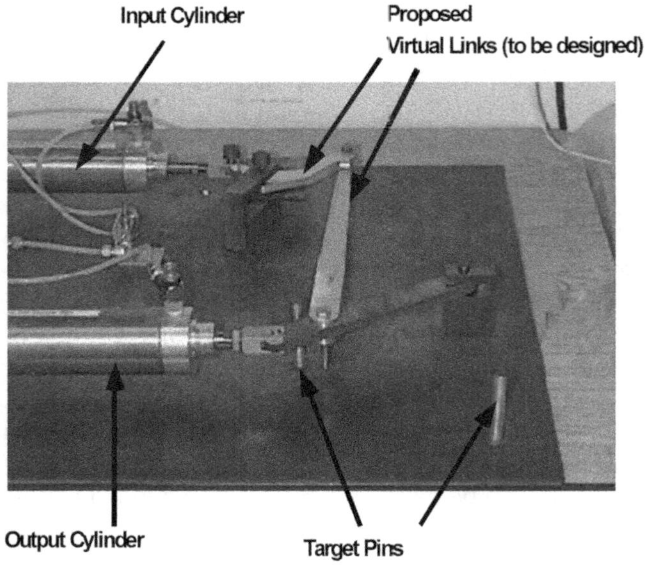

Figure 3.56 An example of the seamless "overlay" of the real (cylinders) and virtual objects (links).

For concurrent engineering, a 3D representation of the product and its production and service environment constitute important information. In order to analyze the interaction between the computer model and the real-world environment, virtual reality-based systems need to create an authentic virtual representation of the real world and then rely on these virtual models to simulate world behavior.

References

[Basmadjian02] Basmadjian, D., *Mathematical Modeling of Physical Systems: An Introduction*, Oxford University Press, Oxford, UK, December 19, 2002.

[Feynman70] Feynman, R.P., *Feynman Lectures of Physics*, Addison Wesley Longman, Boston, MA, 1970.

[Laeng00] Laeng, J., J.G. Stewart, and F.W. Liou, "Laser metal forming processes for rapid prototyping—A review," *International Journal of Production Research*, Vol. 38, No. 16, pp. 3973–3996, 2000.

[Mcmaster07] Mcmaster-Carr Products, Retrieved January 7, 2007, from http://www.mcmaster.com.

[Otto01] Otto, K. and K. Wood, *Product Design*, 1st edition, Prentice Hall, Upper Saddle River, NJ, 2001.

[Shawki66] Shawki, G.S.A. and M.M. Abdel-Aal, "Rigidity considerations in fixture design—Contact rigidity at locating elements," *International Journal of Machine Tool Design and Research*, Vol. 6, pp. 31–43, 1966.

[Wikipedia17] Retrieved on August 26, 2017, from https://en.wikibooks.org/wiki/First_Law_%28Engineering_Thermodynamics%29.

[Yeh00] Yeh, H. and F.W. Liou, "Clamping fault detection in a fixturing system," *SME Journal of Manufacturing Processes*, Vol. 2, No. 3, pp. 194–202, 2000.

chapter four

Material selections and product prototyping

Material selection and design is a critical element in product design and prototyping. One reason material selection is important is that the cost of material may be more than 50% of the total cost. For example, in the auto industry, material cost is about 70% of the total cost. In the shipbuilding industry, material cost is about 45% of the total cost. While the cost of a can of coke may be from 50¢ to $1, the liquid itself is about 2¢, the container material is about 10¢, and the manufacturing cost is about 5¢. The second reason that material selection is important is due to the fact that it will affect the manufacturing process. In other words, when a material is selected, its manufacturing method is selected, and thus may impact many factors, such as product cost, labor, and equipment. For example, when plastic is selected as the product material instead of sheet metal, a manufacturing process such as injection molding is selected instead of a process such as stamping. This chapter focuses on prototyping materials, material properties, and the related material selection processes.

4.1 Prototyping materials

This section will help answer the following questions:

- Why is material selection important?
- Is there any place where guidelines are published for prototype material selection?
- Are there nontraditional materials that can be used for prototyping?
- Are there procedures to select a material for a prototype?
- Where to find resources for the properties of different materials?

4.1.1 Prototyping and material properties

When selecting a material, material properties are critical since they are the link between basic material composition and service performance. Material processing is also critical since it determines part manufacturing processes. Prototyping materials often are different from the final product materials, especially for lower fidelity prototypes, due to the differences in project objectives and time constraints in prototyping. For quick prototyping purposes, there are several materials available.

1. Modeling clay: As shown in Figure 4.1, modeling clay is easy to work with, is useful for visualization and airflow studies, always remains soft, and is available in hobby and craft shops. Each time Congress authorizes a new coin or medal, an artist sketches out ideas for the design. After one design has been approved, the U.S. Mint sculptor-engraver sculpts a clay model as shown in Figure 4.1 [USMint17].

Figure 4.1 Modeling clay is a convenient prototyping material for a new coin [USMint17].

2. Machining wax: Wax can be machined well and is low in cost and useful for prototyping tooling patterns.
3. Foam board: As shown in Figure 4.2, the foam board has a good finish and is easily carved, useful for painting (aesthetic/appearance models), and machinable. Pressurized cans of insulating foam are available that harden quickly and may be cut and formed with a knife and sanding board.

Figure 4.2 Foam board has good finish.

4. Foam core: As shown in Figure 4.3, the foam core is made of sheets of hard paper with internal foam, is useful for mock-ups and layout of square objects, can be used with Bondo/clay for more complex shapes, and is more durable and rigid than cardboard.

Figure 4.3 Foam core is made of sheets of hard paper with internal foam.

5. Rubber and elastomer: As shown in Figure 4.4, rubber and elastomer are useful in energy absorption applications or seals, can be used as a removable mold for castings of other materials, and can be carved.

Figure 4.4 Rubber and elastomers are useful in energy absorption applications or seals.

6. Cardboard, paper, and cloth: Cloth and paper can serve as joints in mock-ups and are very economical.

4.1.1.1 Material selection for high-fidelity prototypes

The previous section summarizes some commonly used prototyping materials. However, these materials are typically used for lower fidelity prototypes. When selecting materials for higher fidelity prototypes, it sometimes helps to know which common material has the "best" properties. Note that they are the highest, or most efficient for commonly used materials, but may not necessarily be the highest or most efficient for all materials. The purpose of the list here is just to provide a general guideline for category selection. For example,

- Graphite epoxy has the highest tensile strength.
- Graphite epoxy has the highest yield strength.

- 30400 Stainless steel has the highest endurance limit.
- Polycarbonate has the highest percentage of elongation.
- Alumina ceramic has the highest modulus of elasticity.
- Polycarbonate has the highest coefficient of thermal expansion.
- 01 Tool steel has the highest melting temperature.
- C268 copper has the highest thermal conductivity.
- C268 copper has the highest density.
- 7075 aluminum has the most efficient cost per pound.
- 01 Tool steel has the most efficient cost per unit volume.
- 4340 Steel has the highest hardness value.

4.1.2 *Material selection methods*

When selecting a material, there are many areas of design concern. One needs to define the objective, such as what is to be maximized or minimized? Does it need to be made economical, lightweight, increase safety, etc., or a combination of these things? Does one need to consider the function of the material, such as what does the component do? Will it support a load, contain pressure, transmit heat, etc.? One needs to consider constraints for the materials, such as what nonnegotiable conditions are to be met, and what are the negotiable but desired conditions? It may be necessary to consider the following application issues in selecting materials:

- Force and load magnitude: Tensile and compressive yield strength of the material should be considered, such as steels, other metal alloys, and plastic composites.
- Creep (slow continuous deformation) applications: When loads are applied over a long period of time and when the temperature is high (30%–40% of the melting temperature of the metal), the creep property becomes critical, and one should use metals and avoid plastics.
- Input load applications: When input loads are the consideration, one can choose high-strength, high-ductility materials.
- Cyclic loads and fatigue applications: When the component is subject to cyclic loads, and fatigue is the issue, the endurance limit should be considered.
- Deformation applications: When structure deformation is considered, metals or thermosets would be good candidates.
- In-service temperature applications: When the in-service temperature is considered, metals or thermosets are good selections.
- Applications exposed to UV light (sunlight): When an application is subject to UV light, nonplastics will be more suitable.
- Applications exposed to moisture: In a moist environment, plastics or aluminum alloys of 1,000, 3,000, 5,000, and 6,000 levels that are resistant to corrosion can be considered.
- Weather applications: When the weather condition is considered, most aluminum alloys, stainless steels, and plastics are good candidates.

Table 4.1 shows a summary of some common material selection properties and characteristics [Dieter99, Ertas96]. As there are thousands of materials available for a prototype or design, the material properties can be used to filter out the unqualified candidates and select the proper ones.

Table 4.1 Summary of common material selection properties and characteristics

Physical properties	• Density, viscosity, porosity, permeability, reflectivity, crystal structure, etc.
Mechanical characteristics	• Strength (ultimate strength, yield strength, shear strength) • Stress–strain curve • Ductility • Modulus of elasticity (tension and compression) • Poisson's ratio • Hardness
Fatigue characteristics	• Corrosion fatigue • High load/low load/extended life • Constant amplitude load • Spectrum load • Fatigue strength • Sample smoothness, notched, etc.
Fracture characteristics	• Fracture toughness • Flaw growth • Crack instability
Thermal properties	• Coefficient of thermal expansion • Emissivity • Absorptivity • Melting/boiling points • Heat transfer coefficient • Specific heat • Thermal conductivity • Thermal shock resistance
Manufacturing	• Producibility • Availability • Processing characteristics (machinability, castability, weldability, moldability, heat treatability, formability and forgeability, hardenability) • Minimum handling thickness • Joining techniques • Quality assurance
Flostile environments	• Moisture • Temperature • Acidity/alkalinity • Salt solution • Ammonia • Hydrogen attack • Nuclear hardness, nuclear half-life
Others	• Electrical, chemical, corrosion, optical properties • Repairability • Reliability • Anisotropy

4.1.3 *Material selection processes for high-fidelity prototypes*

When designing a product or a prototype in which material is more complex than a simple prototyping material, one must use a rigorous material selection method. Just like the design process, material design and selection is also an iterative process. There are two approaches for material selection. One is "material-first," based on choosing the material first and then selecting feasible processes as shown in Figure 4.5. The other is "process-first," based on choosing the process first and then selecting feasible materials, as shown in Figure 4.6.

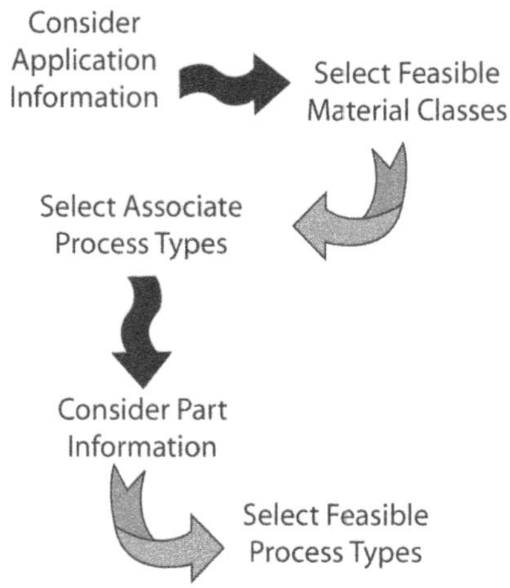

Figure 4.5 Material selection procedure: material-first.

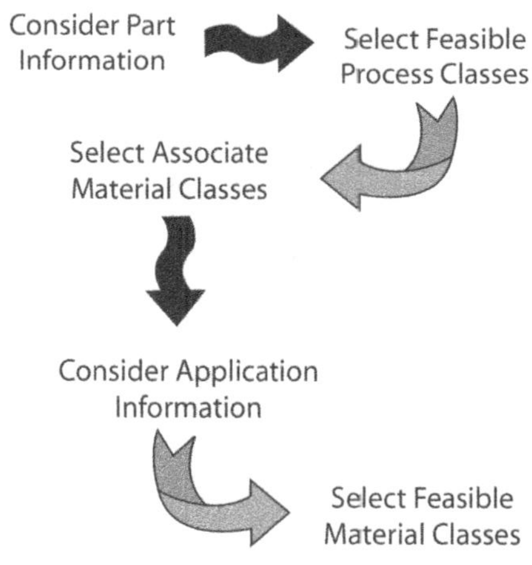

Figure 4.6 Material selection procedure: process-first.

In the material-first method, one can gradually define and then refine the material detail. This is consistent with the design process from conceptual design to detail design, and thus, different level(s) of material specifications can be used in various design stages. There are four different levels to specify a material selection, that is, level I, level II, level III, and level IV. For example, at the level I, one can select between metal and plastic and, based on this selection, design the product or prototype. At level II, one can specify wrought or thermoplastic; at level III, specify between aluminum and acrylonitrile butadiene styrene (ABS); and at level IV, decide between aluminum 6061 and ABS (Dow). The level of detail can proceed when the design process progresses from conceptual design to preliminary and then to detail design.

Level I is the first stage of material selection, and one can choose metal, plastics, ceramics, wood, or concrete. In general, metals have better mechanical properties, whereas plastics are inexpensive and corrosion resistant; have an insulating effect; and improve ergonomics, weight, consolidating parts, etc. Plastics have replaced metals in certain areas, such as automobile bumpers, and fuel tanks. Plastic may be the best option for parts for the following reasons:

1. Decreased piece part prices. For example, the injection molding process has faster cycle times.
2. Eliminate time-consuming and costly secondary operations. As the parts from the injection molding process often do not need secondary operations, this further reduces costs. In addition, the plastic material can be colored with color concentrates before molding—eliminating secondary painting operations. Injection molds can be textured or given various levels of polished surfaces before molding. The costly assembly of several metal stampings or castings fastened together can often be replaced by a single injection molded part incorporating the features of the total assembly.
3. Reduce product weight and improve user ease. Reducing the product weight gives more parts per pound of material and significantly reduces shipping costs and improves the end user's physical ease in utilizing the product.
4. Gain greater product structural strength. Plastic parts can be stronger than metal parts through the use of engineering grade materials and the ability to mold in structural strength, such as ribs, bosses, and gussets.
5. Increase product design options. Metals have design limitations, and plastics increase the design options through the ease of producing complex shapes.

Wood is a commonly used prototyping material. Wood is also a good candidate for prototyping, but some types of wood tend to shrink or deform especially after a long period of time. There are several types of wood that can be used for prototyping purposes. In general, plywood has high strength and is available in large sheets, both hardwood and softwood. Pine is inexpensive and easy to use to make prototypes. Hardwoods, such as oak, cherry, or birch, are higher quality but more expensive, make precise joints, hold fasteners better, and have higher load capacity than other woods.

Level II is the stage to consider more material details. For example, in terms of metals, one can consider cast or wrought. The casting process can produce more complex shapes, hollow sections, and larger parts, and the overall cost can be lower. Wrought products, in general, have simpler geometry, can be produced relatively fast, and are an inexpensive process. In terms of plastics, one can select thermosets or thermoplastics. Thermosets are the plastics which will undergo a chemical reaction by the action of heat, catalysts, or UV light and change into substantially infusible and insoluble materials. Thermosets include epoxy, phenolic, unsaturated polyester, polyurethane, dicyanate, and so on. Thermosets can keep their structural integrity over a wide range of temperatures. Because of their heat resistance and dimensional stability, they are used as a mold for manufacturing products in many areas including the electric industry (electrical connectors, circuit breakers, etc.) and the automobile industry (automotive valve covers, ignition parts, etc.). Thermosets have greater rigidity, good electrical properties, and low absorption in water. The examples of thermosets could include distributor caps, circuit boards, coil caps, and handles for utensils. The cycle time is about 30 s to 5 min per piece. Thermoplastics are faster and easier to process, have a short cycle time (15–60 s), and lower the cost when using injection molding.

Level III is the stage to consider even greater material details. For example, in terms of metals, one can select a nonferrous alloy, such as aluminum, magnesium, copper, and zinc. The nonferrous alloys are more resistant to corrosion, lighter in weight, have better thermal and electrical conductivity, and are easier to fabricate (lower tensile and yield strength) and easier to cast. Ferrous alloys, such as steel, are stiffer and have a high melting point. The low-carbon steels (<0.2% carbon), such as ferrite and pearlite, are used for making bars, sheets, strips, and drawn tubes. Usually, this material is used as it comes from the hot- or cold-forming processes. These steels have high ductility and are acceptable to low-cost production (stamping). The advantages of low-carbon steels include good formability, good weldability (best of all metals), and low cost. The disadvantages include the fact that as the percentage of carbon increases, there is a tendency for the metal to harden and crack and that the hardenability is low. It is rated at 55%–60% machinability as it builds up the heat on the tool.

Increasing the carbon content of steel increases its strength and hardness but decreases its ductility. Mild steel (carbon 0.15%–0.3%) is commonly used in structural sections, forgings, and plates. The advantages include the fact that machinability is 60%–70%; therefore, it cuts slightly better than low-carbon steels, and both hot and cold rolled steels machine better when annealed. Mild steel is also less machinable than low-carbon steel, has good toughness and ductility, and is extremely popular and has numerous applications.

High-carbon steel (>0.8%) is used for hammers, dies, tool bits, etc. The advantages of high-carbon steel include the fact that hardness and wear resistance are high. The disadvantages are that toughness, formability, and hardenability are quite low, and it is not recommended for welding.

Alloy steels (nickel, chromium, manganese) can improve strength, hardness, or resistance to corrosion. In terms of plastics, one can select among composites, polycarbonate, nylon, acetal, and ABS. Composite materials are plastics with additives, fillers, and reinforcing agents, such as glass fibers, carbon, or graphite. Polycarbonate has a very high impact resistance and is easy to produce. It is used for computer parts, peripherals, business machine housings, vacuum cleaner parts, sports helmets, windshields, etc. Nylon (DuPont) is resistant to oils, greases, and most common solvents. It has high strength and high modulus of elasticity, high maximum service temperature, and low friction coefficient. Nylon is commonly used for bearings, bushings, gears, cams, etc. Acetal is strong and stiff, has high resistance to abrasions and chemicals, and has a low coefficient of friction. It is commonly used for brackets, gears, bearings, cams, etc. ABS is good for structural applications, has good thermoformability, has high impact strength, has good fracture resistance, and is inexpensive and available in sheets and rods.

Level IV is a stage with specific materials, such as 1020 steel, 6061 aluminum, and nylon 6/6. Materials are identified with detailed specifications and used in the detailed design.

There are many software packages available for material databases, and some of them can be accessed from the Internet, such as MatWeb [MatWeb17], a searchable database of material data sheets, including property information on thermoplastic and thermoset polymers such as ABS, nylon, polycarbonate, polyester, polyethylene, and polypropylene; metals such as aluminum, cobalt, copper, lead, magnesium, nickel, steel, superalloys, titanium, and zinc alloys; ceramics; as well as semiconductors, fibers, and other engineering materials.

Review problems

1. When a product to fulfill an incident function is designed, why is the material selection very important?
2. Select the material for manufacturing an automobile body by using material-first steps.

3. Select the material to make coffee cups by using process-first steps.
4. Select the material for a condenser (heat exchanger) cooled by brackish water. This condenser is at an early design stage. Its proposed shape is tube sheet.
5. In order to design a product, understanding the properties of materials is very important. Please match the following materials to their "best fit" property. (Each one on the left should match one on the right.)

1. 30400 Stainless steel	a. High melting temperature
2. Alumina ceramics	b. High modulus of elasticity
3. C268 Copper	c. High tensile strength
4. 01 Tool steel	d. High endurance limit
5. Graphite epoxy	e. High thermal conductivity

6. According to application issues in selecting material, what do we need to pay attention to when we design a metal table?
7. Complete the following table for commonly used materials with the "best" properties.

Property	Material
Tensile strength	
Yield strength	
Endurance limit	
Percentage of elongation	
Modulus of elasticity	
Highest density	
Highest coefficient of thermal expansion	
Highest melting temperature	
Highest thermal conductivity	
Most efficient cost per pound	
Highest hardness value	

8. Research the materials associated with fused deposition modeling, selective laser sintering, and stereolithography, and rank the resulting materials in descending order relative to ductility as a function of the percentage of elongation.
9. How are the four levels of material selection related to the stages of product design?
10. Describe the process of material selection for a compact disc using the selection methods outlined in Section 4.1.2.
11. Describe the advantages of a few commonly used prototyping materials, such as insulating foam, fiberglass, and wood.
12. A company has been hired to redesign the cabinet for a line of Sony indoor/outdoor home speakers. Describe some of the things the company must consider when selecting the materials for this cabinet. Choose what must be considered when selecting the material for the speaker cabinet and provide your top selection of material to be used.
13. Select three materials for a fishing pole and discuss some strengths and weaknesses of each material.
14. Based on the materials identified in the previous problem, come up with at least one manufacturing process for making a fishing pole from each material and discuss the strengths and weaknesses of each method.

15. Discuss the results of doing process-first and material-first material selection in the previous problem.
16. Match the material with the possible fabrication process.

Fabrication process	Material
1. Forging	a. Carbon epoxy
2. Stamping	b. Steel
3. Rotational molding	c. Zinc
4. Die casting	d. Acrylic sheet
5. Mandrel layup	e. Titanium
6. Sand casting	f. Polypropylene
7. Thermoforming	g. Aluminum
8. Injection molding	h. ABS

17. List and describe five materials available for quick prototyping purposes.
18. Briefly describe the difference between the two different material selection processes.

4.2 *Modeling of material properties*

> Direct material costs often account for more than half of the total contract cost.
>
> *US Defense Procurement and Acquisition Policy*

This section helps answer the following questions:

- How to conduct material design for a product?
- Are there models that could help aid in material selection?
- Are there simple material models that can be used in product design and prototyping?
- How to model material properties? Are there any equations or graphs that can help to analyze material properties in a model for material selection?

As discussed in Chapter 3, it is possible to use modeling and simulation to perform virtual prototyping of a design or a prototype. It is also possible to use a model and simulation to select a material for a specific application. The purpose of this section is to use examples to illustrate the benefits of modeling material properties to help in product design. It may be necessary to find more related references to gather proper models for a specific application.

The scope of material selection and consideration can encompass more than just the structural makeup. It may involve the five senses used every day: touch, hearing, sight, smell, and taste. For example, one may want to design a coffee cup that feels cool to the touch but keeps the coffee hot inside the cup. Another example is when designing the interior of a car; one may want to absorb road noise but keep the music from the radio in the car.

Proper selection of materials may produce these designs more effectively. Some material properties are easy to model and are very useful in product design and development, such as aesthetic, warmth, abrasion, pitch, resilience, and friction modeling [Ashby02].

4.2.1 *Aesthetic modeling*

One interesting property of the material is its softness. Some products will need to be touched and thus should be designed to be soft. How would one specify a soft material? Material softness, S, can be defined as

$$S = EH \tag{4.1}$$

where
  H = hardness: resistance to indentation and scratching
  E = modulus of elasticity: stiffness of a material

If S is small, the material feels soft. As S increases, the material feels harder. This way "softness" can be defined so that the existing material properties can be used for material selection.

4.2.2 *Warmth modeling*

Sometimes one would like a product to feel warm, and thus, the material should be chosen so that it can let the users feel warm. The material feels "cold" to the touch if it conducts heat away from the finger quickly and feels warm if it resists the heat away from the finger. Assume that heat flows from the finger into a surface such that after time t, a depth x of material has been warmed significantly:

$$x = (at)^{1/2} \tag{4.2}$$

and thermal diffusivity of the material,

$$a = \lambda / \rho C_p \tag{4.3}$$

where
  λ = thermal conductivity
  ρ = material density
  C_p = specific heat

The quantity of heat that has left the area of the finger in time t is

$$Q = x\rho C_p = (\rho \lambda C_p t)^{1/2} \tag{4.4}$$

When Q is small, one feels the warmth of the product.

4.2.3 *Abrasion-resistant modeling*

Abrasion resistance is the ability of a material to withstand mechanical actions such as rubbing and scraping. Abrasion resistance is determined by the material hardness, H. When one material can scratch another material; this material is more abrasion resistant than the other.

4.2.4 Pitch modeling

How would one select a material for a product that requires special sound characteristics? Sound frequency, or pitch, can be heard when an object is struck. Pitch P is related to the modulus of elasticity E and density ρ:

$$P = (E/\rho)^{1/2} \tag{4.5}$$

If P is small, the material's pitch is low. If P increases, the material's pitch becomes higher.

4.2.5 Sound absorption modeling

The proportion of sound absorbed by a surface is called the sound absorption coefficient. The material of 0.8 in sound absorption coefficient will absorb 80% of the sound that hits it, reflecting 20%. Table 4.2 shows the sound absorption coefficients for various materials. It is sometimes important to select a material that can absorb noise in the environment.

Table 4.2 Sound absorption coefficients

Material	At 500–4,000 Hz
Glazed tiles	0.01–0.02
Rough concrete	0.02–0.04
Wood	0.15–0.80
Cork tiles	0.20–0.55
Thick carpet	0.30–0.80
Expanded polystyrene	0.35–0.55
Acoustic spray plaster	0.50–0.60
Glass wool	0.50–0.99

A related topic is how loud is loud? It has been studied that a sound level of above 50 dB can impair concentration, and a sustained sound level above 90 dB can cause damage to hearing. Figure 4.7 shows various sound levels for various activities. Different environments may have different requirements for sound level. For example, to design a quiet car on the highway, it may be necessary to determine how loud the outside noise will be in the car when driving on the highway with windows and doors closed. When the car is running on the road, there are all kinds of noises that can come in, for example, the engine, tires, wind, and other cars. The car interior needs to absorb the noise, and the body should prevent the noise from coming in. In order to feel comfortable, the interior noise should be below 30 dB.

Porous absorbers can be used to retain noise within the engine compartment. The porous absorbers, such as mineral wool, fiberboard, or plastic foams, have an open pore structure. Heat is produced by friction when vibrating air molecules are forced through the pores and interact with the pore walls. These are effective primarily for high frequencies with short wavelengths.

Example 4.1

Assume that a surround sound stereo at a movie theater puts out a sound that is measured at 70 dB. Assume the sound bounces back once. What materials would be needed for sound absorption if an acceptable reflected sound is (A) normal conversation, (B) office background noise, and (C) a whisper?

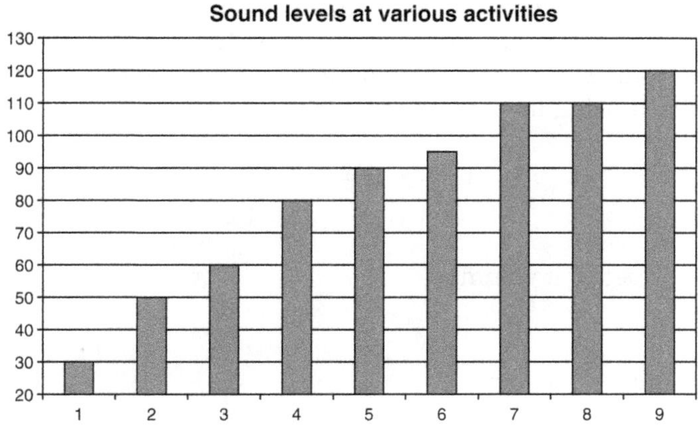

	Threshold of hearing	0
1	Leaves rustling, soft music, whisper	30
2	Office background noise	50
3	Normal conversation	60
4	Road traffic	80
5,6	Subway, shouted conversation	90–95
7	Chain saw, leaf blower	110
8	Pneumatic drill at 1 m	110
9	Jet takeoff at 100 m	120

Figure 4.7 Sound levels at various activities.

Answer:

If the sound that is coming from the first theater is 70 dB, it is 10,000,000 times louder than the threshold of hearing. Normal conversation is 60 dB or 1,000,000 times louder than the threshold of hearing. Office background noise is 50 dB or 100,000 times louder than the threshold of hearing. A whisper is 30 dB or 1,000 times louder than the threshold of hearing.

A. 60 dB/70 dB = 1,000,000/10,000,000 = 0.1% or 10% sound reflected back. This means that something that absorbs 90% of the original sound is needed. There may be a thick carpet out there that will get rather close, but it looks as though glass wool is the best option.
B. 50 dB/70 dB = 100,000/10,000,000 = 0.01% or 1% sound reflected back. This means something that absorbs 99% of the original sound is needed. Glass wool is the only option for this case.
C. 30 dB/70 dB = 1,000/10,000,000 = 0.0001% or 0.01% sound reflected back. This means one needs something that absorbs 99.99% of the original sound. There is no material listed that can achieve this.

This example shows how hard it is to get good sound insulation. One may think that one would come up with a few different materials, but that was not the case.

4.2.6 Resilience modeling

Resilience is the ability to accept large deflection without damage. The ability of a material to absorb energy when deformed elastically and to return to its original shape when unloaded is called resilience. This is usually measured by the modulus of resilience, which is the strain energy per unit volume required to stress the material from zero stress to the yield stress σ. The strain energy per unit volume or modulus of resilience, for uniaxial tension, can be calculated using the following formula:

$$U_r = \tfrac{1}{2}\sigma\varepsilon = \sigma^2/(2E) \tag{4.6}$$

where
U_r = strain energy per unit volume
σ = yield stress
ε = strain
E = Young's modulus

When U_r is large, the material has good resilience. Materials with high resilience can withstand higher amounts of deformation before yielding than materials with low resilience. This is important in many applications such as springs.

4.2.7 Friction modeling

Material friction could be one of the limiting factors of system bandwidth but could also be the major contributor to a system. For example, the road/tire interaction for ground vehicles makes automobiles and other systems work. For simplicity, friction models can be classified into static friction and dynamic friction. The idea of static friction is that friction force opposes the direction of motion when the sliding velocity is zero as shown in Figure 4.8. In this model, the static friction force is equal to the tensile forces until a maximum or minimum is reached:

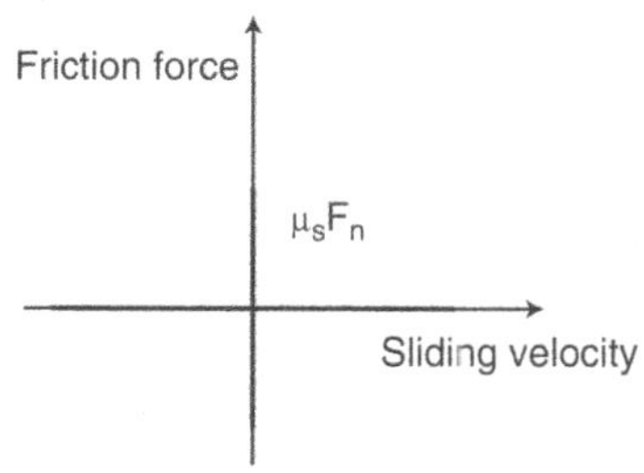

Figure 4.8 Static friction model.

$$F_{fmax} = \mu_s F_n \tag{4.7}$$

$$F_{fmin} = -\mu_s F_n \tag{4.8}$$

where
μ_s = static friction coefficient
F_n = normal force acting on the object

In dynamic friction or Coulomb friction, the friction force is proportional to load, opposes the direction of motion, and is independent of the contact area as shown in Figure 4.9. The friction force can be described as

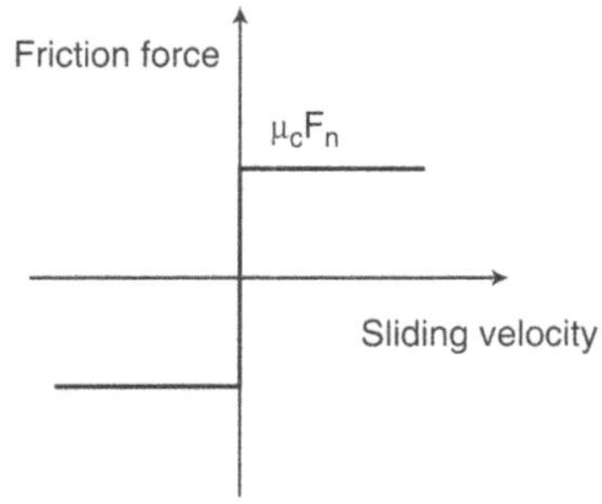

Figure 4.9 Dynamic friction model.

$$F_f = F_n - \mu_c \mathrm{sign}(v) \tag{4.9}$$

where
μ_c = Coulomb friction coefficient
v = tangential velocity between the two objects while sign (v) is taking the sign of object velocity v

When there is fluid between two objects, viscous friction as shown in Figure 4.10 is more effective in modeling the phenomenon. The viscous friction force can be described as

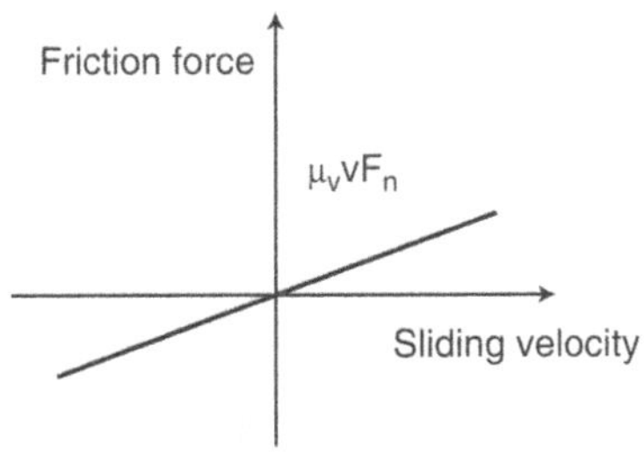

Figure 4.10 Viscous friction model.

$$F_f = \mu_v F_n v \tag{4.10}$$

where μ_v is the viscous friction coefficient.

Friction has a lot of applications in engineering. For example, the friction between the tires of the automobile and the road determines the maximum acceleration and the minimum stopping distance. On dry surfaces, a car might get as high as 0.9 as the coefficient of friction, but on wet roads, it would be dangerous since the wet road coefficient might be as low as 0.1. The design of the tire tread to maximize the friction in wet condition is thus critical to the safety of a driver.

4.2.8 Thermal deformation

The model for object deformation (shrinking or expansion) due to thermal effect is

$$\delta = \alpha \Delta T L \tag{4.11}$$

where

α = coefficient of expansion

ΔT = temperature difference

L = length of the object

Example 4.2

If a 5 ft brass rod is initially at a temperature of 0°F and experiences a temperature increase to a final temperature of 100°F, what is the resultant change in length of the steel? Brass coefficient of expansion is $20 \times 10^{-6}/°C = 11 \times 10^{-6}/°F$.

Answer:

$$\text{Deformation } \delta = \alpha \Delta TL = 11 \times 10^{-6}/°F(100°F–0°F)(60 \text{ in.}) = 0.066 \text{ in.}$$

The thermal deformation often may not have a significant effect on the structure itself. However, if the structure or members of the structure are constrained such that the thermal expansion cannot occur, then significant thermal stress may arise which can affect the structure substantially. This could play a critical role in developing a prototype of a component or a structure that will experience significant and drastic temperature change over a short period. In some applications, the deformation could cause problems in areas in which structure accuracy is critical, and thus, the level of thermal deformation needs to be estimated.

4.2.9 Ductility

Ductility is more commonly defined as the ability of a material to deform easily upon the application of a tensile force or as the ability of a material to withstand plastic deformation without rupture. The percent elongation reported in a tensile test is defined as the maximum elongation of the gage length divided by the original gage length.

$$\begin{aligned} \text{Percent elongation} &= \frac{\text{Final gage length} - \text{Initial gage length}}{\text{Initial gage length}} \\ &= \frac{L_x - L_0}{L_0} = \text{Inches per inch} \times 100\% \end{aligned} \tag{4.12}$$

Reduction of area is the proportional reduction of the cross-sectional area of a tensile test piece at the plane of fracture measured after fracture.

$$\begin{aligned} \text{Percent reduction in area (RA)} &= \frac{\text{Original cross-sectional area} - \text{Minimum final area}}{\text{Original cross-sectional area}} \\ &= \frac{A_0 - A_{min}}{A_0} \times 100\% \\ &= \frac{\text{Decreased area}}{\text{Original area}} \times 100\% \end{aligned} \tag{4.13}$$

These two parameters are often used to represent the ductility of the material. Ductility is one of the most critical factors affecting the building performance. Thus, earthquake-resistant design strives to predetermine the locations where the damage takes place and then to provide good detailing at these locations to ensure ductile behavior of the building.

Review problems

1. Resilience is defined as the capacity of a material to absorb energy when it is deformed elastically and then, upon unloading to have this energy recovered. Which of the following materials has the highest modulus of resilience?

Modulus of resilience for various materials [Keytosteel07]

Material	E (psi)	Yield strength, σ_y (psi)	Modulus of resilience, U_r
Medium-carbon steel	30×10^6	45,000	
High-carbon spring steel	30×10^6	140,000	
Duraluminum	10.5×10^6	18,000	
Copper	16×10^6	4,000	
Rubber	150	300	
Acrylic polymer	0.5×10^6	2,000	

2. Absorption or absorptivity should be considered when the temperature increases of a part when light, such as laser light, irradiates on it. The part should be made of low absorption material if a low-temperature increase is needed. Research and find out how to define absorption or absorptivity. Then find some approximate values of absorption for common materials, such as aluminum alloy and stainless steel.
3. In order to test the strength of the material, we always apply the normal tensile test to acquire its yield strength or tensile strength. However, in many brittle materials, the normal tensile test cannot be performed easily because of the presence of flaws at the surface. Often, just placing a brittle material in the grips of the tensile testing machine will cause cracking. These kinds of brittle materials may be tested using the bend test. By applying the load at three points and causing bending, a tensile force acts on the material opposite the midpoint. Fracture begins at this location. The flexural strength, or modulus of rupture, describes the material's strength. Research and find the model for flexural strength for the three-point bend test.
4. Research and find heat transfer rate of a material under convection.
5. Describe an example that utilizes a graph to model a material behavior so that it may be used in a design.
6. Describe the three different models of friction.
7. By how many times is the sound of normal conversation louder than the threshold of hearing?
8. List and describe three useful material properties that are easy to model for product prototyping. Also, mathematically quantify each property.

4.3 Modeling and design of materials and structures

Rising raw material and energy prices have continued to push up costs and hit profits.

John Cridland

This section helps answer the following questions:

- Is there a way to break the material selection problem down to the basic equations?
- Is there a systematic procedure that can assist engineers in determining material selection?
- Is there a set method that considers both material properties and cost when creating a prototype?

Material selection and structure design often are among the first things to consider in a design. For example, how would one go about prototyping a new bicycle frame? Well, it is possible that it is required to choose a material that is stiff, lightweight, and inexpensive. To do this, it is necessary to find information about Young's modulus, density, strength, and the cost of many different materials. It is unlikely that the cheapest material will also be the lightest and stiffest. Thus, it may be necessary to make some judgments to determine which is the best choice. It will be considerably easier by only choosing from the generic materials; it is a lot easier to decide whether steels are better than aluminum alloys for the bike frame than to choose among the thousands of materials that these generics represent. Since "mild steel" is not a real material, what is its yield strength? The answer is that it will not have a single value, but it will have a range of values. This is because it must represent the yield strength for different types of mild steel, each of which will have its own value.

This section focuses on using the key indexes for material selection in structure design. Material selection for a structure design often is a challenging decision-making process and is sometimes time-consuming as well. This section summarizes a method that demonstrates how to break the structure–material coupling problem down into basic equations. Instead of doing full structural analysis on a part during the material selection phase, it is much easier and, therefore much faster, to utilize only the part of the structural equations that has reference to the requirement [Ashby99].

As shown in Figure 4.11, it is required to design a cylindrical tie rod to carry a tensile force F without failure. To develop equations for material selection, the total mass of the rod can be calculated as

Figure 4.11 A cylindrical tie rod to carry a tensile force F.

$$m = AL\rho \tag{4.14}$$

where

A = cross-sectional area of the rod
L = length of the rod
ρ = material density of the rod

$$A = m/L\rho \tag{4.15}$$

The stress inside the rod should be less than the failure strength:

$$F/A \leq \sigma_f \tag{4.16}$$

where

F = external axial force

σ_f = failure strength

Substituting Equation 4.15 into Equation 4.16, one can obtain

$$FL(\rho/m) \le \sigma_f \text{ or } FL(\rho/\sigma_f) \le m \tag{4.17}$$

Assuming the external force F and rod length L to be the same, the lightest tie to carry F safely is the material with the smallest ρ/σ_f, which is the density-failure strength ratio of a material. In other words, for this particular case, the index ρ/σ_f can be used to select the appropriate material.

4.3.1 *Cost of unit strength*

The case study discussed in Section 4.3 illustrates a procedure to derive a structure problem into an index for material selection. The idea is to use the same structure, under the same load, but select from different materials for the associated structural cross-sectional areas. One approach is called the cost of unit strength, when the structure is designed for strength. For different materials, the associated cross-sectional area will change, the total weight of the structure will change, and thus, the total cost of the material will change.

Considering the yielding of a bar in uniaxial tension as shown in Figure 4.12, the working stress is

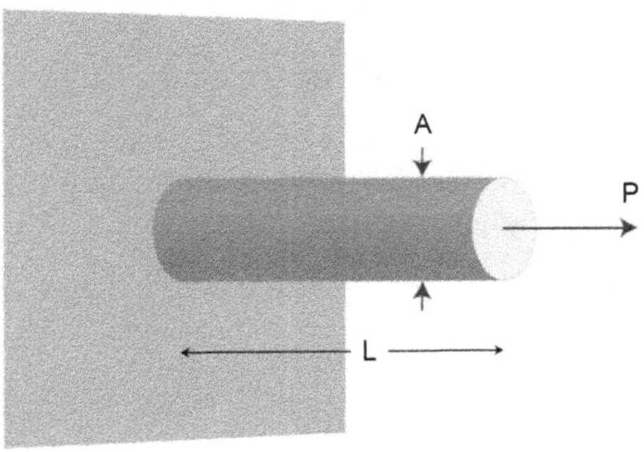

material	cross section	diameter	weight	density
A	A_A	D_A	W_A	ρ_A
B	A_B	D_B	W_B	ρ_B

Figure 4.12 An example of cost of unit strength.

$$\sigma_w = \sigma_{ys}/(\text{Factor of safety}) = P/A \tag{4.18}$$

where

P = axial load

A = cross-sectional area of the structure

Assume equal loading in either material A or material B,

$$P = A_A\sigma_A = A_3\sigma_B \tag{4.19}$$

or

$$\pi D_A{}^2\sigma_A/4 = \pi D_B{}^2\sigma_B/4 \tag{4.20}$$

thus,

$$\sigma_w = \sigma_{ys}/(\text{Factor of safety}) = P/A \tag{4.21}$$

$$\text{The weight of the bar } W = \rho AL = \rho L\pi D^2/4 \tag{4.22}$$

Substituting Equation 4.21 into Equation 4.22, one can obtain

$$\frac{W_A}{W_B} = \frac{\rho_A L\pi(D_A{}^2/4)}{\rho_B L\pi(D_B{}^2/4)} = \frac{\rho_A D_A{}^2}{\rho_B D_B{}^2} = \frac{\rho_A\sigma_B}{\rho_B\sigma_A} \tag{4.23}$$

In other words, the total weight is proportional to ρ/σ = density/strength:

$$W \sim \rho/\sigma \tag{4.24}$$

For this particular case, the index ρ/σ can be used to select the appropriate material. To use the material property chart as shown in Figure 4.13, sometimes it is easier to use the inverse of Equation 4.24 for material selection:

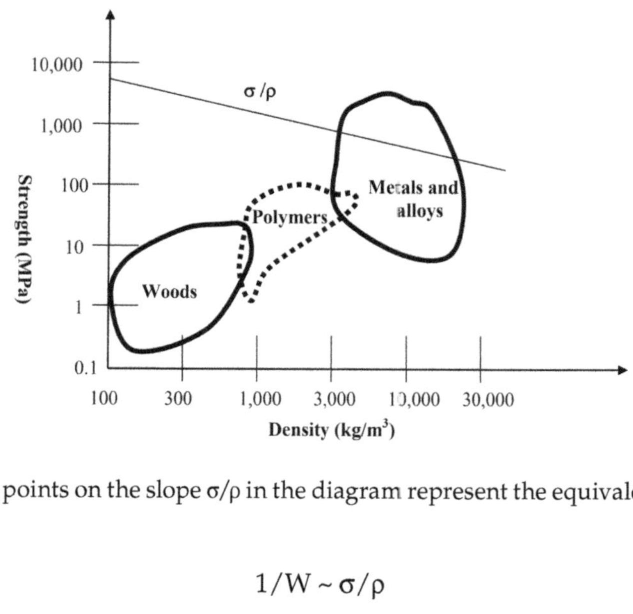

Figure 4.13 All the points on the slope σ/ρ in the diagram represent the equivalent effect of material selection.

$$1/W \sim \sigma/\rho \tag{4.25}$$

However, it is important to remember that 1/W means the inverse effect of W. As shown in Figure 4.13, all the points on a particular slope σ/ρ in the diagram represent the equivalent effect of material selection. The slope line divides the map into two regions, one region represents materials which will have better σ/ρ performance, and the other represents materials which lack σ/ρ performance.

If one is concerned about the total cost of the material, TC, then from Equation 4.23, the following equations can be defined:

$$\frac{TC_A}{TC_B} = \frac{W_A C_A}{W_B C_B} = \frac{\rho_A \sigma_B C_A}{\rho_B \sigma_A C_B} \tag{4.26}$$

where C_A and C_B are the costs per unit weight for materials A and B, respectively.

In other words, T_C is proportional to the index $\rho C/\sigma$. There are associated charts for material selection.

4.3.2 *Cost of unit stiffness*

The case study discussed in Section 4.3.1 illustrates a procedure to derive a structure problem into an index for material selection for strength, and here is an example of stiffness or when the structure must resist deformation.

Considering the deformation of a bar in bending in Figure 4.14, and assuming b, P, and L are unchanged, the deformation

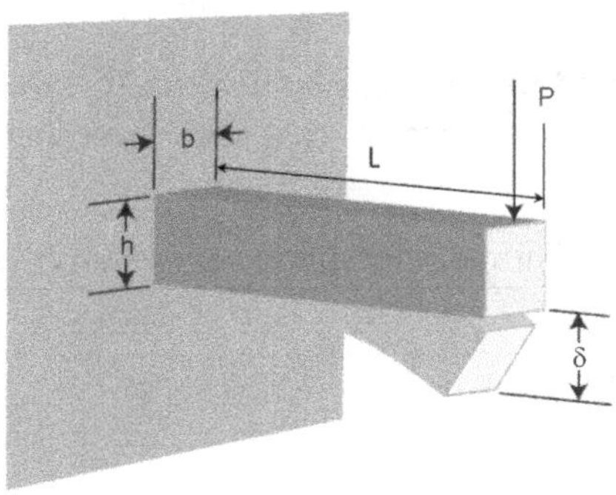

material	Young's modulus	height	weight	density
A	E_A	h_A	W_A	ρ_A
B	E_B	h_B	W_B	ρ_B

Figure 4.14 An example of cost of unit stiffness.

$$\delta = PL^3/3EI \tag{4.27}$$

where

P = bending force

E and I = elastic modulus and moment of inertia of the cross-sectional area of the structure, respectively

$$I = bh^3/12 \tag{4.28}$$

Assume $b_A = b_B$, and assume equal deformation in either material A or B:

$$\delta = PL^3/3E_AI_A = PL^3/3E_BI_B \tag{4.29}$$

It can be found that

$$E_AI_A = E_BI_B \tag{4.30}$$

thus,

$$E_A/E_B = (h_B/h_A)^3 \tag{4.31}$$

The weight ratio

$$W_A/W_B = \rho_A b_A h_A/\rho_B b_B h_B = \rho_A h_A/\rho_B h_B = (\rho_A/\rho_B)(E_B/E_A)^{1/3} \tag{4.32}$$

In other words, the total weight is proportional to ρ/E = density/modulus:

$$W \sim \rho/E \tag{4.33}$$

Similar indexes can be derived using a similar approach. Tables 4.3 and 4.4 show the formulas for minimum-weight and minimum-cost criteria, respectively.

Table 4.3 Formulae for minimum-weight criterion

Components	Strength	Stiffness
Solid cylinder in tension or compression	$\frac{\sigma}{\rho}$	$\frac{E}{\rho}$
Solid cylinder in bending	$\frac{\sigma^{2/3}}{\rho}$	$\frac{E^{1/2}}{\rho}$
Solid cylinder in torsion	$\frac{\tau^{2/3}}{\rho}$	$\frac{G^{1/2}}{\rho}$
Solid rectangular beam in bending	$\frac{\sigma^{1/2}}{\rho}$	$\frac{E^{1/3}}{\rho}$
Thin-walled pressure vessels under internal pressure	$\frac{\sigma}{\rho}$	$\frac{E}{\rho}$
Thin wall shaft in torsion	$\frac{\tau}{\rho}$	$\frac{G}{\rho}$

Table 4.4 Formulae for minimum-cost criterion

Components	Strength	Stiffness
Solid cylinder in tension or compression	$C\frac{\rho}{\sigma}$	$C\frac{\rho}{E}$
Solid cylinder in bending	$C\frac{\rho}{\sigma^{2/3}}$	$C\frac{\rho}{E^{1/2}}$
Solid cylinder in torsion	$C\frac{\rho}{\tau^{2/3}}$	$C\frac{\rho}{G^{1/2}}$
Solid rectangular beam in bending	$C\frac{\rho}{\sigma^{1/2}}$	$C\frac{\rho}{E^{1/3}}$
Thin-walled pressure vessels under internal pressure	$C\frac{\rho}{\sigma}$	$C\frac{\rho}{E}$
Thin wall shaft in torsion	$C\frac{\rho}{\tau}$	$C\frac{\rho}{G}$

These equations will be useful when working with Ashby charts for material selection. One can observe that the key parameters are very similar for various cases, and thus, often the design of a very complex structure can be simplified into some simple parameters in material selection.

Example 4.3

A solid cylindrical bar is in bending with a force P and deformation d. Select a material that minimizes the weight of the bar for the same deformation. Use the material properties given in Table 4.5.

Table 4.5 Example Material Specifications

Material	σ (ksi)	E (GN/m^2)	ρ (Mg/m^3)	Cost ($/ton)
A	50	100	7.8	250
B	30	20	6.5	400
C	40	50	2.5	200

Answer:

For a solid cylindrical bar in bending and concerning stiffness and weight, the following equation should be used:

$$W_A/W_B = (\rho_A/\rho_B)(E_B/E_A)^{1/2}$$

In other words, the weight $W \sim \rho/E^{1/2}$

One can find

$$W_A : W_B : W_C = 7.8/10 : 6.5/4.47 : 2.5/7.07 = 0.78 : 1.45 : 0.35$$

Therefore, the weight would be minimized with material C when designing for deflection. Thus, material C should be selected.

Review problems

1. We want to manufacture a screw to withstand a load of 15,000 N and have a diameter of 10 mm using a factor of safety of 2. What kind of material should be used to manufacture it? The working condition for the screw is shown in Figure 4.15.

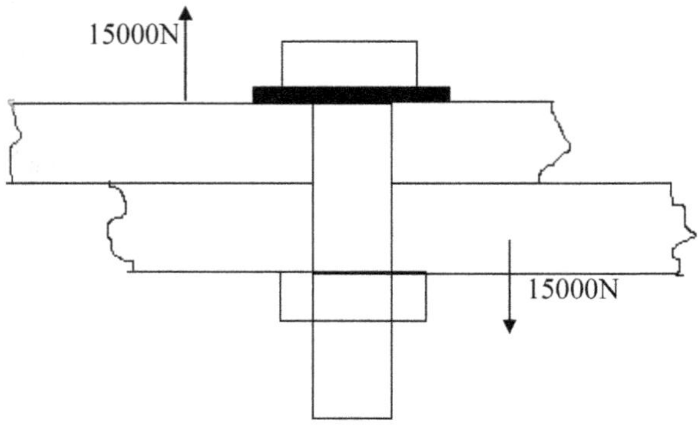

Figure 4.15 To design a screw.

2. Connect the mechanical properties to their proper definition.

Hardness	Stress required to fracture the material
Ductility	Ratio between the lateral and longitudinal strains in the elastic region
Flexural strength	Resistance to indentation and scratching
Poisson's ratio	Stress that corresponds to the maximum load in a tensile test
Tensile strength	Stress that causes a specific amount of permanent strain for the material
Yield strength	Ability to be permanently deformed without breaking when a force is applied

3. Match the following material selection charts listed in Table 4.6 with the material selection criteria listed in Table 4.7.

Table 4.6 Material selection charts

1. Strength–density
2. Young's modulus density
3. Young's modulus–cost
4. Strength–toughness
5. Strength–elongation
6. Strength–maximum service temperature
7. Stiffness–specific strength
8. Resistivity–cost
9. Recycle fraction–cost
10. Energy content–cost

Table 4.7 Material selection criteria

a. Selecting cheap materials that are also stiff
b. Selecting materials that can carry a high load before permanently deforming and can "stretch" a lot before breaking
c. Selecting a material to make a component of low weight while still being stiff enough
d. Selecting materials that are not only strong (carry a high static load before deforming) but are also tough (can withstand impact loads)
e. Selecting strong materials that are also lightweight
f. Selecting cheap conductors and insulators
g. Making trade-off between environmental concerns over excessive waste disposal and the cost penalty
h. Making trade-off between environmental concerns over excess energy use and the cost penalty
i. Selecting light materials that are both strong and stiff
j. Selecting materials that work at high temperatures

4. Find from the materials handbook or Internet the strength–density chart, and the Young's modulus–density chart for material selection, and compare titanium with stainless steel, copper, brass, nickel, and zinc.
5. List the common criteria for selection of materials.
6. Selection of a material for an oar: Mechanically speaking, an oar is a beam loaded in bending. It must be strong enough to carry the bending moment exerted by the oarsman without breaking; it must have just the right stiffness to match the rower's own characteristics and give the right feel, and—very important—it must be as light as possible. Oars are designed on stiffness, that is, to give a specified elastic deflection under a given load. Design the material for an oar.
7. Noise reduction is a major consideration in the design of motor vehicles. State a factor or factors that can be controlled or selected to reduce the noise levels in vehicles.
8. The drive shaft on a typical rear wheel drive vehicle must transfer the power of the motor to the rear wheels. List the criteria for the material of the drive shaft. Which of these criteria may be neglected?
9. Using the criteria stated in the previous problem, use Ashby's material selection charts to pick a suitable material for this application. Explain how each chart was used to arrive at your decision.
10. Using the solution from the previous problem, determine what type of metal would be best suited for this application (i.e., aluminum, copper, 1020 steel, etc.). This may require a Web or library search to find the necessary Ashby charts to find a solution.
11. Choose the suitable magnitude for the relevant sound source.

Sound source	Choice letter	dB
Leaves rustling, soft music, whisper	A	120
Normal conversation	B	30
Road traffic	C	80
Chain saw, leaf blower	D	60
Jet takeoff at 100 m	E	110

12. From the Internet, find a material selection chart of Young's modulus versus cost. Justify the use of wood for building construction.
13. How would you go about prototyping a new skateboard deck?

References

[Ashby99] Ashby, M., *Materials Selection in Mechanical Design*, 2nd edition, Butterworth-Heinemann, Waltham, MA, September 1999.

[Ashby02] Ashby, M. and J. Kara, *Materials and Design: The Art and Science of Material Selection in Product Design*, 1st edition, Butterworth-Heinemann, Waltham, MA, December 2002.

[Dieter99] Dieter, G., *Engineering Design: A Materials and Processing Approach*, 3rd edition, McGraw-Hill Science/Engineering/Math, New York, August 11, 1999.

[Ertas96] Ertas, A. and C.J. Jesse, *The Engineering Design Process*, 2nd edition, John Wiley & Sons, Hoboken, NJ, 1996.

[Keytosteel07] Resilience, Retrieved on May 22, 2007 from www.keytosteel.com/Articles/Art41.htm.

[MatWeb17] MatWeb, Your source for materials information, Retrieved on August 28, 2017, from www.matweb.com.

[USMint17] The United States Mint, Retrieved on August 27, 2017, from www.usmint.gov.

chapter five

Direct digital prototyping and manufacturing

Digital manufacturing is the ability to describe every aspect of the product development process digitally using tools that include digital design, computer-aided design (CAD), office documents, product life cycle management systems, analysis software, simulation, and computer-aided manufacturing (CAM) software. The idea is that the passage of data from one department or discipline to another should be seamless so that the data created is immediately reusable in a different discipline. Direct digital manufacturing (DDM) technology fabricates functional components directly from computer models. Although DDM is generally referred to as an additive process, it includes both additive and subtractive processes in this book as long as a part is fabricated directly from an electronic digital representation. Such a process will lead to dramatic reductions in lead time and manufacturing costs for high-value, low-volume components such as gas turbine engine cases, complex unitized airframes, and other products built from expensive raw materials or requiring high-cost finishing operations. DDM can also be used as an effective prototyping tool.

The core DDM technologies involved in digital manufacturing are CAD and CAM. CAD is the use of computer programs and systems to design detailed two- or three-dimensional models of physical objects, such as mechanical parts, buildings, and molecules. Although it will take some time to model a part in a CAD system, with the current CAD/CAM technologies, it will eventually pay off in the product prototyping process, especially for complex prototypes. As shown in Figure 5.1, the current computer-integrated manufacturing system uses a common database which includes solid model to describe a product's geometry and the relationships among part

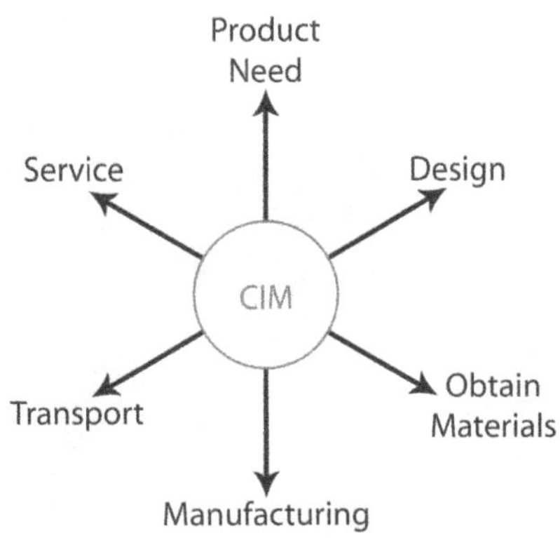

Figure 5.1 A CIM system can use a common database to drive various processes in the product prototyping stages, such as communication with customers to define product need, product design, purchasing of raw materials or part components from vendors, fabrication, packaging for transportation, and product maintenance and service.

components. The trend of current technologies is to use the single database to drive various processes in the product prototyping stages, such as communication with customers to define product need, product design, purchasing of raw materials or part components from vendors, fabrication, packaging for transportation, and product maintenance and service. This is the quickest and best way to build a prototype and eventually design the best product.

This chapter discusses how to use DDM technologies for effective product prototyping and manufacturing.

5.1 Solid models and prototype representation

This section helps answer the following questions:

- What is a solid model?
- Why is a solid model needed in digital manufacturing?
- What is a constructive solid geometry (CSG) model? What is a boundary representation (B-rep) model?
- How is a curved surface represented in a solid model?
- What is B-spline? What is nonuniform rational B-splines (NURBS)? How is a curved surface represented in a B-spline or NURBS?

The language of digital manufacturing and assembly is geometry. Digital manufacturing starts with being able to draw an idea into a computer. Drawing is the art of representing objects or forms on a surface chiefly by means of lines. A part drawing can be used as a communication tool for many tasks in prototype development, including virtual simulation and analysis, such as finite element analysis, flow analysis, and CAD/CAM integration. To have a seamless integration of various prototyping development tasks, a 2D drawing will not do. Solid models are unambiguous representations of the solid parts of an object suitable for computer processing. Other modeling methods include wireframe models, which can be ambiguous about solid volume as shown in Figure 5.2, and surface models that are used extensively in animation as shown in Figure 5.3.

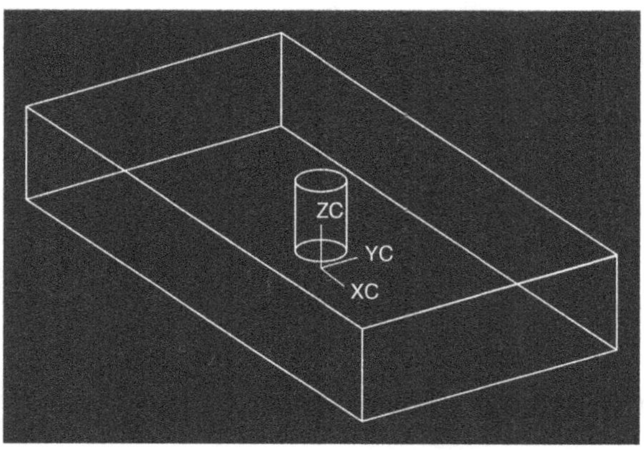

Figure 5.2 A wireframe model of a block with a hole.

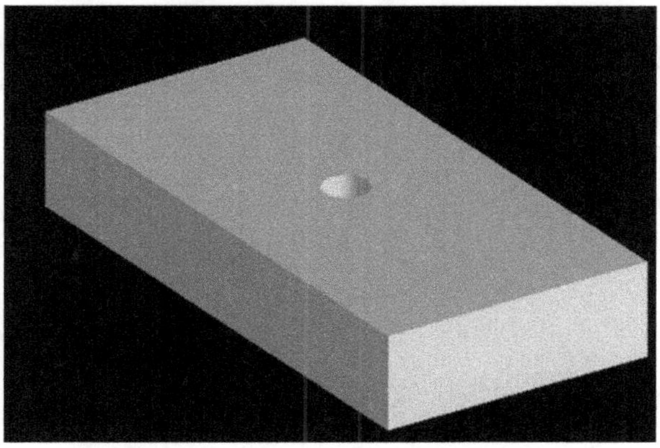

Figure 5.3 A surface shaded model.

5.1.1 Solid modeling

Solid modeling may be the best way for CAD/CAM integration and applications since it can define solid parts well. B-rep represents a solid object by boundary surfaces which are then filled to represent a solid. An example of the previous simple block with a hole can be expressed in the B-rep solid model as shown in Figure 5.4. Because B-rep includes such topological information, a solid is represented as a closed space in 3D space. The boundary of a solid separates the points inside from points outside the solid. B-rep models can represent a wide class of objects, but the data structure is complex and requires a large memory space. It is interesting that this process is analogous to various manufacturing techniques such as injection molding, casting, forging, and thermoforming, in which the boundary is first defined by the tools and then filled with materials.

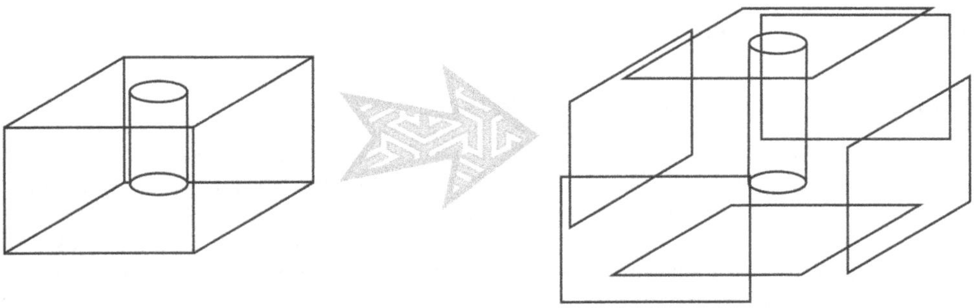

Figure 5.4 A B-rep solid model representation of the block with a hole example.

In CSG representation, an object is specified with reference to a library of simple primitives, such as blocks, cylinders, prisms, pyramids, spheres, and cones, as shown in Figure 5.5. Simple objects are combined using Boolean operators such as union, difference, and intersection. Figure 5.6 shows an example of CSG representation of a block with a hole. The set of allowable primitives may be restricted; for example, curved shapes may be forbidden. In the sweeping process, an area feature is "swept out" by moving a primitive along a path to form a solid feature. These volumes either add to the object called "extrusion" or remove material called "cutter path." It is analogous to various manufacturing techniques such as milling, lathing, and extrusion.

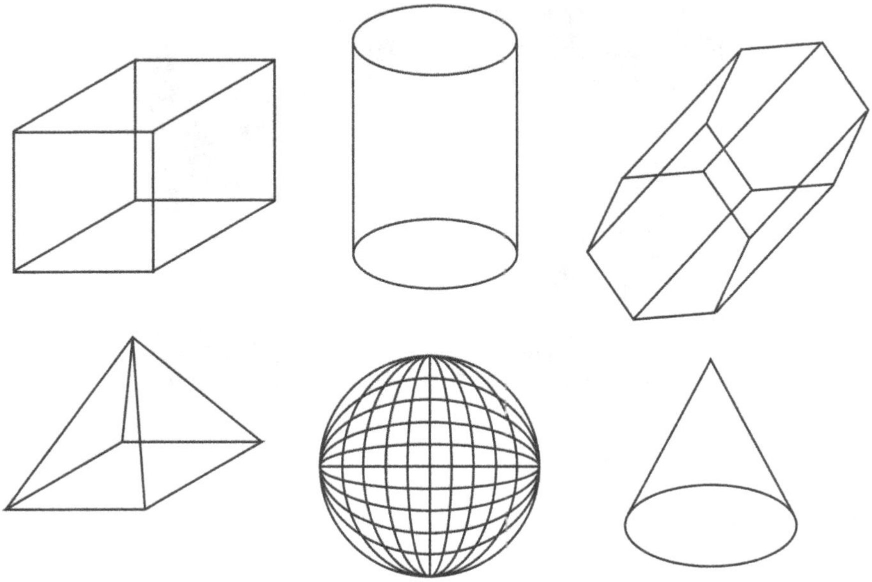

Figure 5.5 A library of commonly used primitives for CSG representation (blocks, cylinders, prisms, pyramids, spheres, cones).

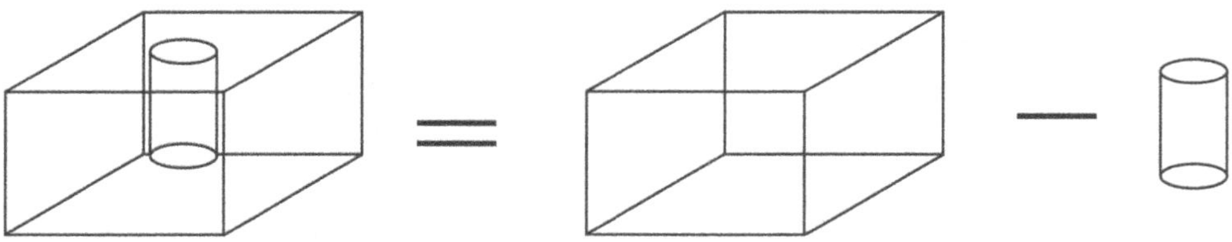

Figure 5.6 A CSG model representation of the block with a hole example.

Example 5.1: Creating a solid model in ACIS

ACIS is a 3D modeling engine. It is a generic graphics library and is used by software developers for applications in CAD/CAM/CAE. ACIS provides software developers and manufacturers with the underlying 3D modeling functionality. ACIS features an open, object-oriented C++ architecture that enables robust, 3D modeling capabilities.

To implement the example shown in Figure 5.6, the following six operations can be performed:

```
Create a block: api_make_cuboid(150,75,25.4,block)
Create a cylinder: api_make_frustum(19.05,12.7,12.7,12.7,cylinder)
Position the block: api_apply_transf(block,translate_
transf(vector(0,0,12.7))) Position the cylinder: api_apply_
transf(cylinder,translate_transf(vector(0,0,6.35))) Boolean subtraction:
api_subtract(cylinder,block)
Save SAT file: save_ent("Block_Example.sat", block)
```

Alternatively, in a C++ program, one can create the solid model with the following statements:

```
// Block_Example.cxx - ACIS 6.0 - 4/15/01 08:30
// This program creates a block and a cylinder,
// subtracts the cylinder from the block
```

```
#include <stdio.h>
#include "kernel/acis.hxx" //Declares system wide information
#include "constrct/kernapi/api/cstrapi.hxx" //Declares constructor
#include "boolean/kernapi/api/boolapi.hxx" //Declares Boolean API's
#include "kernel/kernapi/api/kernapi.hxx" //Declares kernel API's
#include "kernel/kerndata/top/alltop.hxx" //Declares topology classes
#include "kernel/kerndata/data/debug.hxx" //Declares debug routines
#include "baseutil/vector/transf.hxx" //Declares transform classes
#include "kernel/kerndata/lists/lists.hxx" //Declares ENTITY_LIST class
void save_ent(char*,ENTITY*); //Function Prototype
main()
{
api_start_modeller(0); api_initialize_constructors(); api_initialize_
booleans(); BODY* block,* cylinder;
api_make_cuboid(150,75,25.4,block); api_make_frustum(19.05,12.7,12.7,12.
7,cylinder); api_apply_transf(block,translate_transf(vector(0,0,12.7)));
api_apply_transf(cylinder,translate_transf(vector(0,0,6.35)));
api_subtract(cylinder,block);
save_ent("me6659.sat",block); api_terminate_booleans(); api_terminate_
constructors(); api_stop_modeller();
}
void save_ent(char *filename,ENTITY*ent)
{
FILE*fp = fopen(filename, "wr");
if(fp! = NULL){
ENTITY_LIST*savelist = new ENTITY_LIST; Savelist- > add(ent);
api_save_entity_list(fp,TRUE,*savelist); //TRUE for textfile, FALSE for
binary delete savelist;
}
else
printf("Unable to open file!\n"); fclose(fp);
}
```

Clearly, these operations are CSG-like operations, because the model is described as combinations of simpler solids (primitives) in a series of Boolean operations. However, ACIS is actually a B-rep-based technology as shown in Figure 5.7. CSG operations are provided for convenience in human–computer interactions.

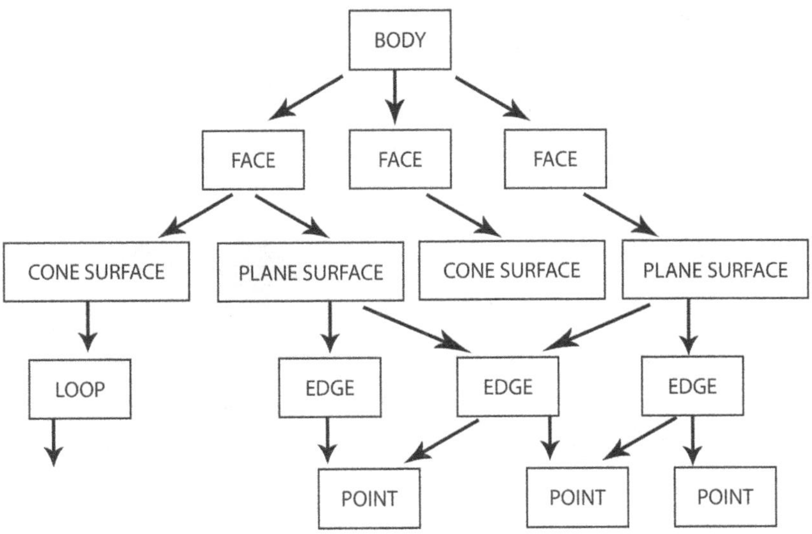

Figure 5.7 The topology of the ACIS SAT file.

5.1.2 CAD data representation

Current CAD tools may include examples as shown in the program above as well as other tools for modeling convenience. Solid modeling software could be used to model solid parts and model assemblies of parts. The immediate use of a solid model is to calculate mass properties of parts and assemblies and reflects the "bill of materials" required to build the product. The solid model can also directly create engineering drawings from the solid models and can also help visualization with shading, rotating, hidden line removal, etc.

The understanding of data representation of a solid model is very important in CAD/CAM integration activities. For example, how is a line represented and stored in the CAD database? It is not possible to use an infinite number of points, but two points should be able to represent a line. Therefore, the two points representing coordinate data, A and B, are stored in the database to represent the line. However, equations, such as the following, will need to be used to show the point P(t):

$$P(t) = A + t(B - A) = (1 - t)A + tB \tag{5.1}$$

where t is a variable.

Included in the database is an algorithm which contains the equation representing the geometry. There may be many different ways to represent the data.

One common question is how to represent free-form curves and surfaces in a CAD system. The free-form curves or surfaces cannot be represented by using the same methods used to represent the basic shapes, such as the CSG primitives discussed above. Monsieur Pierre Bézier was one of the pioneers who developed mathematical representations of free-form curves and surfaces. Some examples of mathematical representations of free-form curves and surfaces are those used in representing the shapes of car bodies and ship hulls. Splines, which were represented by control points lying on the curve itself, are known as Bézier splines. Today, most of the professional computer graphics applications available for desktop use offer NURBS technology.

Simple cubic curves are commonly used in graphics. Curves of lower order commonly have very little flexibility whereas curves of higher order, such as fourth order or above, are usually considered unnecessarily complex, as they may be difficult to solve. Since only second- or third-order polynomials are used, these curves or surfaces are not used individually. Many parametric surface patches are joined together side by side to form a more complicated shape as shown in Figure 5.8.

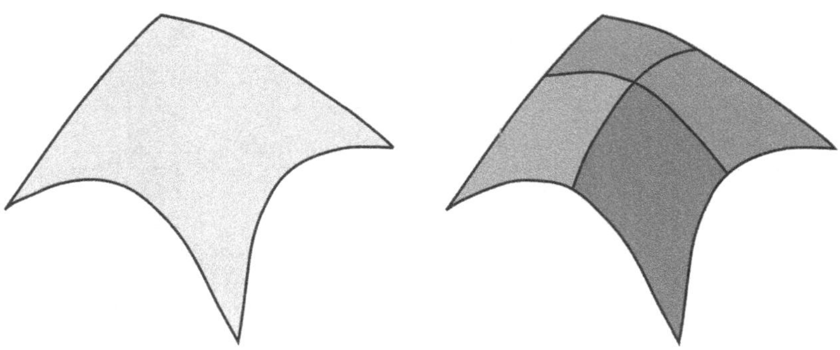

Figure 5.8 Multiple parametric surface patches are used to represent a free-form surface.

As the curves and surfaces are formed by multiple patches, there are several levels of continuity: C_0 curve continuity means simply connected, or when two curve segments join together at an end point. C_1 curve continuity can be recognized when two curve segments join together at an end point and the directions of the two segments' tangent vectors, or the first derivative, are equal at the join point. C_2 curve continuity is commonly referred to as geometric continuity that can be visually recognized as something "very smooth," or when two curve segments join together at an end point, and the two segments' first and second derivatives are equal at the join point.

To illustrate the concept, let us look at curves as they are easier to understand. A quadratic curve can be in the form of

$$P(t) = at^2 + bt + c \tag{5.2}$$

A cubic curve can be in the form of

$$P(t) = at^3 + bt^2 + ct + d \tag{5.3}$$

Both curves can be defined from a series of known points. These curves join through these same points. The cubic curve can also be expressed as

$$\begin{aligned} x(t) &= a_x t^3 + b_x t^2 + c_x t + d_x \\ y(t) &= a_y t^3 + b_y t^2 + c_y t + d_y \\ z(t) &= a_z t^3 + b_z t^2 + c_z t + d_z \\ &1 \geq t \geq 0 \end{aligned} \tag{5.4}$$

where a's, b's,…, d's are 12 shape coefficients. This curve can be defined by four sets of points, or control points. For example, to form a Bézier curve, control points are defined at locations $P(0) = P_0$, $P(1) = P_3$, $P'(0) = 3(P_1 - P_0)$, $P'(1) = 3(P_3 - P_2)$. Thus, by using these formulas, P(t) becomes

$$P(t) = (1-t)^3 P_0 + 3t(1-t)^2 P_1 + 3t^2(1-t)P_2 + t^3 P_3 \tag{5.5}$$

$$\begin{bmatrix} (1-t)^3 & 3t(1-t)^2 & 3t^2(1-t) & t^3 \end{bmatrix} \begin{pmatrix} P_0 \\ P_1 \\ P_2 \\ P_3 \end{pmatrix} = P(t) \tag{5.6}$$

$$\begin{bmatrix} (1-3t+3t^2-t^3) & (3t-6t^2+3t^3)^2 & 3t^2-3t^3 & t^3 \end{bmatrix} \begin{pmatrix} P_0 \\ P_1 \\ P_2 \\ P_3 \end{pmatrix} \tag{5.7}$$

or

$$\begin{pmatrix} t^3 & t^2 & t & 1 \end{pmatrix} \begin{pmatrix} -1 & 3 & -3 & 1 \\ 3 & -6 & 3 & 0 \\ -3 & 3 & 0 & 0 \\ 1 & 0 & 0 & 0 \end{pmatrix} \begin{pmatrix} P_0 \\ P_1 \\ P_2 \\ P_3 \end{pmatrix} = P(t) \tag{5.8}$$

Now, simply input the coordinates (x, y, z) in for P_0, P_1, P_2, and P_3, and the result will be a Bézier curve polynomial. Equation 5.6 can also be written as

$$\begin{bmatrix} B_1(t) & B_2(t) & B_3(t) & B_4(t) \end{bmatrix} \begin{pmatrix} P_0 \\ P_1 \\ P_2 \\ P_3 \end{pmatrix} = P(t) \tag{5.9}$$

where

$$\begin{aligned} B_1(t) &= (1-t)^3 \\ B_2(t) &= 3t(1-t)^2 \\ B_3(t) &= 3t^2(1-t) \\ B_4(t) &= t^3 \end{aligned} \tag{5.10}$$

B_1, B_2, B_3, and B_4 are also called blending function or shape function as shown in Figure 5.9. P_0, P_1, P_2, and P_3 are the control points. The blending functions and control points can define a Bézier curve.

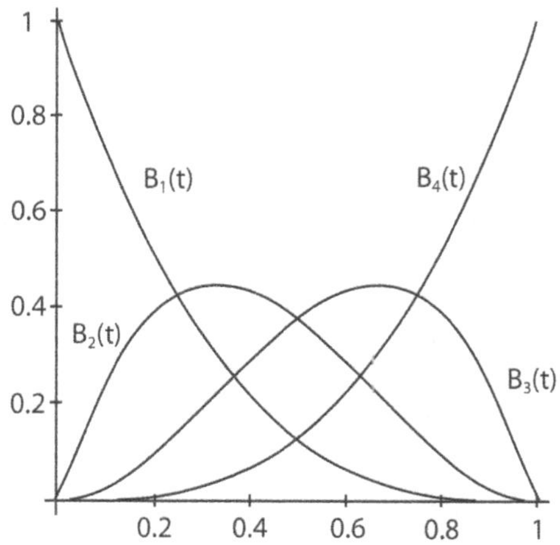

Figure 5.9 Blending function for a Bézier curve.

The blending function of a Bézier curve can also be written as

$$B(u) = \sum_{k=0}^{N} P_k \frac{N!}{k!(N-k)!} u^k (1-u)^{N-k} \quad \text{for } 0 \le u \le 1 \tag{5.11}$$

The corresponding properties of the Bézier curve apply to the Bézier surface. The function remains the same for both the curve and the surface as the formula is the same except that the surface is comprised of two parameters, u and v. The blending function of a Bézier surface can then be written as

$$B(u,v)=\sum_{i=0}^{N_i}\sum_{j=0}^{N_j}P_{ij}\frac{N_i!}{i!(N_i-i)!}u^i(1-u)^{N_i-i}\frac{N_j!}{j!(N_j-i)!}v^j(1-v)^{N_j-j}\quad 0\le u\le 1;0\le v\le 1 \tag{5.12}$$

where P_{ij} is the ith, jth control point. There are N_{i+1} and N_{j+1} control points in the i and j directions, respectively. The surface does not in general pass through the control points except for the corners of the control point grid. The surface is contained within the convex hull of the control points.

There are several different ways of representing curves (or surfaces). A B-spline curve is a piecewise polynomial curve, and the curve is not required to pass through the four control points. The curves are connected by overlapping their control points in C_2 continuity. In other words, the storage of cubic curves is just to store all control data points. A uniform cubic B-spline, or B-spline, is formed when the intervals or knots are equal. The advantages include higher-order continuity, more local control, and invariance under affine transformations.

The cubic form for a B-spline is

$$P(t)=\frac{1}{6}\begin{pmatrix} t^3 & t^2 & t & 1\end{pmatrix}\begin{pmatrix} -1 & 3 & -3 & 1\\ 3 & -6 & 3 & 0\\ -3 & 0 & 0 & 0\\ 1 & 4 & 1 & 0\end{pmatrix}\begin{pmatrix} P_0\\ P_1\\ P_2\\ P_3\end{pmatrix} \tag{5.13}$$

whereas the quadratic form is

$$P(t)=\frac{1}{2}\begin{pmatrix} t^2 & t & 1\end{pmatrix}\begin{pmatrix} 1 & -2 & 1\\ -2 & 2 & 0\\ 1 & 1 & 0\end{pmatrix}\begin{pmatrix} P_0\\ P_1\\ P_2\end{pmatrix} \tag{5.14}$$

Example 5.2: Use B-spline curves to represent a part of an ellipse

To illustrate further, to use the quadratic B-spline curves to represent a part of an ellipse, let us first use four control points to approximate an arc (arc on the same quadrant as point B) in quadrant 1 as shown in Figure 5.10 by a uniform quadratic B-spline. One can select A, B, C, and D as control points and use the quadratic form of B-spline.

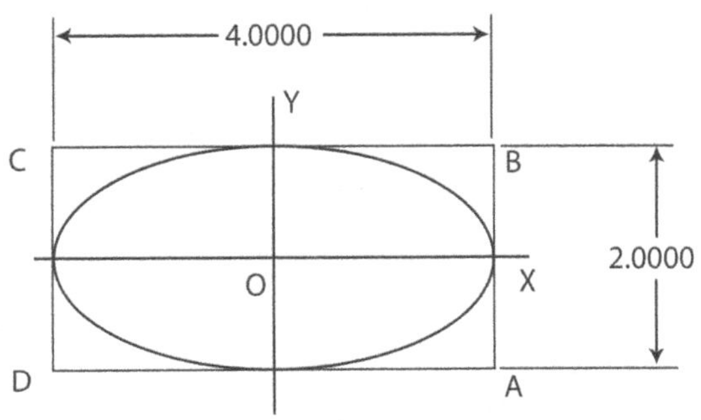

Figure 5.10 One can use quadratic B-spline curve to represent part of the ellipse in quadrant I. A = (2, –1), B = (2, 1), C = (–2, 1), and D = (–2, –1).

To approximate the arc in quadrant 1, one needs to make use of the two lines tangent to the arc, which are defined by points A, B, and C. In Equation 5.14, $P(t) = P_I(t)$, $P_0 = A$, $P_1 = B$, $P_2 = C$. Therefore, the circular arc can be represented by these three points using the quadratic B-spline curve $P_I(t)$:

$$P(t)=0.5\left(\begin{pmatrix} t^2 & t & 1 \end{pmatrix}\begin{pmatrix} 1 & -2 & 1 \\ -2 & 2 & 0 \\ 1 & 1 & 0 \end{pmatrix}\begin{pmatrix} A \\ B \\ C \end{pmatrix}\right)$$

$$P(t)=0.5\left(\begin{pmatrix} t^2 & t & 1 \end{pmatrix}\begin{pmatrix} 1 & -2 & 1 \\ -2 & 2 & 0 \\ 1 & 1 & 0 \end{pmatrix}\begin{pmatrix} 2 & -1 \\ 2 & 1 \\ -2 & 1 \end{pmatrix}\right)$$

$$x=-2t^2+2$$

$$y=-t^2+2t$$

$$t\in[0,1]$$

After synthesizing x(t) and y(t), one can find the approximation model:

$$1-\frac{x}{2}-2\sqrt{1-\frac{x}{2}}+y=0\ \left(0\le x\le 2, 0\le y\le 1\right)$$

For the second, third, and fourth quadrants, curve equations can be derived similarly.

Second quadrant (arc on the same quadrant as point C) using B, C, and D as control points:

$$x=2t^2-4t$$

$$y=-t^2+1$$

$$t\in[0,1]$$

Third quadrant (arc on the same quadrant as point D) using C, D, and A as control points:

$$x=2t^2-2$$

$$y=t^2-2t$$

$$t\in[0,1]$$

Fourth quadrant (arc on the same quadrant as point A) using D, A, and B as control points:

$$x=-2t^2+4t$$

$$y=t^2-1$$

$$t\in[0,1]$$

Since this quadratic B-spline curve is just an approximation of the original curve, there may be some errors at various locations. According to the symmetry, the error only needs to be calculated in the first quadrant. Check the start point and the end point of the approximation curve and the original curve:

When t = 0, $P_I(0) = (2, 0)$, and this point is exactly the start point of the original curve;
When t = 1, $P_I(1) = (0, 1)$, and this point is exactly the end point of the original curve.

5.1.2.1 *Error analysis*

Plot the original curve and approximation curve in the XY plane as shown in Figure 5.11:

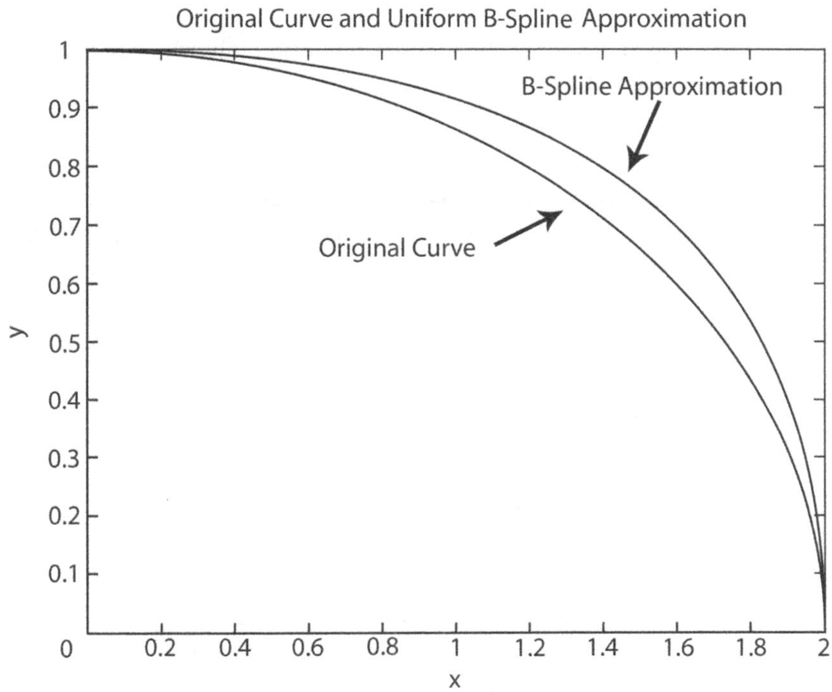

Figure 5.11 The error between the approximation and actual ellipse curves.

Given an ellipse in the XY plane, $\frac{x^2}{2^2}+\frac{y^2}{1^2}=1$,

Taking x as an independent variable, and y as a dependent variable, one can define:

Original curve equation: $y=\sqrt{1-\frac{x^2}{4}}$

Approximation curve equation: $y'=-1+\frac{x^2}{2}+2\sqrt{1-\frac{x}{2}}$

$$\varepsilon=|y'-y|=\left|-1+\frac{x}{2}+2\sqrt{1-\frac{x}{2}}-\sqrt{1-\frac{x^2}{4}}\right|$$

when x = 1.732, $\varepsilon_{max} = 0.0981$. So ε of y-value is varying between 0 and 0.0981.

NURBS has better curvature control and alleviation of roundoff errors in the storage of geometry, and the NURBS equation is

$$p(s,t) := \frac{\sum_{i=0}^{n}\left[\left(\sum_{j=0}^{m} w_{ij}N_{ik}(s)\cdot(N_{ji}(t))\cdot V_{ij}\right)\right]}{\sum_{i=0}^{n}\left[\left(\sum_{j=0}^{m} w_{ij}N_{ik}(s)\cdot(N_{ji}(t))\right)\right]} \tag{5.15}$$

when all $w_{ij} = 1$, it is a nonrational B-spline, and when knot values occur at ends and internal knots are equally spaced, it is uniform.

Review problems

1. Use four control points (a = 8 and b = 6) as shown in Figure 5.12, to approximate an ellipse by a uniform quadratic B-spline. Find its error when t = 0.5, when comparing to the known ellipse.

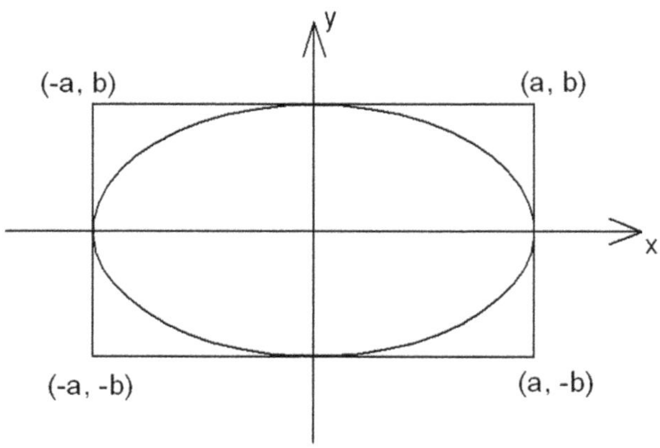

Figure 5.12 Use a uniform quadratic B-spline to approximate an ellipse.

2. Use uniform quadratic B-spline to approximate a cosine curve: $y = \cos(x)$ $(-\pi \le x \le \pi)$ as shown in Figure 5.13.

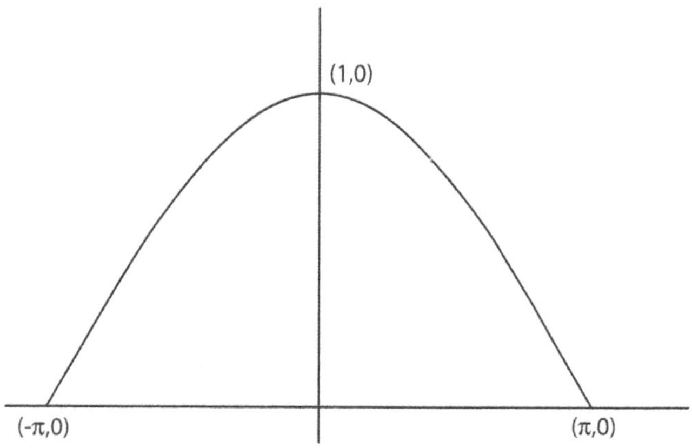

Figure 5.13 Use a uniform quadratic B-spline to approximate a cosine curve cosine curve $(-\pi \le x \le \pi)$.

3. Use three control points $((-\sqrt{3}r, -r)$ $P_1 = (\sqrt{3}r, -r)$ $P_2 = (0, 2r))$ to approximate a circle with radius r as shown in Figure 5.14 by a uniform quadratic B-spline, find curve in region 1.

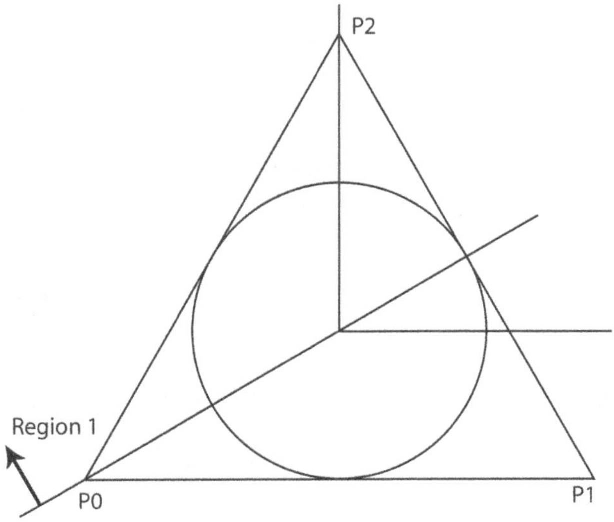

Figure 5.14 Use a uniform quadratic B-spline to approximate a part of a circle.

4. Take B-spline in quadratic and cubic forms to approximate the circle shown in Figure 5.15, and compare them.

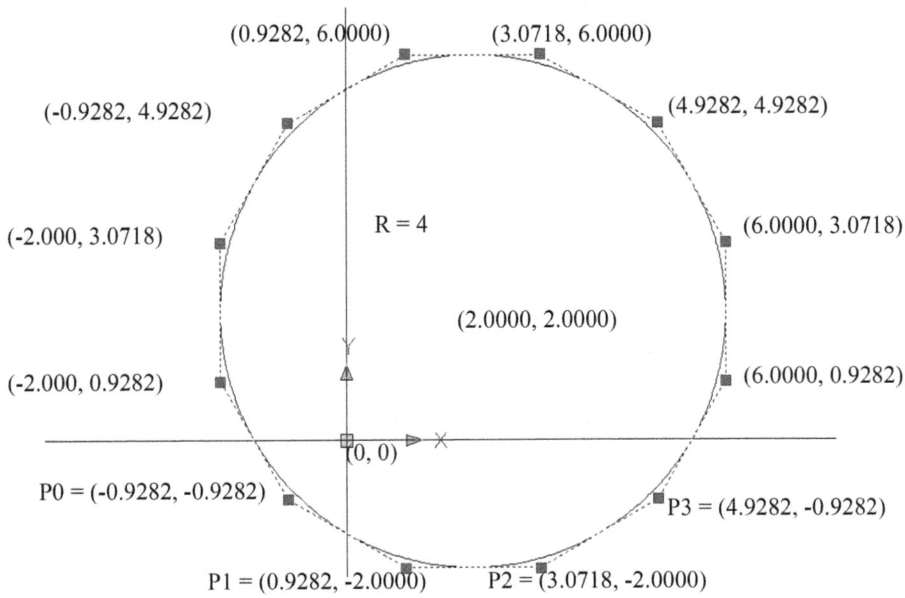

Figure 5.15 Use a uniform quadratic and cubic B-splines to approximate a part of a circle.

5. Use B-spline curves, quadratic form to approximate the "S-shape" curve.

5.2 Reverse engineering for digital representation

The human foot is a masterpiece of engineering and a work of art.

Leonardo da Vinci

This section helps answer the following questions:

- What is reverse engineering? Why is reverse engineering needed?
- How does reverse engineering work?
- What is the procedure for reverse engineering?
- What are the tools for reverse engineering?

In general, reverse engineering is defined as the process of discovering the technological principles of a device, object, or system through analysis of its structure, function, and operation. It often involves taking something apart and analyzing its workings in detail, usually with the intention to construct a new device or program that does the same thing without actually copying anything from the original. Reverse engineering can also be defined as the development of a complete technical data package for an existing component or subassembly. In this section, it is narrowly defined as finding the 3D geometry of an existing object.

5.2.1 Reverse engineering and product prototyping

Reverse engineering is commonly applied to the general process of recreating existing 3D geometry in the computer. This 3D geometry representation could be the shape of a manufactured object, such as a car, or it could be some type of nature shape, like a plant or a human body. Although many manufactured objects are now defined digitally using some type of 3D modeling software, in many instances, the part may not be able to obtain the existing geometry digitally. For example, the digital geometry might be needed for part repair or during product development iterations. For example, engineers may want to put the shape of an old VW Beetle into the computer so that they could construct an oversized sculpture of the car. It could also be an engineer who wants to capture the shape of an airplane to put into a flight simulator program. For nonmanufactured objects, such as rocks, trees, and human beings, there are no existing computer models, and one has no choice, but to re-create the 3D shape on the computer. In general, CAD models are often unavailable or unusable for parts which must be duplicated or modified when

1. CAD was not used in the original design.
2. There is inadequate documentation on the original design.
3. The original CAD model is not adequate to support modification or manufacturing using modern methods.
4. The original supplier is unable or unwilling to provide additional parts.
5. There have been shop floor changes to the original design.

The reverse engineering machines require very high precision, and thus, there are some standard procedures to ensure that the machine is well qualified before it is used. For example, the CMM is an accurate measuring device used to verify dimensions to ensure part quality. A CMM can be used to measure features in 2D and 3D, although 2D features cannot be measured directly. For example, a line is a 2D feature and can be defined by the intersection of a plane to the reference plane. While a plane is a 3D feature, a line can be indirectly defined by defining two planes using a CMM. Most CAD models are made up of simple geometric elements such as planes, edges, cylinders, spheres, and cones, created by machining or forming. Distance, symmetry, intersection, angle, and projection cannot be directly measured, but must be constructed mathematically from measured features. A projection is the reproduction of a part feature on another feature, for example, projecting a circle or a line on a plane or a point onto a line. An intersection is where two existing geometric elements meet and cross each other. For example, the points are created by the intersection of two lines, or of a line and a plane; and the lines are created by the intersection of two planes.

The sequence for programming a CMM with a probe includes the following steps:

1. Home the CMM and establish a global coordinate system.
2. Qualify the tip and compensate for tip diameter. Probe qualification is critical for machine accuracy, especially when the machine starts each day or when a new probe is installed. To qualify the tip,
   A. One needs to measure a reference sphere.
   B. One needs to enter the reference sphere diameter, actual probe diameter, and probe positions into the computer.
   C. The computer calculates the effective probe diameter and location of the center of the probe in the measuring volume.
3. Align the part and establish a local coordinate system on the part. To align a part with its CAD model, one needs to align the following:
   A. The reference plane
   B. The major axis
   C. The part zero
4. Measure the part.

Traditional CMMs typically have better accuracy, but can be limited in the size and complexity of the object to be scanned. Portable CMMs are generally less accurate, but they are portable, with less limitation on the size of an object [Hollister17]. Measurement systems using noncontact technologies, such as various laser scanning probes or laser tracking systems, can scan large and complex surfaces accurately but can be very expensive depending on the type of system. Sensing can be classified into passive sensing and active sensing. Passive sensing is when sensing energy is only received, and no energy is emitted for the purpose of sensing, for example, stereo vision techniques. Active sensing is when properly formatted light or any other form of energy is emitted rather than received once it has interacted with the object to digitize, for example, computerized tomography (CT) and magnetic resonance imaging (MRI).

5.2.2 Reverse engineering process

In reverse engineering of mechanical parts, often, a digitizer is moved along parallel scanning paths, and an NC code is generated to move a cutter along the same 3D path. In effect, no model other than the raw scan data is used. Spline patches are often fit to sampled position points. Geometry is usually represented in terms of surface points or collections of parametric surface patches to describe positional information. However, it is difficult to capture the higher-level structure of the object. There are up to two steps in the process of reverse engineering:

1. Digitizing the parts. This step uses a reverse engineering device such as CMM, laser scanner, faro arms, or other 3D sensors to collect the raw geometry of the object. The data is usually in the form of coordinate points of the object relative to a local coordinate system. A cloud of points as shown in Table 5.1 and Figure 5.16 are taken with a CMM machine. A point cloud is a set of 3D points describing the outlines or surface features of an object.

Table 5.1 Point cloud data of the computer mouse in Figure 5.16

80pts_mouse.pts - Notepad

121.1386	143.6180	-223.9802
106.3888	143.6180	-225.5098
98.1962	143.6180	-227.8520
93.6539	137.3915	-226.5625
105.5783	137.5350	-222.9694
121.0139	137.5350	-221.1369
137.6064	137.5350	-223.0963
148.3482	137.5350	-227.2049
149.4546	132.0477	-224.7795
137.3685	132.4670	-221.0785
121.5080	132.4670	-218.8169
106.0492	132.4670	-220.8041
92.2444	132.4670	-225.2826
90.9163	123.3722	-222.9283
105.6582	123.6270	-217.5460
121.3006	123.6260	-215.0523
138.0862	123.6260	-218.0173
150.2835	123.6260	-222.6006
149.9837	114.3587	-220.1883
138.1530	114.6190	-214.7951
121.8058	114.6190	-211.5010
105.9167	114.6190	-214.1530
92.7343	114.6190	-219.8937
92.6443	105.8103	-218.6253
106.0073	105.9630	-211.5704
121.9213	105.9630	-209.0421
139.4638	105.9630	-213.1160
149.8313	105.9630	-219.9960
149.5491	96.8191	-218.2230
137.2345	97.0249	-210.4700

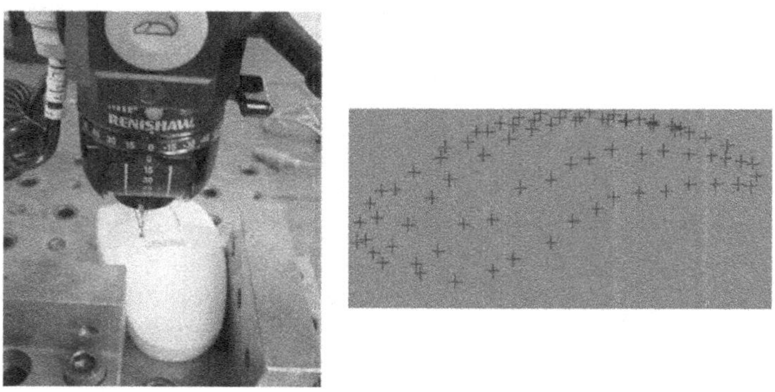

Figure 5.16 Point clouds taken from a computer mouse.

Very often one needs to throw away extraneous or obviously wrong points by visually observing the raw data on the computer. It is advisable to do this before model digitizing is completed so that one can correct any problems by re-digitizing the problem areas.

2. Building CAD models. This step converts the raw point data obtained from step 1 into a usable format. One needs to use a program that has the capability to wrap the cloud of points with 3D, connected polygons. If the point cloud covers several objects, the user of the software may have to split the point cloud into smaller sections before using the polygon wrapping capability. One may also need tools to align point cloud data taken from different views of the object. For example, one can generate a surface by using existing point clouds in NX. Using "Insert\Free form features\Through points\," and follow the procedures as shown in Figure 5.17.

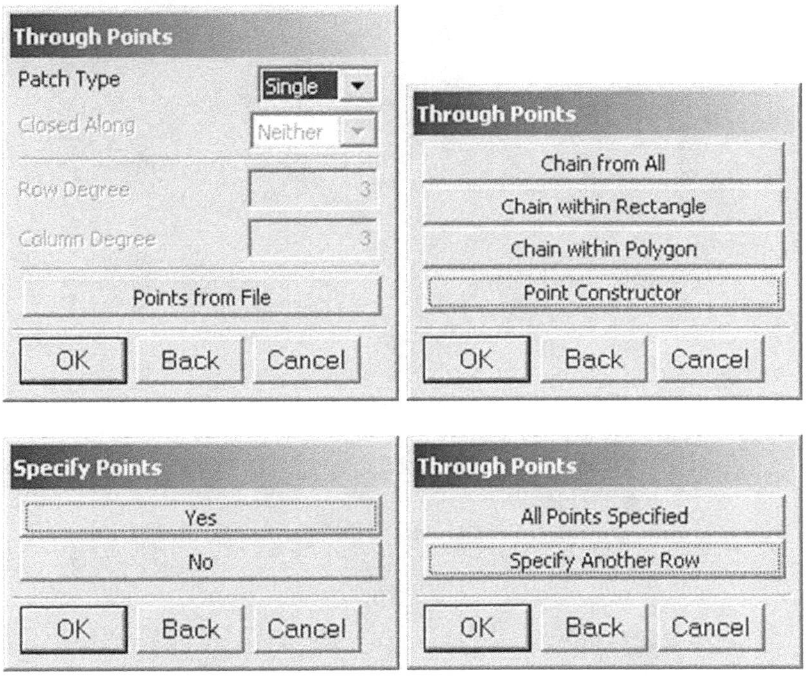

Figure 5.17 The procedures to create free-form surface row by row.

After all the points are selected row by row, click on "All Points Specified" instead of "Specify Another Row," the free-form surface can then be obtained as shown in Figure 5.18. The solid model then can be created as shown in Figure 5.19.

In terms of input devices, CMMs are generally moderate in price, and typically faster and more accurate than manual measurement. The CMM is limited physically by the probe design to scan an object without collisions and interference. Portable types such as the Faro arm are more flexible, but give up some accuracy compared to fixed designs such as gantry CMMs. Laser scanning of a surface is fast and accurate, although slightly less accurate than fixed CMMs. It has the advantage of collecting a large volume of point data very quickly. The speed of operation comes at a relatively high price, however. Other types of noncontact measurement include MRI and CT, which can image internal features of hollow objects as well as external surfaces.

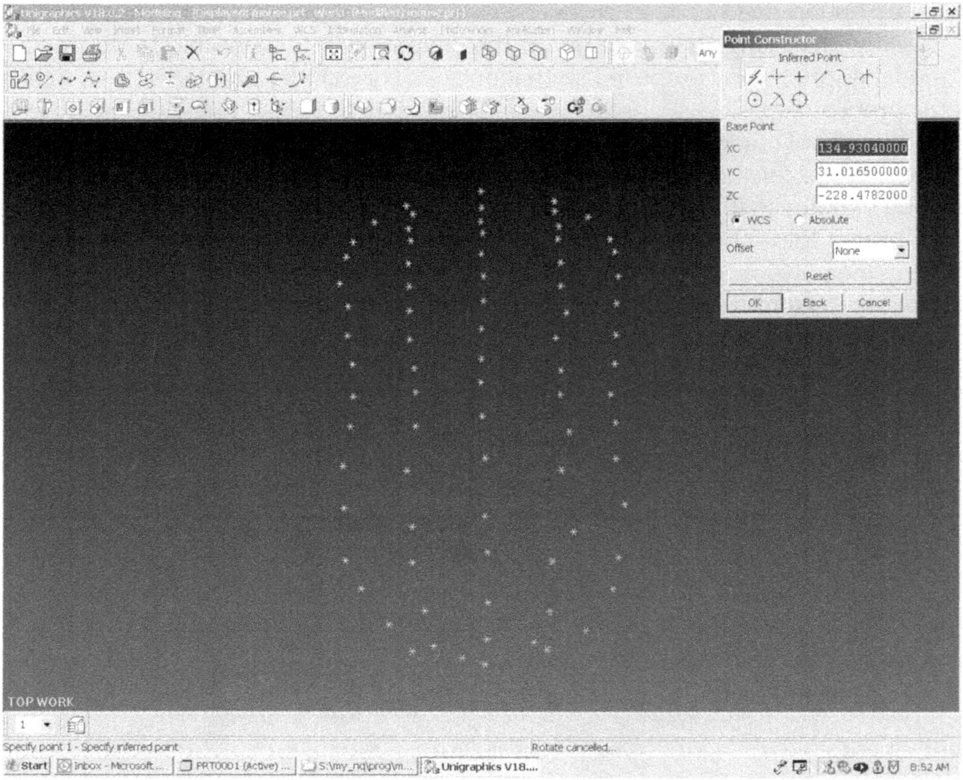

Figure 5.18 Points were selected row by row to create a surface.

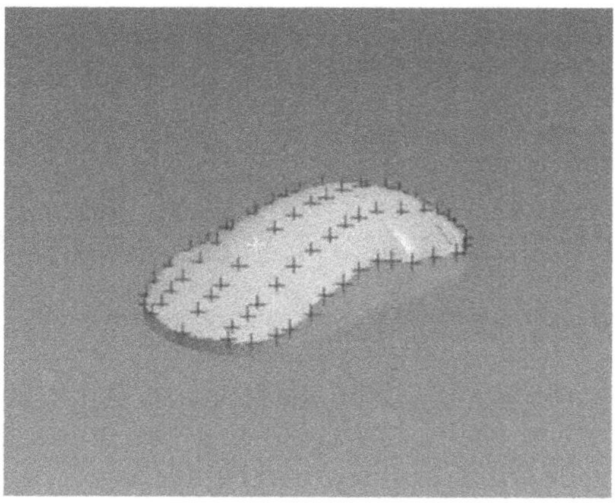

Figure 5.19 A solid model created in a CAD system.

Complex shapes are often divided into zones. While most of the flat or close to flat surfaces can be scanned using a parallel-plane technique or standard features, some areas such as fillets and joints can be scanned with great care and some by the manual method. As an example, one would like to obtain the geometry of a nozzle. From the analysis, one can find that the exterior consists of three segments: an end conical section, a center conical section, and an end cylindrical section as shown in Figure 5.20.

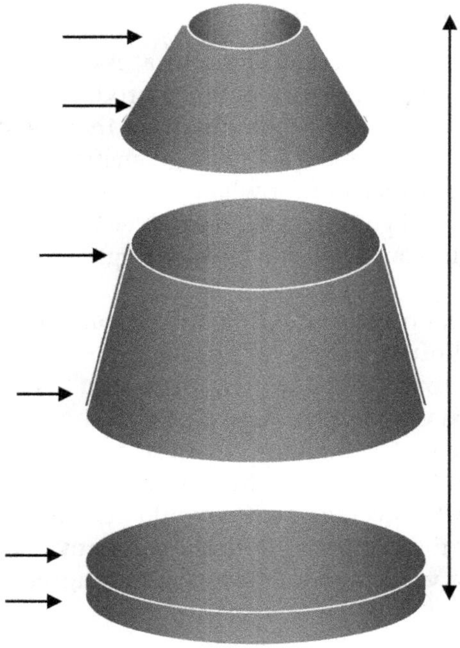

Figure 5.20 The nozzle can be broken into three segments: end conical section, center conical section, and end cylindrical section.

After each section is digitized, the geometric elements can be aligned to construct a section view of one side of the nozzle. A solid model is then constructed as shown in Figure 5.21, and an additive manufacturing model can be made as illustrated in Figure 5.22.

Figure 5.21 The solid model of the nozzle based on reverse engineering.

Figure 5.22 The printed ABS part of the solid model.

5.2.3 *Ethics and reverse engineering*

Many people may have questions regarding ethical issues related to reverse engineering. How can one reverse engineer a competitor's products without infringing on their patents or copyrights? For example, if one discovers that the competitor uses a plastic injection molding process to create the complex geometry of a product and then coats it in rubber for chemical applications or for whatever reason, can one use the same idea in a competitive product without getting in legal trouble? Under regular property law, once one has bought a product, would not one have the right to take it apart and see how it works?

This is a very complex problem, and in many circumstances, it is not legal; in some situations, it is legal, and often, it may be legal, but not ethical. To make proper judgments, an interesting approach called a "newspaper test" is used at Texas Instruments (TI) [TI17]. In short, as TI is very much a large part of Texan industry, bad publicity could severely damage the company's reputation. The test is to ask the TI employee to consider the question, "How would this look in the newspaper?" This simple test encourages engineers to consider how the average person would perceive his or her actions. Would it cause a loss of respect for the company? Both the company and its employees agree that reverse engineering was a valid practice; however, they agree it must be done with care.

Review problems

1. Arrange the sequence for programming the CMM.
   - A. Home the CMM. Establishes a global coordinate system.
   - B. Measure the part.
   - C. Align the part. Establishes a local coordinate system on the part.
   - D. Qualify the tip. Compensates for tip diameter.
2. Define reverse engineering.
3. What is the difference between active and passive sensing?
4. Compare and contrast traditional gantry CMMs with other types of technologies such as portable CMMs or laser scanners.
5. Which of the following is/are false about CMM measurement?
   - A. A line cannot be measured directly.
   - B. A planar surface is a 3D feature.
   - C. 2D features cannot be measured directly.
   - D. Angle and projection cannot be measured directly, but must be constructed mathematically from measured features.
   - E. Distance can be measured directly but must be constructed mathematically from measured features.
6. Why is reverse engineering needed (name at least three reasons)?
7. What is a point cloud?

5.3 *Prototyping and manufacturing using CNC machining*

> Many people see technology as the problem behind the so-called digital divide. Others see it as the solution. Technology is neither. It must operate in conjunction with business, economic, political and social system.
>
> **Carly Fiorina**

This section helps answer the following questions:

- How can one communicate with a Computer Numeric Control (CNC) machine to fabricate a part in a CAD model?
- What is a CNC postprocessor?
- Is it necessary to write a custom designed CNC postprocessor?
- Is there fully automated software to drive a CNC machine right from a CAD model?
- How does CAD interact with CAM?

In the competitive industry, productivity is achieved by guiding a product from concept to market quickly and inexpensively. Advanced CNC machining technology can aid this process by automating the fabrication of a prototype part from a 3D CAD model. While conventional prototyping technologies may take weeks or even months, CNC machining prototyping can be a quicker, more cost-effective means of building prototypes.

Although the CNC machining process does have its limitations when compared to the current additive manufacturing technologies, the major shortcoming is the accessibility to some features such as narrowed corners and internal cavities. CNC machining technology offers low cost of production and flexibility in material selection. Once they are set up, the machines will repeat very accurately, require minimal machinist level intervention, and have a high part-to-part accuracy and consistency. Therefore, fast setup and rapid shipment are possible. Therefore, if accessibility is not an issue, CNC machining is an effective method for prototyping purposes. This section discusses how to use solid models to drive a CNC machine for prototyping and fabrication. Many CAD/CAM packages can help the user to use a solid model to generate CNC codes. However, as there are varieties of CNC machines, often operators will need to be engaged in the process for it to run smoothly. Therefore, it is important to understand some basics of CAD/CAM technology. This section discusses CNC machines only, but the principles can be applied to all other machines using languages that utilize G- and M-codes.

5.3.1 *Machine codes for process control*

The codes used to drive a CNC machine are called G- and M-codes. The definitions of G- and M-codes have been standardized within the specifications of ANSI/EIA-RS-274-D. It is an Electronic Industries Association (EIA) standard that was also adopted by the American National Standards Institute (ANSI). The specifications list no differences between CNC mills and CNC lathes.

G-codes are the preparatory functions. Preparatory functions include the x-, y-, and z-axis movements, thread cutting, radius compensation, canned cycles, circular interpolations, an inch or metric measurement system, dimensional input formats, tool feed rates, and tool spindle revolutions. M-codes are the miscellaneous functions. Many M-codes are defined by the CNC machine manufacturer for their machines' unique operating characteristics. M-codes control options such as program stop, end of program, spindle rotation direction, tool changes, coolant 1 and 2 off/on, clamp and unclamp, and return to program start.

A G-code consists of the letter G followed by two digits that correspond to specifying a control mode of operation. They do things like set the absolute zero for coordinates, change to rapid positioning mode, linear interpolation mode, and circular interpolation modes and so forth. Many of these have to do with movement. Some common G-codes are listed in Table 5.2. M-codes are in a similar format and are miscellaneous functions for

Table 5.2 Some commonly used G-codes

G-Code	Description
G0	Rapid travel (point-to-point)
G1	Linear interpolation
G2	Circular interpolation clockwise (CW)
G3	Circular interpolation counterclockwise (CCW)
G4	Dwell
G15	YZ circular plane with simultaneous A-axis
G17	XY plane selection
G18	XZ plane selection
G19	YZ plane selection
G20	Nonmodel check parameters for inches mode set in SETP
G21	Nonmodel check parameters for metric mode set in SETP
G28	Return to current zero (set home) position
G29	Return from current zero (set home) position
G40	Cutter compensation canceled
G41	Cutter compensation left (climb)
G42	Cutter compensation right (conventional)
G43	Tool length compensation positive
G44	Tool length compensation negative
G45	Tool offset single expansion
G46	Tool offset single reduction
G47	Tool offset double expansion
G48	Tool offset double reduction
G49	Tool length offset cancel
G52	Coordinate system shift
G53	Machine coordinate system
G54	Fixture offset 1 (E1)
G55	Fixture offset 2 (E1)
G56	Fixture offset 3 (E1)
G57	Fixture offset 4 (E1)
G58	Fixture offset 5 (E1)
G59	Fixture offset 6 (E1)
G68	Rotation
G69	Rotation cancel
G70	Modal check parameters for inches mode set in SETP
G71	Modal check parameters for metric mode set in SETP
G74	LH tapping with compression holder
G74.1	LH rigid tapping
G74.2	Prepare for LH rigid tapping
G75	Tapping cycle with self-reversing head
G80	Fixed cycle cancel
G81	Spot drill cycle
G82	Counterbore cycle
G83	Deep hole drill cycle

(*Continued*)

Table 5.2 (Continued) Some commonly used G-codes

G-Code	Description
G84	RH tapping with compression holder
G84.1	RH rigid tapping
G84.2	Prepare for RH rigid tapping
G85	Bore in, bore out
G86	Bore in, spindle off, rapid out
G87	Bore in, bore out
G88	Bore in, dwell, bore out
G89	Bore in, dwell, bore out
G90	Absolute programming
G91	Incremental programming
G92	Programmed coordinate system preset
G93	Rotary axis 1/T feed rate specification
G94	Rotary axis DPM,IPM feed rate specification
G98	Return to initial plane after final Z
G99	Return to RO after final Z

the machine code program. They will start and stop the program, turn the spindle on or off, initiate a tool change, or even control things like coolant flow. Some common M-codes are listed in Table 5.3. CAD is widely used to describe any software capable of defining a mechanical component with geometry, surfaces, or solid models. CAD is a software used to develop NC programs. Integrating CAD with CAM provides the link between software systems that helps streamline the transfer of files between the following important functions: (1) design modeling, (2) manufacturing modeling, and (3) NC programming.

Table 5.3 Some commonly used M-codes

M-Code	Description
M0	Program stop
M3	Spindle CW
M4	Spindle CCW
M5	Spindle stop
M6	Tool change
M7	Coolant 1 on
M8	Coolant 2 on
M9	Coolant 1 and 2 off
M17	End of subroutine (see M30)
M19	Spindle orient and lock
M30	End of all subroutines (sec Ml7)
M60	A-axis brake on
M61	A-axis brake off
M62	B-axis brake on
M63	B-axis brake off
M98	Execute subprogram
M99	End of subprogram or line jump

The key to the success of this process is to develop a software link that integrates the software, necessary for developing the part design, to the machine, used to fabricate the part.

As shown in Figure 5.23, the general CAD/CAM integration process flow starts with the generation of the 3D CAD models by the designer, proceeding through the generation of the machine codes and culminating in part manufacture. No universal NC codes can be interpreted by NC controllers made by various manufacturers. A postprocessor translates actions taken in the CAM software into machine code. The CAM programs often output CL-data files (or cutter location data files) that consist of a generic tool path and critical operations of the CNC machine. CL-data format is designed so that it is machine independent. The postprocessor translates the CL-data into machine-specific G- and M-codes to control the specific machine. The postprocessor writes a machine-specific set of instruction codes and sends it to the CNC machine. With a good postprocessor, one should not have to do any G-code editing, and thus, it is critical to choose a system with good postprocessing capability.

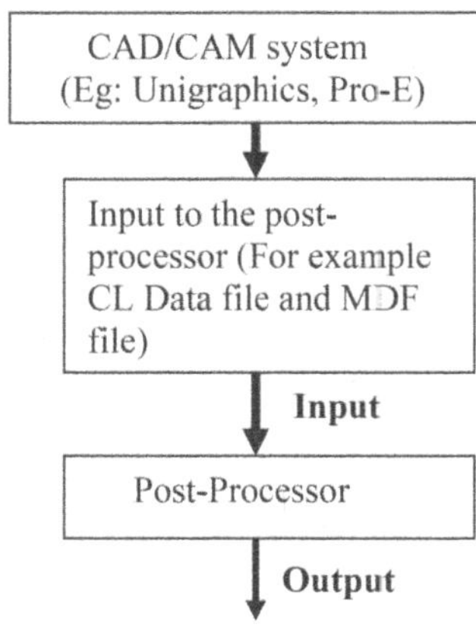

Figure 5.23 CAD/CAM and postprocessor.

The functions of a postprocessor include (1) conversion of a tool position in the part coordinate system to the machine coordinate system; (2) processing of linear interpolation, circular interpolation, etc.; (3) generation of correct spindle speed, feed rate, tooling, and various machine operation code; (4) modify and generate commands that allow for requirements of the NC machine controller; (5) verify the correctness of tool or part travel range, tool motion; and (6) setting the output format as required by the NC machine.

Example 5.3: Comparison between CL-data and CNC codes

Table 5.4 shows an example of CL-data used by quick turnaround cell (QTC). Table 5.5 shows the CL-data generated from a CAD system to make the slot shown in Figures 5.24 and 5.25.

5.3.2 *Using CAD/CAM for digital manufacturing*

Many CAM software packages have the ability to simulate the motion of the cutter paths and the material removed by the paths. These software packages are invaluable in locating problems that may not be apparent until the program is run on the machine.

Table 5.4 The CL-data parameters are described using the following conventions [QTC05]

Code	Term/format	Specification/example
0 Comment	0 S80	Example: 0 begin probing operations
1 x y z **[i] [j] [k]**	Linear interpolation: 1 F4.4 F4.4 F4.4 [F4.4] [F4.4] [F4.4]	x, y, z: location **i, j, k**: orientation unit vector for six-axis machine; Example: 1 25.0250 6.1221 12.0000
2 Plane dir X Y Z X_c Y_c Z_c	Circular interpolation: 2 I1 I1 F4.4 F4.4 F4.4 F4.4 F4.4 F4.4	Plane: 0: Interpolation in the XY plane 1: Interpolation in the XZ plane 2: Interpolation in the YZ plane; Direction: 0: Clockwise interpolation; 1: counterclockwise interpolation; X, Y, Z: location of the circular motion endpoint; X_c, Y_c, Z_c: location of the center of the circular motion arc Example: 2 0 1 17.5840 10.2100 8.7310 17.5840 10.7100 8.7310
3 Rate option	Feed rate: 3 F1.4 I1	Rate: value of specified feed rate; Option 0 For IPM/MMPM feed rate 1 For inverse time feed rate 2 For feed/tooth feed rate Example: 3 9.480 0
4 Option	Rapid traverse: 4 I1	Option 0: Sets rapid positioning OFF 1: Sets rapid positioning ON Example: 4 1
5 Option	Coolant selection: 5 I1	Option 0: For turning the coolant off 1: Specifies exterior flood coolant 2: Specifies exterior mist coolant 3: Specifies through spindle flood coolant 4: Specifies through spindle mist coolant 5: Specifies through spindle pulse coolant Example: 5 1

(*Continued*)

Table 5.4 (Continued) The CL-data parameters are described using the following conventions [QTC05]

Code	Term/format	Specification/example
6 Option tool [retract]	Automatic tool load: 6 I1 [I16] [I1]	Option 0: Prepare the specified tool for loading 1: Load the tool into the spindle Tool: tool number or ID Retract 0: Do not retract when changing tools 1: Retract turret when changing tools Example: 6 0 12
7 Option [rpm]	Spindle operations: 7 I1 [I4]	Option 0: For spindle OFF 1: For clockwise spindle ON 2: For counterclockwise spindle ON rpm Spindle speed value in rpm (used with options 1 and 2) Example: 7 1 319
8 Option diameter	Cutter diameter compensation: 8 I1 [F1.3]	Option 0: Cancels CDC 1: For CDC right 2: For CDC left Diameter Compensation diameter (used with options 1 and 2) Example: 8 1 0.2500

(Continued)

Table 5.4 (Continued) The CL-data parameters are described using the following conventions [QTC05]

Code	Term/format	Specification/example
9 Option [type] [time]	Canned cycle operations: 9 I1 [I1] [F2.2]	Option 0: Canned cycle OFF 1: Canned cycle ON 2: Tap cycle 3: Mill cycle 4: Ream cycle 5: Boring cycle 6: Drill cycle Type 0: Normal (used with option 5) 1: Retract with dead spindle (used with option 5) 2: Bore in, dwell, bore out (used with option 5) 0: Normal drill (used with option 6) 1: Drill in, dwell, retract (used with option 6) Time Amount of time for the dwell option Example: 9 6 1 20.00
10 Option	Measurement unit: 10 I1	Option 0: For U.S. units (in.) 1: Metric units (mm) Example: 10 0
11 Option	Probe cycles: 11 II 11 II 11 II	Option 0: Disarm surface sensing 1: Arm surface sensing 2: Reset register 3: Find and measure surface 4: Test for part presence 5: Compute offset

(*Continued*)

Table 5.4 (Continued) The CL-data parameters are described using the following conventions [QTC05]

Code	Term/format	Specification/example
12 Angle	Index pallet 12 13	Angle To index the table containing the pallet Example: 12 90
13 Option [time]	Program interrupts 13 I1 [F2.2]	Option 0: Program halt 1: Planned program stop 2: Dwell for a number of seconds Time Length of dwell (option 2) in seconds Example: 13 2 10.5
14 Message	Operator message 14 S60	14 Starting milling operations
15 Option [X] [Y] [Z] [B]	Coordinate system operations: 15 I1 [F4.4] [F4.4] [F4.4] [F4.4]	Option 0: Reset coordinates to true machine coordinates 1: Use absolute dimensions 2: Use incremental dimensions 3: Load new absolute coordinate information X, Y, Z, B: New coordinate values for the current tool tip position Example: 15 3 0.0 0.0 0.0 90.0

Note: [], options are enclosed in square brackets; Sn, string of n ASCII characters; In, integer value of n digits; FA.B, float value with a maximum of A digits left of the decimal point and a maximum of B digits right of the decimal point.

Table 5.5 The CL-data and corresponding G- and M-codes in the example

CL-data	Comments	G-codes
0.00	Begin tool setup—comment	G80 G28 G90 G40 G0
6 0 6	Prepare tool T6 for loading	T6
6 1 6	Load tool T6 into the spindle	T6
0.00	Begin machining P-50 slot—comment	
4 0	Rapid positioning	G01 x3.0 Yl.6475 z0.2
7 2 788	Spindle ON CCW with speed 788 rpm	S788 M04
03 6.3087 0	Feed rate of 6.3087 in./min	F6.3087
05 1	Exterior flood coolant selection	M07
09 6 0	Canned drill cycle normal type	G81 X3 Y0 Z–0.46 F6.3087
01 3.0 1.6475 –0.460	Position X, Y, Z, respectively	G01 X3.000000 Yl.647500 Z–0.460
0.00	Begin tool setup flat end mill	
06 0 18	Prepare T18 for loading	T18
06 1 18	Load T18 into spindle	T18
0.00	Begin machining Slot _0	
7 2 1273	Spindle ON CCW with speed 1,273 rpm	S1273 M04
0.00	Begin pocket with 0.625 diameter cutter	
0.00	Level 1 of 3 (workplane $Z_w = -0.153333$)	
4 1	Feed positioning mode	
1 3.0 1.6475 0.2	Position X, Y, Z, respectively	G00 X3.000000 Y1.647500 Z0.200000
1 3.0 1.6475 0.1	Position X, Y, Z, respectively	X3 Y1.647 Z0.1
5 1	Exterior flood coolant selection	M07
4 0	Rapid positioning	
3 7.6394 0	Feed rate of 7.6394 in./min	F7.6394
1 3 1.6475 –0.1533	Position X, Y, Z, respectively	G01 X3 Yl.647 Z–0.1533
3 15.2789 0	Feed rate of 15.2789 in./min	F 15.2789
2 0 0 3 1.3525 –0.1533 3 1.5 –0.1533	CW circular interpolation in XY plane with endpoint X, Y, Z and centered at X_c, Y_c, Z_c	G02 X3 Y1.3525 Z–0.1533 I3 J1.5
1 1 1.3525 –0.1533	Position X, Y, Z, respectively	G01 X1 Y1.3525 Z–0.1533
2 0 0 1 1.6475 –0.1533 1 1.5 –0.1533	CW circular interpolation in XY plane with endpoint X, Y, Z and centered at X_c, Y_c, Z_c	G02 X1 Y1.6475 Z–0.1533 I1 J1.5
1 3 1.6475 –0.1533	Position X, Y, Z, respectively	G01 X3 Yl.6475 Z–0.1533
1 3 1.6875 –0.1533	Position X, Y, Z, respectively	G01 X3 Y1.6875 Z–0.1533
2 0 0 3 1.3125 –0.1533 3 1.5 –0.1533	CW circular interpolation in XY plane with endpoint X, Y, Z and centered at X_c, Y_c, Z_c	G02 X3 Y1.3125 Z–0.1533 I3 J1.5
1 1 1.3125 –0.1533	Position X, Y, Z, respectively	G01 X1 Y1.3125 Z–0.1533
2 0 0 1 1.6875 –0.1533 1 1.5 –0.1533	CW circular interpolation in XY plane with endpoint X, Y, Z and centered at X_c, Y_c, Z_c	G02 X1 Y1.6875 Z–0.1533 I1 J1.5
1 3 1.6875 –0.1533	Position X, Y, Z, respectively	G01 X3 Yl.6785 Z0.2
5 0	Turning coolant OFF	M09
4 1	Feed positioning mode	

(Continued)

Table 5.5 (Continued) The CL-data and corresponding G- and M-codes in the example

CL-data	Comments	G-codes
1 3 1.6875 0.2	Position X, Y, Z, respectively	G00 X3 Y1.6875 Z0.2
0.00	Level 2 of 3 (work plane $Z_w = -0.306667$)	
1 3 1.6475 0.2	Position X, Y, Z, respectively	G01 X3 Y1.6475 Z0.2
1 3 1.6475 –0.0533	Position X, Y, Z, respectively	X3 Y1.6475 Z–0.0533
5 1	Exterior flood coolant selection	M07
4 0	Rapid positioning	
3 7.6394 0	Feed rate of 7.6394 in./min	F7.6394
1 3 1.6475 –0.3067	Position X, Y, Z, respectively	G01 X3 Y1.6475 Z–0.3067
3 15.2789 0	Feed rate of 15.2789 in./min	F15.2789
2 0 0 3 1.3525 –0.3067 3 1.5 –0.3067	CW circular interpolation in XY plane with endpoint X, Y, Z and centered at X_c, Y_c, Z_c	G02 X3 Y1.3525 Z–0.3067 I3 J1.5
1 1 1.3525 –0.3067	Position X, Y, Z, respectively	G01 X1 Y1.3525 Z–0.3067
2 0 0 1 1.6475 –0.3067 1 1.5 –0.3067	CW circular interpolation in XY plane with end point X, Y, Z and centered at X_c, Y_c, Z_c	G02 X1 Y1.6475 Z–0.3067 I1 J1.5
1 3 1.6475 –0.3067	Position X, Y, Z, respectively	G01 X3 Y1.6475 Z–0.3067
1 3 1.6875 –0.3067	Position X, Y, Z, respectively	G01 X3 Y1.6875 Z–0.3067
2 0 0 3 1.3125 –0.3067 3 1.5 –0.3067	CW circular interpolation in XY plane with end point X, Y, Z and centered at X_c, Y_c, Z_c	G02 X3 Y1.3125 Z–0.3067 I1 J1.5
1 1 1.3125 –0.3067	Position X, Y, Z, respectively	G01 X1 Y1.3125 Z–0.3067
2 0 0 1 1.6875 –0.3067 1 1.5 –0.3067	CW circular interpolation in XY plane with end point X, Y, Z and centered at X_c, Y_c, Z_c	G02 X1 Y1.6875 Z–0.3067 I1 J1.5
1 3 1.6875 –0.3067	Position X, Y, Z, respectively	G01 X3 Y1.6875 Z–0.3067
5 0	Turning coolant OFF	M09
4 1	Feed positioning mode	
1 3 1.6875 0.2	Position X, Y, Z, respectively	G00 X3 Y1.6875 Z0.2
0.00	Level 3 of 3 (workplane $Z_w = -0.46$)	
1 3 1.6875 0.2	Position X, Y, Z, respectively	G01 X3 Y1.6875 Z0.2
1 3 1.6475 –0.2067	Position X, Y, Z, respectively	X3 Y1.6475 Z–0.2067
5 1	Exterior flood coolant selection	M07
4 0	Rapid positioning	
3 7.6394 0	Feed rate of 7.6394 in./min	F7.6394
1 3 1.6475 –0.46	Position X, Y, Z, respectively	G01 X3 Y1.6475 Z–0.46
3 15.2789 0	Feed rate of 15.2789 in./min	F15.2789

(Continued)

Table 5.5 (Continued) The CL-data and corresponding G- and M-codes in the example

CL-data	Comments	G-codes
2 0 0 3 1.3525 –0.46 3 1.5 –0.46	CW circular interpolation in XY plane with end point X, Y, Z and centered at X_c, Y_c, Z_c	G02 X3 Y1.3525 Z–0.46 I3 J1.5
1 1 1.3525 –0.46	Position X, Y, Z, respectively	G01 X1 Y1.3525 Z–0.46
2 0 0 1 1.6475 –0.46 1 1.5 –0.46	CW circular interpolation in XY plane with end point X, Y, Z and centered at X_c, Y_c, Z_c	G02 X1 Y1.6475 Z–0.46 I1 J1.5
1 3 1.6475 –0.46	Position X, Y, Z, respectively	G01 X3 Y1.6475 Z–0.46
1 3 1.6875 –0.46	Position X, Y, Z, respectively	G01 X3 Y1.6875 Z–0.46
2 0 0 3 1.3125 –0.46 3 1.5 –0.46	CW circular interpolation in XY plane with end point X, Y, Z and centered at X_c, Y_c, Z_c	G02 X3 Y1.3125 Z–0.46 I3 JI.5
1 1 1.3125 –0.46	Position X, Y, Z, respectively	G01 X1 Y1.3125 Z–0.46
2 0 0 1 1.6875 –0.46 1 1.5 –0.46	CW circular interpolation in XY plane with end point X, Y, Z and centered at X_c, Y_c, Z_c	G02 X1 Y1.6875 Z–0.46 I1 J1.5
1 3 1.6875 –0.46	Position X, Y, Z, respectively	G01 X3 Y1.6875 Z–0.46
5 0	Turning coolant OFF	M09
4 1	Feed positioning mode	
1 3 1.6875 0.2	Position X, Y, Z, respectively	G00 X3 Y1.6875 Z0.2
0.00	End pocket with 0.625 diameter cutter	
0.00	Beg in pocket with 0.625 diameter cutter	
0.00	Level 1 of 1 (workplane Z_w = –0.5)	
4 1	Feed positioning mode	
1 3 1.6475 0.2	Position X, Y, Z, respectively	G00 X3.000000 Y1.647500 Z0.200000
1 3 1.6475 –0.36	Position X, Y, Z, respectively	X3 Y1.6475 Z–0.36
5 1	Exterior flood coolant selection	M07
4 0	Rapid positioning	
3 7.6394 0	Feed rate of 7.6394 in./min	F7.6394
1 3 1.6475 –0.46	Position X, Y, Z, respectively	G01 X3 Y1.6475 Z–0.46
1 3 1.6475 –0.5	Position X, Y, Z, respectively	x3.000000 Y1.647500 z–0.500000
1 3 2 0.5	Position X, Y, Z, respectively	x3.000000 Y2.0 z–0.500000
3 15.2789 0	Feed rate of 15.2789 in./min	F15.2789
2 0 0 3 1.3525 –0.5 3 1.5 –0.5	CW circular interpolation in XY plane with end point X, Y, Z and centered at X_c, Y_c, Z_c	G02 X3 Y1.3525 Z–0.5 I3 J1.5
1 1 1.3125 –0.5	Position X, Y, Z, respectively	G01 X1 Y1.3125 Z–0.5
2 0 0 1 1.6475 –0.5 1 1.5 –0.5	CW circular interpolation in XY plane with end point X, Y, Z and centered at X_c, Y_c, Z_c	G02 X1 Y1.6475 Z–0.5 I1 J1.5

(Continued)

Table 5.5 (Continued) The CL-data and corresponding G- and M-codes in the example

CL-data	Comments	G-codes
1 3 1.6475 –0.5	Position X, Y, Z, respectively	G01 X3.0 Y1.6475 z–0.500000
1 3 1.6875 –0.5	Position X, Y, Z, respectively	G01 x3.0 Y1.6875 z–0.500000
2 0 0 3 1.3125 –0.5 3 1.5 –0.5	CW circular interpolation in XY plane with end point X, Y, Z and centered at X_c, Y_c, Z_c	G02 X3 Y1.3125 Z–0.5 I3 J1.5
1 1 1.3125 –0.5	Position X, Y, Z, respectively	G01 X1 Y1.3125 Z–0.5
2 0 0 1 1.6875 –0.5 1 1.5 –0.5	CW circular interpolation in XY plane with end point X, Y, Z and centered at X_c, Y_c, Z_c	G02 X1 Y1.6875 Z–0.5 I1 J1.5
1 3 1.6875 –0.5	Position X, Y, Z, respectively	G01 X3 Y1.6875 Z–0.5
5 0	Turning coolant OFF	M09
4 1	Feed positioning mode	
1 3 1.6875 0.2	Position X, Y, Z, respectively	G00 X3 Y1.6875 Z0.2
0.00	End pocket with 0.625 diameter cutter	M02

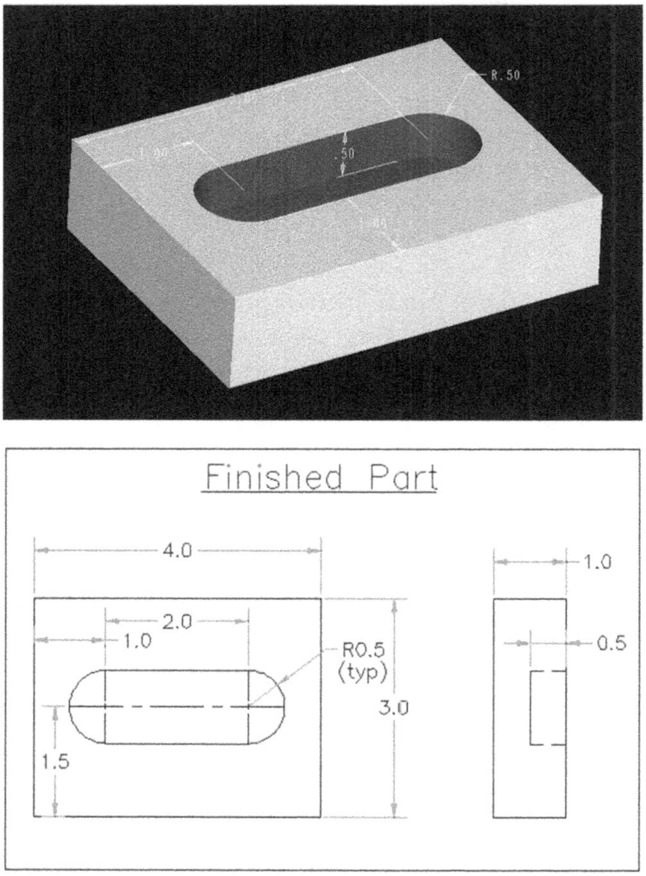

Figure 5.24 A solid model representation of a slot feature.

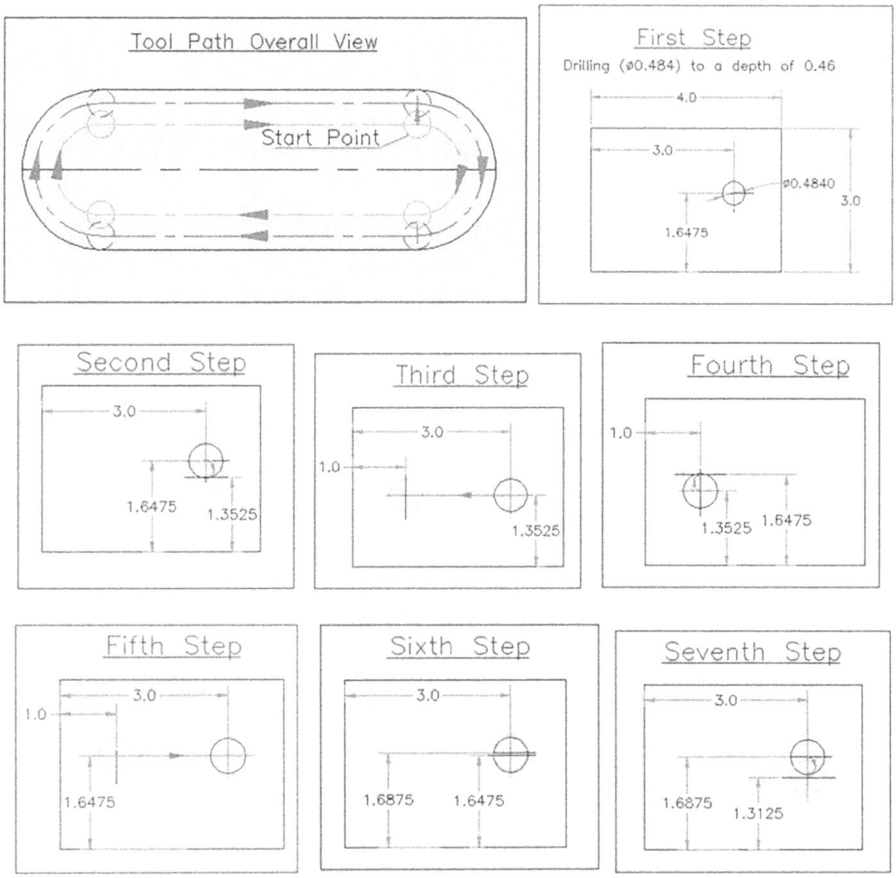

Figure 5.25 Graphical representation of the machining process defined by the CL-data.

For example, in NX, the manufacturing application allows the user to interactively create NC machining programs, generate tool paths, visualize material removal, and post-process. As shown in Figure 5.26, the user can use the CAM application based on a model from CAD (.prt) to prepare the necessary machining operation in the CAM application to generate the tool path and obtain tool path data. After the tool paths and manufacturing operations are defined, the user can output the tool path CLDATA as shown in Figure 5.27. An APT (Automatically Programmed Tool) language-based CL-data file is shown in Figure 5.28.

PostBuilder runs outside of the main NX software as shown in Figure 5.29. The user needs to define the machine configuration (e.g., three-axis mill), machine tool motion limits, machine default unit (in. or mm), G- and M-codes and sequence for a specific machine, etc. Once these steps can be conducted, the output CNC codes then can be used to drive a particular CNC machine to fabricate or prototype the part.

5.3.3 Developing a successful postprocessor

Sometimes, if a machine is not listed as a standard configuration in the CAD software, more effort may be needed to output the proper codes for CNC machining. Summarized

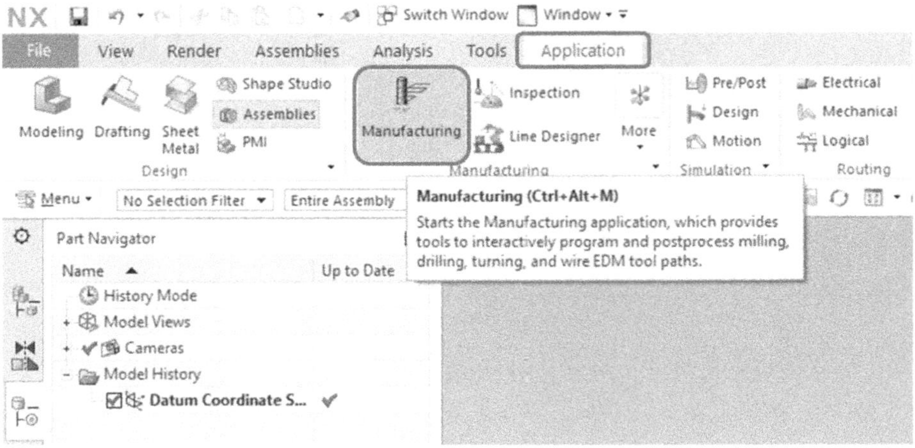

Figure 5.26 Using "Manufacturing" application in NX. (Courtesy of Siemens.)

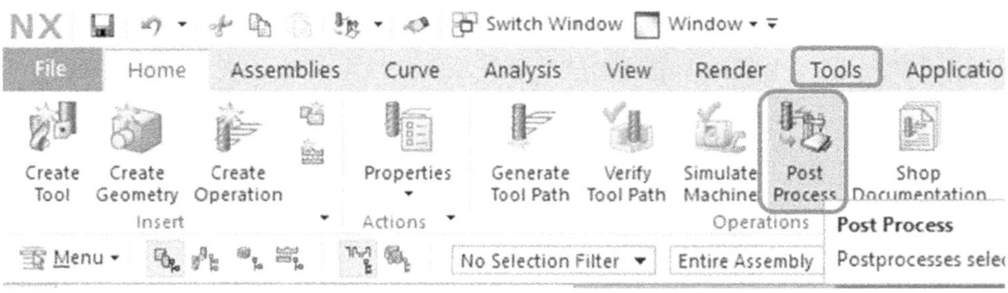

Figure 5.27 Output CLDATA file to a postprocessor. (Courtesy of Siemens.)

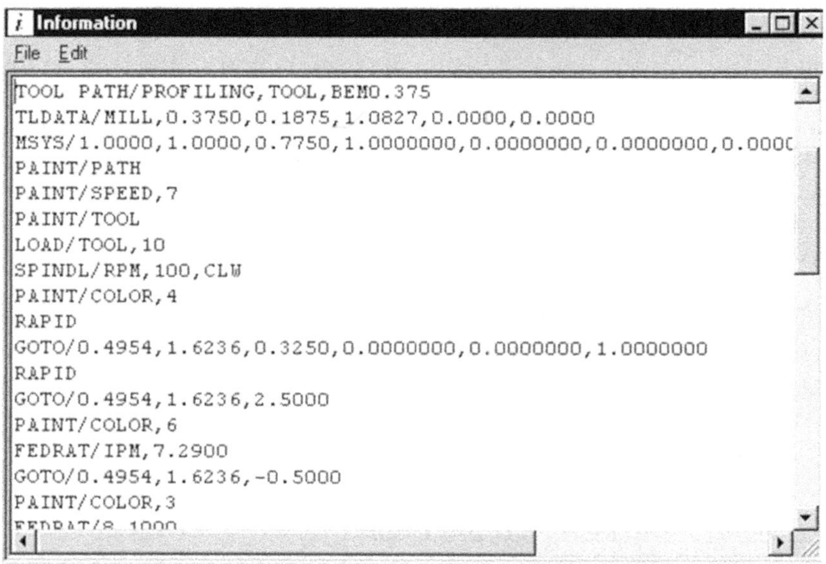

Figure 5.28 A typical NX CLDATA file. (Courtesy of Siemens.)

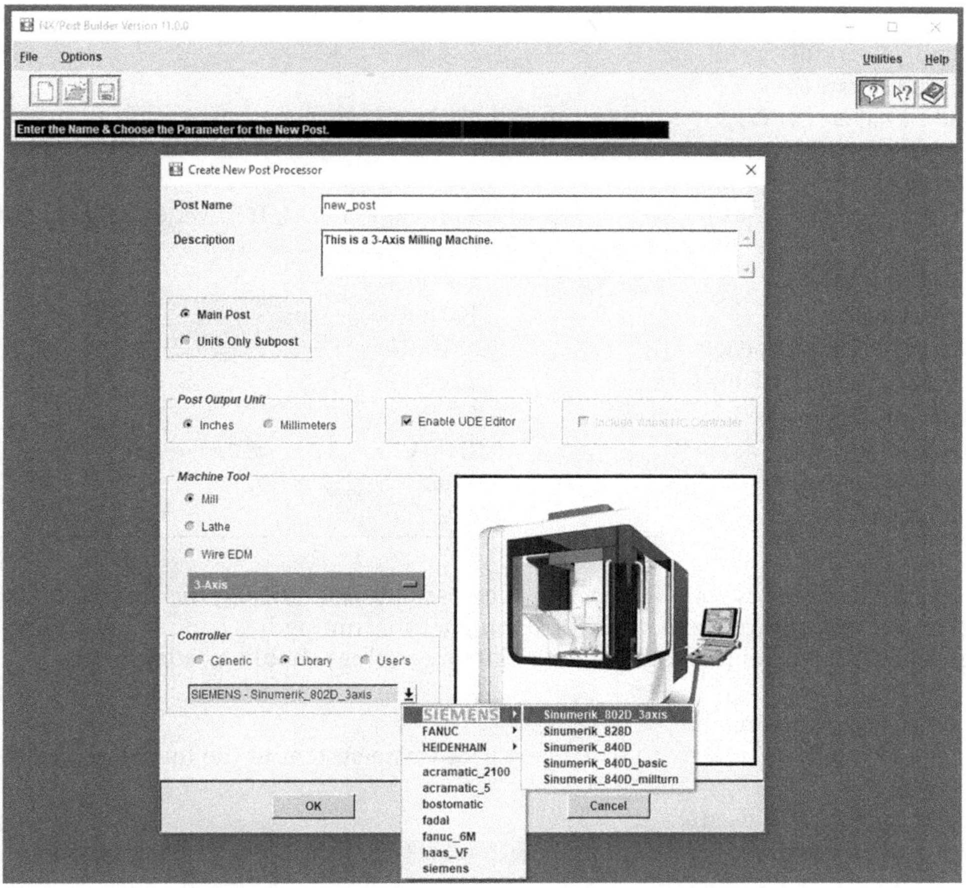

Figure 5.29 Using PostBuilder to transform CLDATA to CNC codes for a particular machine. (Courtesy of Siemens.)

here is an example to output CNC codes from Pro-E to a Fadal five-axis mill, which is a three-plus-two-axis machine. This configuration is not listed in the Pro-E database. Table 5.6 shows the Missouri S&T five-axis machine specifications. Of key importance to the design of the parts was the table size, spindle nose to table height, spindle center to column ways, maximum tool diameter, and maximum tool length. By performing simple coordinated moves in the laboratory, it was determined that this mill functions as a "true" five-axis machine, operating all five axes at once. There are certain limitations to using the axes in tandem, but those limitations have more to do with individual axis limits than with coordination between the axes.

The CNC features that need to be examined to make a successful postprocessor are as follows:

5.3.3.1 *Opening and closing codes*

The opening and closing codes deal with what the machine requires to be done before a machining process can begin, and what it requires to shut the machine down after a process has ended. The beginning processes include starting the program, setting the beginning positioning, setting the units of measure, tool loading and setup, coolant start,

Table 5.6 Five-axis CNC machine specifications

T-slots (No.×width×span)	3×0.562 in.×4.33 in.
Cutting feed rate (X/Y/Z)	0.01–250 in.
Rapid travel rate (X/Y/Z)	500 (X,Y,Z) in./min
Thrust ball screw size (X/Y/Z) 31.75 mm diameter	
X/Y/Z longitudinal (X-axis)	30 in. cross (Y-axis); 16 in. vertical (Z-axis); 20 in. (Opt. 28 in.)
Spindle nose to table	4–24 in.
Spindle torque	160 ft-lbs
Accuracy, axis positioning	±0.0002 in.
Accuracy, axis repeatability	±0.0001 in.
Maximum spindle speed	7,500 rpm
Number of tools	21 ATC
Maximum tool length	15 in.
Tool diameter	3 in.

and turning the spindle on. The ending processes include turning the spindle off, stopping the coolant, returning to a safe position once machining has ended, and ending the program. When starting a program, the following settings should be addressed:

- Start the program.
- Set translational matrix to move part coordinate system to the machine coordinate system.
- Set the unit of measure.
- Load the first tool, set the cutter offset, and set the tool diameter compensation.
- Turn the coolant on.
- Turn the spindle on.

When ending a program, the following settings should be addressed:

- Turn the spindle off.
- Turn the coolant off.
- Return the spindle to a safe position.
- End the program.

5.3.3.2 *Program detail formats*

The Fadal machine allows for two different formats. Format 1 is Fadal style programming and machine operation. This allows the CNC control to reset before restarting machine operation. It also allows the CNC programmer to use minimum commands. Format 2 is a 6M, 10M, 11M style of programming and machine operation. It allows the programmer to completely command the CNC control. The programmer must, therefore, do all resets in programming and in machine operation. For purposes of simplicity, Format 1 was chosen for this explanation. In Format 1 programming, the Fadal programs require line numbers for each line of code, irrespective of whether that line consists solely of comments or of actual commands. Line numbers may be put in manually or may be automatically added when the program is transferred with a tape input command (TA). Renumbering is possible using an NU command.

Line numbers must follow the format N####.###. Decimal points are used to input lines between existing line numbers. For example, if lines N2 and N3 exist, line N2.5 may be added between them.

Dimensions, feed rates, and angles must all be specified using a decimal point. If a decimal is omitted, the controller will add them. For example, X1 will be interpreted as X.0001. For the rotary axes, the sign of the angle determines the direction of rotation. Default, or positive, rotation angle specifies a counterclockwise rotation. To specify clockwise, the programmer must use a negative sign in front of the angle. For example, A100.0 moves the A-axis 100° counterclockwise, whereas A-100.0 moves it 100° clockwise.

The X-axis has a 30 in. limit of movement, the Y-axis 16 in., and the Z-axis 20 in. The A-axis may rotate from 0.001° through 1080°. The B-axis may rotate from 14° below the X-axis in the third quadrant to approximately 75° above the X-axis in the first quadrant. The actual B-axis specifications for these angles are 14° and 255°, respectively. The machine is set up such that the positive X-axis points to the left as the user faces the machine, and the positive Y-axis points toward the user. That makes the 0° datum for the B-axis lay along the positive X-axis, to the left of the user. If the user was standing inside the mill, and looking down, the orientation would be more conventional. The actual orientation makes it difficult to understand immediately. Since the A- and B-axes do not use negative signs to indicate the angle, but rather the direction of travel, it makes it somewhat difficult to understand the limits of the B-axis.

5.3.3.3 Formats of specific G- and M-codes

Tables 5.2 and 5.3 list the basic G- and M-commands that were used with the parts designed for this project, respectively.

5.3.3.4 Transformation matrix

In the CNC machine, there are two coordinate systems that the users will use all the time—the machine original coordinate system and the part coordinate system, which are shown in Figure 5.30. As shown in the figure, the machining center has a machine zero position. This position is typically located at the farthest point of positive travel along the three axes (i.e., X, Y, and Z). Every workpiece requires its own unique program zero, too.

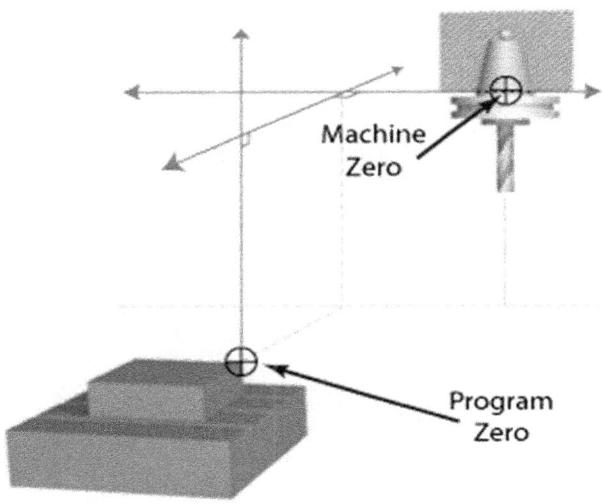

Figure 5.30 Machine zero position in the Fadal machining center.

For most operations, program zero is located in the corner of the part. The primary task deals with the coordinate transformation from the part coordinate system to the machine coordinate system.

Figure 5.31 shows the relation among the program zero users set on the part, the program zero stored in the CNC, and the machine zero. Because the CNC always just remembers where the spindle bottom surface center is, the program zero stored in the CNC is one-tool-length above the program zero on the part.

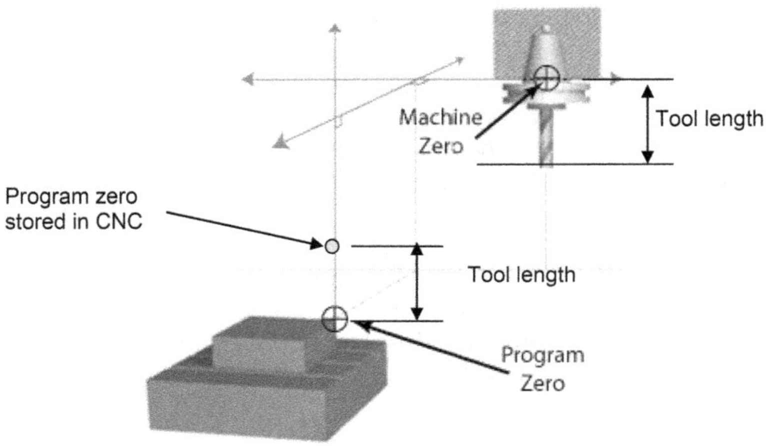

Figure 5.31 Relations among program zero on the part, the program zero stored in the CNC, and the machine zero.

5.3.3.5 *Formation of the transformation matrix for the A- and B-axis rotation*

First, it is necessary to define precisely the position for A- and B-axis rotation with respect to the machine zero.

5.3.3.5.1 Position of the B-axis Figure 5.32 shows the front view of the machine spindle and axis. X_B and Z_B are the position coordinates for B-axis. These data can be found from the manual or by measurement.

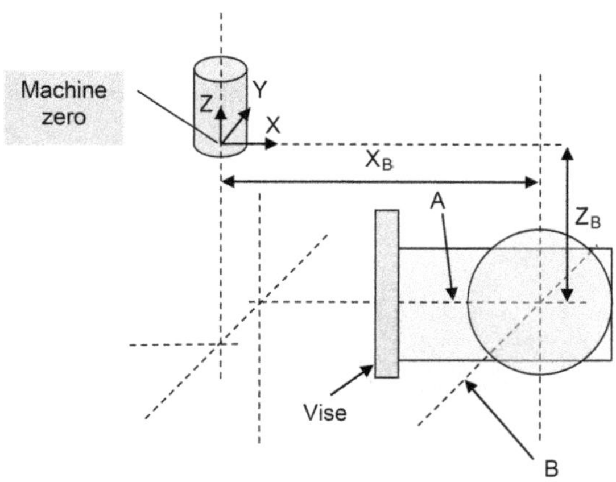

Figure 5.32 Front view of the machine spindle and axis.

5.3.3.5.2 Position of the A-axis Figure 5.33 shows the top view of the spindle and axes, and the coordinate position of the A-axis. As the figure shows, there is an offset between the rotation center and the vise center shown by X_{offset} and Y_{offset}.

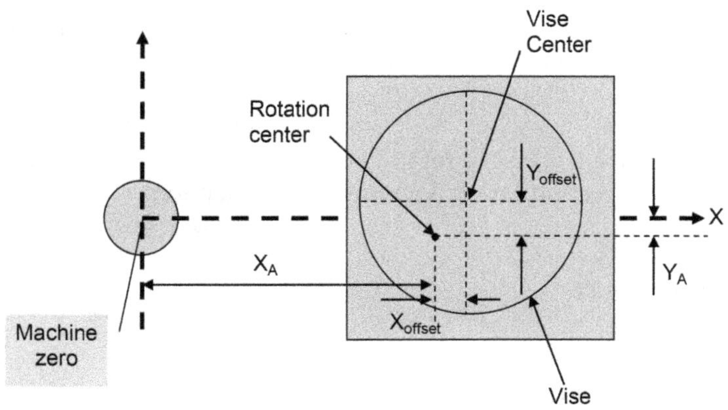

Figure 5.33 Top view of the spindle and axes.

5.3.3.6 Limitation of machine mobility around A- and B-axes

The direction of motion is defined as per ANSI/EIA RS-274-D. Since the vertical machining center (VMC) rotates the workpiece, the rotary head must rotate clockwise to achieve a counterclockwise tool motion and vice versa. The positive sign is assumed. The negative symbol must precede the angular amount as shown in Figure 5.34. Some tilt rotary tables are set up with the B-axis as the tilt portion and others with the A-axis as the tilt portion.

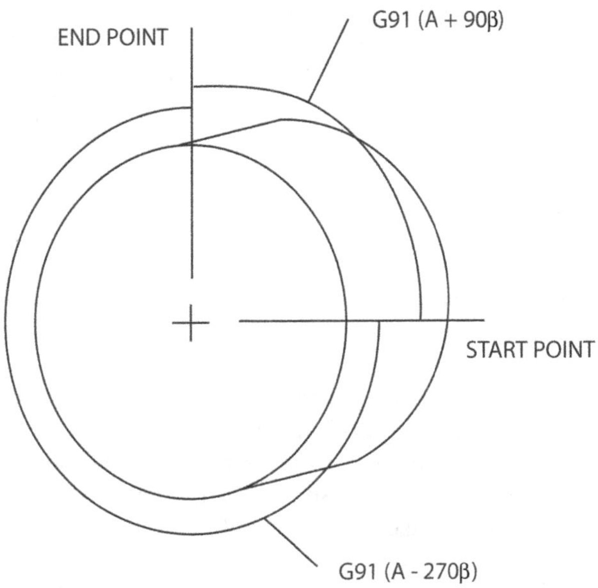

Figure 5.34 The positive sign is assumed. The negative symbol must precede the angular amount. "B-90." is correct, and "-B90." is incorrect.

All rotary table information for a tilt rotary table can be read in the A-axis portion of this section. The VMC tilts must be clockwise to achieve counterclockwise tool motion and vice versa.

5.3.3.7 B tilt table

B+ = Counterclockwise workpiece rotation (viewing in the Y+ direction).
B− = Clockwise workpiece rotation (viewing in Y+ direction).

5.3.3.8 A tilt table

A+ = Counterclockwise workpiece rotation (viewing in the X+ direction).
A− = Clockwise workpiece rotation (viewing in X+ direction).

5.3.3.9 Axis limits

The tilt limits are as follows:

1. B tilt table as shown in Figure 5.35
   - 105° – from the cold start position
   - 15° + from the cold start position

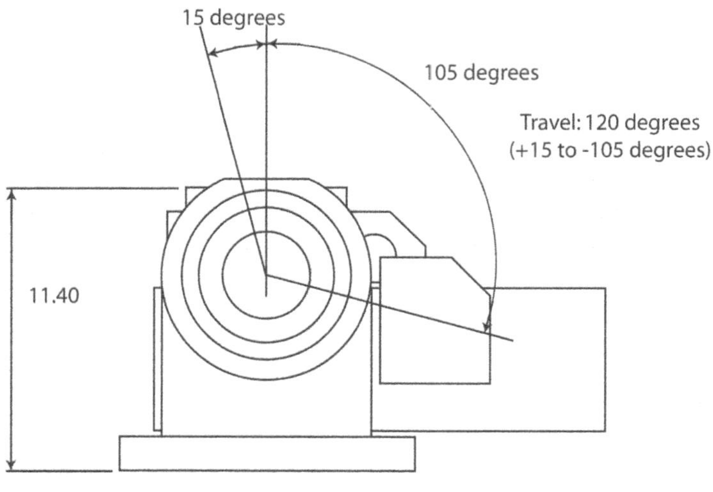

Figure 5.35 B-axis as seen from the back of the machine.

2. A tilt table
   - 105° + from the cold start position
   - 15° – from the cold start position

It is very vital to understand where the machine zero is as this is the reference point considered by the machine controller and all the information in the machine controller is referenced only with respect to the machine center. Also, the position of A- and B-axes should be clearly defined in the postprocessor for the machine. The offset between the vice center axis and the A-axis should also be incorporated.

5.3.3.9.1 CAM/CAM modeling and postprocessing A postprocessor for five-axis Fadal machining has been implemented for the chosen part as shown in Figure 5.36. The generated G- and M-codes were initially 75,000 lines with a resolution of 1 μm. As this was

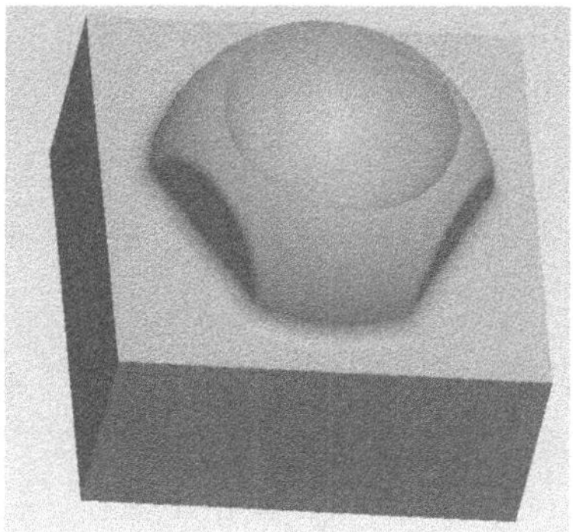

Figure 5.36 A part to be fabricated to illustrate the CAD/CAM procedure.

just a prototype, the code was cut short to 3,000 lines with a resolution of 1,000 μm from the earlier 75,000 lines to increase the operating speed. Since this was the first test, a slow feed rate was used, and thus, the present codes that give rough machining took about 2 h on the Fadal machine. One constraint observed on the machine was the CNC machine memory. A maximum of approximately 1,500 line of code can be loaded and run successfully. This makes the task of ensuring continuity between batches of codes for one big program very tedious. The user needs to be extra careful in doing this job that may otherwise lead to tool crashing. The part in Figure 5.37 had a crash in the center when the codes were broken into three parts, which created problems in running the machine. Distributive numerical control (DNC), which allows a single computer to be networked with one or more machines that use CNC, is recommended to process complex parts. The DNC network is typically a broadband version of Ethernet.

Figure 5.37 Part machined with Fadal five-axis CNC machine using codes generated using CAD software and a postprocessor.

Review problems

1. Explain the difference between the command G00 and G01.
2. What is a postprocessor? Why are postprocessors necessary in CNC machining?
3. What is the relation between postprocessor and CL-data?
4. Which of the following are needed for the development of NX postprocessors?
   A. Toolpath data
   B. G-code
   C. Machine specifications
   D. Postprocessor program

5. Arrange the sequence of the processes of generating tool path data.
   A. Generating the CL-data
   B. Define machine
   C. Define operations
   D. Define the sequence of different machining processes
   E. Define tools
6. Match the following code with the corresponding categories:

X, Y, Z; U, V, W	Feed word
I, J, K	Circular dimension
a, b, c, d, e	Speed word
S	Tool word
F	Distance dimension
T	Angular dimension

7. Write the CNC program to mill 0.625 in. a radial slot having a depth of 0.6 in. as shown in Figure 5.38.

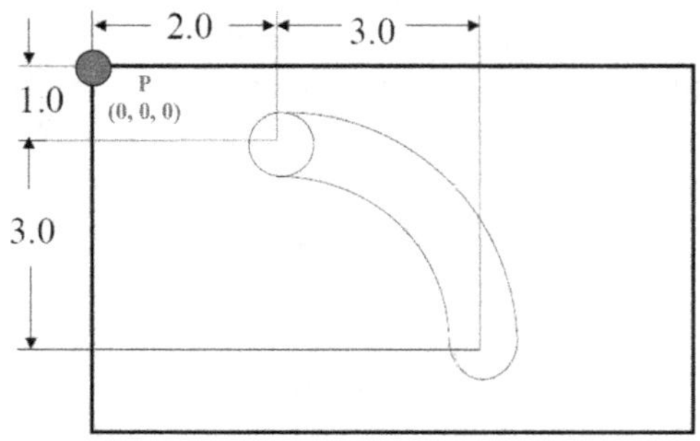

Figure 5.38 Write the CNC program to mill 0.625-in. radial slot (all dimensions are in inches).

8. To mill around the outline of the given part with a CNC mill, the cutter path has been calculated, as shown in Figure 5.39. The cutter path involves linear and circular interpolation and drilling of three holes. The tools include a 0.5 in. cutter and a 0.25 in dia. drill bit assembled in their collets and collet chunks in the CNC laboratory. The depth of the cutter is 0.5 in. Spindle speed is 2,000 rpm, and the feed rate is 16 in./min. Write the part program for CNC milling machining.
9. Finish NC codes to conduct the operations as shown in Figure 5.40. Note: This is a milling process. Processes 1, 2, 3, and 4 are cutting material. The diameter of the cutter is 0.5 in.

N10 G90 G70 G00 X-0.25 Y-0.25 Z 0.1 S1000 T1 M03
N20 G01 Z-0.5 F4
N30 _____________ F8
N40
N50
N60_
N70 G00 X-1 Y-1 Z2

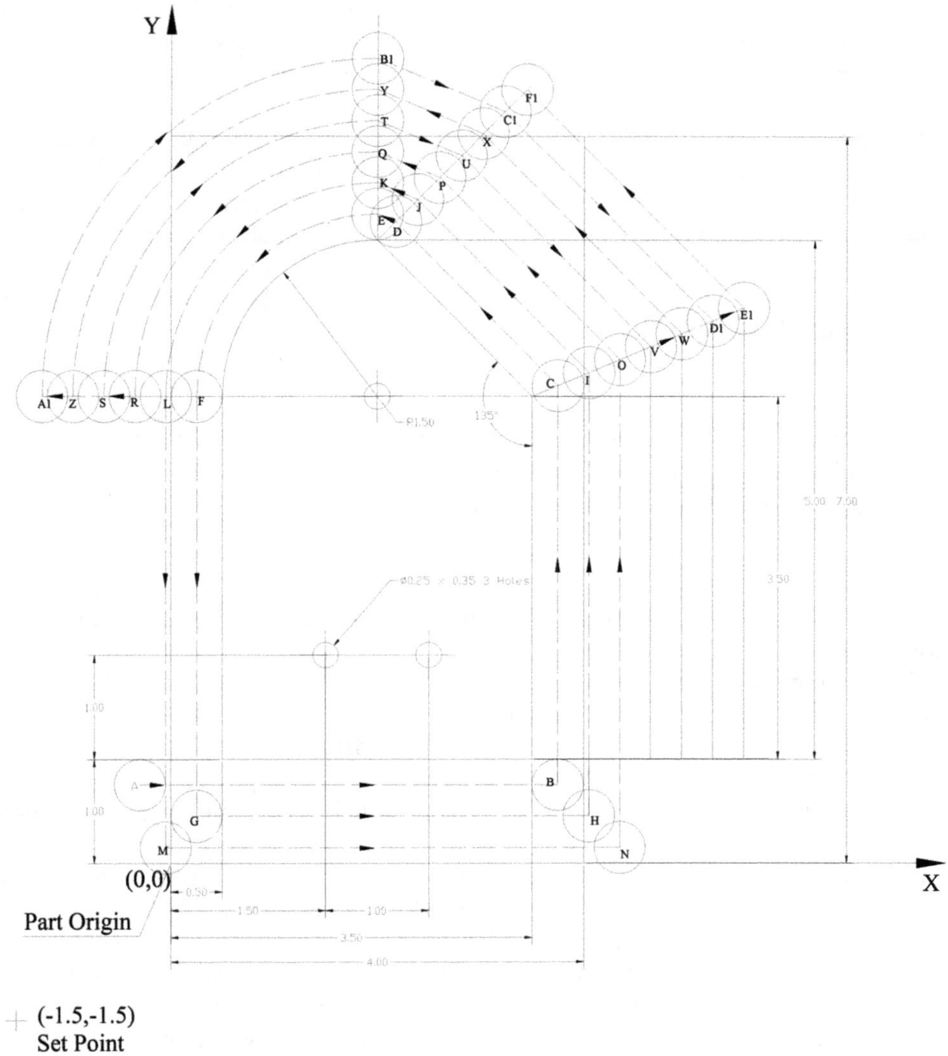

Figure 5.39 The cutter path to mill around the outline of the given part.

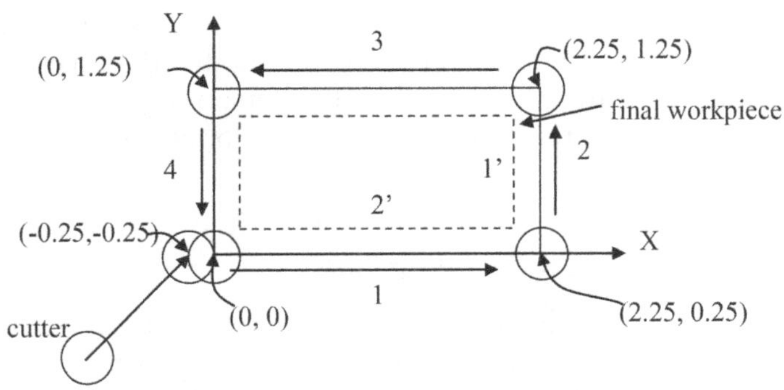

Figure 5.40 Finish NC codes to conduct the operations.

10. Write a G/M-code N/C program that drills a hole at location (4.5, 5.67, 0) 2 in. deep. Use a spindle speed of 700 rpm and a feed of 3 in./min with coolant. Assume the drill bit is already chucked into the milling machine, and its current location is up and away from the part.
11. Write a G/M-code N/C program that creates a 4 in. diameter boss 2 in. deep on a 4 in. × 4 in. × 4 in. block. The center of the boss is in the center of the block. Home point (0, 0, 0) is located in the lower left corner of the top surface. Each cutting pass may take 1.0 in. deep (Z-axis) of material because of the softness of the material. Use a spindle speed of 1,000 rpm and a feed of 5 in./min with coolant. Assume that the 1 in. diameter required end mill is located in the tool holder position 01. Assume the machine head starts well away from the part.

5.4 *Fully automated digital prototyping and manufacturing*

Putting a computer in front of a child and expecting it to teach him is like putting a book under his pillow, only more expensive.

This section helps answer the following questions:

- What would it take to provide a fully automatic digital prototyping and manufacturing system?
- What is computer-aided process planning (CAPP)? What level of part complexity can be handled by current CAPP software?
- What is feature-based design? Why is it important in digital prototyping and manufacturing?
- Is it true that feature-based design will only be suitable for some simple solid models?

In the previous section, the use of CAD/CAM technologies to generate G- and M-codes for CNC machine part fabrication was discussed. This session discusses an even more advanced system, a fully automated CAPP system or generative CAPP system. Process planning translates design information into the process steps and instructions to efficiently and effectively manufacture products. As the design process is supported by many computer-aided tools, CAPP has evolved to simplify and improve process planning and achieve more effective use of manufacturing resources. Process planning can roughly be classified into three classes—manual, variant, and generative. Manual process planning is based on a manufacturing engineer's experience and knowledge of production facilities, equipment and its capabilities, processes, and tooling. A skilled individual examines a part drawing to develop the necessary instructions for process planning. A process plan for a similar part might be retrieved and modified. Process planning is very time-consuming, and the results vary based on the person doing the planning. One can make use of variant and generative CAPP technologies for effective product prototyping and fabrication.

5.4.1 *Process planning and digital fabrication*

Process planning encompasses the activities and functions to prepare a detailed set of plans and instructions to produce a part. The planning begins with engineering drawings, specifications, parts or material lists, and a forecast of demand. The following are the results of planning:

- Routings that specify operations, operation sequences, work centers, standards, tooling, and fixtures. This routing becomes a major input to the manufacturing resource planning system to define operations for production activity control purposes and define required resources for capacity requirements planning purposes.
- Process plans that typically provide more detailed, step-by-step work instructions including dimensions related to individual operations, machining parameters, setup instructions, and quality assurance checkpoints.
- Fabrication and assembly drawings to support manufacturing.

Variant CAPP is based on a group technology (GT) coding and classification approach to identify a larger number of part attributes or parameters. These attributes allow the system to select a baseline process plan for the part family and accomplish most of the planning work. The planner will add the remaining effort modifying or fine-tuning the process plan. The baseline process plans stored in the computer are entered using a super planner concept, which develops standardized plans based on the accumulated experience and knowledge of multiple planners and manufacturing engineers. In other words, this uses a knowledge-based system to quickly prototype a part.

Generative CAPP is a later form of process planning, in which process plans are developed automatically for each new workpiece. It starts with a CAD model and includes a manufacturing database and logic for decision making. Artificial intelligence has also been added to these systems. A fully automated generative CAPP system will be able to prototype or manufacture a product with great benefits. First, as not much user interaction is needed, the prototype or part can be fabricated quickly right after the part's solid model is completed. Second, as design is fully tied to prototyping, manufacturing, and other product development activities, the information from the latter phases can be online providing feedback to the solid modeling stage. For example, when designing a feature of a product, the information such as manufacturability and manufacturing cost can be directly shown to the designer. The user can then choose the optimal design to lower the prototyping or manufacturing cost. This way, many process optimization issues can be resolved in the product design stage and thus will greatly accelerate the product development process. Therefore, a generative CAPP system will be one of the future prototyping systems as it will have the capability of bringing all product development phases into the product definition stage and finding the optimal solution.

5.4.2 Feature-based design and fabrication

CAD systems generate graphically oriented data and may go so far as graphically identifying metal, etc. to be removed during processing. To produce NC instructions for CAM equipment, basic decisions regarding equipment to be used, tooling, and operation sequence need to be made. To automate CAD/CAM integration, two schools of approaches are being developed. One is called a feature-based approach, and the other is a generative CAPP, which uses a standard CAD database such as B-rep or CSG to directly produce NC instructions. A feature is a physical constituent of a part. However, the following information about a part can be considered as a feature: (1) shapes such as drilled holes, ribs or bosses in castings, and grooves in shafts; (2) the required material surface finish of the part; (3) the function of a part; (4) dimensional tolerances; and (5) assembly relationship between one part and another.

The feature-based approach has been successfully implemented. However, general CAD/CAM integration is still under investigation. The feature-based approach and a case study to implement a user-assisted feature-extraction system are discussed below.

For many feature-based modelers, each feature is constructed from a sketch on a datum plane that is extruded, revolved, or swept. That feature contains information for the datum, sketch, and creation properties that are stored and can be recalled to modify the feature. The modifications can be made quicker and easier than topological boundaries. Each of these features can be added or subtracted from the part design. Some feature-based modelers utilize common features or libraries that can be quickly dropped into place and defined, which reduces the amount of development time. Having a feature-based model design enables the use of feature-based CAM interaction.

CAM software relies on features to define operations and if each feature is defined, then determining and locating the features is easier for the software. Since the feature-based design has the features defined, these features can also become the manufacturing data for the fabrication of the detail part. An attempt can be made to generate process plans automatically from the solid model. If it is developed and set up properly, it can provide an effective bridge between CAD and CAM.

The concept behind feature-based modeling is to create geometry based on attributes of the part or object being modeled. For example, the part shown in Figure 5.41 could be generated in two steps. First, the user could create a solid block. Second, the user could remove a defined area, in this case a slot, from the block. For this example, the part has two features: a block and a pocket (slot-shaped). Alternatively, the same geometry could be generated by simply defining the exact coordinates of the corners of the block and the exact coordinates of the geometry required for the slot. Feature-based modeling techniques more closely resemble the attributes considered during fabrication. In this case, during fabrication, the slotted pocket will be removed from a block similarly to how the geometry was defined during the CAD of the part using feature-based modeling.

Feature-based modelers allow operations such as creating holes, fillets, chamfers, bosses, and pockets to be associated with specific edges and faces. When the edges or faces move because of regeneration, the feature operation moves along with it, keeping the original relationships. Figure 5.42 shows some machining features used in QTC [Albuquerque00].

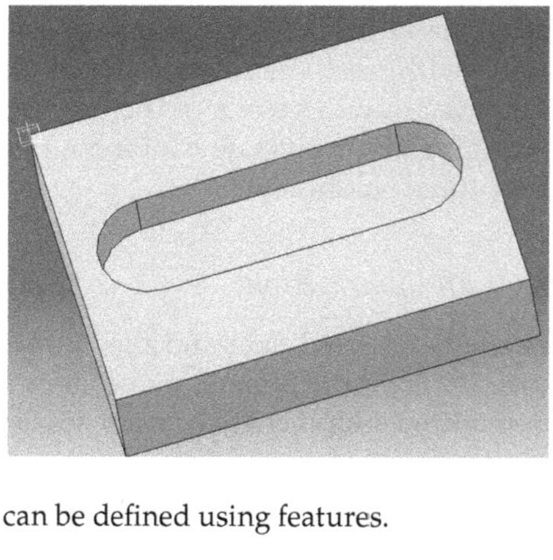

Figure 5.41 A part like this can be defined using features.

Note that these features are defined based on the capability of a machining tool so that when a part is designed using these features, they can be machined. Also, some of the machining information, such as tool type, diameter, and tool path, can be directly or indirectly obtained once a design is completed.

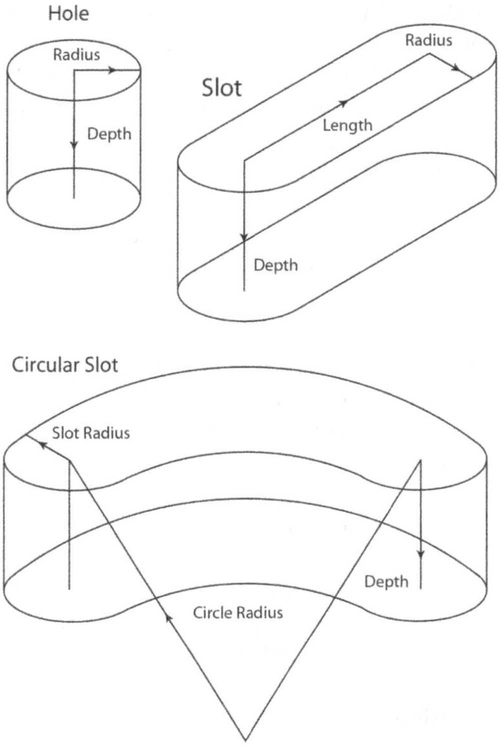

Figure 5.42 Some machining features used in QTC [QTC05].

5.4.3 *User-assisted feature-based design*

The feature-based approach has immediate success in some areas as features can provide an effective bridge between CAD and CAM. However, limitations do occur. The critical step in transforming a CAD drawing into manufacturing information is called feature recognition. There is a lot of literature reported in the area of feature recognition [Chang90, Subrahmanyam95]. Most of the feature recognition algorithms use some kind of pattern matching to determine the machining sequence for some previously existing entity, to divide the cavity into laminae, or to decompose the part into primitives of arbitrary shapes, etc. The obvious limitations of these methods are that they apply to a domain of specific features. However, there are challenges in automated feature recognition, especially when dealing with intersecting features.

The B-rep in its raw form provides the low-level information that is not directly usable for feature recognition. Additional information regarding the type of face adjacencies and relationships between sets of faces need to be explicitly available to assist in feature recognition. To facilitate the recognition process, the concept of an adjacency graph built on the underlying B-rep can be used. The low-level B-rep can then be converted into features that form a higher-level structural entity whose primitives are the elements of a B-rep. A typical application of features during design is to support operations such as repositioning and removing related geometric elements or adjusting their dimensions. This requires the capability of recognizing an intersecting subset of the geometry as a feature.

Recognition can be made a lot easier if the user can tell the program what to look for and pick one or more entities of the feature to get the process started on the right track. This is easy to do when a graphical image of the part is available to the designer. However, picking all the entities belonging to a feature is cumbersome. Instead, one can automate the process of finding any features that a selected entity is part of. Therefore, use of a hybrid system, which combines interactive and automatic feature recognition capability, is the preferred way to provide robust feature recognition. Once a feature is recognized, an interference check needs to be performed. There should be an option for the user to reject the selected feature. The following are some of the salient features that can be employed in the work:

1. The basic assumption made here is that the human eye is the best feature-recognizer. The human brain has a very advanced set of heuristics that help in distinguishing features. Hence, it would be appropriate to defer to the discernment of the user whenever the feature recognition algorithm is unable to provide the desired results.
2. The time taken for feature recognition can be greatly reduced with some "hints" from the user.
3. The approach here is one of the model growing. From the given B-rep model, known as the original model or OM, features are recognized and another model is created or grown, known as the grown model or GM. The algorithm quits once both the models are totally similar.
4. Once a feature is recognized, the dimensions of this feature are calculated.
5. There could be three modes of model growing, manual, semiautomatic, and automatic.
6. In the automatic mode, the user picks a machining plane. The algorithm tries to detect all the faces resulting due to machining on this face. It then tries to group faces together to detect features. If the algorithm is unsuccessful in finding a feature, either of the other two modes can be used.
7. At any given instance, the user can reject the feature selected by the automatic feature recognition module. The user can tell the module to select an alternate, if possible.
8. In the semiautomatic mode, the user picks a set of faces. The algorithm then tries to fit a feature to these faces. Once again the user can reject the feature given by the algorithm.
9. In the manual module, the user picks surfaces of OM and picks the feature that these faces are part of. One need not pick all the faces making up the feature. The minimum sets of faces must be selected to allow for dimensional extraction.
10. If the dimension extraction algorithm gives incorrect dimensions for a feature, the user can undo the previous pick and start over.
11. If the problem of incorrect dimension still remains, then that feature is ignored.

Review problems

1. What is a feature?
2. Which of the following information about a part can be considered a feature?
   A. Shapes such as drilled holes, ribs or bosses in castings, and grooves in shafts
   B. Material, required surface condition, etc. of a part
   C. Function of a part
   D. Dimensional tolerances

   E. Assembly relationship between one part and another
3. Match the CAPP approach and its characteristics.

CAPP Approach	Characteristics
Manual approach	Identify the similar parts and plan the similar process
Variant approach	Automatically generate a process plan from engineering specification of the finished part
Generative approach	Relies on the knowledge of the skilled individual

4. Which of the following is/are true about approaches to automation of process planning?
   A. The manual approach does not rely on the knowledge of the planner.
   B. In the variant approach, similar parts will have similar process plans using GT.
   C. Generative approach automatically generates a process plan from engineering specification of the finished part.
   D. Generative approach systems rely heavily upon the methods of artificial intelligence or very complex algorithms.
5. Define CAPP and list two shortcomings of today's CAPP systems.
6. What are the commonly used machine features?

References

[Albuquerque00] Albuquerque, V.A., F.W. Liou, and O.R. Mitchell, "Inspection point placement and path planning algorithms for automatic CMM inspection," *International Journal of Computer-Integrated Manufacturing*, Vol. 13, No. 2, pp. 107–120, 2000.

[Chang90] Chang, T.C., *Expert Process Planning for Manufacturing*, Addison-Wesley Publishing Company, Inc., Boston, MA, 1990.

[Hollister17] Hollister, S.M., Reverse engineering, Retrieved on August 30, 2017 from http://pilot3d.com/Reverse%20Engineering.htm.

[QTC05] QTC. Brief QTC user's guide, Retrieved on July 19, 2005 from www.cadlab.ecn.purdue.edu/qtcplm/3.html.

[Subrahmanyam95] Subrahmanyam, S. and M. Wozny, Rensselaer Polytechnic Institute, "An overview of automatic feature recognition techniques for computer-aided process planning," *Computers in Industry*, Vol. 26, pp. 1–21, 1995.

[TI17] Code of Conduct (Rev. A)—Texas Instruments, Retrieved on August 30, 2017 from www.ti.com/lit/szza06.

chapter six

Additive manufacturing processes

> Developing a prototype early is the number one goal for our designers, or anyone else who has an idea, for that matter. We don't trust it until we can see it and feel it.
>
> **Win Ng**

In the past, constructing a physical prototype has been seen as a very time consuming and costly activity. However, additive manufacturing (AM) technologies have changed such prospective for a great deal. Today's highly intense competition pushes manufacturing companies within different industries to apply world-class manufacturing practices. However, these practices require more investment in new technological knowledge to bring high quality and sophisticated products satisfying user requirements not only faster but more economical. To bring the products to market swiftly, many of the processes involved in the design, test, manufacture, and marketing of the products have been squeezed, both in terms of time and material resources. In recent years, new tools and approaches which enable the efficient use of these materials and resources have evolved. These tools mostly involve computer technologies. In product development, time-to-market pressure is the main reason behind these technologies which evolved rapidly in the last few years. AM is a free-form manufacturing process that allows users to fabricate a real physical part directly from a computer-aided design (CAD) model. The CAD model is "sliced" into many layers by any number of software packages that can also prepare the part for whichever layered manufacturing machine is to be used. The part is then built layer by layer without extraneous tools. This process allows us to quickly build geometrically complex parts.

The AM processes have many benefits and applications [Hanson13]. In addition to reducing cycle time, AM processes allow for highly complex structures with internal features, small and intricate shapes, overhangs, and freeform structures; economical production of small-size near-net-shape parts, with complex geometries and wide-ranging sizes; and parts-on-demand production that may lead to mass customization. AM processes also promote design freedom and innovation without adding to the time or cost constraints and enable the fabrication of digital materials, in that materials can be digitally designed and distributed across its volume based on functional need.

This chapter covers the general AM procedure and various AM processes.

6.1 Additive manufacturing overview

This section helps answer the following questions.

- What is AM?
- How has AM technology been revolutionizing the manufacturing industry?
- What are the different types of AM methods?
- What are the benefits and risks associated with the various types of AM? What determines if a method of prototyping can be called AM?

- Why is AM helpful in the industry?
- When would it be better to use CNC (computer numerical control) machines versus AM techniques?

6.1.1 What is AM

As discussed previously, the role of prototypes in today's industries includes but is not limited to the following:

- Experimentation and learning
- Testing and proofing
- Communication and interaction
- Synthesis and integration
- Scheduling and makers

Prototypes enable the product development team to think, plan, experiment, and learn the processes while designing the product. For example, in designing an appropriate elbow support for an office chair, several physical prototypes can be developed to learn about the feel of the elbow support when performing typical tasks in the office chair. Since physical prototypes in AM can be built with speed and accuracy, many of these roles can be accomplished quickly and effectively, together with other productivity tools.

AM is the physical modeling of a design using a special class of machine technology. It involves adding and bonding materials in layers to form objects and thus is also called "layered manufacturing" or "solid freeform fabrication." The advantages of AM include the fact that objects can be formed with any geometric complexity or intricacy, reducing the construction of complex objects to a manageable, straightforward, and relatively fast process. Recently, AM technologies have been used not only for rapid prototyping but also for rapid manufacturing. In some AM processes, materials can even be varied in a controlled fashion at any location in an object, called digital materials.

6.1.1.1 AM applications

Although AM can be applied in almost every industry, some typical AM applications are listed as follows.

1. Communications

   AM applications are commonly used for communication purposes. Most people tend to learn more in a shorter amount of time from a physical model than from drawings. For example, many companies reported a savings of 30%–50% or more in vendor quotes when using AM models to communicate with the vendors. This is due to the fact that when a vendor understands more about an actual part, the quotation tends to be lower as they feel more confident about the bid.
2. Receiving input from tool makers and suppliers

   Another application of AM is to receive direct input from tool makers and suppliers. AM parts will be a much more effective communication tool than plain drawings because a tool designer often can provide useful suggestions on minor changes to the product designer to reduce tooling effort and cost.
3. Tooling applications

   Tooling industry can also directly benefit from AM technology. Just as a part can be designed in a CAD system to make the physical part, the mold of the part can also be made in the CAD system and then produced using AM. Therefore, instead of making the AM part, one can directly make the molds and dies, which can be used for mass production. Another area of tooling application is to repair a worn tool, such as molds and dies.

4. Verify CAD data

   An AM part can also be used to verify CAD databases, especially misaligned holes, interferences, structured ribs in the wrong place, improper mating of parts, and whatever was forgotten in creating the model. Sometimes these errors are difficult to detect in the CAD model but can easily be spotted with a physical AM part.

5. Styling and ergonomics studies

   An AM process is also useful in styling and ergonomics studies. For example, many of the products must fit in some way to the human body such as a helmet, breathing apparatus, gear for the military, and driving masks. These products need trial and error to ensure fit and comfort, and thus, the AM process can provide immediate feedback to accelerate the development process. Recognizing that many design decisions are initially incomplete or wrong, and assuming that design errors will need to be quickly detected or corrected, one will be able to better manage the changes using AM processes. In other words, AM processes can help shorten the overall design and development process by encouraging changes before serious product development begins.

6. Rapid manufacturing

   Since AM does not require tooling, it may produce the final products directly, and thus has the capability to be a rapid manufacturing tool. Manufacturers have now turned to AM to directly produce the final parts when the product batch is small, such as in aerospace and biomedical industry. Many products now require a small volume, and as the demand to obtain these parts quickly continues to rise, AM processes are positioned to leverage their advantages for this purpose. For low-volume applications, AM offers much more responsive production than traditional manufacturing methods. With advancements in material and increases in part strength, AM can directly offer to build more and more functional parts.

7. Design freedom

   Due to the fully automated manufacturing process, highly complex structures can often be fabricated without extra cost. Therefore, AM parts can be designed to be extremely lightweight by optimizing the part structure. Since AM does not require tooling, it brings design freedom without time or constraints of the traditional production environment. AM-enabled designers are not constrained by traditional manufacturability. Thus, they can focus their energy on more creative designs.

6.1.2 *What are the alternatives to AM processes?*

Before the alternatives of AM processes are discussed, it is essential to understand the fundamental automated processes or fabricators. There are three basic fabricators: subtractive, additive, and formative. A subtractive operation, as shown in Figure 6.1, starts with a single block of solid material, and the material is removed until the desired shape is obtained. The examples include machining and grinding processes.

An additive fabricator, as shown in Figure 6.2, is the exact reverse of subtractive where the material is manipulated so that successive portions of it are combined to form the desired object. The examples include welding, soldering, brazing, and most AM processes.

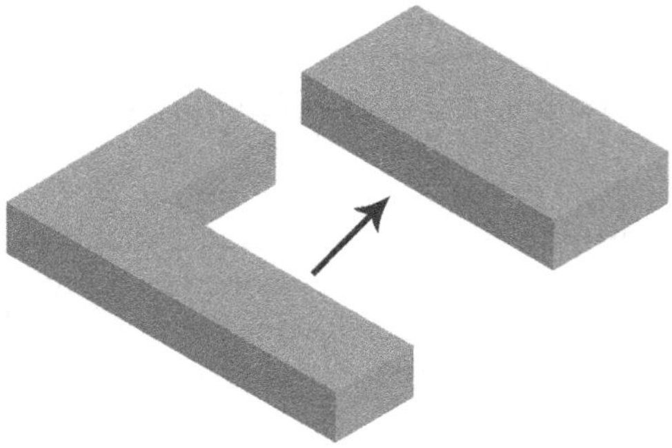

Figure 6.1 A subtractive fabricator starts with a single block of solid material, and the material is removed until the desired shape is obtained.

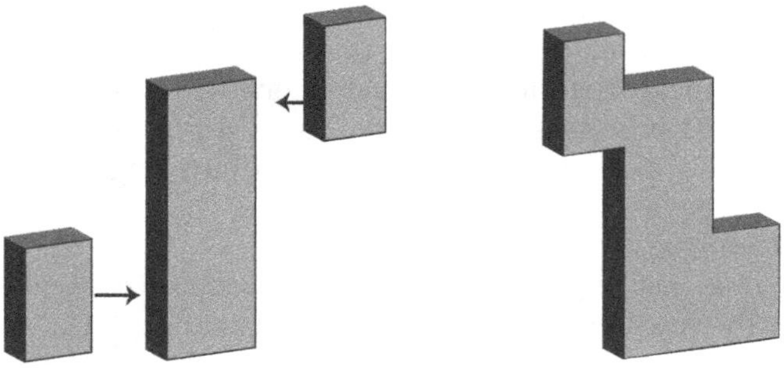

Figure 6.2 An additive fabricator is a process where a material is manipulated so that successive portions of it are combined to form the desired object.

In a formative fabricator, as shown in Figure 6.3, mechanical force or restrictive force is applied to materials so it can be formed into a desired shape. The examples include bending, forging, and injection molding processes.

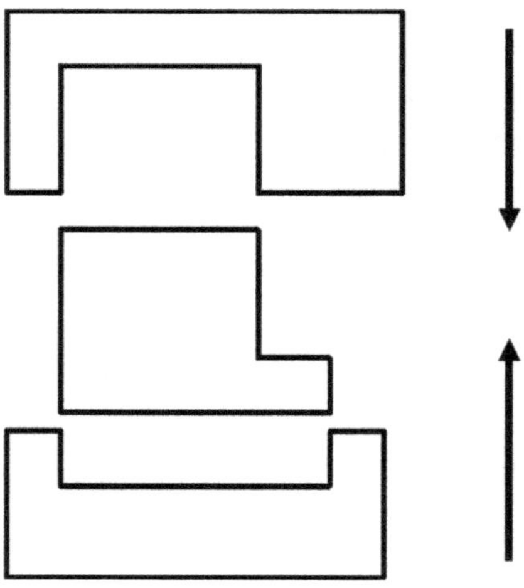

Figure 6.3 In a formative fabricator, mechanical force or restrictive force is applied to a material so it can be formed into a desired shape.

Although most of the current AM processes are additive fabricators, it does not limit the manufacturing processes to just additive fabricators. Future manufacturing processes can be either one of the processes or any combination of the fabricators.

In general, the CNC machining process can be thought of as a prototyping process to rapid prototype parts. A CNC machining process is generally not considered to be a rapid prototyping technology for the following reasons: (1) it still requires skillful human intervention to help plan the operations, (2) custom fixturing and special tooling is often required, and (3) machining has inherent geometric limitations. However, although it is difficult, it is possible to use intelligent CAD to do automatic process planning, tool selection, fixture, and machine setup. This way, the process can be automated and can thus be used for physical prototyping applications. The advantages of using a CNC process for prototyping and manufacturing include almost no limit on materials, excellent dimensional control, and good surface finish. The major limitation is the possible accessibility of machine tools.

In some cases when the part geometry is complex, the AM process can offer faster turnaround time than the traditional CNC process. By adapting the slicing procedure of layered manufacturing processes, a material removal system, such as a machining center, can also be used for physical prototyping. Instead of adding material layer by layer, the material is machined layer by layer to form the final shape. The basics of machining are very straightforward, including cutting off small chips. The process is a reverse layered manufacturing process. This traditional computer-aided manufacturing (CAM) software needs a trained CAM specialist, who can correctly interpret and set the many parameters and must also be able to check the resulting toolpath, as errors are possible that must be prevented by changing the parameter settings. The subtractive approach is aimed at product designers, who do not know much about machining and do not want to be bothered with such know-how. The CAM software for prototyping would work like a "black box," making the prototype creation process as automatic as possible. This does solve the problem that at that moment the time needed to generate the CNC toolpath was too long for real manufacturing. Figure 6.4 shows an example built by a commercial CNC machining software called DeskProto [DeskProto18]. DeskProto is a 3D CAM program. It can import STL (stereolithography) files from a 3D CAD program, calculate CNC toolpath,

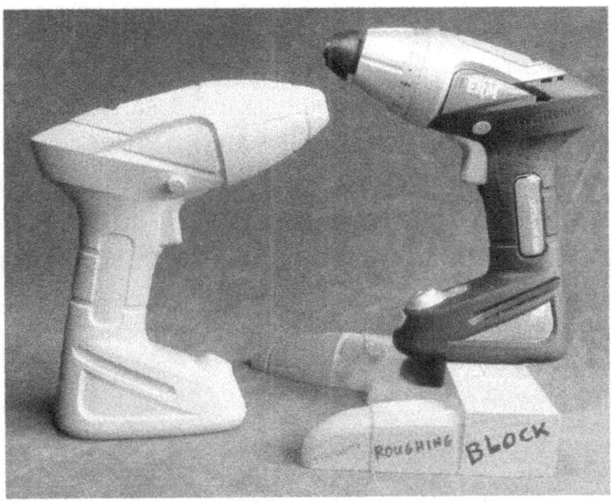

Figure 6.4 An example of a part milled using CNC machining and DeskProto software. (Courtesy of Brandes en Meurs industrial design [Brandes18] and DeskProto[DeskProto18].)

and then write NC (numerical control) program files for most brands of the CNC milling machine (three to five axes).

The advantage of this approach is to enable CNC machining as a fully automatic process or an inverse layered manufacturing process, and thus, it yields CNC machining benefits such as the choices of accuracy in various areas and the choice of materials. The material choices are very broad. Another advantage of the subtractive system is the possibility of making much larger parts.

6.1.3 Producing functional parts

As discussed in the previous section, rapid manufacturing is a trend in AM. Many AM processes are limited by the types of part materials that they can produce. Parts with functional metals are especially the bottleneck. In order to make parts with a functional material, there are direct and indirect methods.

1. Indirect method

   The indirect method is to use AM to make molding and die casting for product volume production. This process creates an AM model as a pattern, uses the AM pattern to make a durable pattern, and then uses the durable pattern for volume production. This way, the AM process can be indirectly used to fabricate metals or parts with other materials. The drawback to this indirect approach is the fact that during the transition, some accuracy may be lost.
2. Direct methods

   The direct fabrication methods for metal AM parts could include powder bed fusion (PBF) processes or directed energy deposition (DED) processes. The PBF processes include direct metal laser sintering (DMLS), electron beam melting (EBM), selective heat sintering, selective laser melting (SLM), and selective laser sintering (SLS). DED processes use a laser or EBM to fuse materials, powder, or wire, together. PBF processes can produce parts with complex geometry, but depending on applications, its part density is limited to about 99%. For some applications, such as fatigue critical structures, it may be limited.

   The laser-based deposition process is a DED technique, which can produce fully functional parts directly from CAD models and eliminate the need for intermediate steps. However, such a process is currently limited by the need of supporting structures—a technology commonly used in all the current AM systems. Support structures are not desirable for high-strength and high-temperature materials, such as metals and ceramics, since these support structures are difficult to remove. Multi-axis systems can offer more flexibility in building complex objects. Laser-aided AM is advancing the state of the art in the fabrication of complex, near-net-shape functional metal parts by extending the laser cladding concept to AM. A hybrid process combines laser deposition and machining processes to build functional metal parts as shown in Figure 6.5. The additive and subtractive processes allow direct production of CNC-level precision metal parts.

Review problems

1. What is AM?
2. What are the advantages of using AM in the design process?
3. How is the AM process different from the conventional machining or forming processes?

Figure 6.5 A metal part fabricated with DMG/Mori LASERTEC 65 3D hybrid. (Courtesy of DMG/Mori.)

4. What is the basic working principle of AM machines?
5. List some of the disadvantages of AM.
6. Identify and give a short description of the three fundamental automated processes (or fabricators) for AM.
7. Which statement(s) is (are) correct about AM?
   A. Involves adding and bonding materials in layers to form objects
   B. Reduces the construction of complex objects to a manageable, straightforward, and relatively fast process
   C. Objects can be formed with almost any geometric complexity or intricacy
   D. Helps shorten the overall design and development process by encouraging changes before serious coding begins
   E. None of the above
8. What are the differences between rapid tooling and conventional tooling?

6.2 *Additive manufacturing procedure*

> Productivity is never an accident. It is always the result of a commitment to excellence, intelligent planning, and focused effort
>
> **Paul J. Meyer**

Expectations:
This section helps answer the following questions.

- How can a CAD solid model be used to drive an AM process?
- How can an AM machine translate the STL file into signals that perform the motion necessary for manufacturing on a prescribed precision?
- What is the significant importance of slicing the geometry into triangles rather than rectangles or circles?
- What are the major differences between creating a prototype on an AM machine and creating one on a CNC machine?

The recent improvements of AM technologies can be closely linked to the developments of computer technologies. The declining costs of computer technologies and advancements in many computer-related areas including CAD, CAM, and CNC machining tools and approaches have completely changed today's factory functions. The existence of AM systems would not have been possible unless these computer technologies evolved with a reduction in costs. Moreover, many other technologies and advancements in other fields such as manufacturing systems and materials have also played pivotal roles in the development of AM technologies.

6.2.1 *Why is AM process faster?*

In conventional CNC machining techniques, a part is manufactured by removal of the material, which is the opposite of AM. Hence, there are many machining preparation steps which have to be taken into account, such as tooling design, tool path determination, and feeding the machine with tool path information. These steps may involve setting up the workpiece for machining, tedious programming, and possibly postprocessing. These steps require extensive planning and preparation in advance. AM is a faster process because it takes virtually no human effort to run the machine. After the virtual model of a part is created, only a few simple buttons can begin the automatic AM processing. By AM processing, since the model is built in layers, almost all structures can be built without an accessibility issue. By machining a physical prototype, the process may be limited by the tools available or space needed for the tools.

6.2.2 *A typical AM process*

There are many different AM processes, but the basic operating principles are very similar. Figure 6.6 shows the data flow diagram of the basic process. It includes the following steps:

1. Construct the CAD model.
2. Convert the CAD model to STL format.
3. Check and fix STL file.

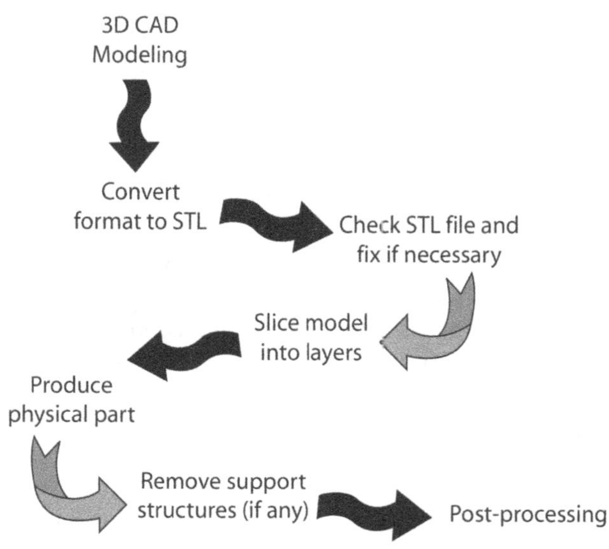

Figure 6.6 The data flow of the basic AM process.

4. Generate support structures, if needed.
5. Slice the STL file to form layers.
6. Produce physical model.
7. Postprocess the physical model.

The AM input can be described as the electronic information required to specify the physical object with 3D data. There are two possible starting models, that is, a computer model and a physical model. A computer model created from a CAD system can be either a surface model, such as an STL model, or a solid model. A physical model can be obtained by digitizing or scanning the geometry of a physical part. The 3D data from digitizing a physical part is not always straightforward. It generally requires data acquisition through a method known as reverse engineering, often with a CMM (coordinate measuring machine) or laser digitizer.

The industry standard for AM is the .STL file, a file extension from **ST**ereo**L**ithography. It is a format that uses a mesh of triangles to form the shell of the solid object, where each triangle shares common sides and vertices. The CAD software generates a tessellated object description. In the STL format, the file consists of the X, Y, and Z coordinates of the three vertices of each surface triangle, with an index to describe the orientation of the surface normal. In many AM processes, the support structure is generated before slicing operation to anchor a deposition or hold overhanging surfaces during the build.

Most CAD packages can export a CAD file in .STL file format, and good .STL files will assure good quality AM models. The STL format can be an ASCII (American Standard Code for Information Interchange) or binary file used in the AM process. It is a list of triangular surfaces that describe a computer-generated solid model. A binary file has a smaller file size compared to an ASCII file, thus is commonly used for AM processing. The facets define the surface of a 3D object. As such, each facet is part of the boundary between the interior and the exterior of the object. The orientation of the facets (which way is "out" and which way is "in") is specified redundantly in two ways, which must be consistent. First, the direction of the normal is outward. Second, the vertices are listed in counterclockwise order when looking at the object from the outside (*right-hand rule*) as shown in Figure 6.7.

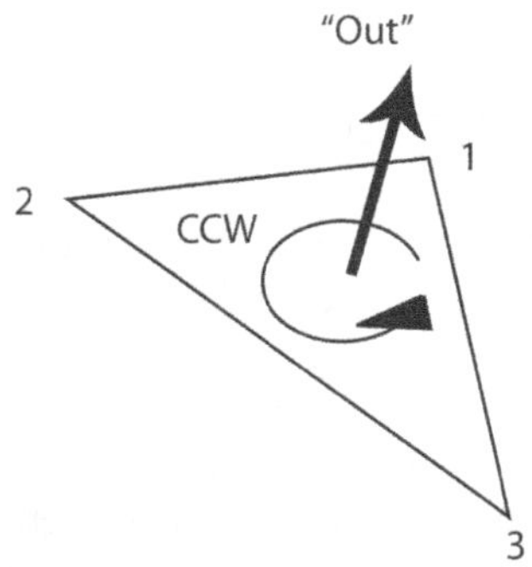

Figure 6.7 The triangle with three vertices. The sequence of the storage of the vertices indicates the direction of the triangular face.

6.2.3 *Why STL files?*

STL files translate the part geometry from a CAD system to the AM machine. All CAD systems build parts and assemblies, store geometry, and generally do many things in their own independent and proprietary way. Instead of having a machine that has to communicate with all of these different systems, there is a single, universal file format that every system needs to be able to produce so that an AM machine can process what a part looks like for "slicing" operation. This is the STL file.

Why is STL format used? The reason is due to the fact that slicing a part using STL format is easier compared to other formats, such as boundary representation (B-rep) and constructive solid geometry (CSG), which will need complex geometric reasoning and data conversion. Figure 6.8 shows the representation of a cube in B-rep. The right-hand side of the figure shows the data structure of the geometric entities. To calculate the interaction between the geometry and a plane that represents the slicing operation is not very efficient. The slicing operation is computed by "intersecting" a ray of virtual lines with the object of interest. In other words, it is necessary to compute the intersections between many lines and the object. The STL format allows us to transfer the slicing operation into a routine of finding the interactions between lines and triangles. This operation judges whether the intersection point is within or outside the triangles, and there are very efficient codes to do just that.

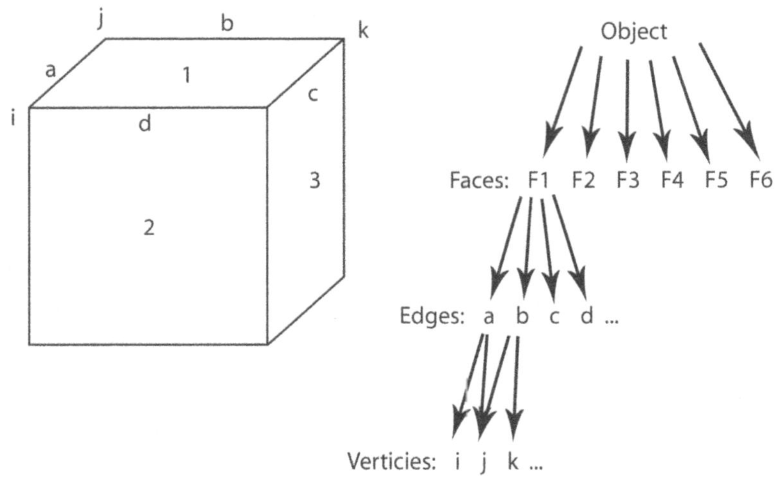

Figure 6.8 B-rep of a cube and its data structure.

The reason that the STL format is the industry standard is due to the fact that it was the first AM data format, it can make the process robust and reliable to get the correct result the first time and the fact that high-end data processing tools, such as surface and STL repair and translation tools, are available in market.

The AM model representing the physical part should be with closed surfaces defining an enclosed volume. The meaning of this is that the data must specify the inside, outside and the boundary of the model. A solid model will automatically satisfy the requirements, as it ensures that all horizontal cross sections essential to AM are closed curves. The internal representation of a CAD model as shown in Figure 6.9 can be in B-rep or CSG representations, whereas its STL representation is shown in Figure 6.10. The STL representation is often used as the standard format to interact with the CAD model and an AM machine. The STL representation approximates the surfaces of the model by triangles meaning that STL files for curved parts can consist of many triangles in order to represent the original geometry well.

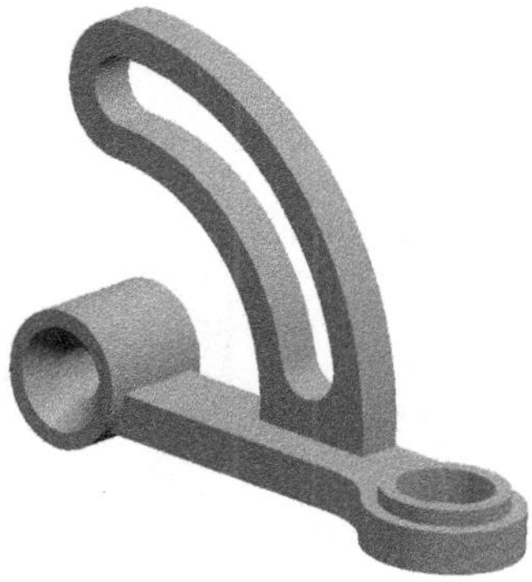

Figure 6.9 An example of a CAD model.

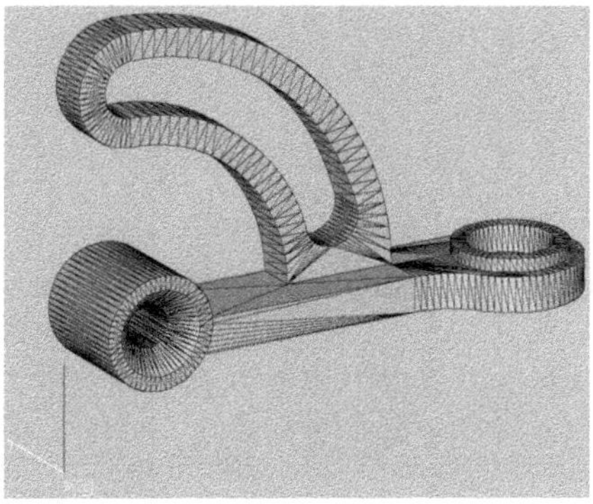

Figure 6.10 An example of an STL triangulation model.

In other words, the CAD models can have smooth curved surfaces, but the AM process must have the model broken down into discrete volumes to build the part. To have a continuous smooth curved surface, the volumes for each discrete piece would have to be close to zero which would require the number of entities to be infinite which makes for large file size in the real world. In order to minimize the file size to something that is more manageable, the system makes the volumes of the discrete pieces larger. The larger these volumes, the fewer are needed to approximate the part. Triangulation, as shown in Figure 6.10, is breaking the model into these discrete pieces, and balancing the number and size of these pieces can make a practical file size without sacrificing too much accuracy.

6.2.4 Converting STL file from various CAD files

Most CAD software can directly output an STL file from a CAD model, but the actual command may change depending on various versions. When generating the STL files, triangular surfaces are used to express the real surfaces of the part. To explain the concept of STL tolerance, Figure 6.11 is used. The left figure shows a square representing the circle and the right figure shows an octagon representing the circle. In order to measure the closeness between each of them, two techniques are used. The first involves measuring the distance between the tangent to the circle and the side of the polygon. Another technique is to find the angle made by the tangent to the circle and the side of the polygon. The latter method serves as a good measure of the degree to which the polygon represents the circle well.

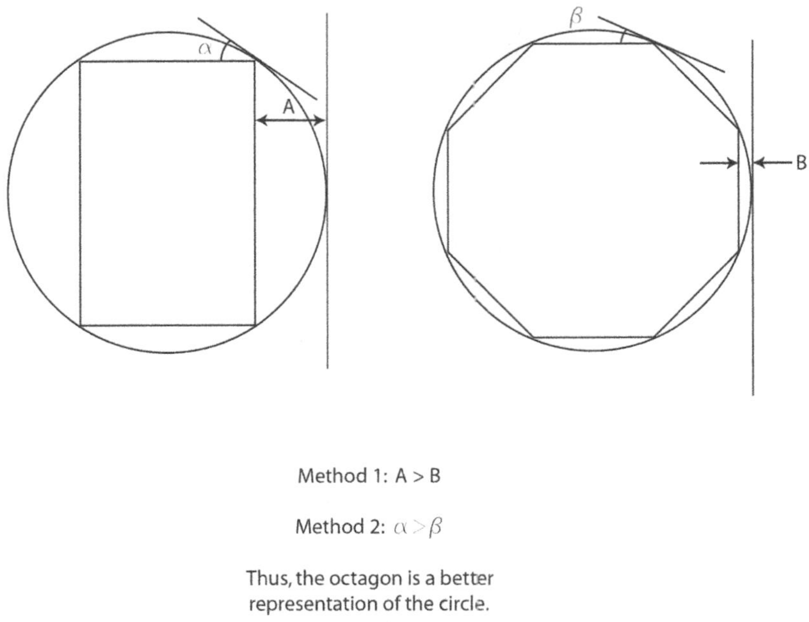

Figure 6.11 The tolerances between a circle and an octagon representation of the circle.

There are two ways for users to control the tolerance of the triangulated model, which can also be explained by the following example as shown in Figure 6.12. The users can input the maximum acceptable angle between the model line and the tangent of the original curve. The users can also input the maximum acceptable distance between the model line and the original curve. As the two figures show, it is easy to find that the smaller the maximum value input by users is, the closer the model line will be to the original curve, and consequently, the smaller the tolerance will be.

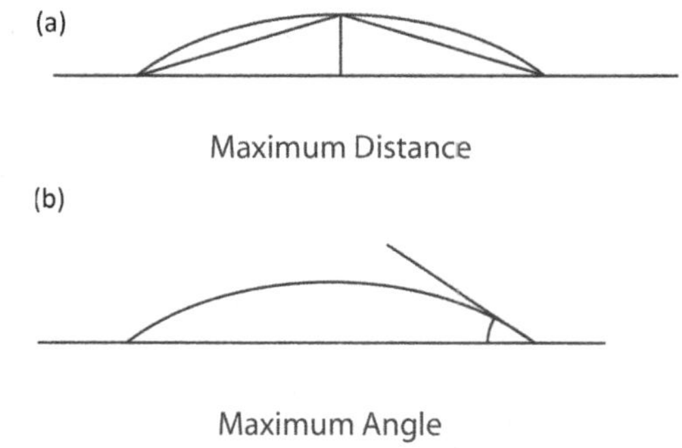

Figure 6.12 Two ways of STL approximation: (a) chord height control—maximum acceptable distance between the model line and the original curve; (b) angle control—maximum acceptable angle between the model line and the tangent of the original curve.

6.2.5 Controlling part accuracy in STL format

There are commonly two options, angle control and chord height, used to specify the precision of a .STL representation. A part as shown in Figure 6.13 was used to illustrate

Figure 6.13 A CAD model used to make STL files.

the impact on the number of triangles and the actual part by varying angle and chord height as shown in Table 6.1. In this table, the first model has the default settings. The file contained a total of 1,280 triangles as shown in Figure 6.14. The representation is not as smooth as the model, but it is a good estimation. By making the chord length smaller, the number of triangles was increased by two times resulting in 2,736 triangles as shown in Figure 6.15. With this file, no noticeable difference was detected from the actual model.

Table 6.1 STL file configurations

Model number	Chord height	Angle control	Number of triangles	Figure number
1	0.261532	0.5	1,280	6.14
2	0.05	0.5	2,736	6.15
3	0.261532	1	1,352	6.16
4	0.261532	0.05	1,296	6.17
5	10	0.5	208	6.18
6	1	0.5	2,208	6.19

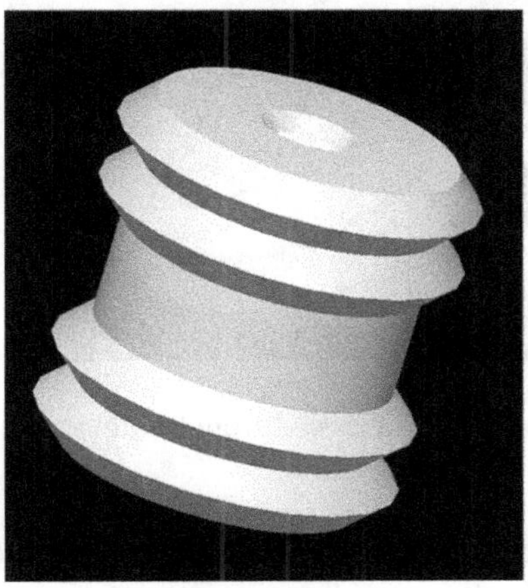

Figure 6.14 Model #1's default values for parameters (STL format); chord height = 0.261532 and angle control = 0.5.

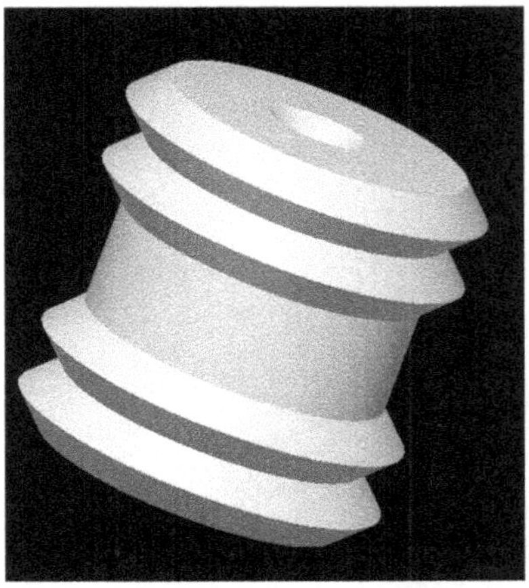

Figure 6.15 Model #2's chord height is decreased from model #1; chord height = 0.05 and angle control = 0.5.

For the third model, the angle control was doubled. This netted similar results to the first model with 1,352 triangles as shown in Figure 6.16. To see what effect the angle control had, it was set at the other extreme of 0.05 in model 4. This file looked the same as model 3 and contained 1,296 triangles. Therefore, in this case, the angle control does not have a major effect on the model as shown in Figure 6.17. In the fifth model, the angle control was left at default, and the chord length was changed to 10. This resulted in 208 triangles and a round part looking square as shown in Figure 6.18. The chord length is set to 1 for the sixth model. This produced 2,208 triangles and a slight circular distortion as shown in Figure 6.19. It can be seen that the chord length is a significant factor in the number of triangles computed.

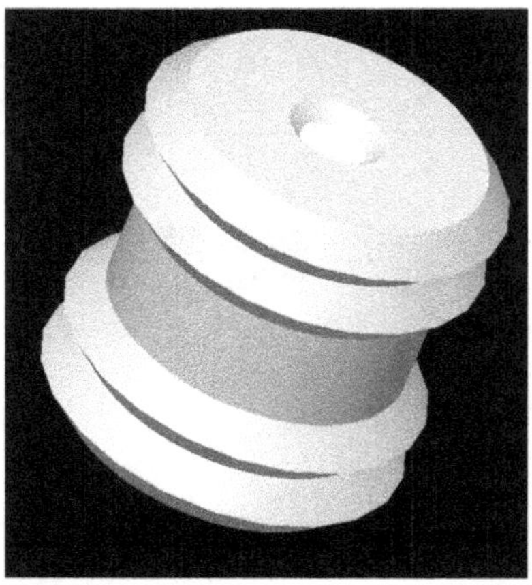

Figure 6.16 Model #3's angle control is increased from model #1; chord height = 0.261532 and angle control = 1.

Sometimes there are errors when generating an STL file, and the errors will often need to be corrected before further processing. For example, all facets in an STL data file should construct one or more non-manifold entities according to Euler's Rule for legal solids: $F - E + V = 2B$, where F, E, V, and B are the number of faces, edges, vertices, and

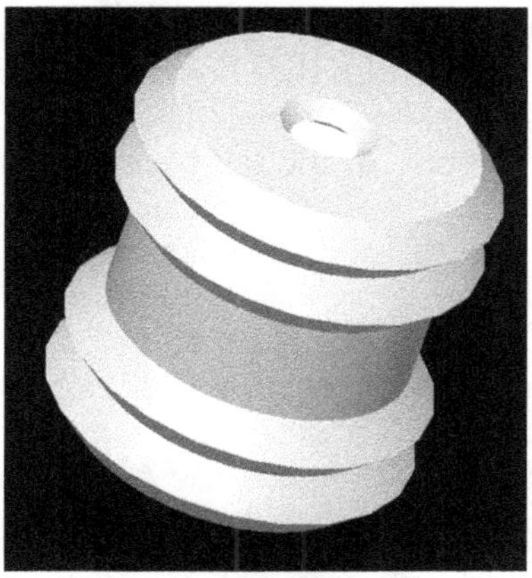

Figure 6.17 Model #4's angle control is decreased from model #1; chord height = 0.261532 and angle control = 0.05.

Figure 6.18 Model #5's chord height is greatly increased from model #1; chord height = 10 and angle control = 0.5.

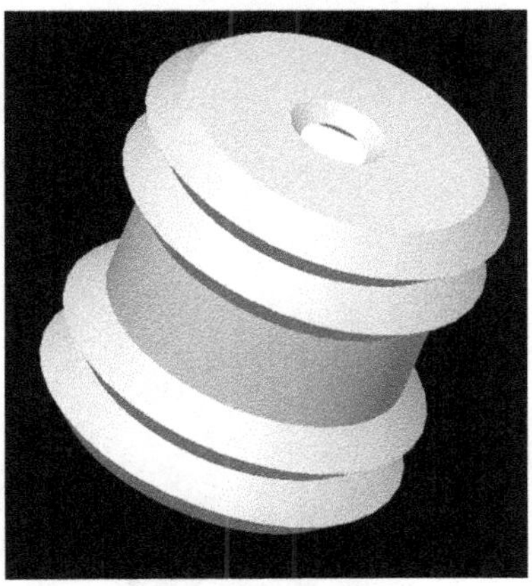

Figure 6.19 Model #6's chord height is increased from model #1; chord height = 1 and angle control = 0.5.

separate solid bodies, respectively. If the relationship does not hold, the STL model is "leaky," and this will cause it to produce slice boundaries that are not fully closed. Other examples can be seen in "degenerated facets." The type of degenerated facets include the following: (1) The three vertexes of the facet are colinear or become colinear when the previously non-colinear coordinates are truncated by the algorithm of importing the application; and (2) the three vertexes of the facet that coincide with or become coinciding when the previously non-coincided coordinates are truncated by the algorithm of importing the application. Another type of error is called model error. These type of errors are not generated during the STL conversion process but rather inherited from the incorrect solid models themselves, often due to the designer's human errors. These errors may cause inconvenience to the AM process. In this case, the AM machine may build a bad part or may even stop building the part during the AM processing.

Once the STL file is obtained, the next step is to select the build direction. Often, a surface of the largest area of a part may be chosen as the bottom surface, and the part is then built along the z-direction. However, it may be different depending on part complexity and support structure design. In most AM processes, the z-direction is usually the weakest axis for tensile strength and a rougher surface finish. The user may rotate the part to fit within the build chamber, keep the higher required strength, and/or better surface finish of the part in the x–y plane of the build. The build time is also dependent on the total length of the tool path and how much support material is deposited. To save support material and/or build time, the user usually needs to keep the support material to a minimum, at least a minimum height in the z-direction. On the other hand, one cannot sacrifice the quality of the part in order to reduce support material height; otherwise, one may end up with a pile of filament or bad surface finish.

To illustrate the process, Figure 6.20 shows the .STL file of a die to be loaded into a software package used to convert an .STL file into a file that can be downloaded onto an AM machine. The first thing to do is to set the build orientation. Often from the STL menu, the user can specify the build orientation, or one can select a preferred orientation from the options it suggests. For example, one can select the build ordination based on either the "maximum horizontal area" or "minimize supports" method for most parts. "Rotate to fit envelope" is useful for particularly large parts that barely fit inside the machine.

Figure 6.20 A tooling (die) example.

6.2.6 Slicing the STL file

The generation of a series of closely spaced 2D cross sections of a 3D object is known as "slicing." Depending on the type of AM machine, the user can generally specify the Z-thickness of the slice. This is often an approximate process. The primary error associated with this is the staircase error as shown in Figure 6.21 because of the fact that each layer is often considered as the same height and width.

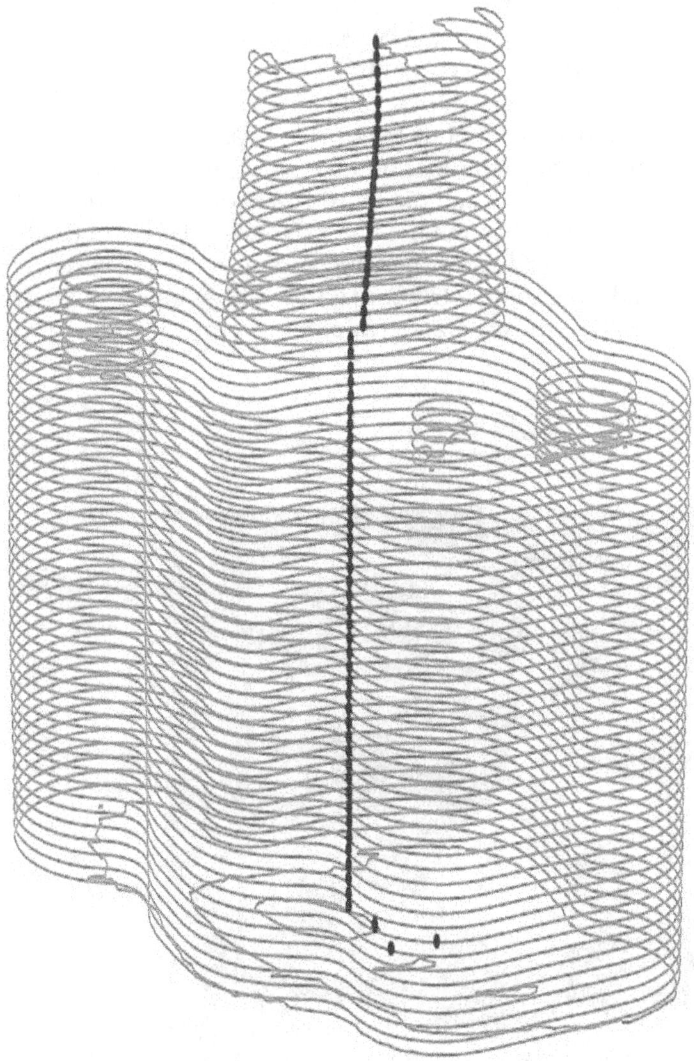

Figure 6.21 The die is sliced into layers.

After the .STL part is properly oriented and positioned, the user then "slices" the part into layers. Figure 6.21 shows the sliced file for the first orientation of the die shown in Figure 6.20. After the part is sliced, it automatically generates the support structure, as most AM software does. For the first orientation of the cone, some layers of support material were generated in AM software under the base of the part.

After the orientation and the support decision have been made, the user then determines the toolpath that the machine will use to build each layer. The toolpath allows a great variety of options to the user depending on finished part strength, build time, or weight and surface finish. One of the biggest variables is the orientation of the toolpath. The software allows the user to set the orientation in each layer. For a more isotropic part, it automatically sets the layers to a ±45° orientation with the exterior boundary curve. It also has the capability to generate a "hollow" part by creating a toolpath with larger air gaps in the middle that forms a mesh-like interior. This sparse fill, as it is called, dramatically decreases the material used, which in turn increases the build speed of the part.

Figure 6.22 shows an airplane example to be sliced with Insight™. The upright orientation will produce better surface quality on the wing structures but will result in a longer build time. Figure 6.23 shows the result of slicing the STL file at the selected layer thickness. The STL file disappears but may be redisplayed if necessary. Each layer is a cross section of the interior and exterior boundary, and one can view a single layer, a group of layers, or all layers in any orientation. Figure 6.24 shows two different build patterns of the cross section of the plane through the wing section, solid build (body cross section of the airplane), and sparse build (below the wing and up above the body cross section of the airplane). The part and the support structure will be sliced together. Figure 6.25 shows the result of the part including the generated support structure.

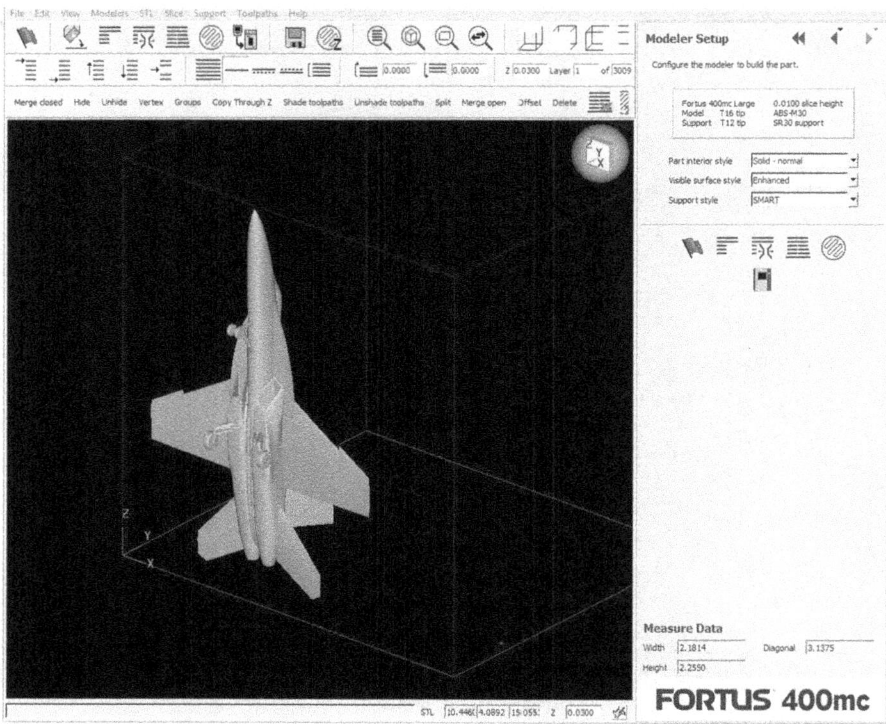

Figure 6.22 The STL view of an airplane example. (Courtesy of Stratasys.)

The default toolpath widths may result in unfilled areas as shown in Figure 6.26. As this is a printer part that was to be used as a functional prototype, the hole was to be reamed out to a precise size, but with the gap, the center would break out making it

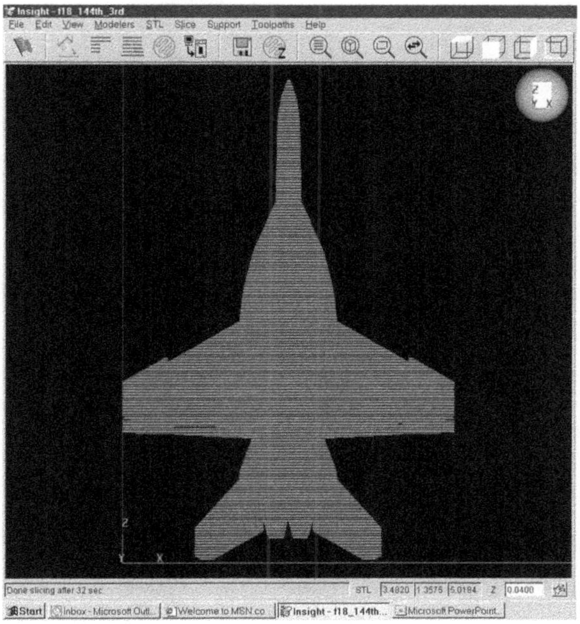

Figure 6.23 The sliced model of the airplane example. (Courtesy of Stratasys.)

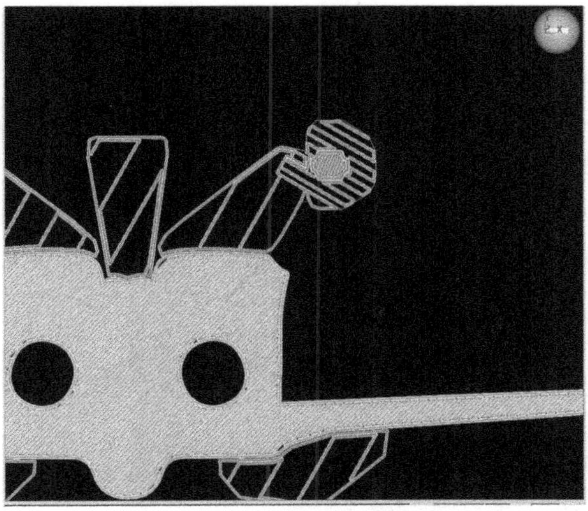

Figure 6.24 Solid and sparse build toolpath of the cross section of the airplane. (Courtesy of Stratasys.)

unusable. Within software, such as Insight™, the user can create custom groups of toolpath parameters as shown in Figure 6.27 to achieve the proper fills enabling them to ream the hole and use the part as a functional prototype.

After the support and toolpath are defined, the location and orientation of the model to be built should be planned. Figure 6.28 shows the location of the airplane to be built. The user can relocate the position and orientation of the model within the build chamber. Multiple parts can be built simultaneously. If the parts are compatible, we can bring in multiple parts

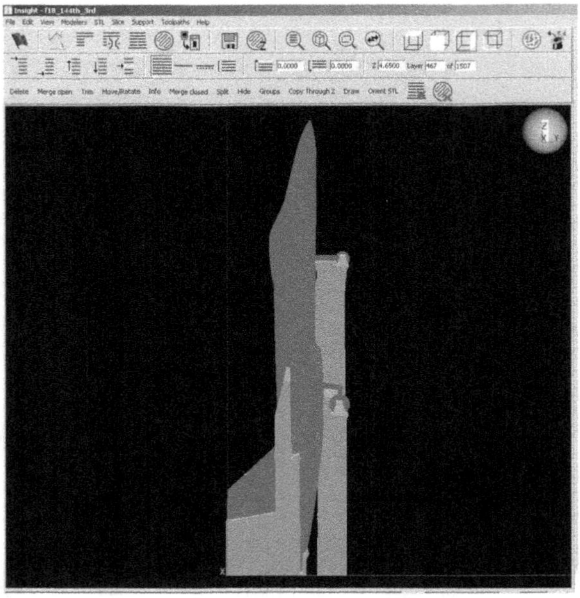

Figure 6.25 Generated support structure is also sliced. (Courtesy of Stratasys.)

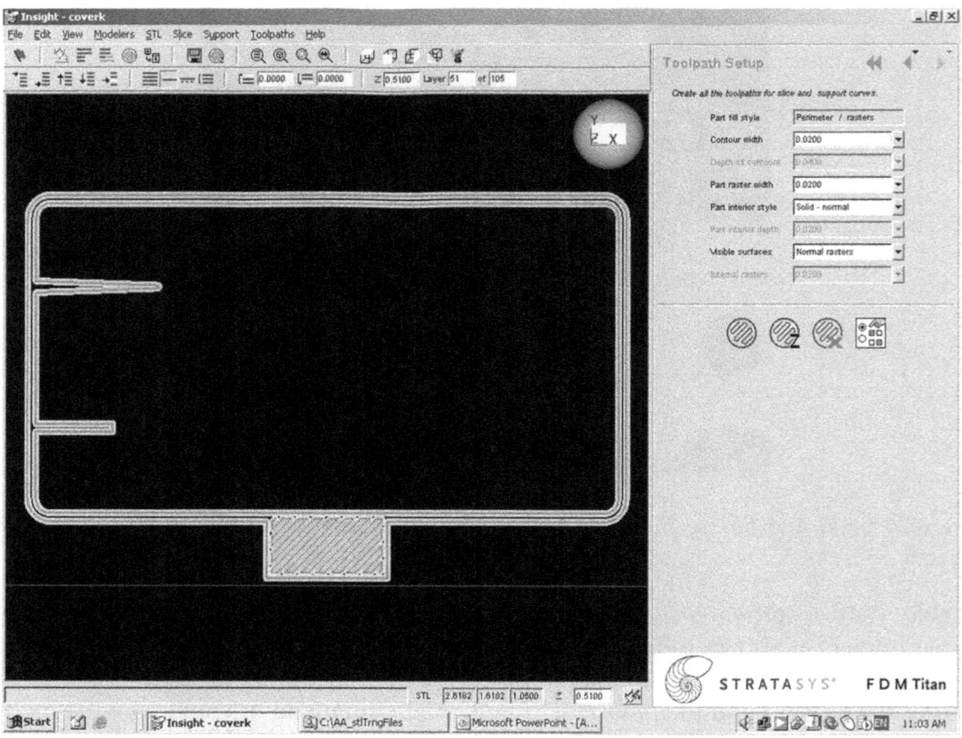

Figure 6.26 Just using the default toolpath widths may result in unfilled areas. (Courtesy of Stratasys.)

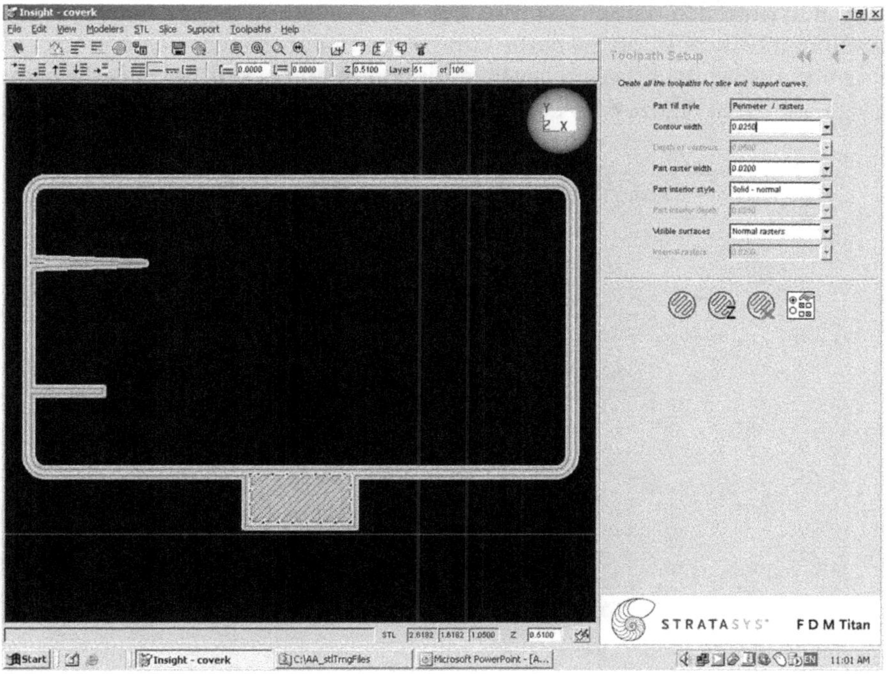

Figure 6.27 Creating custom groups to achieve the proper fills. (Courtesy of Stratasys.)

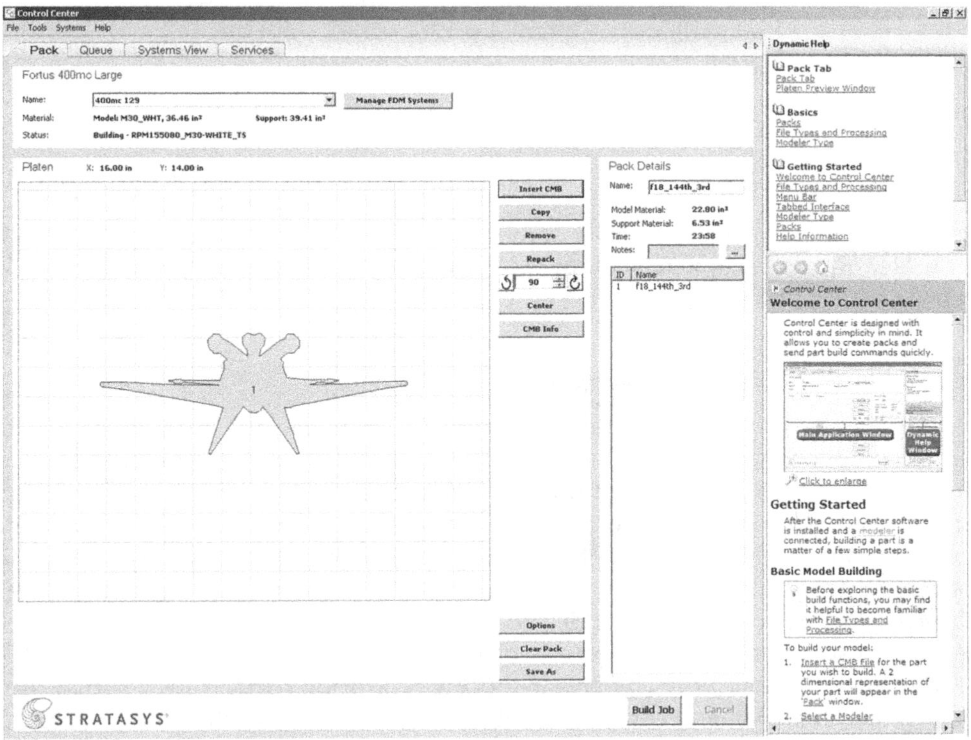

Figure 6.28 Positioning of the airplane to be built. (Courtesy of Stratasys.)

and fit them into the machine. Some software provides a button to rotate items, and a repack button to automatically position the parts within a quarter of an inch of each other. One needs to make sure that slice thickness, or layer resolution of all packed parts, is the same as the warning sign shown in Figure 6.29. One can also click and drag parts to move them around.

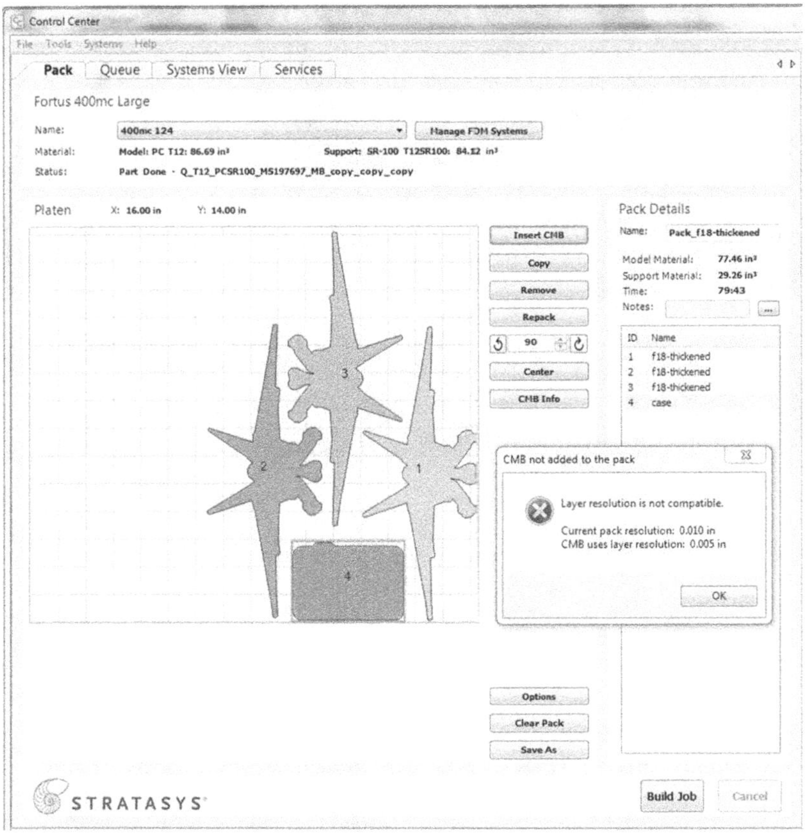

Figure 6.29 A warning sign to remind the user that all packed parts should have the same layer resolution (Courtesy of Stratasys.)

6.2.7 Building an AM part using an STL file

Once the STL file is sliced and transferred to an AM machine, the build process is automated. The support material can then be removed from the build material either by force or by increasing the temperature of the solution to melt the support material but not the build material which has a higher melting temperature. The method of support material removal can vary from process to process. For example, some processes will require support breakaway and/or grinding operation to remove the support material. There are other alternatives. For example, in PBF processes, powder may be used as a support, but some support structures may still need to be removed.

AM processes can be very effective in quickly producing parts. However, they are not without limitations. Some general AM cautions include being cautious of possible limitations in building thin walls and small features, as well as ensuring the multiple part assemblies fit together accurately by considering the shrinkage effect.

6.2.8 AM file format

Up to now, STL format still is widely used for rapid prototyping and CAM. An STL file contains information only about the surface geometry of a 3D object, and there is no information about color, texture, material, substructure, and other properties of the fabricated target object. As AM technology is quickly evolving from producing single-material, homogenous shapes to fabricating multimaterial geometries with functionally graded materials (FGMs) and microstructures, there is a growing need for a standard interchange file format that can support these features.

Current research indicates that several efforts have been made toward the standardized representation of heterogeneous objects to address this issue. Using the information modeling methodology developed for STEP (Standard for the Exchange of Product data), an information model for heterogeneous objects is proposed. AM file format (AMF) standard is jointly managed by ASTM (American Society for Testing and Materials) and ISO (International Organization for Standardization). AMF is an extended format of STL and can describe the color, texture, material, and structure information of the desired object. AMF utilizes the curve triangular mesh to describe the model more accurately and efficiently [Lipson18].

AMF is using the format of XML (Extensible Markup Language), a markup language that defines a set of rules for encoding documents in a format. XML is both human readable and machine readable. XML allows each specific industry to develop its own tag sets to meet its unique needs without forcing everyone's browser to incorporate the functionality of tag sets and without forcing the developers to settle for a generic tag set that is too generic to be useful. By defining the tags, the specific markup language in terms of our desired format is set.

An AMF (.amf) can represent one object or multiple objects arranged in a constellation. Each object is described as a set of nonoverlapping volumes. Each volume is described by a triangular mesh that references a set of points (vertices). These vertices can be shared among volumes. An AMF file can also specify the material and the color of each volume, as well as the color of each triangle in the mesh as shown in Figure 6.30. One can specify the material and the color of each volume and the color of each triangle in the mesh.

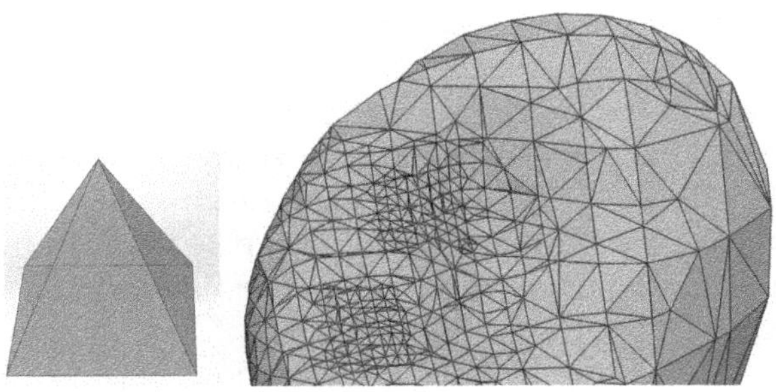

Figure 6.30 The volume structure of an AMF format (left). An object consists of several AMF unit volumes (right).

Using the three new peak points and normal as shown in Figure 6.31, one can split the triangle into four subtriangles as shown in Figure 6.32 and obtain a representation with the better precision. One can repeat that step for each triangle until the desired display or manufacturing precision is reached or until no significant improvement is obtained.

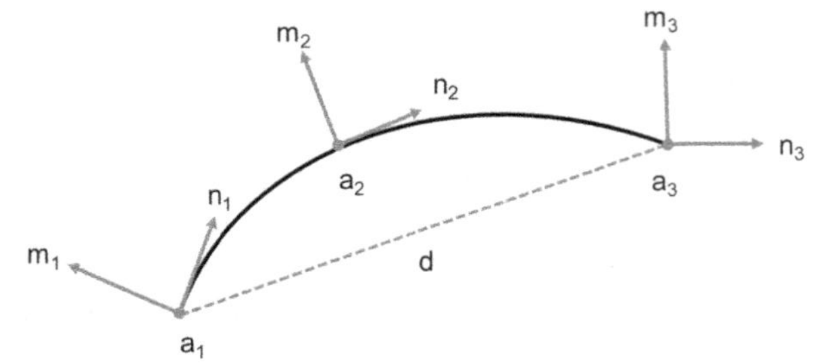

Figure 6.31 Using three new peak points and normal, a triangle can be subdivided into four subtriangles in AMF.

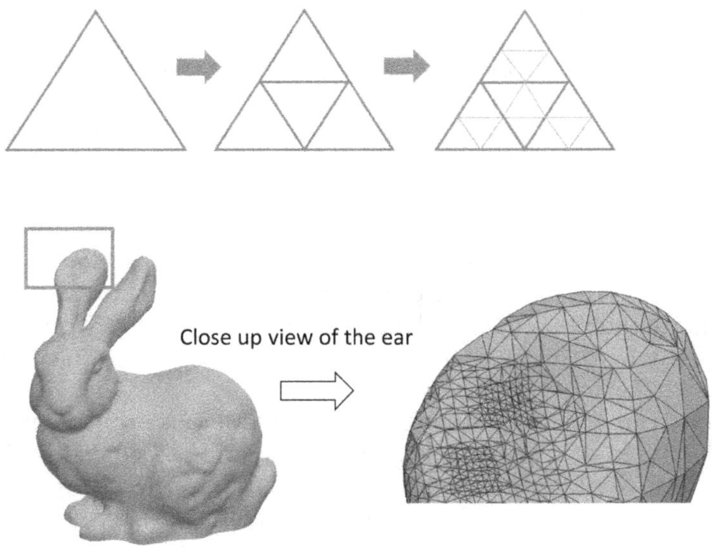

Figure 6.32 Each triangle is split into four subtriangles to provide better precision representation.

The basic AMF file structure includes the following information:

1. Object (parts): defined by volumes and materials. The volumes are defined by a triangular mesh, and the materials are defined by properties, such as <metatdata type="Elastic Modulus"> 2GPa, or by names, such as <metatdata type="Name"> ABS. The material can be graded materials or can be lattice or mesostructure.
2. Texture: color and texture mapping. Color and texture can be assigned to a material, a region, or a vertex.
3. Constellation (optional): specifies the position and orientation of objects to increase packing efficiency and to describe large arrays of identical objects. Constellation can include information such as print orientation, duplicated objects, sets of different objects, efficient packing, and hierarchical.
4. Metadata (optional): specifies additional information about the object(s), such as a name, authorship, copyright information, and special instructions.

AMF format is a volume-based method. Although the volume mesh reduces the amount of data needed to represent an FGM object compared to the voxel representation, it is more complex and not as easy to work with. Moreover, most of the AM machines are compatible with STL files, but not AMF files; thus, this limits the AMF usage. However, as the AM processes are getting more sophisticated, it is expected that AMF file may become more useful. Figure 6.33 showed some representative FGM models using AMF format.

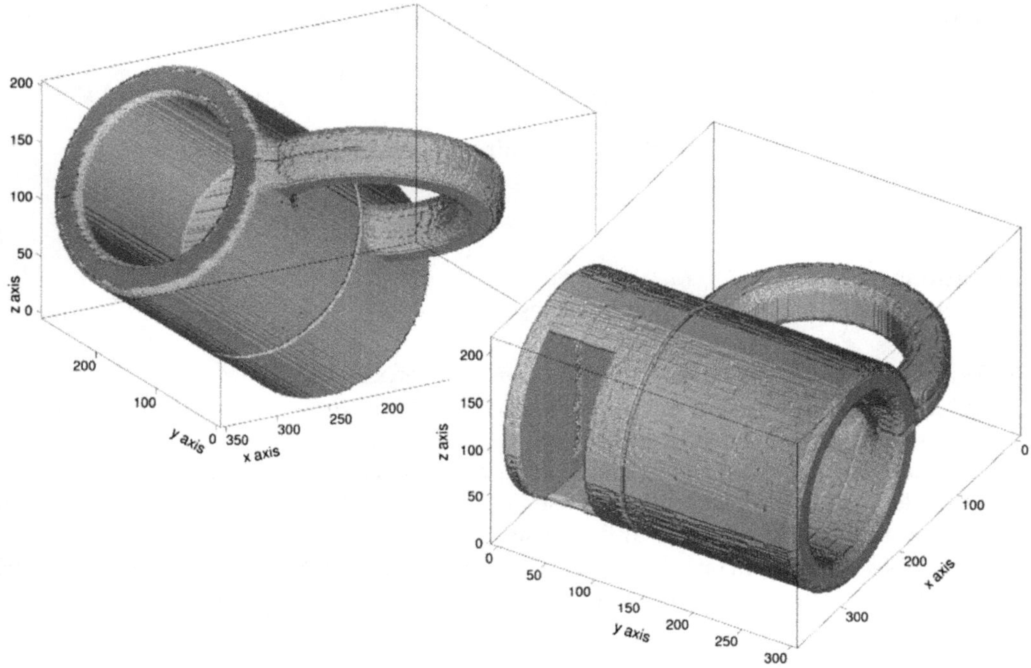

Figure 6.33 AMF representations of the FGM models (in grey level).

Review problems

1. What is an .STL file? How is it sliced?
2. Explain how the user controls the tolerance of a triangulated model by inputting important data.
3. Explain binary format and ASCII format for the STL standard.
4. Use a CAD solid modeling tool to make an STL file.
5. Given the procedures for AM, arrange them in the right sequence:
   A. The physical model is produced.
   B. Check and fix STL file.
   C. A CAD model constructed.
   D. Part postprocessed.
   E. Converted to STL format.
   F. Support structures are removed.
   G. Slice the STL file to form layers.
6. Why is the AM technique more flexible and faster in producing prototypes than that of the traditional fabrication methods, such as machining or casting?
7. Make a solid model from CAD software which can generate an STL file. Generate the STL file (unit: mm) from the CAD software with various resolutions (or various control parameters). Read the STL file and prepare the model in ModelPress Reader (or any other STL viewer) for processing. View the STL files with an STL toolkit. Summarize

your findings on the relationship between resolution control and the smoothness of the solid model. Explain how your software defines the resolution and/or control parameters to output the STL format. Document this example in Word.

Note: There are many STL file viewers on the Internet. For example, one can download a demo version of ModelPress Reader from: www.modelpress.com/download.htm

8. What is AMF file format? How is the AMF format different from the STL format?

6.3 Liquid-based AM processes

Scientists investigate that which already is; Engineers create that which has never been.

Albert Einstein

Expectations

This section helps answer the following questions.

- What are the various types of liquid-based AM processes?
- What are the differences between the various liquid-based AM processes? How to choose among the AM processes?
- What are the advantages and disadvantages of each of the AM processes?
- How do various AM technologies use liquid to build a part? How difficult would it be to maintain the thickness and porosity of the deposited layer?
- How does the raw surface finish of the different types of AM compare to each other?
- Are there any possible issues when using the liquid-based approach? (Any parts that cannot be built?)
- What are the material choices in liquid-based processes?

The objective of this section is to help understand the concept of liquid-based AM processes, such as stereolithography (SLA®) and inject-based processes. Liquid-based AM has its initial form of materials in the liquid state. In general, liquid-based processes can take advantage of smoothness of liquid surface in steady state and result in parts with a quality surface finish. However, its limitation is on the type of materials that can be solidified. Most of the liquid-based processes use a heat source to scan a 2D profile within a vat of heat-sensitive epoxy resin. Through a curing process, the liquid can be converted into a solid state. The heat source, such as UV laser, is chosen to control the curing in a tiny spot to gain good part accuracy.

In general, there are many other ways to solidify the liquid. For example, for the liquid to be solidified, a cold source to cool the liquid point by point is also possible. However, currently, not all of them have been used in AM processes, such as electrical or chemical approaches to change the liquid state to solid state. These represent various opportunities to invent new processes.

There are many liquid-based AM processes, such as 3D Systems™, stereolithography apparatus (SLA®, [3DSystem07]), mask-based process, such as Cubital's Solid Ground Curing (SGC, [SGC07]), and inject-based process. Some of the commercially available processes are introduced in the following subsections.

6.3.1 Stereolithography process

SLA process is the first commercialized AM process and is the representation of the STL process. Patented in 1986, STL started the AM revolution. It works on the principle of solidifying a photosensitive resin using UV laser light, layer by layer to develop a 3D object. STL uses a photocurable resin that can be classified as an epoxy, a vinylether, or an acrylate. Acrylics only cure to about 75% or 80% since curing stops as soon as the UV light is removed. Epoxies continue to cure even after the laser is not in contact. The device as shown in Figure 6.34 consists of a platform that is moved down as each layer is formed in the tank containing the resin. The laser light is moved in the XY plane by a positioning system. In some cases, a support structure has to be created to support the overhanging parts.

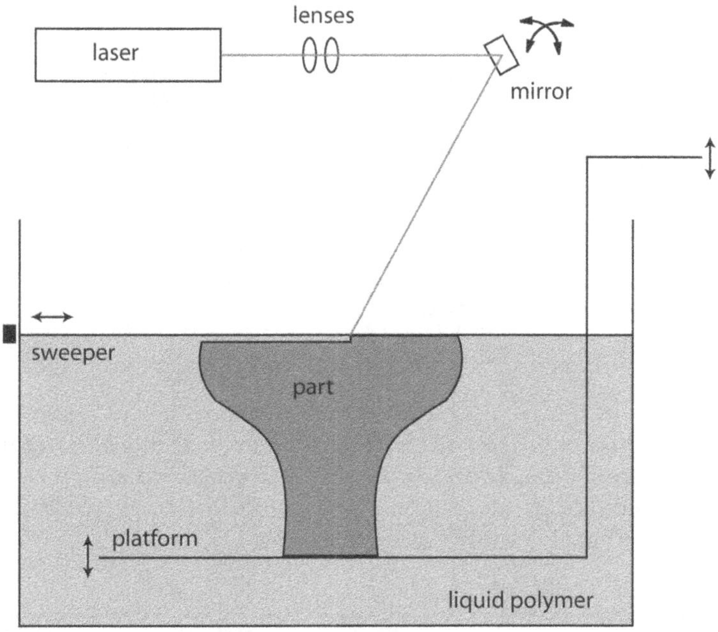

Figure 6.34 An illustration of the SLA process.

The STL process converts 3D computer image data into a series of very thin cross sections, much as if the object were sliced into hundreds or thousands of layers. A vat of photosensitive resin contains a vertically moving platform. The part under construction is supported by the platform that moves downward by a layer thickness (typically about 0.1 mm/0.004 in.) for each layer. A laser beam then traces a single layer onto the surface of a vat of liquid polymer as shown in Figure 6.35. The UV light causes the polymer to harden precisely at the point where the light hits the surface. As shown in Figure 6.35a, the model is built upon a platform situated just below the surface in a vat of liquid epoxy or acrylate resin. A low-power highly focused UV laser traces out the first layer (see Figure 6.35b as the laser is tracing in the middle of the layer), solidifying the model's cross section while leaving excess areas liquid. The UV laser is controlled by a galvanometer scanner to generate X–Y motion, and thus, the table does not need to move in the x and y directions.

Next, an elevator incrementally lowers the platform into the liquid polymer as shown in Figure 6.35c. The laser is tracing from the left. A sweeper recoats the solidified layer with liquid, and the laser traces the second layer atop the first. This process is repeated until the prototype is complete (Figure 6.35d and e). Afterward, the solid part is removed from the vat and rinsed clean of the excess liquid as shown in Figure 6.35f. In all the cases, when a part is built, there is a small structure attached to the bottom called the "supports."

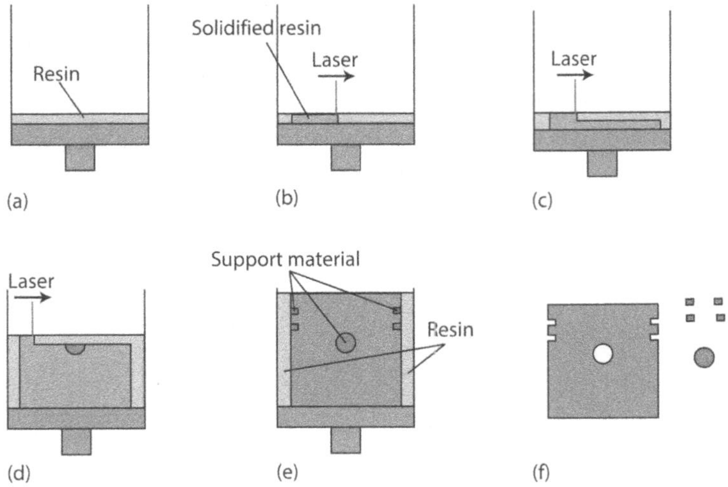

Figure 6.35 A step-by-step STL process: (a) a layer of resin to be solidified on a platform; (b) UV laser selectively traced out the first layer; (c) second layer with laser tracing from the left; (d and e) repeat to build the rest of the layers; (f) the final part after the support structures are removed.

Their purpose is to raise the part off the platform and provide a "bridge"-type structure that only touches the part by small points. This structure is removed after the part is completed. Supports are broken off, and the model is then placed in an UV oven for complete curing. Some SLA systems have adjustable beam size to accelerate part building speed. An example of a dashboard part is shown in Figure 6.36.

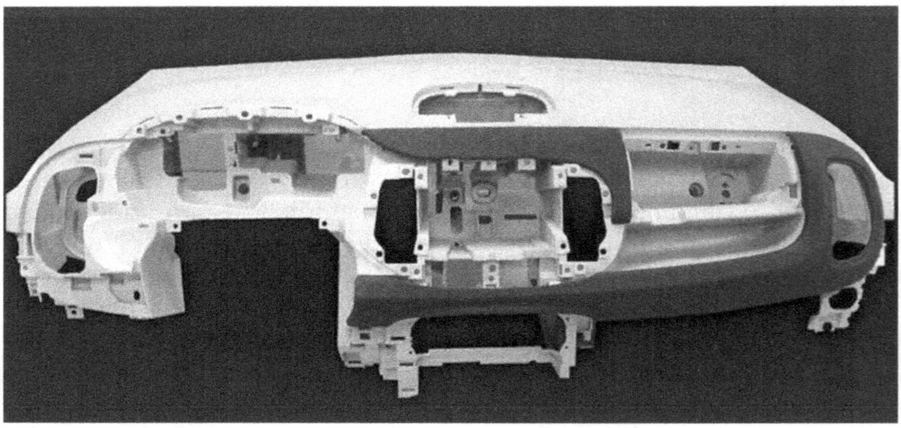

Figure 6.36 An SLA® model of a dashboard part. (Courtesy of 3D Systems.)

Figure 6.37 shows a part using upside-down STL by Formlabs. This is a small machine but much less expensive for SLA processing. One advantage of this upside-down method is more flexible build volume, which can substantially exceed the tank volume. There are many good day-to-day examples that people can readily associate with the benefits of today's and future potentials of the technology.

Figure 6.37 A part being build using an upside-down STL by Formlabs. (Courtesy of Formlabs.)

The uniqueness of this process is its resolution and accuracy. The end product is a very close physical model, or prototype, of the 3D drawing—giving designers, engineers, manufacturers, sales managers, marketing directors, and prospective customers the opportunity to handle the new product or prototype. In this way, design iterations can be made quickly and inexpensively, guaranteeing companies the best product possible, in the shortest time possible. It has become a much-used technology in many industries. Examples are aerospace, armaments, automotive, consumer electronics, consumer products, toys, industrial equipment, medical equipment, surgical applications, and dental applications. Because it was the first technique, SLA is regarded as a benchmark by which other technologies are judged. Early SLA prototypes were relatively brittle and prone to curing-induced warpage and distortion, but recent modifications have largely corrected these problems, and there are several material choices.

Like many other AM processes, STL is an unattended building process—once it is started, the process is fully automatic and can be unattended until the process is completed. It also has good dimensional accuracy. The process is able to maintain the dimensional accuracy of the built parts to within ±0.1 mm. Due to liquid properties, the product produced has good surface finish—glass-like finishing can be obtained on the top surfaces of the part. Although stairs can be found on the side walls and curved surfaces between build layers, 3D Systems Inc. have developed a software called Quickcast™ for building parts with a hollow interior, which can be used directly as a wax pattern for investment casting. This is an excellent example of extending AM processes into tooling applications.

One disadvantage is that resin solidification over time in thin areas will result in curling and warping. The materials available are only photosensitive resin. The parts in many cases have not been fully cured by the laser inside the vat. This is due to the fact that when a laser is curing a spot, the energy is a cone shape as shown in Figure 6.38, and during processing, there are some uncured regions throughout the part, and thus, a post-curing process is normally required.

The major advantage of AM processes is that any geometrical shape can be made with virtually no limitation. However, like most AM processes, liquid processes also have process limitations. For instance, parts with enclosed and hollow structures are problems as the liquid may be trapped inside the enclosed body during building. Unless the liquid can be drained during the process or somehow cleaned up after the build, it remains a process limitation. The areas of applications are restricted due to the given material properties.

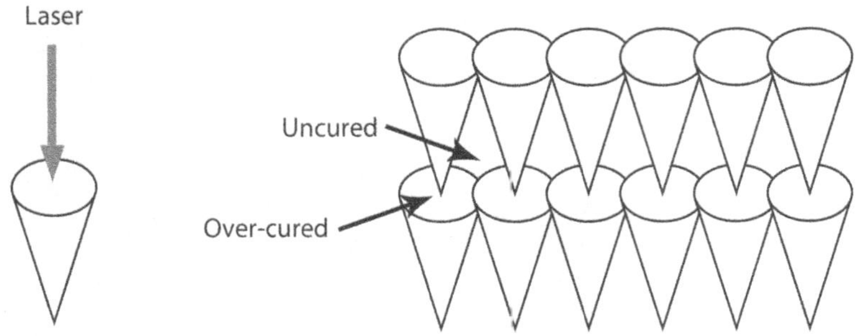

Figure 6.38 The cone features generated by the laser curing process resulting in uncured regions throughout the part.

6.3.2 *Mask-based process*

A representation of the mask-based process is SGC, a process that was invented and developed by Cubital Inc. of Israel. The schematics of a mask-based process are shown in Figure 6.39. The primary material used for this process is resin, and the secondary material used is wax. It works on a principle similar to the previous STL process, but in this case, a whole layer is produced at a time. A mask-based process analyzes a CAD file and renders the object as a stack of slices. The image of the working slice is "printed" on a glass photomask using an electrostatic process similar to laser printing.

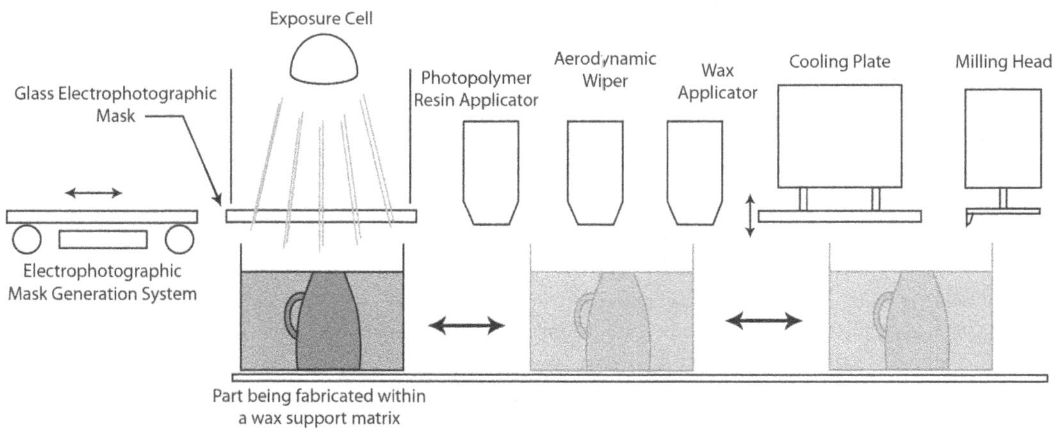

Figure 6.39 The schematics of mask-based process prototyping a coffee cup.

To illustrate further, as shown in Figure 6.40, this is a two-cycle process having a mask generation cycle and a layer fabrication cycle. It takes about 2 min to complete all operations to make a layer:

1. Spray a photosensitive resin on the platform (see Figure 6.40a): The object under construction is given a coating of photopolymer (photosensitive) resin as it passes the resin applicator station on its way to the exposure cell.
2. Prepare photomask (Figure 6.40b): A mask is generated by electrostatically transferring toner in the required object cross-sectional image pattern to a glass plate at a specified resolution. In this process, an electron gun writes a charge pattern on

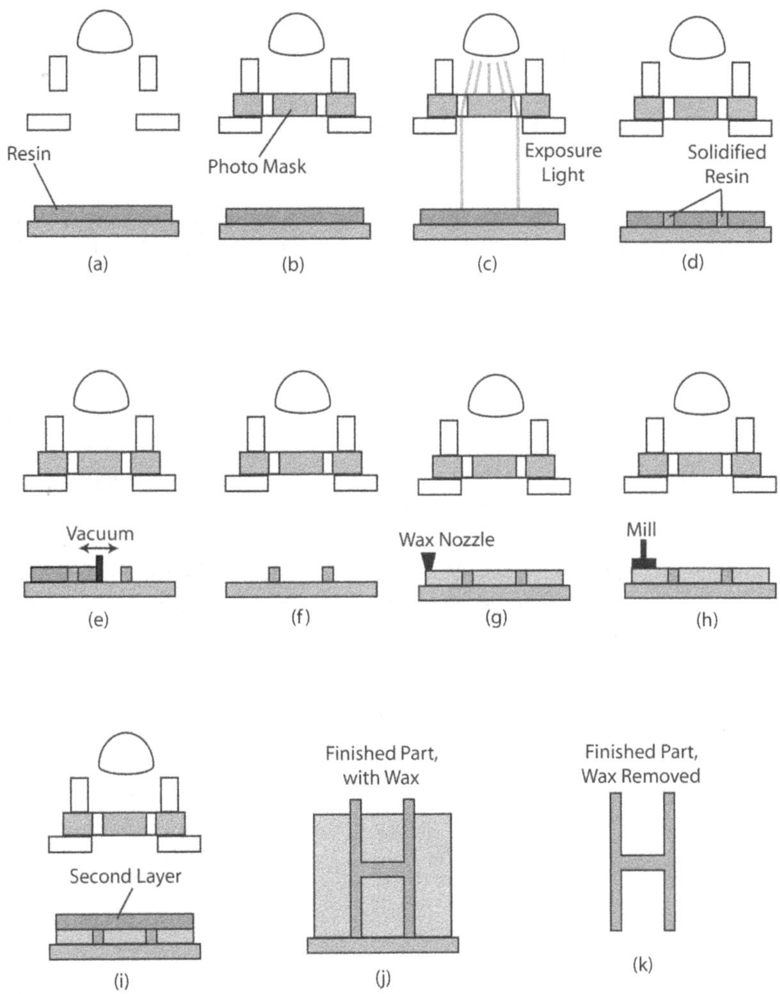

Figure 6.40 The mask-based process.

the plate which is developed with toner. This forms a pattern, which is a negative image of the cross section. The machine will then present black electrostatic toner that adheres to the ion-charged portions of the plate. The transparent areas reflect precisely the cross section of the part(s). The glass plate then moves to the exposure cell where it is positioned above the object under construction.

3. Cure patterned resin (Figure 6.40c): A shutter is opened allowing the exposure light to pass through the mask and quickly cure the photopolymer layer in the required pattern. In other words, the photosensitive resin layer is exposed to UV light through the photomask, and only this exposed portion is solidified, and the remaining area is liquid (Figure 6.40d). Because the light is so intense, the layer is fully cured, and no secondary curing operation is necessary, as is the case with STL. The glass mask is cleaned of toner and discharged. A new mask is electrophotographically generated on the plate to repeat the cycle.
4. Vacuum off liquid resin (Figure 6.40e): The object moves to the aerodynamic wiper where any resin that was not hardened is vacuumed off and discarded (Figure 6.40f).

5. Fill with wax (Figure 6.40g): A thin layer of liquid wax is then spread over the entire layer, filling the areas that before held liquid polymer. The wax, which goes through a cooling process, surrounds and supports the part.
6. Mill the surface (Figure 6.40h): Once the wax solidifies, a milling cutter is used to machine the extra wax from above the level of the photosensitive resin. The layer is milled to the correct thickness and produces a flat surface ready for the next layer. The chips, thus, formed are simultaneously removed by a vacuum pump. Both the wax and photopolymer are milled to a uniform thickness, and the cycle is repeated until the object is completely formed within a wax matrix.

This process is repeated for each layer (Figure 6.40i) and by virtue of the wax. Secondary operations are required to remove the wax. It can either be melted away or dissolved using a dishwashing-like machine. The object is then sanded or otherwise finished. The wax matrix makes it unnecessary to generate extra support structures for overhangs or undercuts (Figure 6.40j). This, along with the large volume capacity of the system, also makes it easy to nest many different objects within the build volume of high throughput.

In contrast to many traditional AM processes, the mask-based process is considered a high-throughput production process. The high throughput is achieved by hardening each layer of photosensitive resin at once. Many parts can be created at once because of the large workspace and the fact that a milling step maintains vertical accuracy. The multipart capability also allows quite large single parts (e.g., 20 in. × 20 in. × 14 in.) to be fabricated. Wax replaces liquid resin in nonpart areas with each layer, so that model support is ensured. In this process, no warping or curling of the part takes place, and the use of the milling operation helps in undoing a wrong operation. However, the cost of the process is relatively high, and it is rather complicated which requires skilled workers to monitor it. Therefore, while the method offered good accuracy and a high fabrication rate, it suffered from high acquisition and operating costs. This led to poor market acceptance.

6.3.3 Inject-based process

One representative inject-based process is a hybrid of material jetting or printing and STL. As engineers are very familiar with the desktop printing process, the process uses printing technology to deposit supports and build material combined with photo or UV curable materials. Unlike some 3D printing machines, it is capable of producing results similar to those from STL processes. Figure 6.41 shows the schematics of Stratasys PolyJet system. In this process, the jetting head slides back and forth along the X-axis, depositing a single super thin layer of photopolymer onto the build tray. Immediately after building each layer, UV bulbs alongside the jetting bridge emit UV light, immediately curing and hardening each layer. This step eliminates the need for additional post modeling curing, as required by other technologies. The internal jetting tray moves down with extreme precision, and the jet head begins building the next layer. This process is repeated until the model is complete.

The process software manages the process, which uses multiple jetting heads. Each head can be individually replaced by the user. The software tools enable the multi-heads to work in parallel, to synchronously jet identical amounts of resin on the tray. This results in an even and smooth surface. Two different materials are used for building; one is used for the actual model, whereas another gel-like photopolymer material is used for support. Similar to other AM processes, the geometry of the support structure is preprogrammed to comply with complicated geometries, such as cavities, overhangs and undercuts, or delicate features and thin-walled sections. It relies on a build platform that moves down in the Z-direction

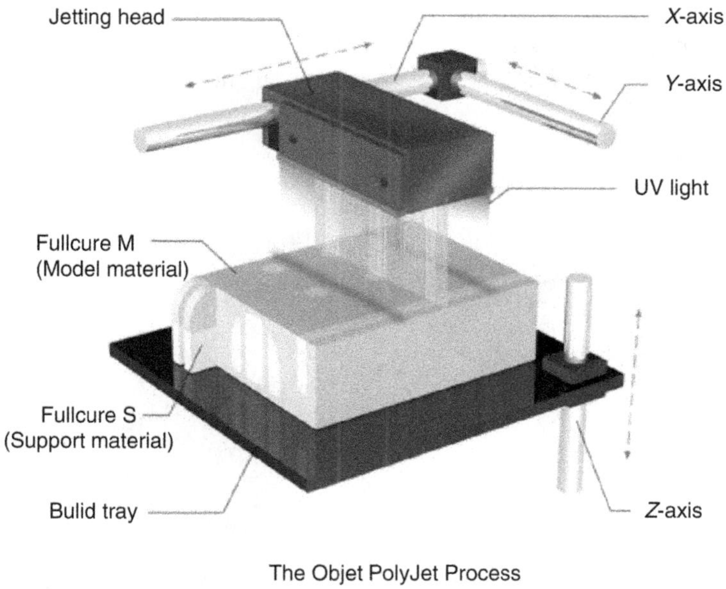

Figure 6.41 The schematics Stratasys PolyJet system. (Courtesy of Objet.)

(vertical), with moving "print heads" which pass over the platform and print both the model material and the support material to build each layer. Once built, each layer is then cured and hardened by exposure to UV lighting. The next layer is then built on top of that and so on. When the build is finished, a water jet easily removes the gel-like support material.

Several materials can be used in the process as shown in Figure 6.42. It shows that digital materials, produced by combining two or more base resins, can be printed. Some

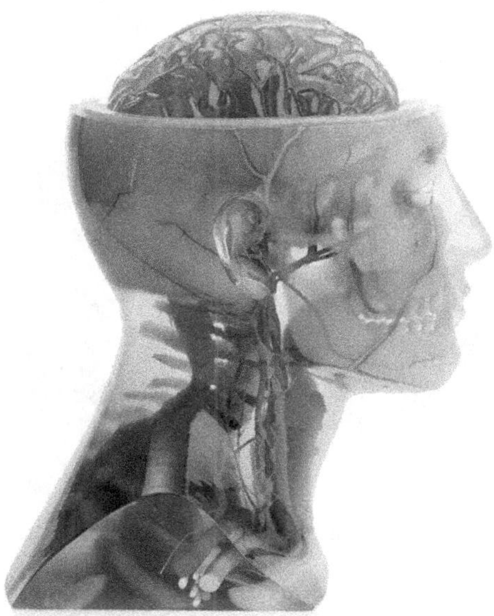

Figure 6.42 Digital materials, produced by combining two or more base resins. (Courtesy of Stratasys.)

materials are rubberlike flexible materials, which enable models that closely resemble the "feel" of flexible products as shown in Figure 6.43. The applications include consumer electronics applications, medical, shoes, toys, general industrial applications, and rapid tooling.

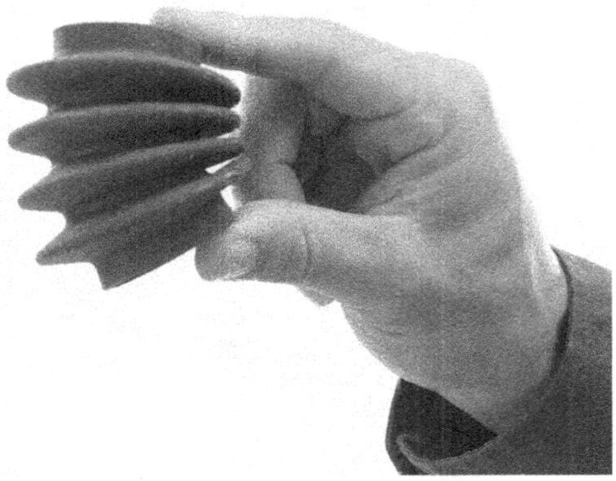

Figure 6.43 Some materials can provide good elasticity with hardness. (Courtesy of Stratasys.)

Part resolution is as good as the inject resolution, such as 600 dpi (dots per inch) in X- and Y-axes, and the resolution in Z-axis could be better, such as 1,600 dpi, which is then translated into jetting a 16 μ layer thickness onto a build tray. This means that the stair effect on parts is reduced, so there is likely to be less of a need for hand finishing. This also results in smooth surfaces for simple to complex geometries. However, this generally slows down the speed. Separation of support and model materials is made with a high-pressure waterjet or by hand. This process allows finishing of most parts within a reasonable amount of time. Figures 6.44 and 6.45 show further example parts built from inject-based machines. The prototyped part closely resembles the target as it has the capability of producing smoother curves, and it significantly reduces the model development time and almost eliminates the post-molding polishing process.

Figure 6.44 A prototype shoe model using inject-based process. (Courtesy of Stratasys.)

Review problems

1. What are the advantages and disadvantages of using liquids in prototyping?
2. Generally, describe the materials used in the STL process.

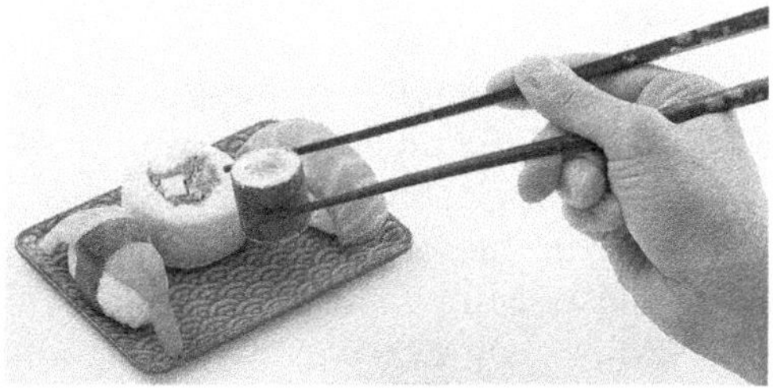

Figure 6.45 A 3D-printed sushi model using inject-based process. (Courtesy of Stratasys.)

3. Compare the advantages and disadvantages of the STL process and mask-based process.
4. Use an example to explain the limitation of the STL process.
5. Why are support structures not needed in the mask-based process?
   a. The material used for the parts in the mask-based process is self-supporting.
   b. Milling removes the need for support structures.
   c. Parts with intricate shapes are not fabricated by the mask-based process.
   d. Wax supports the model, filling the uncured resin areas.
   e. All the above.
6. Are there any problems or geometrical limitations local to the liquid-based AM approach?
7. Name (four) liquid-based AM processes. Briefly explain how each process transfers from liquid to form a part.

6.4 *Solid-based AM processes*

> Sometimes when you innovate, you make mistakes. It is best to admit them quickly, and get on with improving your other innovations.
>
> **Steve Jobs**

Expectations
This section helps answer the following questions.

- What are the various types of solid-based AM processes?
- What are the differences between the various solid-based AM processes?
- What are the advantages and disadvantages of each of the solid-based AM processes?
- How do various AM technologies use solid to build a part?
- How are the overhang structures in solid-based AM processes supported? How is a support structure removed after build?
- How to choose among the AM processes?
- How to compare liquid- and solid-based processes?
- Are there any possible issues when using the solid-based approach? (Any parts that cannot be built?)
- What are the material choices in solid-based processes?

Solid-based AM processes are meant to encompass all forms of materials in the solid state. The solid form can include the shape in the form of a wire, a roll, laminates, and pellets. The presentations of extrusion-based, contour-cutting, and UC (Ultrasonic Consolidation™) processes are introduced in the following sections.

6.4.1 Extrusion-based process

Unlike a liquid-based process, which changes the material from a liquid into a solid state, an extrusion-based process feeds material in wire form and then melts it into a shape and forms a solid. The Fused Deposition Modeling (FDM™) process was originally developed by Advanced Ceramics Research in Tucson, Arizona, but the process has been significantly advanced by Stratasys, Inc. of Minneapolis, Minnesota. FDM™ is a non-laser filament extrusion process that utilizes engineering thermoplastics, which are heated from the filament and extruded in layers to build each model from the bottom up. The models can be made from ABS, polycarbonate, PPSF (polyphenylsulfone), and various versions of these materials. Furthermore, the models are tough enough to perform functional tests.

The FDM processes use a CAD model to produce physical prototypes by taking the STL file created by the CAD model and first converting it into machine code. The machine code is created through the STL file to generate a toolpath for each layer or slice. The machine then uses the toolpath to maneuver ahead to the determined locations to deposit material.

The extrusion-based process utilizes filaments of molten thermoplastic that are extruded from a heated tip to build up layers comprising the physical model. Figure 6.46 is a depiction of the head and process in which the material is pulled off the spool, heated just above the melting temperature, and deposited at the desired location. The key steps are as follows:

1. Start with the filament being fed into the drive wheels.
2. The drive wheels force the filament into the liquefier.
3. The heater block melts the filament.
4. The solid filament is used as a piston.
5. The melted filament is forced out through the tip.

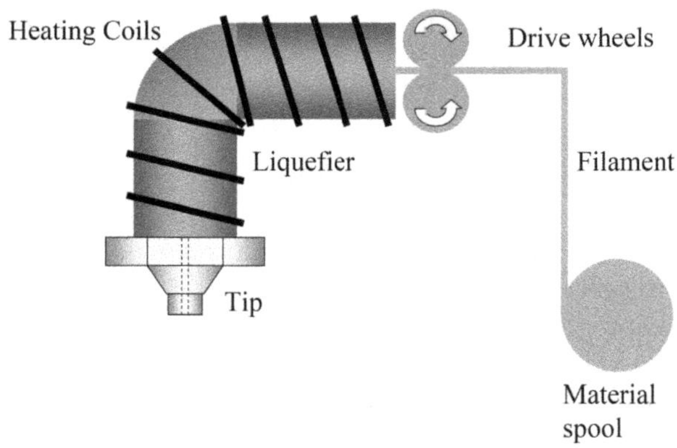

Figure 6.46 FDM drive system. (Courtesy of Stratasys.)

The material used is fed into the head in solid wire form and then liquefied in the head and deposited through a nozzle. The extrusion head can move in the x–y plane and is controlled to deposit thin beads of molten material onto the build platform to form the first layer. The platform is maintained at a lower temperature to ensure the deposited thermoplastic hardens quickly in 0.1 s. After the platform is lowered, the extrusion head deposits the second layer on the first. The material then cools and solidifies in place. The speed of the drive wheels can determine the width of the extrusion path, which is controlled using the software.

Figure 6.47 shows the two FDM heads, one for build material and one for support material. Figure 6.48 shows the schematics of two feeders and nozzles. The build process lays down both the modeling and the support material in separate steps for one layer at a time. To switch between modeling and support material, one nozzle will raise up so it would not interfere with the material being laid down. The appropriate amount of "Z" axis movement is determined by setting within the software. The heads are moved in the X–Y plane by a set of linear motors to improve resolution that hang from the machine ceiling.

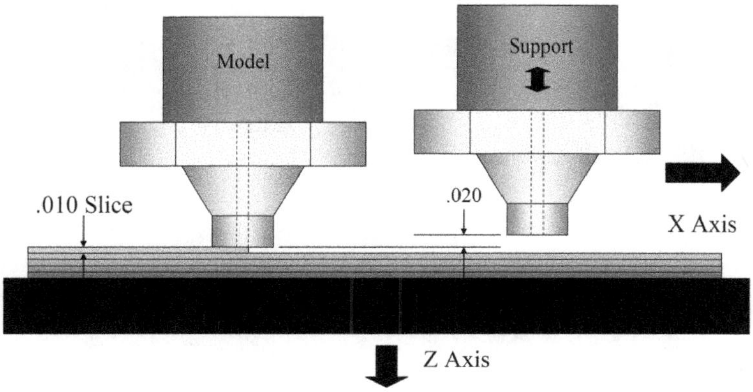

Figure 6.47 The two nozzles (model and support) of an FDM process. (Courtesy of Stratasys.)

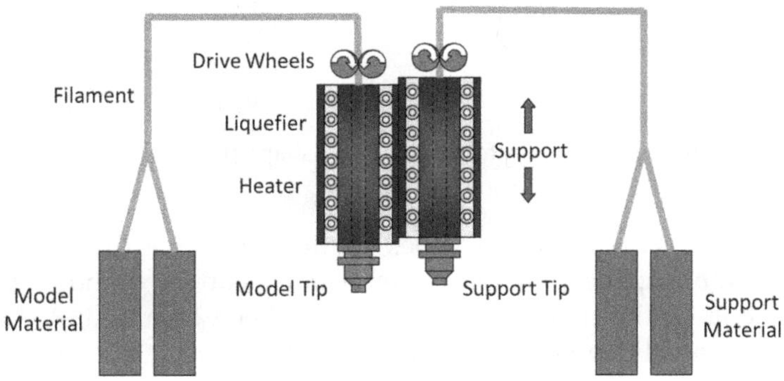

Figure 6.48 Titan dual nozzles and a part being built. (Courtesy of Stratasys.)

Figure 6.49 shows a typical extrusion-based process, where the plastic filament is used as the material for deposition and is supplied from a coil. This plastic filament passes through an extrusion nozzle that is maintained at a temperature high enough to melt the plastic filament as shown in Figure 6.49a. This nozzle is attached to an X–Y positioning system or the stage system. As the extrusion nozzle melts the plastic film, the stage moves the nozzle, depositing the material in a layer. Once a layer is complete, indexing is done in the Z-direction, and the process is continued as shown in Figure 6.49b. The support material, also thermoplastic, is melted and then extruded through a specially designed head onto a platform to create the support as shown in Figure 6.49c. This process is repeated and alternates between the build and support materials until the part is completed as shown in Figure 6.49e. The support material is then removed, and the part is cleaned as shown in Figure 6.49f.

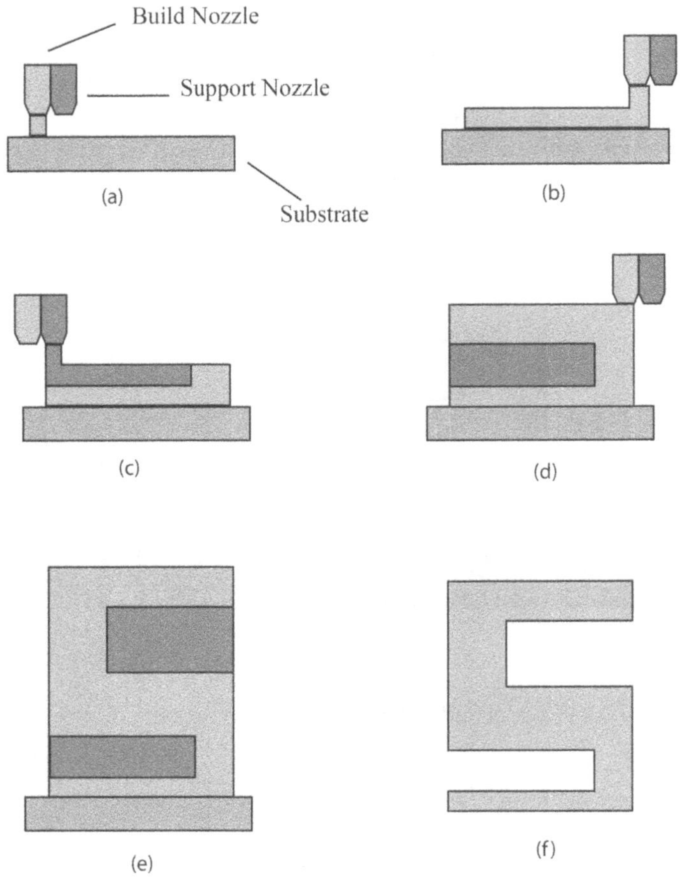

Figure 6.49 An extrusion-based process to make an "S" shape part.

This FDM process is office friendly, does not make noise during the build, produces parts with comparable speeds, and very small parts can be manufactured. In recent years, improvements in the surface finish have also been obtained. Although this process is classified as a solid process, as shown in Figure 6.50, due to the nature of the material, the process actually lays down ribbons of material creating a secure bond from layer to layer.

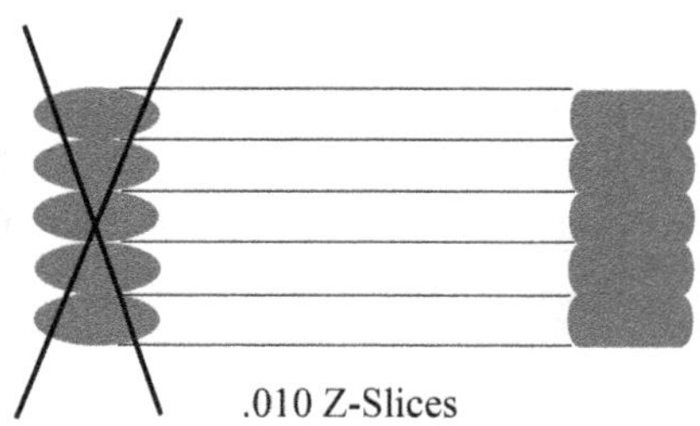

Figure 6.50 Extrusion-based process does not lay down ropes of material to build a wall but lays down ribbons of material creating a secure bond from layer-to-layer. (Courtesy of Stratasys.)

The support in FDM printing may result in marks or damages to the surface in contact with the support. The support material can be difficult to remove from intricate, internal features. The selection of the material for support is important. It is important to note that not all overhang structures will need support. Depending on a specific material, in general, an overhang of, say 45° or less (measured from the vertical), may not need support material. Some FDM processes are designed with a single nozzle (without support material) by using the concept of such angle to build support structures of the same material. The support structures then can be removed.

The support material may be removed in different ways. For some processes, such as the sigle nozzle process, the support can be broken away manually, and the model is then cleaned with sandpaper and tools. The use of the water-soluble material is very convenient as it is easy to remove those structures. For some processes, the support material can be removed in a liquid-based solution with either another energy source, such as ultrasonic vibration, or controlled temperature to melt the support material. The ultrasonic based water-soluble material may be noisy and, thus, should be placed in an isolated room. However, the automated support removal process can free up time, and it also results in a model with greater surface finish smoothness and feature detail.

The extrusion-based process is widely used in almost every industry, including automotive, aerospace, business, commercial machines, medical, consumer products, and architecture. The advantages of the extrusion-based process include functional materials, including ABS (as shown in Figure 6.51), medical ABS, investment casting wax and elastomer, as well as multiple material colors, no exposure to toxic chemicals, and thus, it can operate in an office environment. It can also build with polycarbonate, a durable, high-performance plastic, and PPSF, a durable, high-performance plastic, typically used for medical applications.

Figure 6.51 A wheel part built from an extrusion process.

Another advantage is that the extrusion-based process does not waste much material during or after producing the part. It is also easy to use and safe, uses inexpensive materials, fits on a desktop, produces nonfragile parts, and even uses water-soluble support material. But, it should be noted that the water requires heat, circulation, and also sodium hydroxide in order for the support structure to dissolve. The ABS materials used are very cost-effective. It is also easy to change materials and uses relatively inexpensive binders. Due to nozzle diameter, accuracy can be limited compared to the liquid-based processes and is difficult to build parts with complicated details. The disadvantages are limited materials, limited size, and unpredictable shrinkage. Currently, there are FDM processes that can be used to deposit other materials, such as metals and ceramics [Gibertia16, Desktopmetal18], as shown in Figure 6.52. This can be accomplished either by mixing the metal powder with polymer following a sintering process or by directly extruding the molten metal.

Figure 6.52 An FDM printed part with 80% copper.

6.4.2 Contour-cutting process

A contour-cutting process utilizes the layered manufacturing technique for AM. It, however, differs from other AM systems in that, rather than building up a part by adding materials to a stack through a forming process, layers of sheet materials, such as paper, plastics, or composites are attached to a stack, and a laser cuts away the unused portions. A representative contour-cutting process is Laminated Object Manufacturing (LOM™) developed by Helisys Inc. in 1985. The process begins with a computer slicing a 3D solid model of the part into 2D cross sections. Input data is in the STL file format. The thickness of the computer-generated cross section corresponds to the thickness of the sheet material. The sheets can be 0.001–0.005 in. thick, and a winding and an unwinding roll provide a ribbon of the material. Figure 6.53 shows the schematics of a contour-cutting process. It is capable of producing parts that can be machined or sanded to provide good surface finish, but since the parts are made from paper, they must be sealed from moisture.

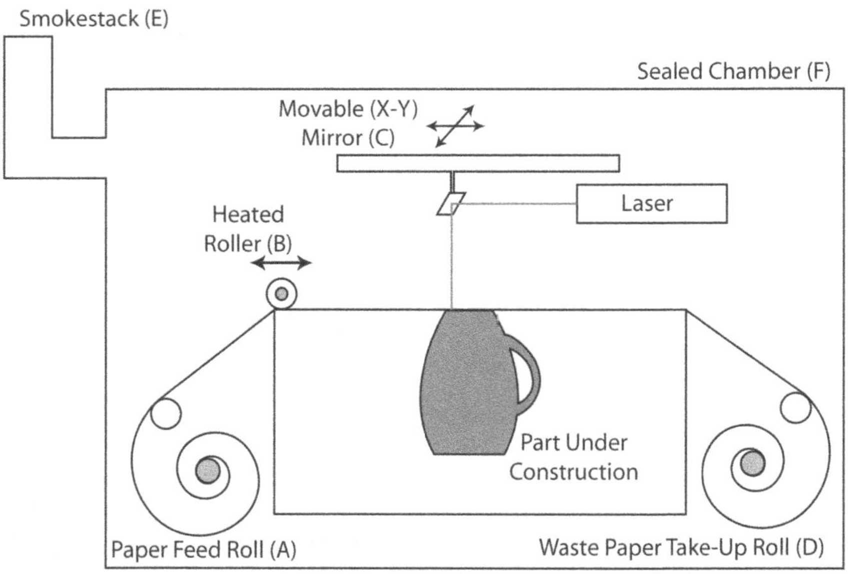

Figure 6.53 The contour-cutting process.

Profiles of object cross sections are cut from paper using a laser as shown in Figure 6.53. The paper is unwound from a feed roll (A) onto the stack and bonded to the previous layer using a heated roller (B). The roller melts a plastic coating on the bottom side of the paper to create the bond between layers. The profiles are traced by an optics system that is mounted to an X–Y stage (C). The process generates considerable smoke. Either a chimney or a charcoal filtration system (E) is required, and the build chamber (F) must be sealed.

After cutting, the geometric features of a layer are completed, and the excess paper is cut away to separate the layer from the web. The extra paper of the web is wound on a take-up roll (D). The method is self-supporting for overhangs and undercuts. Areas of cross sections which are to be removed in the final object are heavily crosshatched with the laser to facilitate removal. It can be time consuming to remove extra material for some geometry.

6.4.2.1 The process

In a contour-cutting process, the material is positioned onto the building platform, and a heated roller moves across the surface of the material, bonding it to the stack as shown in Figure 6.54a. An x–y positioning table with mirrors and optics reflects and focuses the CO_2 laser beam, which cuts a profile of the part as shown in Figure 6.54b and c. The area of material surrounding the part profile is cut in a crosshatch pattern to facilitate its removal later (Figure 6.54d). The excess material left on the building block acts as a support structure for the next layer. Adjustments in the laser power and cutting speed enable it to cut through only one layer at a time. The platform lowers to accommodate another layer of sheet material as shown in Figure 6.54e, in which the rolls advance, and the thermal roller bonds to the stack as shown in Figure 6.54f. A new layer of material is then bonded to the

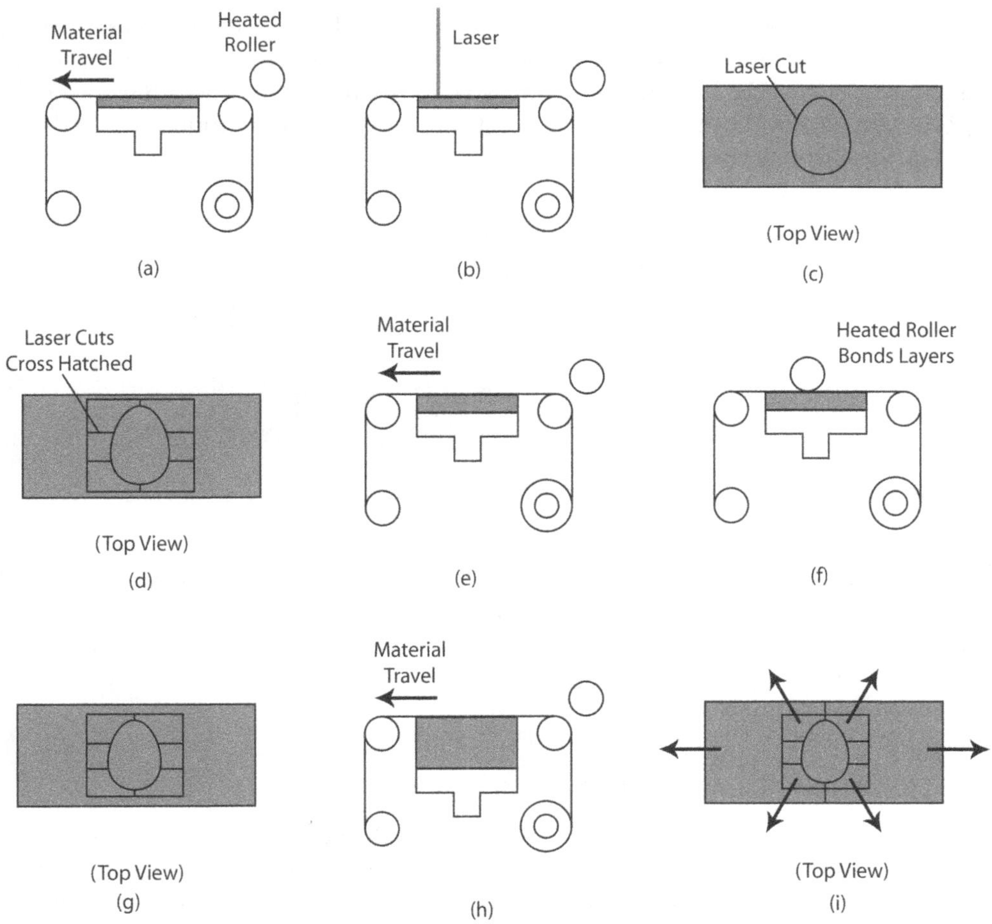

Figure 6.54 Using a contour-cutting process to make an "egg" shape part.

top of the previously cut layer. The laser cuts another cross section as in Figure 6.54g, and this process continues until the part is complete as shown in Figure 6.54h.

Upon removal from the platform, the extra material surrounding the part must be removed as shown in Figure 6.54i. This process involves separating the part from the "crosshatch" of waste cut by the laser. Since the excess material is not removed until after the building process is complete, parts with hollow cores or internal cavities cannot be fabricated as a single part. Furthermore, the materials used in the process are in sheet form, and the system does not subject them to either physical or chemical phase changes, and the finished parts do not experience shrinkage, warpage, or other deformations.

Some sample LOM™ parts are shown in Figure 6.55. There is no need for a contour-cutting process to produce special support structures because the support structure is the portion of the sheet that was not used as a portion of the part. Another benefit to the process is that it is capable of producing large parts. Since the process cuts the contour-sectional area in the top layer instead of processing the entire area of the top layer, it is quick. The finish and accuracy are not as good as with some methods. However, the objects have a look and feel of wood, but the material is estimated to be about three times stronger. The produced parts can be machined, sanded, polished, coated, and painted. The process works well as a one-time investment casting or even as stretch sheet metal tooling. On the other side, the postprocessing and the surface finish are a couple of the disadvantages of the process. The produced parts require sealing in order to protect the part from moisture. Also, since it uses paper, there are no environmental concerns for material waste or part destruction, but the parts can also be a fire hazard for the same reason. Like many other processes, the enclosed hollow features are difficult to produce when they are trapped inside the material.

Figure 6.55 Some sample LOM™ parts.

The applications of the process include sand castings, investment castings, plaster castings, silicone RTV (Room-Temperature-Vulcanizing silicone) molds, vacuum forming, and injection molding. In particular, the process is very effective when the application involves larger prototypes, cheaper production cost, and fast production time, and one-time investment casting is required.

6.4.3 UC process (Ultrasonic Consolidation™)

The UC™ process is a solid-state manufacturing method invented and patented by Solidica Inc. [White02] for AM and direct-to-metal manufacturing. At the heart of the UC process is an ultrasonic metal welding system (UMW) that creates true metallic bonds between thin layers of metal. By alternately bonding and selectively machining, the UC process can be used to build solid metal objects with complex internal geometries that would otherwise be impossible by conventional manufacturing methods. Solidica integrates this technology with a CNC platform, thereby bringing about the *Formation*™ machine tool system.

In the UC process approach, with its unique ability to grow parts in a "cold" solid-state fashion, holds perhaps the greatest promise of all of the metal deposition manufacturing systems, especially if the base metal capability can be expanded to meet new market demands. The system inherently lends itself to producing metal parts with excellent dimensional control and materials' properties due to several unique process characteristics:

- Solid-state fabrication—there is no liquid to metal transition in the UC process; thus, dimensional accuracy easily meets engineering requirements that are far out of reach for many other direct metal processes. This allows rapid fabrication and real-time feature generation. In addition to this, the "cold" deposition process lends itself to the incorporation of support materials that must be used in temperature ranges well below the melting point of the material so that solidification occurs, thereby leveraging the added stiffness and rigidity for complex part manufacturing.
- Integrated three-axis milling center—UC is a hybrid additive metal process in that the deposition stages are periodically followed by precise material removal stages using conventional CNC technology. This is made possible in large part to the point noted above.
- Ultrasonic bond quality—the UC process utilizes ultrasonic joining of thin foil layers for a solid-state process. The resultant interlaminar bond quality is that of a metallurgical bond and is characterized by fine grain structures unlike the coarse recast layers common to other processes.
- Multimaterial laminates—the UC enables bonding and joining of dissimilar materials, such as glass/metals, Al/Ti, and Cu/Al.

In the UC process, a sonotrode is a textured cylinder that clamps the additive layer or foil (or tape) to the substrate and vibrates at ultrasonic frequencies and micron-scale amplitudes. In the ideal case, the sonotrode and new layer move together, and the vibration energy is transferred from the sonotrode/tape interface zone into the tape/substrate interface. This tape/substrate interface deformed region, called deformation affected zone, experiences plastic deformation that breaks and disperses surface oxides and contaminants, thereby bringing atomically clean metal surfaces into contact and creating a true metallurgical bond.

These combined attributes of the system enable such designs as a fully integrated conformal cooling system without the use of bolts and sealants and multiple parts. This significantly reduces costs in materials, labor, and time-to-market.

The UC process can be further illustrated in Figure 6.56. Figure 6.56a indicates two strips of materials to be welded. Figure 6.56b shows the materials are welded using a rolling sonotrode. Figure 6.56c shows the machining operation after the materials are welded. Figure 6.56d shows the circular shape after machining. Figure 6.56e shows the welded part with the third strip machined to a slot in the center. Figure 6.56f shows the part with the fourth strip. In this way, the UC can be used to produce some internal cavities. By using ultrasonic welding and machining, complex geometries and optimal designs can now be realized where conventional machine tool processes would not be able to produce such features.

In summary, the advantages of the UC process include the following:

- Directly build metal products from cheap foil feedstock
- Incorporate complex internal geometries, cavities, and channels

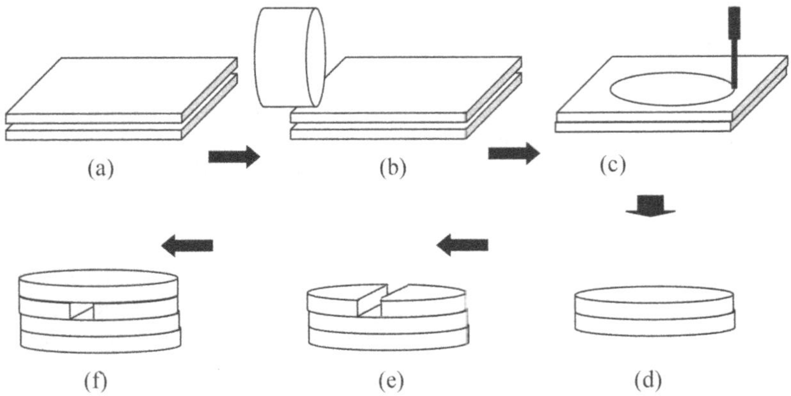

Figure 6.56 An example to illustrate the UC process.

- Functionally graded composite manufacturing capabilities
- High feature-to-feature accuracy (±0.002 to ±0.005 in.)
- High fabrication speed
- Low process heat
- Low cost compared to using a high-energy source to produce metals
- Reduced secondary machining operations
- Stop and restart to build at any time

Although some related research is being conducted to extend this technique to a wide variety of metals, the present limitation for UC is that it needs to work on low temperature melting alloys such as aluminum alloys. The range of the material selection may be its main disadvantage.

The applications of the UC process include investment casting, vacuum forming, blow molding, injection molding, and direct aluminum prototype parts. In addition to forming metallurgical bonds, plastic flow generated by ultrasonic excitation is strong enough to encapsulate a variety of reinforcing materials. Fiber composites can be assembled by placing the fibers between layers of metal foils. High strength fibers, such as silicon carbide and boron, can be embedded to create stiff panels, whereas 2D meshes can promote isotropy. Shape memory alloys, such as nitinol, can be built into an aluminum matrix to make intelligent movable parts. Sensors, such as thermocouples or strain gages, can also be inserted. This process is a good automated alternative to glue, screws, or snap-fit designs, but it is usually used in applications for small pieces, such as watches, cassettes, toys, and medical tools. The applications of UMW are rapidly expanding, and it has already been proven applicable to automobile aluminum chassis, such as the all-aluminum Jaguar.

Review problems

1. What advantages and disadvantages does an extrusion-based process have? What are its application areas?
2. What advantages and disadvantages does a contour-cutting process have? What are its application areas?
3. Which of the following AM techniques does not use a laser?

A. Contour-cutting process
B. Powder-bed sintering process
C. Extrusion-based process
D. STL process
E. None of the above

4. What is cross-hatching? How is it useful in a contour-cutting process?
5. Match the appropriate characteristic with its corresponding AM technique.
   Extrusion-based processes
   a. Non-laser based
   b. Capable of producing large parts
   c. No need for UV cure chemicals
   Contour-cutting process
   d. Multiple material colors are possible
   e. No extra support required
   f. No cleanup required
   g. Parts often need to be machined, sanded, or painted
   h. Must be sealed from moisture
6. Compare the liquid- and solid-based processes by performance metrics or other ratable characteristics.
7. Compare the FDM™ and SLA® processes and discuss their advantages and disadvantages.
8. You want to produce a glass that looks like the picture shown in Figure 6.57. What method of solid-based prototyping (FDM™ or LOM™) would you use to produce this object to be used to test the fit to a person's hand and the volume of liquid it can hold? (Assume the wall thickness never gets smaller than 0.060 in.)
9. Among the solid-based (not powder) AM processes, which process can produce metal parts such as aluminum? How does it work?
10. Name (three) solid-based AM processes. Briefly explain fundamentally how each process transfers from solid to form a part.

Figure 6.57 A glass to be prototyped.

6.5 *Powder-based AM processes*

Creativity is thinking up new things. Innovation is doing new things.

Theodore Levitt

Expectations:

This section helps answer the following questions.

- What are the various types of powder-based AM processes?
- What are the differences between the various powder-based AM processes?
- What are the advantages and disadvantages of each of the powder-based AM processes? How to choose among the AM processes?
- What are the material choices in solid-based processes?
- How do various AM technologies use powder to build a part? How difficult would it be to maintain the thickness and porosity of the deposited layer?
- When is a mold needed for a small batch, for a smooth, hard surface, which AM process should be used?
- When a part needs multiple colors, which AM process should be used?
- When a part has convoluted internal spaces, for easy support removal, which AM process should be used?
- How to compare powder-based processes with liquid-based processes?
- How does electron beam sintering compare to laser sintering?
- One of the most severe limitations of AM has always been that the materials used in most processes leave a lot to be desired in terms of their physical, thermal, and chemical properties. How can one overcome this limitation?

PBF and direct laser deposition () processes belong to the category of powder-based AM techniques. These techniques, in a basic sense, are similar except for the type of materials used. They use lasers or EBMs for the purpose of manufacturing. In general, the principle used is to fuse a powdered material and deposit it on the required regions. The following sections discuss various powder-based processes.

6.5.1 PBF processes

The earliest powder-bed laser sintering process is SLS® patented in 1989, using a laser to bond or sinter powdered material into the solid part. This process is similar to the STL process, but the photosensitive resin is replaced with a powdered thermoplastic or high-temperature material with a thermoplastic binder. As shown in Figure 6.58 in a build chamber, a laser with a galvanometer is used to scan and fuse layers of powdered materials together to form solid 3D objects. The powder needs fine grains and thermoplastic properties so that it becomes viscous, flows, and then solidifies quickly. Its main advantages revolve around material properties. A variety of materials are possible, and these materials can approximate the properties of thermoplastics, such as polycarbonate, nylon, or glass-filled nylon. This process was further developed utilizing powdered materials, such as nylon, polycarbonate, and investment casting wax which are transformed into solid objects, one thin cross-section at a time, using a modulated laser beam. It can be utilized to generate models and prototypes, patterns, molds, and tools required for short-run production. Later, the laser sintering process developed into SLM (also DMLS) and EBM powder bed processes, and they are called PBF processes here. PBF is an AM process where a point heat source selectively fuses or melts a region of a powder bed.

The PBF processes primarily consist of a laser (or EBM) and three powder beds. Two beds hold the feed powder, and the third bed holds/builds the part. The part bed is placed in the middle of the beds with the laser acting directly to this bed. A roller or wiper is used

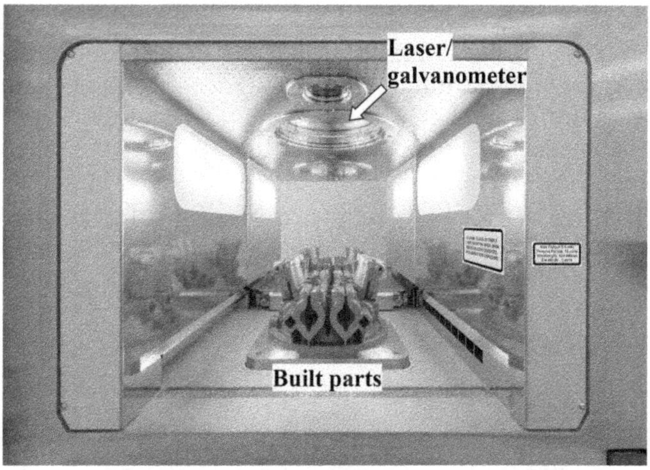

Figure 6.58 A build chamber with a laser/galvanometer to melt layers of powder to form parts. (Courtesy of Renishaw.)

to push and level the powder over the part bed. The beds are preheated to a certain temperature to minimize residual stress buildup and also minimize the laser power required for the process. The process itself is a very simple repeatable one. Figure 6.59 shows a powder bed and the sintering/melting of the part. The process is somewhat similar to STL in principle. The thermoplastic powder is spread by a roller over the surface of a build cylinder. The piston in the cylinder moves down one object layer thickness to accommodate the new layer of powder. A laser beam (or electron beam) is traced over the surface of a compacted powder (A). The powder is spread by a roller or wiper (B) over the surface of a build cylinder (C). A piston (D) moves down one object layer thickness to accommodate the layer of powder.

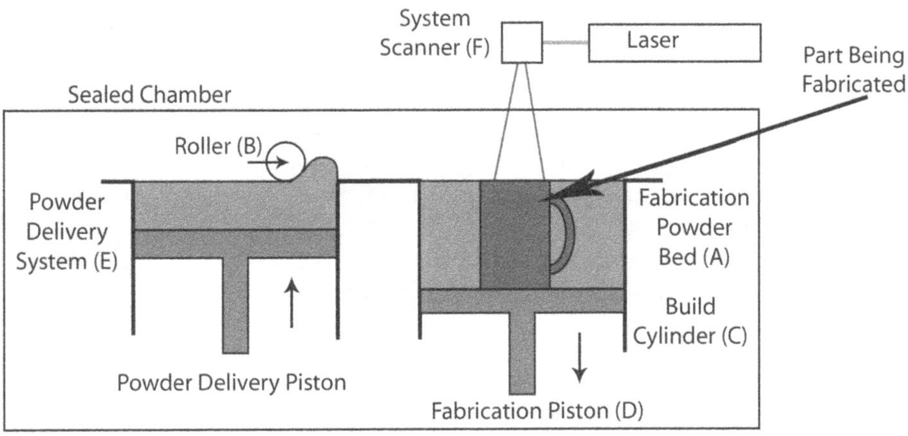

Figure 6.59 A laser sintering/melting process.

The powder supply system (E) is similar in function to the build cylinder. It also comprises a cylinder and piston. In this case, the piston moves upward incrementally to supply the powder for processing.

In the laser-based PBF process, a laser beam is traced over the surface of a tightly compacted powder made of thermoplastic material. The laser melts the powder when it strikes under the control of the scanner system (F). The fabrication chamber is sealed and maintained at a temperature just below the melting point of the powder. Thus, heat from the laser only needs to raise the temperature slightly to sinter the thermoplastic powder, greatly speeding the process. An argon atmosphere is also maintained in the fabrication chamber which prevents the possibility of an explosion during powder handling.

After the object is built, the piston is raised to elevate the object. Excess powder is removed, and manual finishing may be carried out. It may take some cool-down time before the part can be removed from the machine. Support structures for PBF processes are often needed to support or anchor some overhang or base structures.

6.5.1.1 PBF process steps

In a PBF process, parts are created by sintering or melting powdered materials with the heat from a laser (or electron) beam, directly from 3D CAD models. As discussed previously, in the building process, the powder is spread over an area with a roller or wiper. The piston at the bottom of a powder delivery container moves up so that the roller can catch a layer of powder. In the meantime, the piston at the bottom of the fabrication container lowers the same amount.

The part building is a repeatable two-step process. In the first step, a roller/wiper is positioned alongside one of the feed beds. The feed bed then raises a set amount (usually less than 0.1 mm) as the roller passes over the device. The excess powder from the delivery container is spread over the top of the fabrication container as shown in Figure 6.60a. The excessive powder is then deposited on the other side of the container as shown in Figure 6.60b. In the next step, with the layer of powder present, the laser starts to deposit the specified shape of the part in the powder in effect sintering/melting the powder as shown in Figure 6.60c and d. During this step, the laser beam passes over the area of the part and sinters/melts it to the layer of powder below it. Depending on the process and material, the container is kept at an elevated temperature (below the melting point) of the powder so that the laser can quickly increase the temperature of the powder to sinter/melt it. Once this step is completed, the part bed lowers the set amount, and the process continues from the opposite side with the other feed bed raising as shown in Figure 6.60e and f, and the roller distributing another layer of powder over the part bed and the laser etches out the shape as shown in Figure 6.60g. The part is built up in slices with each layer of powder representing a single slice of the part, and the laser melts/sinters the powder as each layer fuses together to yield a solid part. These steps are repeated until the part is completed as shown in Figure 6.60h and i. Once the prototype is finished, it is raised out of the container, excess powder is blown off, and a final finishing pass is made. Figure 6.60j shows the top view of the completed part, and Figure 6.60k shows the part after the excessive powder is removed. Because a part is made up of slices, very complex shapes and designs can be manufactured that would otherwise be impossible by conventional means.

In order to process metal in high temperature at the melt pool, the PBF process is normally operated in an inert gas environment. Figure 6.61 shows the inert gas chamber. Before a part is built, build chamber is prepared with vacuum removal of air (oxygen level, say below 0.1% or 1,000 ppm, particle per million) and replaced with argon. Eventually, the chamber is filled with inert argon gas before the part can be built. Figure 6.62 shows the powder spreading, laser fusion, and process repeating the process. Figure 6.62a shows that one layer of metal powder is uniformly distributed on the substrate; Figure 6.62b

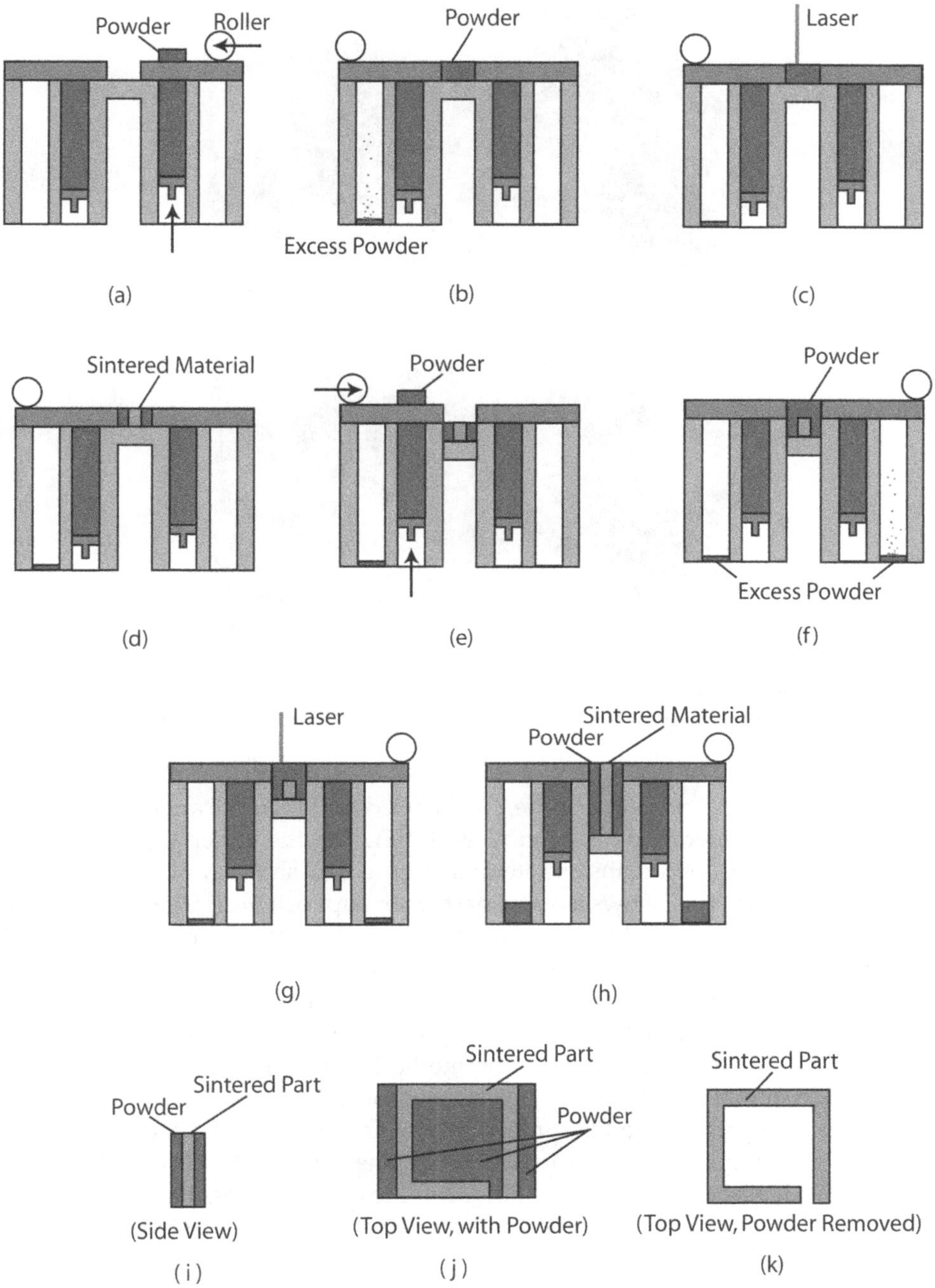

Figure 6.60 An SLS machine to make an "a" shape part as in figure (k).

shows the laser scanning on the powder. After the laser scan, the build plate moves down, and the next layer is built up as shown in Figure 6.62c.

The PBF process can offer the key advantage of making functional parts in essentially final materials. The parts are created in an accurately controlled environment, and thus, the processes provide good part stability. A variety of thermoplastic materials, such as

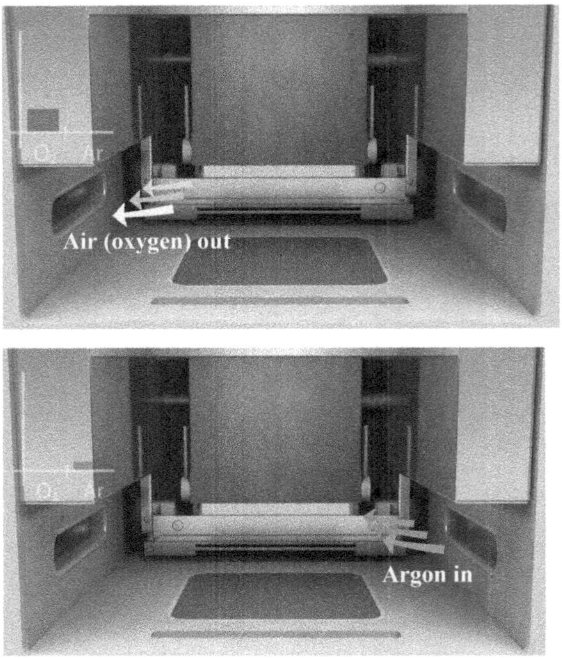

Figure 6.61 Renishaw AM250 build chamber inert gas environment. Before a part is built, build chamber is prepared with vacuum removal of air and replaced with argon. (Courtesy of Renishaw.)

nylon, glass filled nylon, and polystyrene, can be used for SLS fabrication. Surface finishes and accuracy are in general not as good as with STL, but the material properties can be rather close to those of the intrinsic materials. Figure 6.63 shows a manifold made from the SLS process. Figure 6.64 shows a metal part (with support) built from an SLM process.

The SLS method has also been extended to provide direct fabrication of metal and ceramic objects and tools. However, since the objects are sintered, they are porous. For certain applications, it may be necessary to infiltrate the object with another material to improve mechanical characteristics. Much progress has been made over the years in improving surface finish and porosity. The method has also been extended to provide direct fabrication of metal and ceramic objects and tools.

The surface of a part from a PBF process is powdery, such as the base material whose particles are fused together. Thus, the surface finishes are pretty good but not as good as those from STL. Other issues of the PBF processes include possible porosity of parts, the first layers to require a base anchor to reduce thermal effects, extended overhang to need support structures, and part density to vary. Material changes require cleaning of the machine; thus, cross-contamination is still a possible issue.

For some applications, there is one problem with the technique: the end product generally has a density of 80% to 99+%, while a density of 100% is desired, especially with metal parts. Many efforts are in progress for the creation of full-density, non-permeable, metal objects. SLM and EBM processes are very similar to SLS in terms of equipment, but use a much higher energy density to melt the powders. Therefore, the fabricated parts exhibit a density very close to the theoretical one. The process can be used to process a variety of materials, such as zinc, bronze, stainless steel, tool steel, and Ti64. Since these powder materials need to be in spherical shapes, such as those from a gas atomized process, the

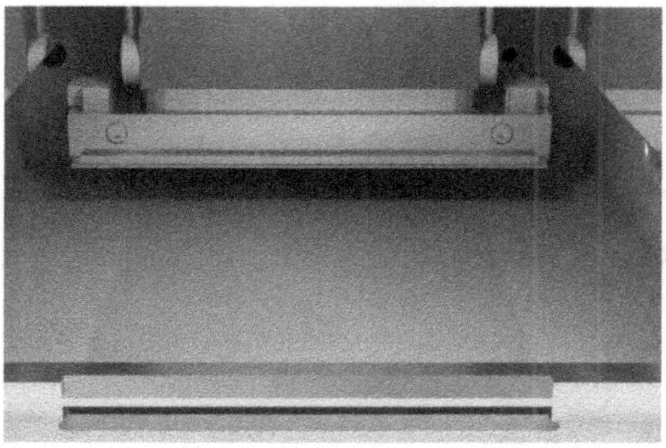

(a) Powder uniformly spread on the build plate.

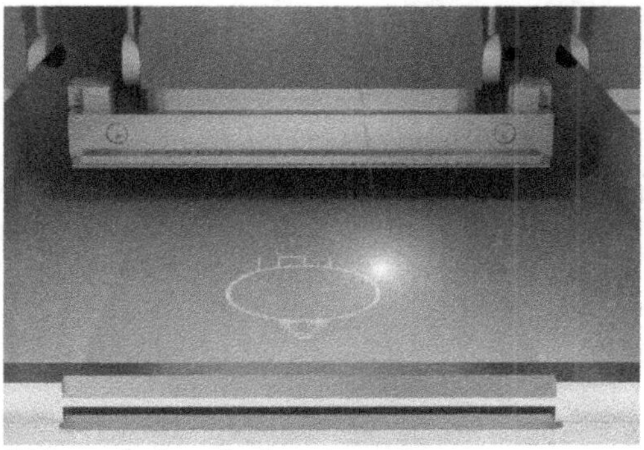

(b) Laser fusion of selected powder

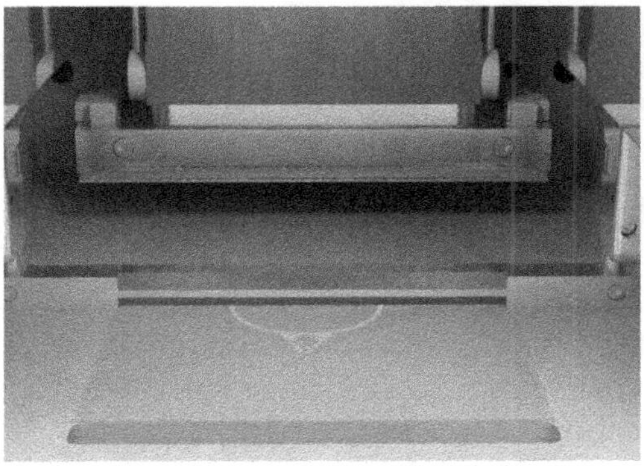

(c) After the laser scan in (b), the build plate moves down and next layer is built up. The process is repeated by spreading the powder on the next layer.

Figure 6.62 Powder spreading, laser fusion, and process repeating the process. (Courtesy of Renishaw.)

powder tends to be costly. The spherical powder can provide good flowability, so that the powder can provide good support for complex part. Figure 6.65 shows an example build of 80 specimens built in vertical orientation and 60 specimens built in 45° orientation. In many cases, most powder is used to support; thus, they are not sintered or fused after the build. Recycling of the unused powder is critical to bring the cost down. Thus, treatment of recycled powder to yield certain standard for recycling is also important.

The properties of AM products, as well as the PBF processes, can be very different from the feedstock material. This is due to the fact that the complex mechanics of a PBF process often leads to very complicated and nonequilibrium microstructure. Evidence

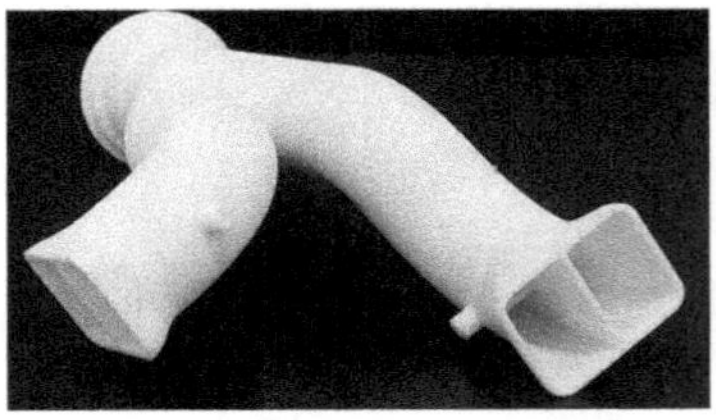

Figure 6.63 A manifold from an SLS system. (Courtesy of 3D Systems.)

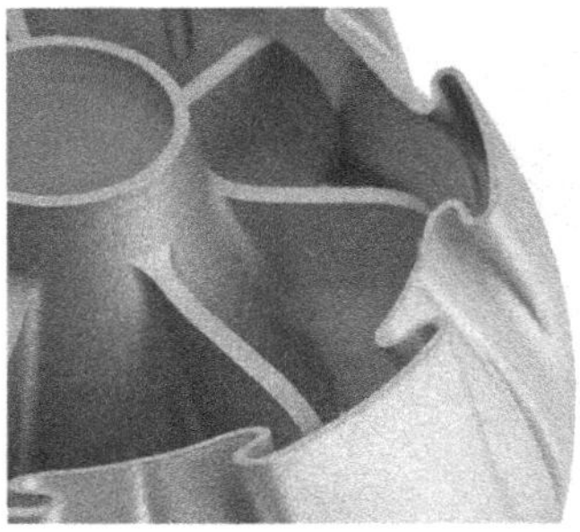

Figure 6.64 A metal part built from a metal 3D printer. (Courtesy of Renishaw.)

Figure 6.65 PBF process can provide good support as shown of parts built with various angles due to powder support.

suggests that the properties of the fabricated material could be anisotropic. In general, the performance of PBF processed materials can be sensitive to fabrication conditions and are also influenced by important process attributes, such as part shape, part size, and orientation in the build [Karnati17]. Anisotropies consequent of varying the orientation have been observed by researchers for several materials.

SLM allows the manufacture of functional components with high structural integrity. The smaller size range of powders used in the SLM process compared to say powder feed processes makes it feasible to fabricate parts with dimensions close to design, while attaining good surface finish. Like most AM processes, SLM material consists anisotropic properties that are different from materials fabricated from conventional methods. The rapid solidification of material and complicated implementation of the process results in nonequilibrium properties. Substantial thermal gradients occur during fabrication in

AM processes, the resultant thermal stresses cause distortion and residual stresses. The cooling rates as large as 10^3–10^4 K/s have been recorded during AM fabrication. The heat flow is directional and the layer-by-layer build approach causes thermal cycling. The high cooling rates, directional heat flow, and thermal cycles have complicated effects on the microstructure. These characteristics of fabrication cause various types of anisotropy and residual stresses. The scale of these effects can vary depending on the input power, build strategy employed material used, etc. Material properties, such as thermal conductivity, the coefficient of thermal expansion, elastic modulus, and yield stress, are all affected by the residual stresses.

6.5.2 3D inject printing process

One representative of the inject printing process is Three-Dimensional Printing technology (3DP™), originally developed at the Massachusetts Institute of Technology (MIT) in 1993. 3DP™ is the basis of Z Corporation's prototyping process. Z Corp was acquired by 3D Systems in 2012. One of the products the technology is commercialized by 3D Systems was ProJet CJP (Color Jet Printing). Inject printing technology creates 3D physical prototypes by solidifying layers of deposited powder using a liquid binder. This technology is very similar to laser sintering technology, but instead of using a laser, it uses the printing technology to bind the powder together and thus can operate at very high speeds and low costs.

Similar to other AM processes, in the inject printing process, after exporting a solid modeling file from a 3D modeling tool, users can open the file and virtually section the solid object into digital cross sections, or layers, creating a 2D image of each slice along the z-axis and, then, send 2D images of the the slices to the 3D printer via a standard network.

The inject printing process uses the inkjet printing technology to make parts layer by layer by depositing a liquid binder onto each powder layer. Instead of feeding a paper under the print heads, such as a 2D printer, the inject printer moves the print heads over the powder bed to print the cross-sectional area. The process requires the powder to be distributed accurately and evenly across the build platform. As shown in Figure 6.66a, inject printers realize this task using a feed piston and platform that rises incrementally for each layer. A roller mechanism spreads the powder fed from the feed piston onto the

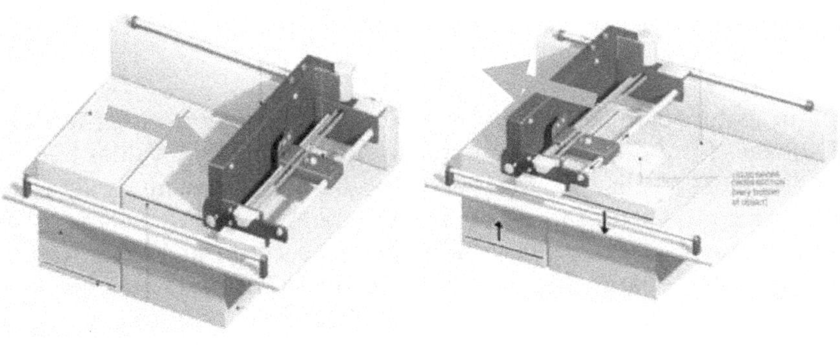

(a) (b)

Figure 6.66 (a) Spread a layer of powder; (b) print binder on the cross section. (Courtesy of 3D Systems.)

build platform; intentionally spreading certain percent of extra powder in each layer to make sure a full layer of packed powder can be formed on the build platform. The extra powder falls down an overflow chute, into a container for reuse for the next deposit. Once the powder is spread, the inkjet prints the cross-sectional area of each layer of the part, binding the powder together as shown in Figure 6.66b. A piston then lowers the build platform by a layer thickness, and a new layer of powder is spread on top. The print heads apply the data to the next cross section onto the new layer, which binds itself to the previous layer. This process is repeated for the rest of the layers of the part. The inkjet printing process creates the physical model of the geometry represented by the digital data. Since multiple inkjet heads can be used, the processing time can be greatly reduced.

When the inkjet printing process is complete, the unbind powder surrounds the part in the build chamber. These loose powder also served as a support material. Users can then remove the part from the build chamber after the materials are set and return unprinted powder back to the feed platform for the next build. Users can use forced air to blow the excess powder off the printed part. This technology does not require other solid supports during the printing process, and the unused material is reusable.

The technology uses inkjet print heads with good resolution, such as 600 dpi (dots per inch), and focuses on a drop-on-demand approach. The inkjet technology allows printing of multiple parts simultaneously, while only adding a small amount of time to print each part. The accuracy of the models depends on the jet size and motion control.

The software processes data in parallel with the printing of the part. Unlike the other processes, while the 3D printer deposits the first layer, the software slices and processes, say, the fifth layer. Figure 6.67 shows the example of a bearing model built using 3D inject printing process.

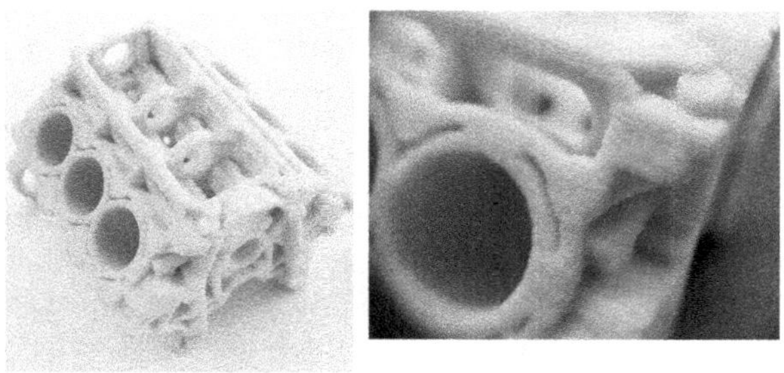

Figure 6.67 An engine model built by a 3D Inject printing process with zoomed view on the right. (Courtesy of 3D Systems.)

One unique feature of inkjet printing technology is the color printing capability. When printing 2D images from digital files, computers convert the RGB values (red, green, and blue) to CMYK colors (cyan, magenta, yellow, and black). Typically, a 2D color desktop printer has a print head with three of the color channels, CMY, and another for black. Using these inks, the printer combines several dots in each printed pixel through ordered patterns to create the appearance of thousands of colors. The same principle applies to 3D printing. Inkjet printers use four-colored binders: CMYK, to print colors onto the shell of the part. The software communicates the color information to the printer within the slice data. Users can also use different colors to represent analysis results on the model or to annotate design changes to enhance the communication value of the model as shown in Figure 6.68.

Figure 6.68 These examples of models if in colors can provide meaningful and creative applications for color 3D printing including product labeling, topographical analysis, and production planning. (Courtesy of 3D Systems.)

Inject printing technology printers produce very little waste. The unprinted powder surrounds and supports complex parts during printing. Users can reuse all unused support powder. Thus, printed-part volume becomes the basis for all part-creation costs. The use of an off-the-shelf print head allows for inexpensive, quick replacement of the system's primary consumable component. The application of modular design techniques to the printer's electronics, printing, and maintenance components makes the printers efficient to maintain with minimum downtime, further reducing costs.

A similar process, Direct Shell Production Casting (DSPC), is a patternless casting process for metal parts in which the casting molds are generated automatically, directly from 3D CAD data. With DSPC there is no need for physical patterns, core boxes, or any other tooling and no part-specific setup. The only "pattern" is the CAD design itself. Another similar process, ProMetal™ technology, formerly owned by Extrude Hone, is an application of MIT's 3DP™ process to the fabrication of injection molds. ExOne Company was formed by spinning out several disparate technologies from Extrude Hone in 2005.

DCPC is slightly different from the systems, such as SGC or SLA®, in that a cavity is produced to the dimensions of the CAD model required. The shell is built in a ceramic allowing the direct production of investment shells from CAD data.

A CAD file is used to define the required cavity, the system modifies this file to suit the investment casting process by introducing fillets and removing machined features such as holes. The number of castings required is then entered into the computer controlling the system and the model of the shell generated.

As shown in Figure 6.69, this process is similar to 3D printing. A layer of alumina powder is laid down onto a cylindrical bed. A print jet projects a fine stream of colloidal silica in the required plan of the slice onto the powder coating. This solidifies and also adheres to the previous layer. Once the slice of shell has been drawn, the machine bed lowers and the process is repeated until the entire shell has been formed. As with the process, the excess powder acts as a support while the shell is being produced. On completion, the shell is removed from the bin and excess powder is removed from within the cavity.

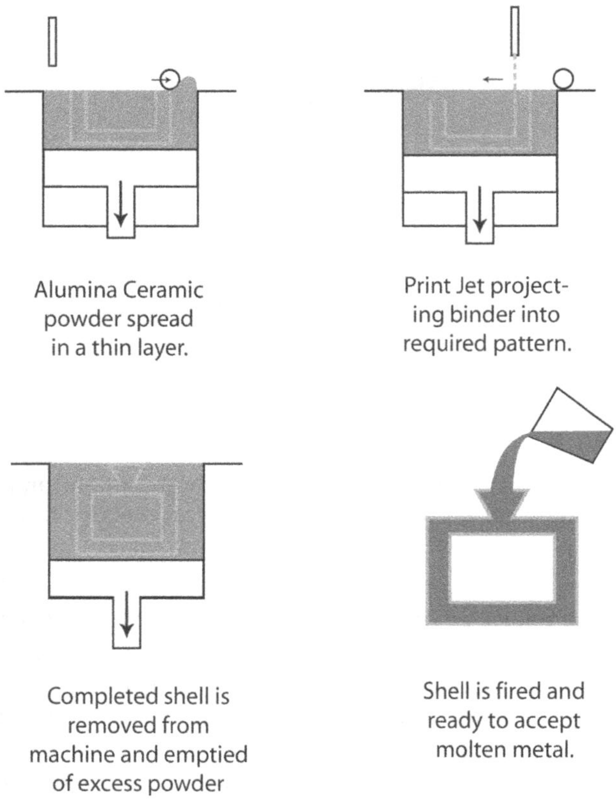

Figure 6.69 Direct shell production casting.

After a mold is printed, it is fired to create a rigid ceramic mold. One can pour any molten metal into these molds, thereby eliminating several steps required in investment casting. Plus, these molds are more accurate than those from standard sand casting. Tolerances for lengths smaller than 1 in. are ±0.021 in.; for lengths greater than 6 in., accuracy is ±0.031 plus 0.003 in./in. over 6 in. (i.e., when past is longer, the tolerances will be larger).

6.5.3 Direct laser deposition

A DLD process, a class of DED process, also called blown powder process or powder feed process, is a laser-assisted direct metal manufacturing process that uses computer-controlled lasers in which, in hours, weld air blows out streams of metallic powders into custom parts and manufacturing molds. Figure 6.70 shows a blown powder (or powder feed) process with side nozzle. Some DED processes use wire instead of powder, but the concept is similar. A representative process is called the Laser Engineered Net Shaping™ (LENS®) process. It uses CAD file cross sections to control the forming process by Optomec Inc. The DLD process can be used throughout the entire product life cycle for applications ranging from materials' research to functional prototyping to volume manufacturing. An additional benefit is its unique ability to add material to existing components for service and repair applications. Powder-metal particles are delivered in a gas stream into the focus of a laser to form a molten pool of metal. It is a layer-by-layer AM process. The DLD process allows the production and repair of parts, molds, and dies that are made out of the

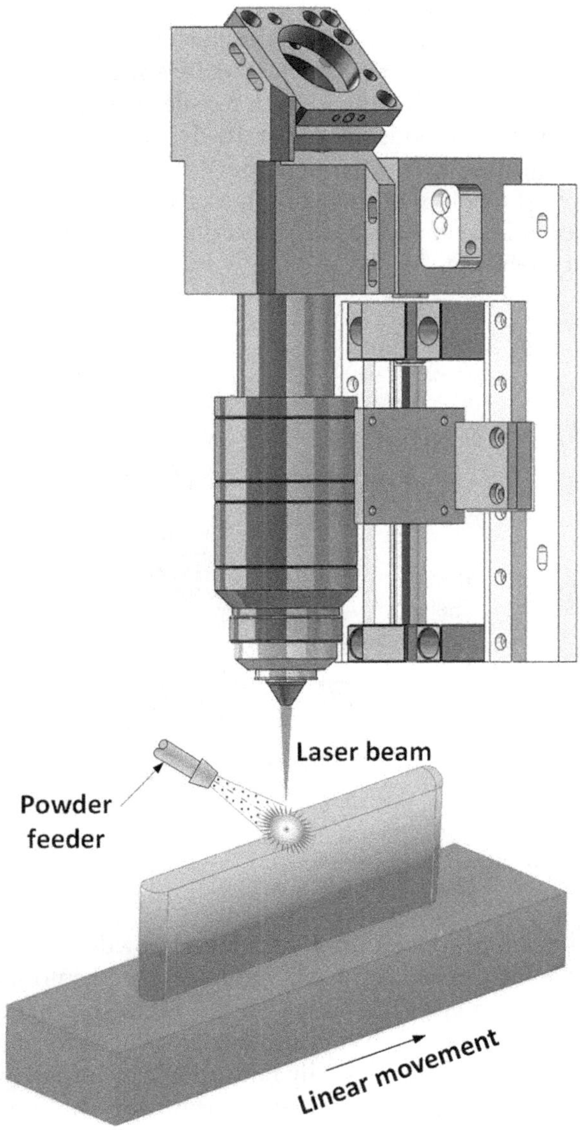

Figure 6.70 A blown powder (or powder feed) AM process with side nozzle.

actual material, such as aluminum or tool steel. In other words, this produces the high-temperature materials, which are difficult to make using the traditional AM processes.

The laser beam is moved back and forth across the part and creates a molten pool of metal where a precise stream of metal powder is injected into the pool to increase its size. This process is the hybrid of several technologies, including lasers, CAD, CAM, sensors, and powder metallurgy. This process also improves on other methods of metalworking in that there is no waste material or subtractive processes necessary. It can also mix metals to specific standards and specifications in a manner that have never been possible before.

The build rate of a DLD process is dependent on laser power and material. For example, the build rate for a 500 W laser power on steel can be about 1 in.^{3}/h. The DLD uses a laser deposition process with the possibility of having an oxygen-free environment. The basic process is to use a laser to create a melt pool within the substrate, and power stream can be injected into the melt pool, point by point, and layer by layer, to form parts. The benefits of the DLD system are the fine microstructure it creates when comparing to a traditional welding process. This results in high tensile strength that meets or exceeds a metal made with traditional methods. However, its deposited material is, in general, less ductile. As shown in Figure 6.71, the main components of a DLD system include a laser, an x/y table to generate 2D motion of the workpiece, a method of z-axis motion of laser, a powder feeding device, a laser focusing device, and a control system. The resulting part surface finish is about 40–50 μm.

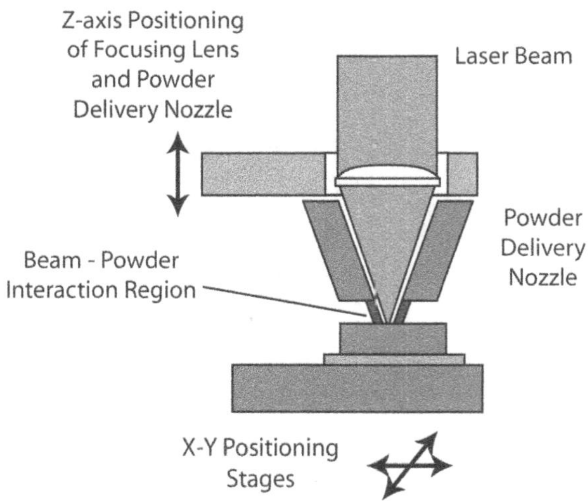

Figure 6.71 The schematics of the DLD process.

A DLD system uses a high-power laser with metal powder to build dense structures directly from a CAD solid model. The CAD model is sliced into a 2D toolpath, which instructs the machine how to print the part. The part is constructed layer by layer under the control of a software that monitors various parameters to ensure geometric and mechanical integrity. The process is generally housed in a chamber, which is purged with argon such that the oxygen level stays below a certain threshold to ensure there is no impurity pickup during deposition. The laser beam typically is focused to a small spot by one or more lenses. The X–Y table travels in raster fashion to fabricate each layer of the object. The head is moved up vertically as each layer is completed. The metal powder is fed to the process by a powder-feed system, which is able to flow small quantities of powder very precisely. Metal powders are delivered around the circumference of the head either by gravity or by using an inert, pressurized carrier gas. Even in cases where it is not required for feeding, an inert shroud gas is typically utilized to shield the melt pool from atmospheric oxygen for better control of material properties and to promote layer-to-layer adhesion by providing better surface wetting.

When complete, the part is removed and can be heat-treated, hot-isostatic-pressed, machined, or finished in any other manner. Several materials can be processed, such as stainless steel, Inconel 625, and Ti-6Al-4V. Material composition can be altered dynamically and continuously, leading to parts with material properties that might be mutually exclusive using classical fabrication methods.

Parts fabricated are near-net shape but often will require finish machining. They are fully dense with fine grain microstructure and have properties similar to, or even better than, the intrinsic materials. DLD processes can also be used to repair parts as well as fabricate them.

The key process parameters for the DLD process include (1) laser processing diameter, (2) process speed, (3) laser power, (4) powder feed rate, (5) laser beam path width or path overlap, (6) layer thickness, and (7) energy density. The typical resulting material parameters considered include (8) surface roughness, (9) hardness, and (10) grain size. The mechanical performance of the part includes the parameters, such as tensile strength and fatigue strength. Other process parameters, such as nozzle-to-surface distance, or standoff distance, nozzle gas flow rate, absorptivity, and depth of focus with respect to the substrate, also play essential roles.

1. Laser processing diameter

   This laser processing diameter is one of the most important parameters because it determines the power density which is the laser power divided by the cross-sectional area of the laser processing beam. In laser deposition, laser beam density is the critical parameter to account for the laser input. It is difficult to precisely measure for high-power laser processing diameter. This is partly due to the fact that the position of the standoff distance could dynamically change depending on the processing geometry.
2. Process speed

   It is the relative speed between the laser beam and the workpiece spot being heated. In general, decreasing process speed increases the layer thickness and increases deposition diameter. However, there is a threshold to reduce process speed since too much energy may cause previous layer tempering or secondary hardening. Process speed for the DLD process should be well chosen since it has a strong influence on microstructure. Depending on the servo of the motion system, often when the deposition toolpath changed the direction, the speed may not remain constant, thus often makes an undesirable deposition.
3. Laser power

   As mentioned previously, the layer thickness during the laser cladding process is directly related to the power density of the laser beam and is a function of incident beam power and laser processing diameter. Generally, for a constant laser processing diameter, the layer thickness increases with increasing laser power provided with adequate powder feed rate. It was also observed that the deposition bead diameter increased with increasing laser power.

4. Powder feed rate
   The smallest diameter of the powder "beam" is dependent upon the design of the nozzle. If the powder beam diameter becomes too large compared to the laser beam diameter, much of the supplied powder may never reach the cladding melt pool. Thus, there will be unacceptably low powder utilization rate. It is desirable that the nozzle be designed so that the powder beam diameter is smaller than the laser beam diameter to yield high powder utilization rate.
5. Beam path overlap
   It is the overlap between two adjacent deposition paths. As the deposition profile is not uniform, beam overlaps as shown in Figure 6.72 are used to smooth out the deposition surface. Beam overlap has a strong influence on top surface roughness. The reason for this decrease in surface roughness is depicted in Figure 6.72. As the cladding pass overlap increases, the valley between passes is raised due to the overlap, therefore, reducing the surface roughness. Since each material has certain bead profile due to different surface tension, the beam overlap may be different. For example, the beam overlap of about 40% is used for steel deposition, and titanium is smaller. The hatch distance, as shown in Figure 6.73, can determine the percent of overlap between the two deposition tracks [Trosch16]. When the hatch distance is large shown in Figure 6.73a, there may be a cavity between two tracks. Thus, the surface may be rough. When the hatch distance is reduced to the proper distance shown in Figure 6.73b, the cavity may be greatly reduced near flat. Thus, the surface may be smooth. When the hatch distance is further reduced shown in Figure 6.73c, the material may be accumulated to a lump. Thus, the overall surface may become rough again.

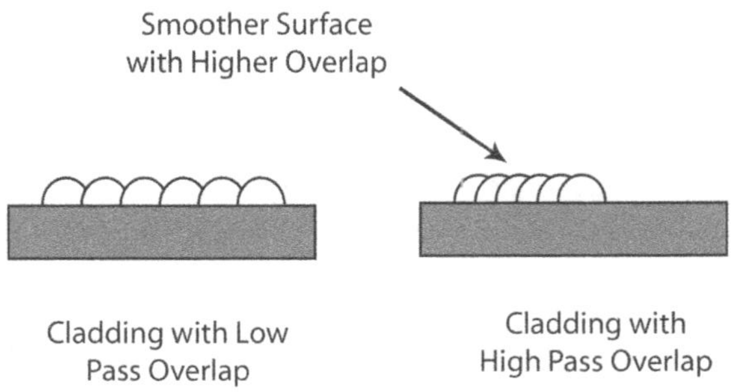

Figure 6.72 Surface profile change from increase in percent overlap.

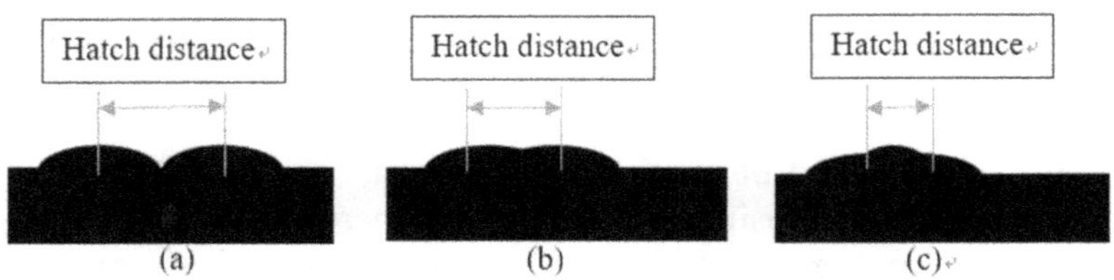

Figure 6.73 Illustration of the effect of various hatch distances on the surface roughness.

6. Layer thickness
   Deposition layer thickness is not easy to control in DLD process due to the nature of the deposition profile as shown in Figure 6.72. There is a large range of layer

thickness as well as deposition rates that can be achieved using laser deposition. The layer thickness is critical, especially for true 3D parts. The layer thickness may cause inaccuracy in the 3D part building, unless in a hybrid process in that a substrate process can remove the excessive material and control the layer thickness. In general, monitoring and control of deposition layer thickness is needed to achieve good part quality.

7. Energy density

Both the layer thickness and the volume deposition rates are affected predominately by the energy density. Energy density is one of the critical driving parameters toward this material density. As a means of standardization, it is theorized that if the energy density during manufacturing is maintained over a certain threshold, near fully dense parts can be expected. Finding this energy density threshold is important to ensure that the detrimental effects of overheating do not affect part quality. In general, the energy density supplied to the powder layer is represented by the following equation.

$$\text{Energy density} = \frac{P'}{V \times W \times H} \tag{6.1}$$

where,
P' = absorbed laser power
V = process speed
W =deposition bead width, or beam path overlap
H = layer thickness or deposition bead height

Also it has been known that the actual laser power absorbed in the melt pool is not the same as nominal laser power measured from the laser power monitor due to reflectivity and other plasma-related factors depending on the materials. The use of adjusted energy density is thus preferable. Considering the factors, it was reported that there is a positive linear relationship between the energy density and deposited material density.

8. Surface roughness

Similar to the layer thickness, the surface roughness is not easy to control. Surface roughness was found to be highly dependent on the direction of measurements with respect to the cladding. In checking the surface roughness, at least four directions should be tested from each sample; the length and width direction on the top surface and the horizontal and vertical directions on the walls. Since the largest roughness on each sample is of primary interest, measurements should be only taken perpendicular to the clad direction on the top surface and in the vertical direction on the walls. The surface roughness is also a function of beam path overlap or hatch distance, as discussed in #5. "beam path overlap" of this section.

9. Hardness

It is the degree of how resistant a solid material is to shape change when a compressive force is applied. Many hardness tests exist, in which the material is indented until an impression is formed. These tests can be performed on a macroscopic or microscopic scale. It is a local material property and normally is not considered as a fundamental material property. However, since hardness is dependent on many other material properties, such as elastic stiffness, plasticity, strength, ductility, strain, toughness, viscoelasticity, and viscosity, if measured with proper procedure, it may reveal a lot of material information. The surface finish of an object does not have an effect on the hardness measurement if only the indentation is large compared to the surface roughness. Since hardness test is a nondestructive evaluation method, it can provide a very effective and economical first assessment of the material properties for AM. In general, hardness is related to microstructure, composition, grain size, and porosity.

10. Grain size

    Laser-based AM processes tend to produce a part with finer grain size compared to that of the traditional manufacturing processes. This is because the metal AM processes, in general, involve the cooling rate. The greater the temperature difference, the quicker the phase transformation tends to happen. Thus, the higher cooling rate will result in finer grains. Resistance to plastic deformation is an important grain size-related property of metals. One of the simple measures of this resistance is the hardness test. The Hall–Petch relationship [Hall51, Petch53] described the linear relationship between hardness H and grain size d as follows:

$$H = H_0 + K_H d^{-1/2} \tag{6.2}$$

    where H_0 and K_H are hardness constants. This equation expresses the fact that the finer the gain, the harder the material unless the grain size is too smaller (less than 10 nm or so).

    Gain size can be used to predict the properties of several different materials, such as metals, ceramics, intermetallics, and polymers. There is a direct grain size effect on ductile fracture, though there is certainly an indirect effect on the effect of grain size on strength.

6.5.3.1 Advantages of DLD process

The strength of DLD lies in the process' ability to fabricate fully dense metal parts with good metallurgical properties at reasonable speeds. DLD is an efficient approach that reduces production costs and speeds time-to-market for high-value components. In addition to economic advantages, the DLD systems enable the fabrication of novel shapes, hollow structures, and material gradients that are not otherwise feasible. DLD can even be used to add features to cast or forged parts.

The DLD process can be used for prototyping, production tooling, and even rapid manufacturing in a variety of industrial applications. Figure 6.74 shows some example parts produced by the LENS® process. Manufacturing companies can use this process to produce rapid metal components. Using DLD, it is possible to modify or add material to existing aluminum, steel, or other metallic parts, or make a fully functional prototype directly from the CAD design. In die repair and refurbishment, downtime costs can mount quickly when mold or die cracks or becomes worn. Replacement can represent a significant expenditure with traditional repair processes often being substandard. These processes are also time-consuming and can warp or damage the mold or die. The DLD repair process can produce a durable, high-quality repair with a small heat-affected zone that does not weaken or damage the part. This results in molds or dies with tool life, strength, and heat resistance comparable to or even better than the tooling produced by conventional manufacturing methods.

The other capability that opens up a whole new world to the imagination is the possibility of changing from one type of material to another within a part. This would allow parts to be very specifically designed by possibly adding a high strength material

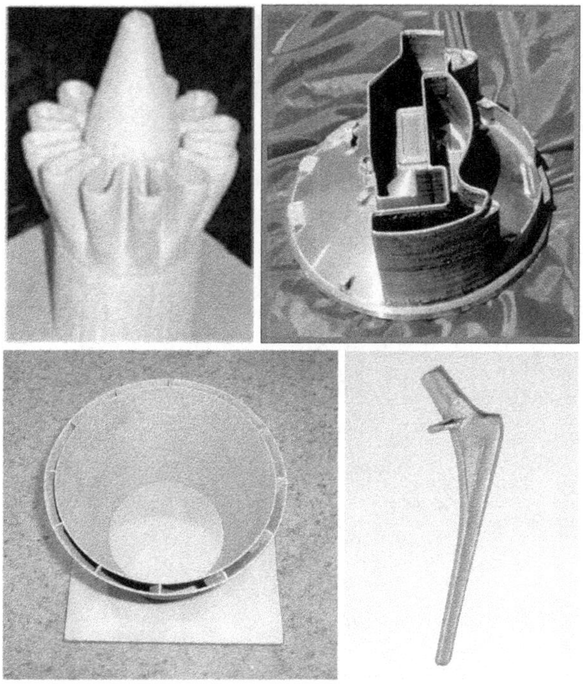

Figure 6.74 Some example parts produced by the LENS process. (Courtesy of Optomec.)

in an area of high stress or a stress concentration. One could possibly add an area of material that could be tapped in a part that is mostly made from a material that cannot be tapped.

6.5.3.2 Limitations of DLD process

The primary limitation is the complexity of parts that can be manufactured. Since DLD is a free-form process, there is a limit to the overhang angle that can be built. The limit is material, and process dependent, typically greater than 30°–45° from vertical, which presents a challenge. There are other issues to be improved, such as the microstructure control, dimensional accuracy, surface finish, and residual stress built into components. Most AM processes are using three-axis tables, and thus, support structures are very often needed in building overhang parts. These structures are not desirable in laser-based processes involving metals. One could use a high melting-point material to build the support structures and use other processes, such as chemical etching, to remove the support material afterward. However, such a secondary process may take a long time to remove the support material, and even after etching, the remaining surface may still require postprocessing to be considered acceptable.

6.5.4 EBM process

The EBM process was developed initially at Chalmers University in Sweden in the 1990s. It is a powder-based method having a lot in common with SLS® but replaces the laser with a scanned 4 KW electron beam to produce fully dense parts, although porosity cannot easily be avoided due to the nature of the laser sintering process. This method projects electrons at half the speed of light onto the powder surface, and the kinetic energy of the electrons induces melting. The electron beam is, in general, more energy efficient than a laser for conductive materials.

In an EBM-melting process, as shown in Figures 6.75 and 6.76, parts are fabricated in a vacuum and at about 1,000°C to limit internal stresses and enhance material properties. The vacuum chamber is typically 10-5 torr for titanium deposition but can be set down to 10-2 for other types of metals. A layer of metal powder is spread over a platform in a vacuum chamber. To reduce residual stresses that cause distortion in a fabricated part, an electron beam gun preheats the powder layer. After the preheating is finished, the layer is selectively melted by increasing the beam power or decreasing the speed. In the melting process, electrons are emitted from a filament which is heated to over 2,500°C. The electrons are accelerated through the anode to half the speed of light. A magnetic lens

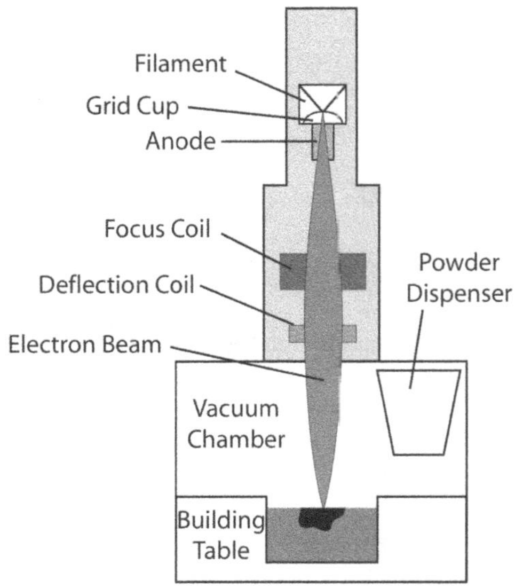

Figure 6.75 EBM melting the powder using the speed of the electrons.

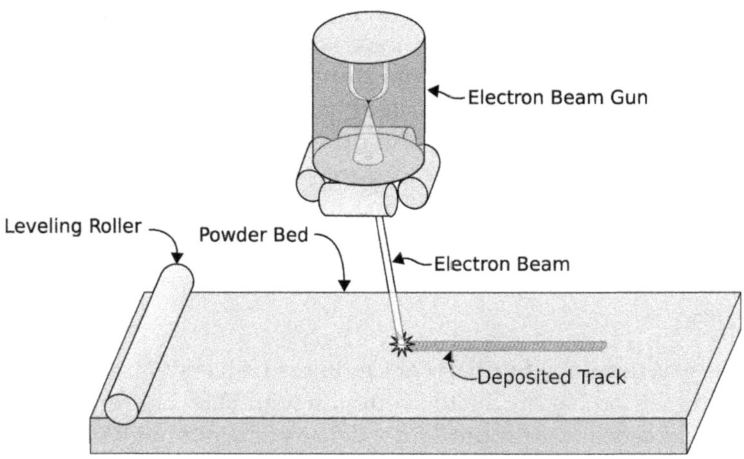

Figure 6.76 Another view of the EBM process.

brings the beam into focus, and another magnetic field controls the direction of the beam. When the electrons hit the powder, kinetic energy is transformed into heat which melts the metal powder. The power is controlled by regulating the number of electrons in the beam. The cooling process is also controlled to produce well-defined hardening. As with other processes, the parts require some final machining after fabrication. The processing in a vacuum provides a clean environment that improves metal characteristics.

Once the first layer is complete, the platform is then lowered, a new layer of powder is uniformly distributed, and the process is repeated until the part is complete. The resulting part is close to fully dense. Finish heat treatment or machining may be used to improve the surface finish. Extremely complex internal geometry is the limiting factor of the EBM process. The EBM process may ultimately be applicable to a wider range of materials than competitive processes and also has the potential to offer much better energy efficiency. Some of its benefits include the ability to achieve a high-energy level in a narrow beam, vacuum melt quality to yield high strength properties of the material, and vacuum environment to eliminate impurities such as oxides and nitrides, and the ability to permit welding in refractory metals and combinations of dissimilar metals. Compared with laser melting, additional benefits include higher efficiency in generating the energy beam resulting in lower power consumption, lower maintenance and installation costs, the high actual overall power resulting in high build speeds, and the deflection of the beam to avoid moving parts resulting in high scanning speed and low maintenance.

The unique operating environment of electron beam technology requires a vacuum. This could be an advantage when it is in space, but could be a disadvantage as it costs more, thus, limiting the part size it can build. Some disadvantages include producing gamma rays while in operation and requiring electrically conductive materials.

6.5.5 *Hybrid material deposition and removal process*

For more than a decade, AM has given industry an approach to achieve the goal of providing products in a shorter time and at a lower cost. Most of the current AM systems are built on a 2.5D platform. Among them, the DLD process is a potential technique that can produce fully functional parts directly from a CAD system and eliminate the need for intermediate steps. However, such a process is currently limited by the need for supporting structures—a technology commonly used in all the current AM systems. Support structures are not desirable for high-strength and high-temperature materials such as metals and ceramics since these support structures are very difficult to remove. Multi-axis systems can offer much more flexibility in building complex objects. Laser-aided AM is advancing the state-of-the-art in the fabrication of complex, near-net-shape functional metal parts by extending the laser cladding concept to AM. Laser-aided manufacturing process (LAMP) developed in the LAMP Laboratory at the Missouri University of Science and Technology (Missouri S&T) combines laser deposition and machining processes to develop a hybrid rapid manufacturing process to build functional metal parts. This section summarizes the research and applications of such a hybrid process for fabrication and repair of metallic structures.

As discussed in Section 6.5.3, DLD is a free-form fabrication process that has the capability of direct fabrication of metal parts. The process uses a focused laser beam as a heat source to create a molten pool on an underlying substrate. Powder material is then injected into the molten pool through nozzles. The incoming powder is metallurgically bonded to the substrate upon solidification. The part is fabricated in a layer-by-layer manner in a shape that is dictated by the CAD solid model, which is sliced into thin layers. After the slice data is then translated into laser scanning paths, an outline of each feature of the layer is generated, and then the cross section is filled using a rastering technique to fabricate a single layer. After the deposition of a single layer, the subsequent layers are deposited by incrementing the nozzles and focusing the lens of the laser in the z-direction, until the 3D part is completed.

In order to expand the applications of metal deposition processes, multi-axis capability is often needed. A multi-axis rapid manufacturing system can be hardware-wise configured by adding extra degrees of mobility to a deposition system or by mounting a laser deposition device on a multi-axis robot. The configuration could also be a hybrid system in which a metal deposition system is mounted on a multi-axis CNC machine. With the addition of extra mobility, the support structures may not be necessary for the deposition process in order to build a complicated shape. Figure 6.77 illustrates the process to build an overhang structure on a 2.5D and multi-axis deposition system. Similar to the previous deposition processes, it is driven by a slicing procedure, which uses a set of parallel planes to cut the object into layers.

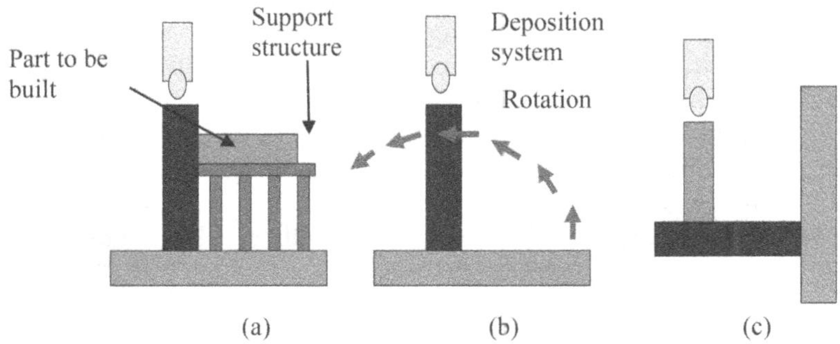

Figure 6.77 A multi-axis deposition system: (a) build a part with the support structure; (b) with multi-axis capability, after building the column, the table can be rotated; (c) after rotation, continue to build the component from another direction.

This hybrid process uses laser deposition for material deposition and CNC milling for material removal. It includes two major systems: a laser deposition system and a CNC milling machine system. The laser deposition system and CNC milling machine work in shifts in a five-axis motion mode. The laser deposition system consists of a laser and a powder feeder. With a five-axis deposition process integrated with five-axis machining, many obstacles can be removed. The bottleneck is the automated coding of the multi-axis machine's movement and planning of the material additive and subtractive processes.

Process planning, simulation, and toolpath generation for such a hybrid system allow the designer to visualize and perform the part fabrication from a desktop. The LAMP Planning uses B-rep models as input and generates a description that specifies contents and sequences of operations. The objective of the process planning is to integrate the multi-axis motion system and deposition-machining hybrid processes. The results consist of the subpart information and the build/machining sequence [Ruan05]. The basic planning steps involve determining the base face, extracting the skeleton, decomposing a part into subparts, determining build sequence and direction for subparts, checking the feasibility of the build sequence and direction for the machining process, and optimizing the deposition and machining.

1. Skeleton computation

   An algorithm for computing the skeleton of a 3D polyhedron as shown in Figure 6.78 is needed. The algorithm is based on a classification scheme for points on the skeleton computation in which the continuous representation of the medial axis is generated with associated radius functions. Because it is used as a geometric abstraction, the skeleton is trimmed from the facets that touch the boundary of the object along every boundary edge for which the interior wedge angle is less than π rad [Ruan05].

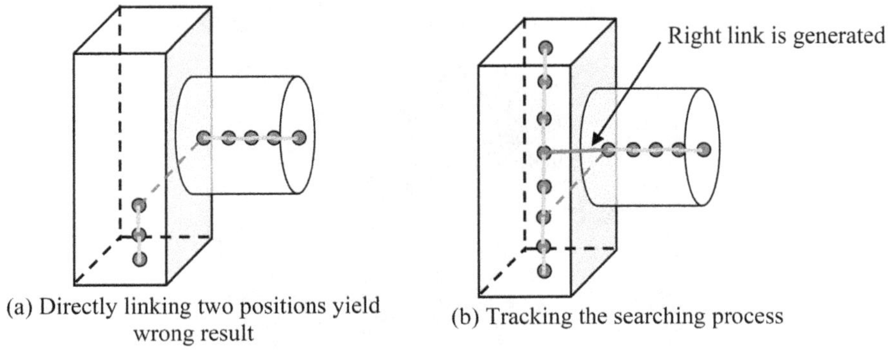

Figure 6.78 Topology link in skeleton computation.

2. Part orientation

   The determination of the base face from which the building process of the part starts is very important. The base face functions as the fixture in the machining process. Therefore, when in the machining process, it must provide enough resistance against the cutting force. The maximal resistance force depends on the area of the base face. The base faces have to satisfy the following conditions: (1) it should belocated on the convex hull of the part, and (2) it should have certain amount of contact area.
3. Part decomposition and building direction

   The objective of part decomposition is to divide the part into a set of subparts, which can be deposited and machined as shown in Figure 6.79. The topology of the part can be obtained from the skeleton. Each branch of the skeleton corresponds to a subpart. One of the partitions preformed is along a nonplanar surface. Therefore, close to the partition area, 3D layers are needed to build the connection between two subparts. The build direction of a subpart may not be constant. It changes when the part is built layer by layer so that for two adjacent layers, the later layer can be deposited based on the earlier layer without support structures. To achieve the nonsupport build, the build directions need to be along the skeleton.
4. Building sequence

   The results of the decomposition are recorded in an adjacency graph where nodes represent subparts and edges represent the adjacency relationship between connected nodes. After considering part building order, a directed graph that represents the precedence relationship among subparts can be constructed. From the precedence graph, one can identify in what order the subparts can be built. With the precedence graph, a set of alternative building plans can be generated. Each plan

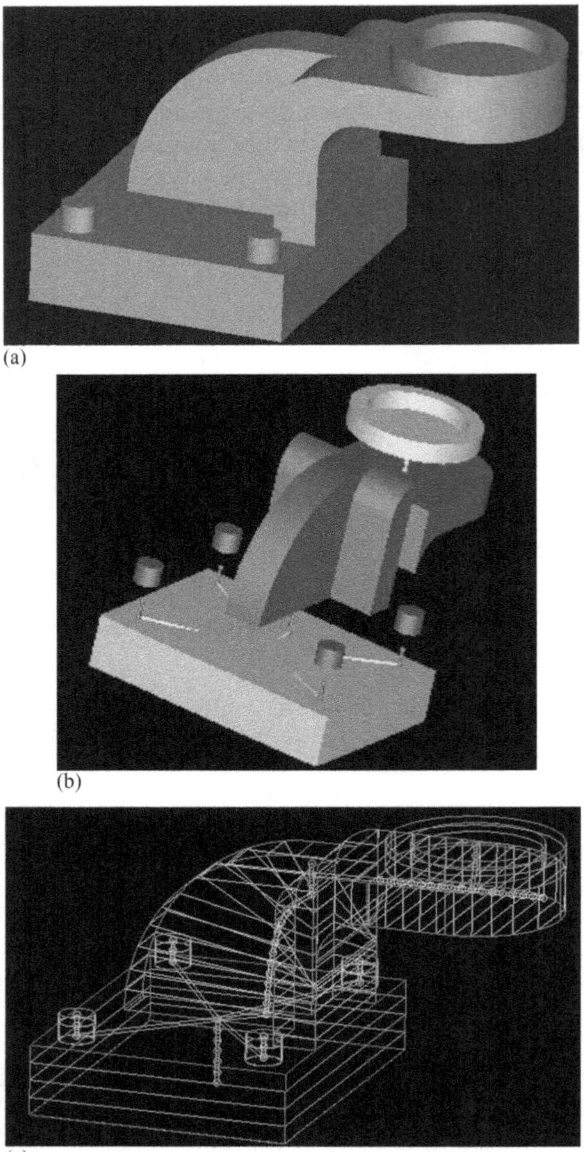

(a) (b) (c)

Figure 6.79 Example of planning a bearing seat deposition.

represents a possible building sequence on the decomposed geometry and can be chosen optimally depending upon machine availability or other criteria such as minimum building time (Figure 6.80).

5. Machinability check

   The main purpose of a machinability check is to choose an optimal building sequence from the sequence set. Local and global collision checks are operated first to choose acceptable sequences since the building direction is different in each sequence. If any kind of collision happens or an undercut plane appears, the corresponding sequence will be discarded. For the rest of the building sequences in the

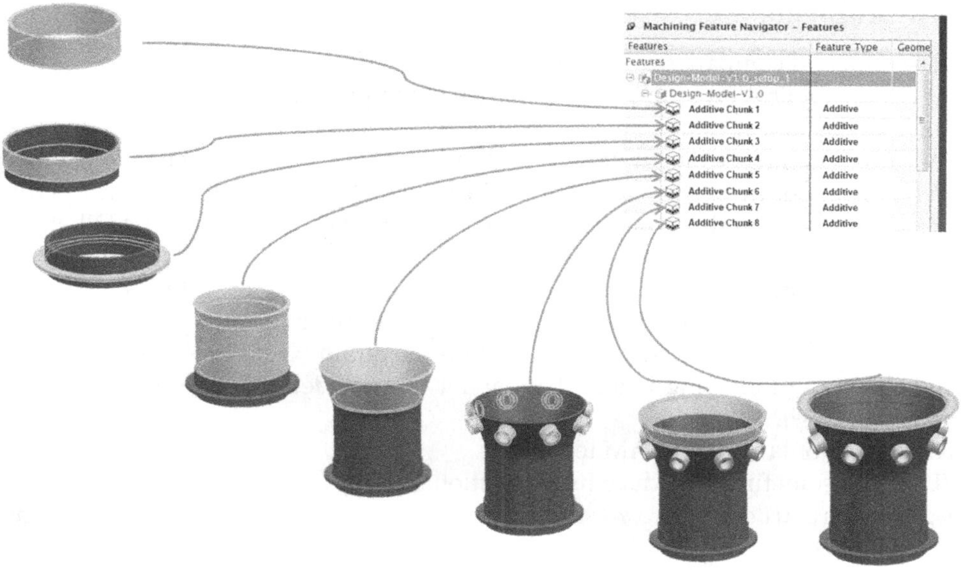

Figure 6.80 Process planning for an example part. (Courtesy of Siemens PLM.)

set, the buildability check and machining time computation are performed to find an optimal building sequence.

During process planning, a part may first be computed to find the skeleton diagram. Based on the skeleton diagram, it is then decomposed into subcomponents for slicing. Figure 6.81 shows the part fabricated using DMG/MORI's hybrid additive and subtractive machine, LASERTEC 65 3D hybrid.

The advantage of this process is the ability to produce parts with sound microstructure, including fabrication of large complex net-shape parts. A useful application of this process is to add components to existing structures, such as bearings. As this hybrid process has both material deposition and machining functions, it can be used to repair high value and high precision components. Also due to the excellence of local heat control, it can be used to effectively fabricate net-shape components with desirable microstructures and also 3D components and structures with FGMs.

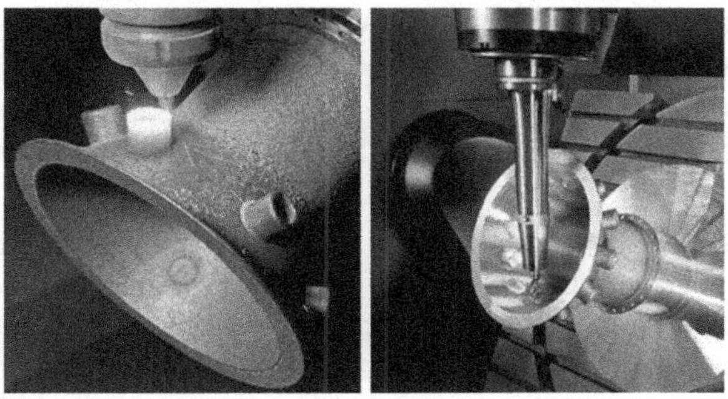

Figure 6.81 Fabrication of the example part in DMG/MORI's LASERTEC 65 3D hybrid. (Courtesy of DMG/MORI.)

Review problems

1. Research and compare the various AM technologies.
2. Give two application examples to illustrate the importance of prototyping a part with different colors.
3. What are the geometric limitations and main application areas of the DLD process?
4. What kinds of materials can be used in the DLD process? How about the dimensional accuracy and surface roughness of the parts processed by DLD before any postprocessing?
5. List the advantages and disadvantages of a powder-bed laser sintering process.
6. What are the advantages and disadvantages of 3D inject printing process?
7. Compare the powder-based processes with the liquid-based processes.
8. Which of the following is (are) true about the powder-bed laser sintering process (such as SLS®)?
   A. It is a nonlaser-based AM technique.
   B. It can sometimes produce fully functional parts.
   C. Support structures are washed away and broken with a knife, after building the part.
   D. The part produced could shrink disproportionably on cooling.
   E. Very limited materials can be used in the powder-bed laser sintering.
9. Select which characteristic describes an appropriate process: STL, laser sintering, 3D inject printing, DLD, UC, and contour cutting
   a. Uses a laser to bind or sinter a powder thermoplastic material together
   b. Injects a binder onto a thin layer of powder using inkjet technology
   c. Can mix metals to get specific properties
   d. Uses ultrasonics as layer bonding method
   e. Paper is the build material
   f. Uses a liquid raw material and requires a post cure process

6.6 *Summary and future AM processes*

There are various AM processes, and more processes are being developed. As AM is an emerging area, each process is improving rapidly and becoming more competitive. Although there are often questions on which process or which material should one use, the choice will highly depend on the specific application. End-use material performance can be impacted by specific part design, conditions, test conditions, etc. Actual values may vary with build conditions, and as mentioned, each process is improving quickly.

As presented, although AM technologies have significantly impacted on the product development process, most of the AM processes being used in industry are still limited to specific materials, and the technologies have not yet been fully utilized. Based on the needs of industry, the following summarizes the future trends of the AM technologies:

- Reasonably inexpensive. Just like the current printer industry, the cost of AM machines is expected to be lower, and general engineers will be able to have good access to the technology. More sophisticated AM machines may become common with reasonable price.
- More varieties of materials. As discussed previously, many AM processes can produce metal and ceramic parts or are still in the development or at initial commercialization stages, and thus, the machines to make parts with a good selection of

materials, including high-performance alloys and ceramics, and multi-materials will be available and more robust.

- More accurate. As AM machines are becoming more efficient to make parts with higher resolution and the trend is to integrate multiple processes to produce quality parts, many AM processes will be able to produce parts with better accuracy and precision.
- Much larger parts. When pushing the technologies to the larger side, some AM processes will be able to produce large full-scale parts, and some parts may be redesigned from the integration of several parts. An AM process to build a car, a house, or even a community of various houses is not impossible.
- Much smaller parts. Due to advances in micro/nano technologies, some AM processes will be able to make micro- or nanoscale 3D prototypes.
- Rapid manufacturing. When the technologies advance and overall cost can be lowered, instead of performing prototyping and tooling, many more processes will be able to directly produce functional parts for use and compete with the traditional manufacturing processes. The niche of AM processes will be for smaller quality products, for parts that cannot easily or efficiently be produced using traditional methods.
- Extensive medical applications. Many medical applications require custom-made AM parts. When AM processes can produce a variety of materials for medical application at a reasonable cost, AM technologies will be a natural match for medical applications.
- Repair and reuse. As some additive processes can effectively bind two materials together, including high-temperature materials, application of AM technologies for part repair, and reuse will be very cost effective. The applications will be in repairing and reusing medium and high-value parts. For example, one good application will be to repair a worn-out mold or to modify and reuse an existing mold for different applications.
- Functionally gradient materials. Some AM processes involve point-by-point material deposition. Thus, it is possible to control the material composition of the part. In other words, these processes will be able to produce a part which "smoothly" combines two or more materials together, metal to metal, metal to ceramic, optical material and metal, etc. There will be an infinite number of combinations to create parts with new material properties, and the applications are beyond imagination.

References

[3DSystem07]. 3D Systems, Retrieved on January 2, 2007 from www.3dsystems.com/.

[Brandes18]. Brandes en Meurs industrial design, Retrieved on July 31, 2018 from www.brandesenmeurs.nl/.

[DeskProto18]. DeskProto, Retrieved on July 31, 2018 from www.deskproto.com.

[Desktopmetal18]. Desktop Metal, Retrieved on June 1, 2018 from www.desktopmetal.com/.

[Gibertia16]. Gibertia H., M. Strano, and M. Annoni, An innovative machine for fused deposition modeling of metals and advanced ceramics, *MATEC Web of Conferences*, Milan, Italy, Volume 43, 2016, Article Number 03003.

[Hall51]. Hall E.O., "The deformation and ageing of mild steel: II characteristics of the L ders deformation," *Proceedings of the Physical Society. Section B*, Vol. 64, p. 747, 1951.

[Hanson13]. Hanson, J., Top five benefits of additive manufacturing (you never considered), 2013 July, www.pddnet.com/article/2013/02/top-five-benefits-additive-manufacturing-you-never-considered.

[Karnati17]. Karnati, S., J. Hoerchler, F. Liou, and J. Newkirk, Impact of specimen dimensions on miniature tensile characterization of powder bed fabricated 304L stainless steel, *Proceedings of the 28th Solid Freeform Fabrication Symposium*, Austin, TX, August 6–9, 2017.

[Lipson18]. Lipson, H., Additive manufacturing file formats, Retrieved on November 23, 2018 from fab.cba.mit.edu/classes/S62.12/docs/formats.ppt.

[Petch53]. Petch, N.J., "The cleavage strength of polycrystals," *Journal of the Iron and Steel Institute*, Vol. 174, p. 25, 1953.

[Ruan05]. Ruan, J., K. Eiamsa-ard, and F. Liou, "Automatic process planning and toolpath generation of a multi-axis hybrid manufacturing system," *SME Journal of Manufacturing Processes*, Vol. 7, No. 1, pp. 57–68, 2005.

[SGC07]. Solid ground curing, Retrieved on January 2, 2007 from http://home.att.net/~castleisland/sgc.htm.

[Trosch16]. Trosch, T., J. Strößner, R. Völkl, and U. Glatzel, "Microstructure and mechanical properties of selective laser melted Inconel 718 compared to forging and casting," *Materials Letters*, Vol. 164, pp. 428–431, 2016.

[White02]. White, D.R., Object consolidation employing friction joining, United States Patent 6,457,629, October 1, 2002.

chapter seven

Building a prototype using off-the-shelf components

Often, the quickest way to realize a physical prototype is to build the prototype using off-the-shelf components. Most prototyping projects will require acquiring components to make a prototype, and in fact, the more off-the-shelf components that can be purchased and used, the quicker the prototype can be realized. However, many engineers do not have the knowledge to systematically proceed with component acquisition. This chapter presents a procedure for prototyping a system using component catalogs for a portion or all of the components. The procedure presented for choosing components should also be useful in most engineering projects. The issues regarding choosing components from a catalog apply to any design project and can be extremely frustrating and costly. Many engineers have been involved in many time-sensitive projects where the wrong component or material was specified, ordered, or delivered. Usually, there is some type of fire drill that will occur to order the proper component/material or find an acceptable substitute. This chapter defines what is needed to solve the problem, and the steps include the following:

- How to decide what to purchase?
- How to find the catalogs which have the needed components?
- How to ensure that these components will work together?

Most of the prototyping components can be found by following these steps.

7.1 How to decide what to purchase?

This section helps answer the following questions:

- How to decide what to purchase?
- What to consider in order to decide what is needed?
- How to come up with a bill of materials so that the system is complete?
- Why and how to prioritize a list of items to purchase?

To decide what needs to be purchased for a product prototype, one can follow the product development process steps:

1. Define the requirements.
2. Produce a conceptual design.
3. Produce a final design.
4. Manufacture the machine or system.

The question is, in which step does one decide that one will need a component from a catalog? There is no absolute answer as it depends on each particular case. It is normally not practical to decide whether to purchase a component when initially defining the product requirements, unless it is a major component and a decision is made at an early stage. In general, one seldom decides to purchase a component in the stages of producing a conceptual design, but sometimes off-the-shelf components could be purchased when a prototype or a product is developed. For example, when conceptually designing an aircraft, sometimes it is known whether an engine will need to be purchased as the engine is a major component. However, one may decide to purchase a particular joining device at the last minute of prototype fabrication. What this says is that the decision could be made at any stage, but often it is made in steps 2 and 3. The reason is due to the fact that it may be necessary to know many part requirements in order to make a decision.

The analysis of part requirements should include many aspects:

- Functionality
- Working environment, temperature
- Volume, size, weight
- Safety
- Cost
- Schedule
- Make/buy

7.1.1 Purchasing decision for a prototype

Make/buy decision is used to decide whether a part should be made in-house or should be purchased. Many of today's products are so complex that not many companies have all the necessary knowledge about either the product or the required processes to completely design and manufacture them in-house. As a result, most companies are dependent on others as part suppliers. Most companies design and make only a portion of what makes up their products, buying the rest from a complex multilink chain of suppliers. An initial understanding of automotive supply chains was achieved when it was found that the most successful Japanese car companies design and make as little as 30% of the components that go into their cars [Clark89].

The reasons for buying instead of making a component may be due to one or more of the following facts:

1. The company cannot make the component or easily acquire such a capability and must seek a supplier.
2. The supplier has a lower cost and/or faster availability.
3. The supplier's version of the item is better for any number of possible reasons.

On the other hand, the reasons for making a component in-house could be due to one or both of the following facts:

1. The item is crucial to the product's performance, or the skill in producing it has been judged critical to the company's technical competence.
2. In-house has lower cost or faster capability.

A company should make what matters most to the customer or what differentiates the product in the marketplace. The make/buy decision should also consider customer's needs.

The make/buy decision may also be impacted by whether the product is designed to have modular architecture or integral architecture. A product with a modular architecture has components that can be mixed and matched due to the standardization of function to some degree and standardization of interfaces to an extreme degree. Home stereo equipment has a modular architecture as one can choose speakers from one company, a CD player from another, a tape deck from a third, and all the parts from the different manufacturers will assemble together into a system. IBM-compatible computers are also modular with respect to the CPU, keyboard, monitor, printer, software, etc. A product with an integral architecture, on the other hand, is not made up of off-the-shelf parts, but rather comprises a set of components and subsystems designed to fit with each other. Functions typically are shared by components, and components often display multiple functions. Airplanes are an example. The product must be developed as a system, and the components and subsystems are defined by a design process exerted from the top-down, rather than the bottom-up design process that may be used by a computer manufacturer.

After the needs and requirements have been defined, a conceptual design of the system or machine can be developed. This allows the engineer to integrate all of the users' requirements with currently available components and technology. This should create an overview of what the system may look like. The overview may be more of a list of specifications for each component within the machine or system than an actual picture. From these specifications, changes or trade-offs may be made before the final design and manufacturing stages. These changes or trade-offs include the following:

- Changing power sources
- Increasing or decreasing power requirements
- Changing material selection
- Changing controller systems
- Modifying operator interface requirements
- Modification of ambient operating environment

7.1.2 What to purchase?

No matter what kind of product, to decide what kind of components are needed, the functional efficiency technique discussed in Section 2.5 can be used. It is a top-down approach to define and understand product function and how the function relates to others. The first step is to start with a list of the derived functions for a new product with a noun and a verb. The second step is to find the relationships among the functions by arranging the functions into a logical orderly fashion and finding the subfunctions and its related units. When the function can be replaced by a physical object, the process can be stopped, and thus, the bottom components are physical objects, needed for the design. The physical objects that cannot be made in-house are the components that will need to be purchased.

Once the physical objects or components have been defined, one needs to investigate the details of the prototype, such as cost, functionality, and other important points of the prototype or product before making or buying. For example, if the product is a type of artwork or furniture that is advertised as handcrafted, then automation may negatively affect the bottom line of the sale of the product to the customer who is willing to pay extra for something that is handmade. If a company promotes itself as offering affordable products, then any cost reduction, including automation, should be considered.

These changes and trade-offs require some technical information and expertise, both of which can often be found in component catalogs. Component catalogs often have a plethora of information about the components listed within them. This information may be overwhelming at first due to the fact that there is so much information. This is mainly due to the fact that the component suppliers are trying to answer every question that some engineer at some point will ask in one catalog. The list of specifications produced during the conceptual design step was created to help wade through the sea of information.

For example, let us say that the list of specifications calls for a remotely activated flow valve. Also, in the specifications, it mentions that this valve is to control flow from an existing hydraulic power source. These two specifications greatly reduce the number of valves and valve suppliers which need to be reviewed. To limit the number even more, the designer should look at the list of specifications again. What pressure will the hydraulics operate at and with what flow rate? Also, the specifications should list how the valve is going to be controlled. If an existing PLC (programmable logic controller) program were to control the valve, then the designer could limit the search to electrically activated valves.

Assume that from the specifications list, one has limited the valve to an electrically controlled hydraulic valve with a known pressure and flow rating. The component catalog probably will still list several options including voltage and AC- or DC-powered valves. Again, the specification should help limit the choices but perhaps not directly. Since the designer did not know what the options from the component supplier were, some specifications may have been overlooked. Perhaps, the designer does not know what voltage the PLC outputs or controls currently, but a call to the supplier should help answer the new specifications questions.

This process of limiting choices by looking at the specifications list is very easy, but it may cause some original specifications to change slightly or be added to. By limiting the choices for each component so that only one or two components fit the specifications, the designer or engineer can list all of the components needed for the new machine or system. If there are some components that the designer cannot decide on, it never hurts to call the components' vendor and ask for his opinion. The vendor usually has an excellent technical support line that will know more about the products than what is published.

When all the components are specified, list each component's specification and review the new list of specifications. Make sure there is nothing missing or perhaps out of place. If, for example, hydraulic components are listed, and all of them are rated for 3,000 psi, except one, which is rated for 2,000 psi, investigate why one component is rated differently and whether it will cause the system to fail or work improperly. If it is a problem, then some changes or trade-offs must be made so the system can work.

When designing an automated system from catalogs, the main problem that is presented is the inability to build or order custom components. This may not seem that big of an issue, but remember how integral all of the signals and components are with each other. If one sensor or actuator is not receiving the proper signal, the whole system could fail. With a large risk, like the one presented, it is best to develop a plan that will segment the decisions that need to be made. This step-by-step process will allow one to make an informed decision inside a small scope and then later apply those smaller scopes to the larger scope. Computer-aided design (CAD) files of the components sometimes can be found on the vendor's website. These are great resources to help layout the machine or system. A general CAD drawing may show some problems with component sizes, locations, or other specifications.

Example 7.1: Sensor purchasing

This case study includes an oxygen-detecting system in a newly developed chamber. Assume that one does not have much knowledge about sensors and does not have the capability to make a sensor, but needs to purchase the system. The question is where and how to start? Maybe, one would think that since it is going to be purchased, a phone call would be needed to order the oxygen sensor. Is it that simple? Sometimes, it may not be that straightforward, and it is necessary to consider the complexity in the available options and price issues. If a project can start with a more systematic selection process, it can be planned well in the decision-making process:

1. Objective: To design, purchase, and integrate a sensor in an argon chamber to detect the oxygen level during the entire manufacturing process when the chamber air is replaced with the argon gas.
2. Goal: In addition to maintaining less than 500 ppm (parts per million) oxygen level for processing, it is necessary to maintain a safe environment for the operator outside of the chamber.
3. Constraints/conditions:
   a. The air contains about 21% oxygen and 78% nitrogen.
   b. The argon-filling process will take several hours to bring the oxygen level from 21% to 500 ppm.
   c. Monitoring the oxygen level inside the chamber during these hours is necessary.
   d. The chamber will occupy about half of the volume of the room, and thus, an oxygen sensor and an alarm system are needed outside the chamber to make sure that the operator can operate in a safe condition.
4. Sensor system design and specifications:
   e. Overall: There should be three sensors: the first sensor to detect the oxygen level in the chamber, the second sensor to detect the level of oxygen outside the chamber, if possible, and the third sensor to act as a backup to the second sensor for the safety of the people inside the lab, but outside the chamber.
   f. Inside sensors: The inside sensor should measure the oxygen level in ppm. If possible, it should have the capability to measure from 21% to 500 ppm.
   g. Outside sensors: The two outside sensors should measure the oxygen level in percentage of oxygen. They should be independent of the system controller or system PLC.
   h. Life: Oxygen sensors normally have a certain life span as it involves a chemical reaction, and thus at certain time intervals some chemical components should be replaced. The sensor should have a considerably long life.
   i. Integration: The sensors will be integrated in a chamber controlled PLC and thus should be compact, easy to handle, and mount.
   j. Cost: The sensor should fit in the estimated budget.
   k. Accuracy: Accuracy is the relationship between the readout and the true concentration. This relationship is indicated by an error factor (indicated by "±," e.g., ±0.5%). The lower the number, the greater the instrument's accuracy. The outside sensors do not need very high accuracy, but they need to be reliable. The inside sensor needs to have good accuracy.

   Figure 7.1 shows a sensor design result based on the functional efficiency technique. This design process results in four sensors (sensors nos. 1–4) to be purchased, and one alarm tied to the outside sensor.
5. Sensors selected: Based on the previous design and requirements, the sensors in Table 7.1 were selected. As it was difficult to find an inside oxygen sensor at the time of ordering which can measure a wide range of oxygen levels, two separate sensors were purchased for inside usage. The outside sensors are not too expensive as they only need to provide one setting (warn the user when the oxygen

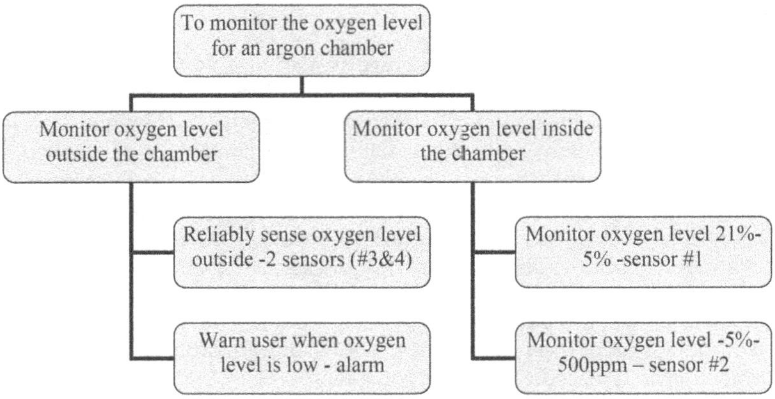

Figure 7.1 The sensor design result based on the functional efficiency technique.

Table 7.1 Sensors selected

Product	Company/ specifications	Unit	Unit price		Comments
Series 1000 oxygen analyzer	Alpha–omega; Accuracy: ±1% of full scale	1	$1,744		Oxygen sensor to be used inside the chamber for use in the range 21%–1%
Series 3000 trace oxygen analyzer	Alpha–omega; Accuracy: ±1% of full scale	1	$3,018	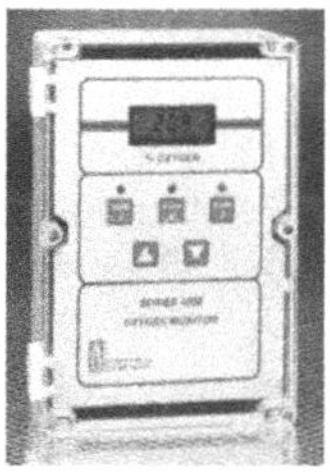	Sensor to be used inside the chamber in the range 5%–500 ppm or 500–50 ppm
Fixed oxygen monitor	Analox: two alarm set points at 19.5% and 18%	2	$595	 The Analox 1 main sensor and remote repeater unit	Sensors to be used outside the chamber to warn the operator when the oxygen level is low. One extra sensor is used as a backup

level is low), and they come with an integrated alarm. Also, they are independent and not integrated with the PLC so that in case the PLC is down, they will still work. There is also an alarm repeater which provides four status indicators, which mimic the status indicators on the main Analox 1 enclosure. The alarm repeater can be installed in a separate room so that the operator can monitor the oxygen level in the chamber room before actually entering into the chamber room.

7.1.3 *Draw a flow diagram of signals and components*

It is a good practice to summarize the system components before purchasing them. The first step is to create a basic flow diagram of the system as shown in Figure 7.2. This will help identify the nature of the system. What types of motion will be created? What types of events will need to be sensed? What types of output signals must be sent? This flow diagram does not need to be very detailed. All it needs is to allow one to visualize the process and the information flow through the system.

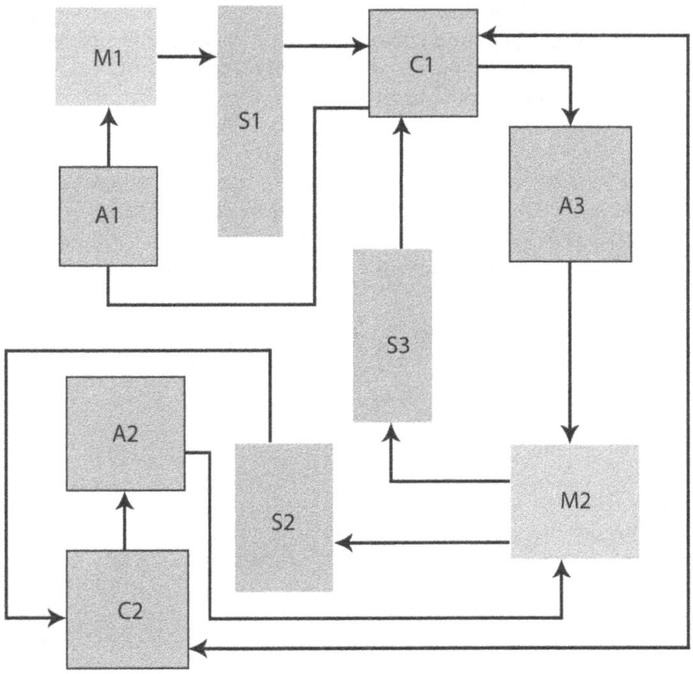

Figure 7.2 Signal flow diagram.

To create this flow diagram, start by drawing each of the mechanisms; one will need to perform the necessary tasks. Then, add the necessary actuators to power these mechanisms. Next, add the main control station. Then, add the sensors needed to monitor the process. The diagram may be quite cramped, but it will soon make sense of things.

To identify each of the components, number them in any desired logical manner. For example, the mechanisms can be identified with an M1, M2, M3 designation and likewise for the actuators (A), sensors (S), and control stations (C).

Now draw the signals that will need to travel between each component. While doing so, log the signals in a spreadsheet, showing from which component the signal originated, where it is being sent, and the type of signal as shown in Figure 7.2. This spreadsheet will help later when selecting what methods and hardware will be necessary to automate the process.

The flow diagram should now be complete, and look something like what is shown in Figure 7.2. This diagram relates all major components in a logical manner. For example, on the upper position of the diagram in Figure 7.2, it shows that device C1 (e.g., a computer) provides a voltage to device A1 (e.g., an actuator), device A1 outputs a force to device M1 (e.g., a mechanism), M1 outputs the speed received by device S1 (e.g., a sensor), and S1 outputs a voltage to device C1. It is still a basic diagram and can change throughout the process, but will be a great launch point for the remainder of the decisions.

Example 7.2: Selecting and purchasing sensors for a friction stir welding robot

In friction stir welding, a robot with three nonparallel telescopic translational joints and three rotational joints is used. The robot can handle very large force. Thus, an external sensor with analog outputs needs to be applied to measure the forces and moments from the friction spindle. Considering both the work requirements (high load, high velocity, low control frequency) and the cost, a multi-axis force/torque sensor system is selected. Figure 7.3 shows the schematic diagram of the robot motion joint and the force sensor. This is a simple feedback control example in which the controller (C) controls the motor (M) with a voltage; the motor (M) moves the robot (R) joint with a certain rotating speed; the robot (R) joint processes the friction stir device (G) with a controlled depth; the force sensor (S) receives the force exerted in the friction stir process; and the sensor sends the voltage back to the controller. Figure 7.4 shows the system's signal diagram. Although only the sensor needs to be purchased, the signal flow diagram can be used to check the compatibility of the surrounding components.

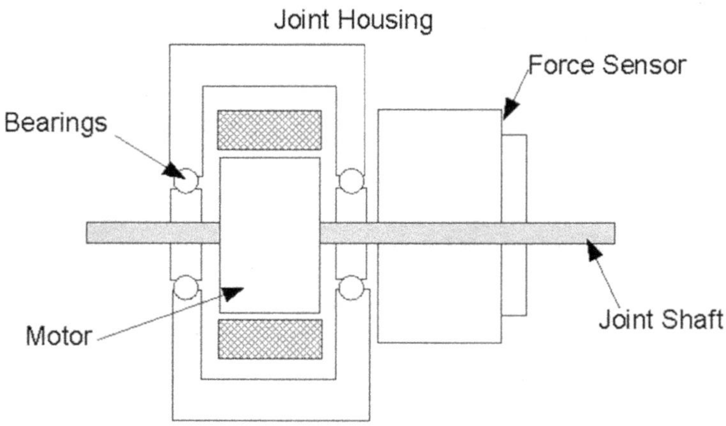

Figure 7.3 Schematic diagram of the robot motion joint and the force sensor.

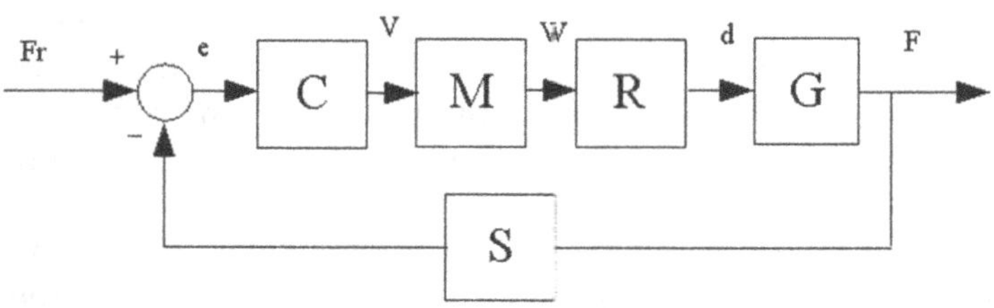

Originator	Receiver	Signal
C (controller)	M (Motor)	V: Voltage
M(Motor)	R (Robot joint)	W: Rotation speed
R (Robot joint)	G (Friction Stir Process)	d: Plunge depth
G (Friction Stir Process)	S (Force Sensor)	Force

Figure 7.4 Signal flow diagram of the sensor.

7.1.4 Prioritize the precision of the system

Once a basic flow diagram has been created, determine which components are the most important. Obviously, most of them are important, but in this instance, which signals need to be most accurate, or which device needs to be the most precise (and which ones are not as critical)? The accuracy of the signal is a very important characteristic of the component. When selecting standard parts from a catalog, options are limited; therefore, one wants to make the decision of selecting the most important component(s) at the time when the most options are available. Because after each decision for a component, the options for the other components are reduced, the selection of the most critical component should be the first decision.

To decide which components are most important to the whole process, one will have to look at the individual processes. Each process will need to be treated differently. Note that not all components or systems fall into this prioritization. For example, certain mechanisms may not be as important as the sensors. Each component and system should be evaluated independently. The spreadsheet created in the first step could be used at this time as well as later in the process to identify which components provide the most communication or are "parents" of the most processes.

Example 7.3: Development of a temperature controller for CPU fans

CPU coolers have been getting more and more efficient over the last few years. However, this higher performance usually has a price: more noise. The reason is that in many PCs, the fans are always spinning at maximum speed—no matter if the CPU is running at full load or is idle, and no matter if the outside temperature is 308°C or 168°C. Existing temperature-controlled fans have some problems. A major problem is that the user cannot adjust the temperature. Another problem is that each fan needs its own sensor, so it is necessary to make it user-accessible and provide an economic temperature controller for CPU fans.

1. Requirements

   Based on the problems discussed above, the requirements of this temperature controller are as follows:
   - Temperature setting can be adjusted by the user
   - Fans are switched off if the temperature is low enough
   - Several fans can be controlled with just one sensor and temperature control
   - Low cost
2. Function analysis

   The functional analysis is summarized in Figure 7.5.

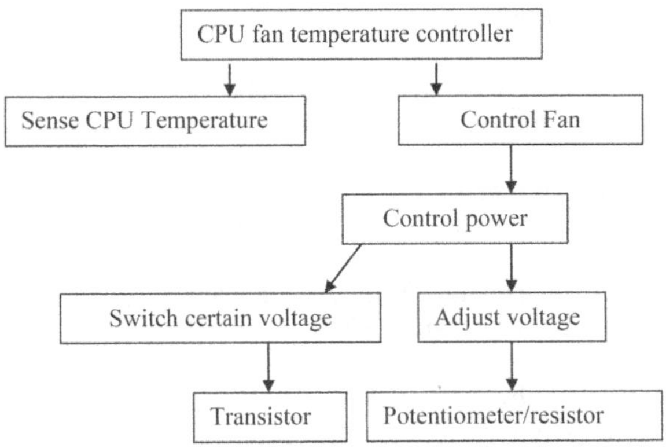

Figure 7.5 The functional diagram of the temperature controller for a CPU fan.

3. Planning

   Objective: To build a temperature controller that costs less than $20. The user can adjust the temperature at which fans should be turned on. Several fans can share one sensor and temperature control.

4. Purchasing components as shown in Figure 7.6
   - Power transistor
   - Temperature sensor
   - Potentiometer or resistor
   - Heat-contractible tubing
   - A little printed circuit board (PCB) (for mounting such parts)

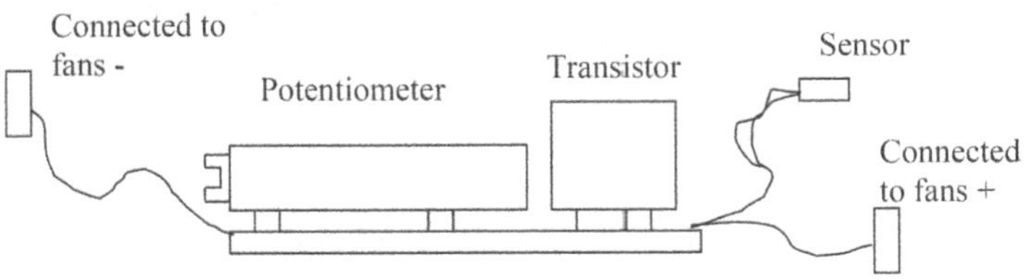

Figure 7.6 Assembly of the system components.

5. Signal flow diagram as shown in Figure 7.7

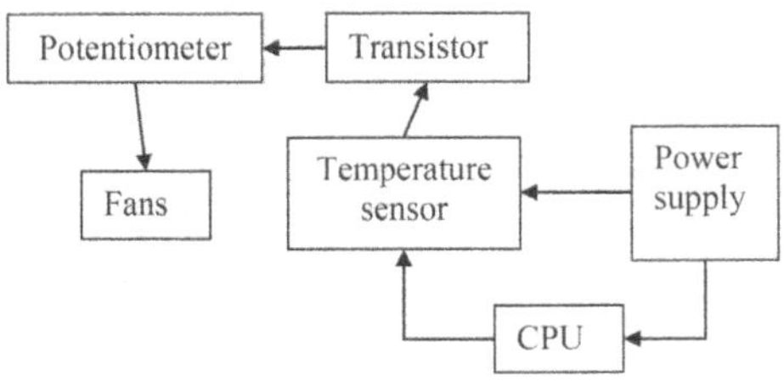

Originator	Receiver	Signal
CPU	Sensor	Temperature
Sensor	Transistor	Voltage
Transistor	Potentiometer	Voltage
Potentiometer	Fans	Voltage
Power	CPU	Voltage
Power	Sensor	Voltage

Figure 7.7 Signal flow diagram.

6. Prioritize the precision of the system. The precision from high to low is sensor > transistor > potentiometer
7. Main components selection. The main selection criteria are as follows:
   - The size should be small
   - Inexpensive (less than $5)
   - Resistance should be at least 10 kΩ
8. Final choice: NTC Thermistor (from www.jameco.com) at $1.59
   - Operating temperature range: –30°C to +130°C
   - Thermal dissipation factor (@ 25°C): 6.5 mW/°C
   - Resistance value allowable difference: 10%
   - Resistance (@ 25°C): 10 kΩ
   - B Constant 25°C/50°C (10%): 4,100
   - Temperature coefficient (at 25°C): –4.6 (%/°C)

9. Secondary components selection
   a. Power transistor
      - Requirement: it can handle the 12 V voltage and the current the fan requires
      - Final choice: IRFZ24N MOSFET power transistor at $1
      - IRFZ24N MOSFET power transistor can usually handle over 50 V and over 10 A
   b. Potentiometer
      - Requirement: not less than 10 kΩ
      - Final choice: 10 kΩ spindle trimming potentiometer at $1.59

Review problems

1. Use functional efficiency technique to design a nozzle used in laser deposition. This nozzle should allow a laser to pass through the center and allow for concentration of powder into a small region.
2. List five key things that should be done before you go to a catalog and begin to order parts.
3. What should be considered in selecting a temperature sensor?
4. Define the signal diagram and detail components needed to build an XYZ 3D motion system. The motion should be precise; hence, the system needs position, velocity, and acceleration data. The velocity and acceleration can be derived from the first and second derivatives of the position, and thus, only the position data is sufficient, but sensors are needed to detect the position of the axes. To identify the absolute position of the motors compared to the slides of the gantry, position limit triggers are needed. To control the motion of the entire three axes, a controller is essential for the system. Hence, we need the following:
   A. X, Y, and Z orthogonal axes
   B. Motors and amplifiers
   C. Position sensors
   D. Home triggers and position limit triggers
   E. A controller
5. A company wants to implement a simple and inexpensive robotic arm as shown in Figure 7.8 in one of their assembly lines. The material needed to make the arm has been defined, and we need to purchase the motion components. The arm is rotated around the base and has one pivot in the middle (total of two rotation points). Other information can be assumed. A motor will be placed at each pivot point. The end

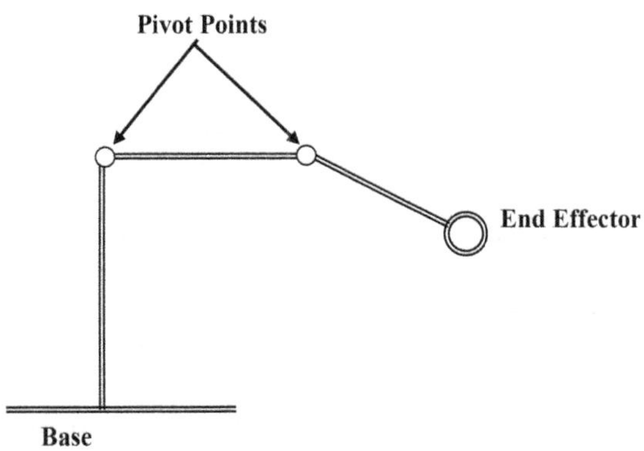

Figure 7.8 The schematic diagram of a simple and inexpensive robotic arm to be implemented.

effecter will be made of magnetic material to pick up ferrous objects. Draw the signal flow diagram and make recommendations on the motors, magnet, and controller.

7.2 *How to find the catalogs that gave the needed components?*

The artist is the only one who knows that the world is a subjective creation, that there is a choice to be made, a selection of elements.

Anais Nin

This section will help answer the following questions:

- Where to find the components needed to purchase?
- How to know that a vendor or a product is a good choice?
- Very often there will be many choices when purchasing a part. How to select the right one?

In the current Internet age, finding information and products is easier than ever before. If a company is implementing automated systems, one may want to start from a catalog from one of the large component suppliers, such as Grainger or McMaster-Carr. If for nothing else, these catalogs serve as a good reference manual for many different automated devices. They can be used to help generate automation ideas by looking through the specifications for different equipment. They can also be used as a price reference when deciding on the best device for a particular application. Instead of waiting for a catalog, refer to the websites listed in Table 7.2.

Table 7.2 Some commonly used websites for off-the-shelf components

Company	Website
FastenerNet	www.fastenernet.com/
Carr Lane	www.carrlane.com/
Grainger	www.grainger.com/Grainger/wwg/start.shtml
iCrank	http://icrank.com/cgi-bin/pageman/pageout.cgi?path=/directory.htm&t=1
McMaster-Carr	www.mcmaster.com/
Information Handling Services	www.ihs.com/catalogxpress/
Global Spec	www.globalspec.com/
Thomas Register	www.th omas net.com/
Wiha Quality Tools	www.wihatools.com/distlink 2.htm
Part Solutions	www4. cade nas.de/wunschkatalo g.asp?sid=2
Omega	www.omega.com/
Fisher Scientific	wwwl.fishersci.com/index.jsp

In addition to paper and electronic catalogs, automation devices are often stocked at manufacturing parts distributors that can be found in any major U.S. city. Simply look in the local yellow pages or go online to www.yellowpages.com and look under industrial equipment, industrial supplies, manufacturing part distributors, machine tools, or other related topics to find local distributors. It may also be applicable to find automation services that, for a fee, will help plan and install automated equipment in the company.

7.2.1 Evaluating companies and products

To narrow the vendor company and product list, the designers need to make some judgments which cannot be addressed completely because of almost an infinite number of issues. However, some critical issues will be discussed. If a designer already owns equipment from a company on the list, he or she may want to continue with that company—repair parts and maintenance might be similar, and technical support is readily available. If that is not the case, the designers will have to weigh several factors:

- Cost—it is the cost to the life cycle of the product.
- Quality—it is balanced with cost but is important to the final product.
- Technical support—it is helpful both during the design while the product is being integrated into the system and during use when it requires repair or maintenance.
- Features—more features upfront may mean the designers will have to do less integration into the automation system.
- All the requirements—power usage, noise, space, etc.—that were placed on the black box solution earlier, also have to be placed on these components.

The vendor/component selection process should have a system perspective and always have integration in mind. For example, selecting a machining center with a pneumatic chuck would work best with a system with pneumatic actuators and vice versa.

With the operation centers chosen, the movement between them must be detailed. As discussed previously, modeling the system in a CAD environment will greatly reduce the cycle time for the design–analyze–evaluate–modify the iterative process. Tools such as kinetics analysis, which can be used to visualize the movement of systems, may be included with available CAD systems. The analysis tools may also be capable of analyzing forces at nodes to determine if a joint will withstand the stresses on it.

7.2.2 Component selection

Ordering components can sometimes be just as difficult as trying to figure out which components are to be used in the first place. Some of the components needed are back ordered or are no longer available. If time is not overly important, then a backorder is not a problem, but an item which is no longer available may be a concern. As long as the reasonable amount of time is allowed for either case, a lot of last-minute panic or cost can be avoided. In either case, first, ask the vendor or check with other sources if there is another component that is equivalent in specifications but perhaps made by another manufacturer. Often this is the case, and no redesigning is required. Also, ask if the component has been replaced with a newer component of the same specifications from the same manufacturer. Again this often is the case. The worst case scenario can happen when no substitute components can be allocated; then, the designer will need to jump back to redesigning or explore other alternatives.

Sometimes, the correct part is ordered, but the wrong part is shipped. The wrong part may have to be shipped back before the vendor will ship the correct part, so correction should be processed as soon as possible. If the part was ordered online, save the screenshot of the order and print a copy. This simple step may help prove that the vendor made the mistake and not the person that ordered the components. If the order is placed over the phone, write down the day, time, the person that took the order, and the part numbers requested. Many times, if one can name the operator that took the order, the vendor is much more cooperative. These are common practices, but are often ignored in the purchasing process especially when trying to meet a deadline.

Most importantly, do not forget to make detailed drawings of how all of the components are to be assembled to fit together. This will speed up the assembly time greatly. In the decision-making process, many tools, such as the house of quality method as discussed in Chapter 2, can be used.

Example 7.4: Designing and implementation of a transfer chamber

The objective of this project is to design and implement a removable transfer chamber for an argon chamber. To achieve an oxygen-free environment in the main chamber, an inert gas will have to be used, which in this case is argon. A transfer chamber can transfer the specimens from outside to inside or inside to outside with a minimal amount of gas loss. The transfer chamber is designed to be removable so that it allows movement in front of the chamber to be easier and reduces the risk of a collision with the transfer chamber, thus causing a leak in the main chamber that can possibly harm someone. The total budget is estimated to be at $1,000. To start the designing process, the product definition and requirement were used to start brainstorming solutions to the problem. Multiple ideas surfaced from the brainstorming. With these multiple ideas, there were two ways to go. Either make a chamber or purchase a chamber.

To purchase a transfer chamber, one option was to purchase an inflatable chamber or to purchase a box style transfer chamber. The options of making a chamber are building a box style transfer chamber as shown in Figure 7.9 or building a tube style transfer chamber. From looking at the cost of buying a system, $8,000 for the box style and $1,800 for an inflatable chamber at the time, it was determined to look further into making a transfer chamber. To assist in cost estimation, paper prototypes of the tube design and the box design were created. Initially, the tube design consisted of a 10 in. clear PVC pipe with two gate valves and rollers to transfer the parts. After completing a simple cost estimation, it was found that a gate valve would run around $400–$500. That amount would need to be doubled to close both ends. The valves alone drove the price up to the point where another method should be considered. The concept of placing bulkhead type doors would reduce the price significantly, but there was wasted space and not much room for bigger parts. To better utilize space, a box chamber was considered.

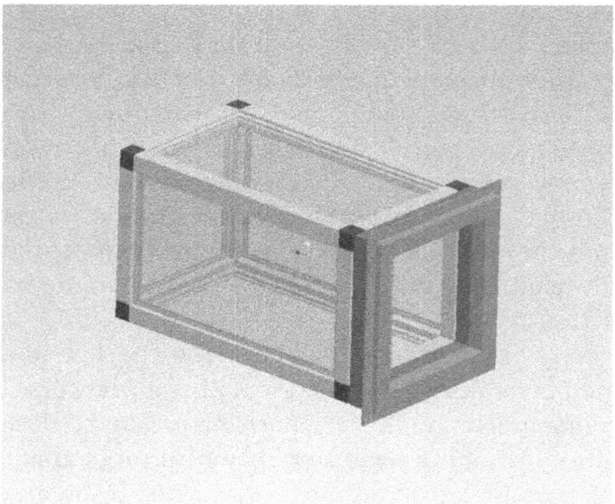

Figure 7.9 Option to make a box transfer chamber.

Similar types of doors, like on the tube design, were initially planned to be used.

Quotes were received from the suppliers of the commercially available transfer chambers. For an inflatable chamber, the cost would have been $1,789.50, and for the box transfer chamber, the cost was $8,000.00. To obtain cost estimation for the other two transfer chambers, an estimated bill of materials was made. Then, the components were looked up in catalogs to get a price. The estimated price for the tube chamber was $1,300.00. For the box transfer chamber to be made, the price was $290.00 (Table 7.3).

Table 7.3 Estimated cost for made box transfer chamber

Description	Price ($)	Quantity	Total ($)
Polycarbonate (box) 24 × 24 × 0.25	27.51	1	27.51
Polycarbonate (hatch) 12 × 12 × 0.375	13.90	1	13.90
Aluminum extruded channel (box no. 7) 8 in.	29.62	1	29.62
Aluminum extruded channel (box no. 7) 4 in.	15.40	1	15.40
Aluminum extruded channel (chamber no. 2) 4 in.	15.40	1	15.40
3-Way corner pieces	8.95	8	71.60
Aluminum 6063-T5 square tubing (6 ft) 0.0625 wall	8.33	3	24.99
Lift-off hinges (left 2 in. width × 2 in. leaf height)	1.96	2	3.92
Ball valve for purge (3/8 pipe size) female/female	4.38	2	8.76
Barbed × pipe fitting (3/8 pipe size)	6.88	1	6.88
Aluminum 6063 T-52 angle bracket (1.5 in. × 1.5 in. × 0.0625) 8 in.	10.00	1	10.00
90° pipe bend (male × male)	3.67	2	7.34
3/8 pipe lock nut	0.81	5	4.05
Sealing materials for transfer chamber	1.58	7	11.06
Tubing	4.00	1	4.00
Latches	8.00	1	8.00
Screws	4.00	1	4.00
Internal hinges	8.00	1	8.00
Internal latches	8.00	1	8.00
Drawer slides	5.95	1	5.95
		Total	288.38

To help select which transfer chamber to use, the house of quality method was used. The factors that were used to compare were if it could accept the part size and withstand the maximum weight; how well the volume is used; the storage of the chamber; ease of use; if it can be customized; the price; ease of manufacturing; and if any additional parts needed (i.e., a table or stand). This information was put into Table 7.4, and each criterion was weighted so that the sum of the weights equaled one. The most important factor was the price of the chamber followed by the part size and weight capacity. Each requirement was rated on a scale of one to five, with five being the best. That score was multiplied by the weight to get its weighted score. The weighted scores were summed up, and it was found that the self-made box chamber was the best way to go, and the items to be purchased are in Table 7.3.

Vendors try to give all the information they can, so that the engineer, designer, and end operator can work to make sure that a successful prototype can be built. It is up to everyone involved in producing a new system or machine to read each component's specifications to see if it will work within the design parameters. When in doubt, look to see if the vendor has a technical support line. Often what they sell is their expertise, and they will be an invaluable resource for the system design.

Table 7.4 House of quality for comparing the ideas for absolute importance

5 = Meets requirement/best 1 = Fails requirement/worst		Type							
		Purchase box style chamber		Self-made PVC pipe		Purchase inflatable chamber		Self-made box style chamber	
Requirement	Weight	Raw	Weighted	Raw	Weighted	Raw	Weighted	Raw	Weighted
Part size	0.15	5	0.75	5	0.75	5	0.75	5	0.75
Weight capacity	0.15	5	0.75	5	0.75	5	0.75	5	0.75
Usable volume	0.1	4	0.4	2	0.2	3	0.3	4	0.4
Storage	0.1	1	0.1	3	0.3	5	0.5	3	0.3
Ease of use	0.1	4	0.4	4	0.4	4	0.4	4	0.4
Customizable	0.05	1	0.05	3	0.15	1	0.05	4	0.2
Price	0.2	1	0.2	2	0.4	1	0.2	4	0.8
Ease of manufacturing	0.1	5	0.5	3	0.3	5	0.5	2	0.2
Additional accessories	0.05	4	0.2	4	0.2	3	0.15	3	0.15
Total	1	30	3.35	31	3.45	32	3.6	34	3.95

Review problems

1. The purpose of a diagram of signal flows and components is to (choose all that apply)
   A. Help identify the nature of the system.
   B. Identify what types of motion are required.
   C. Identify what types of signals are needed.
   D. Identify what properties must be sensed.
2. Give the proper sequence for what we have to do before we begin to order parts.
   A. Decide which operating system or control software will be used.
   B. Select the main components from the catalogs.
   C. Prioritize the precision of the system.
   D. Draw a flow diagram of signals and components.
   E. Research the main components in the catalogs and select a few options for each.
   F. Select secondary components.
3. Choose the correct answers in "How do you find the catalogs which have the components you need."
   A. We can search the components on some common search engines.
   B. We can search on a company website if we know their websites.
   C. We need to develop a requirement specification, such as functional requirements, acceptance tests, and documentation.
   D. We need to draw a flow diagram of signals and components. Then, we can choose what types of controllers, actuators, or sensors we want to use.
   E. We need to complete an analysis of task requirements, such as functionality, safety, and cost.

7.3 How to ensure that the purchased components will work together?

> Creativity in science could be described as the act of putting two and two together to make five.
>
> **Arthur Koestler**

This section helps answer the following questions:

- How to make sure that the items purchased can be used in the project?
- What are the tools available to ensure proper assembly of components?
- In addition to the assembly issue, are there other factors to be considered?

To make sure that the ordered components will work, a lot of homework needs to be done before purchasing, including prototype assembly planning, using a CAD system for virtual simulation, tolerance analysis, and effective communication with the components' vendor. Functional compatibility and geometric compatibility are the keys. To ensure functional compatibility, in addition to the tasks specified in Section 7.1, clearly written specifications for the needed capability are critical. It is just as critical to be able to judge that the supplier can indeed deliver the required items that meet the specifications. In general, it is important to read the specifications in the component catalog carefully when purchasing the items. Make sure that the device, wiring/hose, and connectors will all fit together before the purchase.

Writing prototype specifications is critical, and it should start from defining prototype goals as shown in Figure 7.10. It is an evolving process at various prototype development stages. The prototype design goals will be matured into prototype requirements to further define component requirements and specifications.

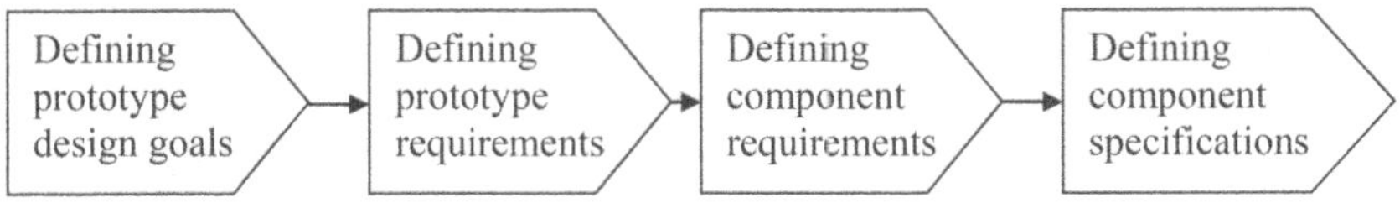

Figure 7.10 Defining prototype components specifications.

In other words, component specifications are the matured development from the design goals. Similar to the product design specifications (PDS) as discussed previously, the key specifications when defining prototype design goals may include performance, environment, service life, cost, shipping, quantity, aesthetics, materials, standards and specifications, quality and reliability, and safety.

To ensure geometric compatibility, two steps need to be carried out: prototype assembly planning and tolerance analysis. CAD prototyping is an effective way for prototype assembly planning. Besides its capability to simulate the assembly of various components, a CAD prototype has a number of benefits. Some representatives are listed as follows:

- Allows for quick design–evaluate–modify iterations, reducing cost and time.
- Improves accuracy and quality.
- Less or no need to produce a physical prototype.
- Capable of simulation and virtual prototyping, including computational fluid dynamics, finite element analysis, motion analysis/kinetics, and human interaction.
- Measures tolerances and clearances.
- Defines multiple configurations.
- Determines volume and weight.
- Parts can be readily produced from the prototype model.

A CAD prototype model was also critical for the prototype development team. The budget was limited, and producing physical prototypes would not have been practical. Time, raw materials, and resources were only available to produce some components only once. They had to work the first time with as little postproduction modification as possible.

Used or refurbished components can greatly cut down the prototyping cost, as often a prototype is used to demonstrate a concept, and thus, durability may not be a main issue. However, extra care and planning in terms of timing and scheduling should be made to adapt a used or refurbished component to ensure successful prototype development. The following example is a case study to develop a 2D plotter assembly within a very limited amount of funds. It may not be applicable for all projects. However, it would be a good engineering example especially when the project budget is very limited.

Example 7.5: Prototyping a cost-effective 2D plotter assembly

Computer numerical control (CNC) router tables, water jetting, and engraving machines are automated devices that are essentially two dimensional. This case study used an example conducted by a prototype design group to build a device of a similar function, capable of drawing simple paper plots. The goal was to demonstrate the ability to design and integrate an effective automated system within a very reasonable budget. The finished machine is capable of movement in the X–Y plane with limited movement in the Z direction for lifting a pen from a drawing surface. A controller board operates the motors and is controlled through the parallel port of a computer. During the 2D plotter development, components to be used in the construction were researched and virtually assembled to determine proper fit.

The process of implementing this project was broken into three main sections: development, construction, and integration. During the development process, components to be used in the construction were researched and virtually assembled to determine proper fit, both physically and mathematically. These selected components then were used in the second phase of construction. Other aspects of the project were also developed in this first phase. The initial logic of the postprocessor and electronic circuitry was developed during this phase. The components were mostly ordered online using web catalogs and company websites. Other noncritical components were purchased from local hardware and electronic supply shops.

In this case study, some details such as PDS are omitted as they have been discussed in Chapter 2, and the focus of this case study was on the CAD integration of the system components to assist in purchasing. The primary concern when selecting components was cost. The group compared the concepts of a moving pen head, a moving table, or a combination of both. A moving head was selected as it would require less structure and smaller motors. A concept that used timing belts and a system of rails was developed, and then, large-scale component research began.

LINEAR RAILS

Most of the linear rail systems were costly. This was mostly due to the price of linear bearings. One member of the group had attended manufacturing week in Chicago and was familiar with the igus company which specializes in polymer bearings. The group visited their website at www.igus.com and also found a rail system, the Drylin N40 series, which was lower in cost compared to any other solution. Solid models of the rails were downloaded from the site and used to model the plotter.

MOTOR

To obtain good control of the proposed system, plain DC motors, which were unlikely to produce the resolution needed, were avoided. The group had to decide between stepper motors and servomotors. Small, radio-controlled servomotors were easily accessible, but most did not provide for more than 3,608 of rotation. Some servos can be purchased or modified as a continuous rotation servo. These servos have had the feedback loop removed. The motors were less than ideal since they would have to be timed like a DC motor solution. Stepper motors seemed to be the cost-effective and accurate choice.

Three main requirements drove the purchase of the stepper motors. The first was cost. The second was that the motors had enough power to move the rail assembly. That requirement was difficult to determine since it relied on the friction and alignment of the system. The third was that it could be integrated into the system readily. Since the design used belts and pulleys, a nominal-sized motor shaft that would mate with a common pulley inner diameter was necessary.

While extensively searching Internet electronic catalogs, a surplus of old stepper motors was found at All Electronics Corp. They were labeled as a Hansen brand motor from 1984. The shaft diameter was 0.25 in. which was better than the more common 0.20 in. The torque was unknown, but was comparable in specs to a current Hansen motor which seemed strong enough. The step size was 7.58 per step which was acceptable for the resolution. Most importantly, the price was only $3.75 apiece.

MOTOR DRIVER

The parallel port had eight data lines to control. If the stepper motor was controlled directly, with the parallel port supplying voltage for each phase, the control would require a multiplexer and programming to accurately control the phases. During the research phase, the team came across integrated circuit chips which could drive stepper motors with fewer inputs. The stepper motor chosen had six wires and therefore had unipolar windings. All Electronics had a driver IC, the EDE1200, which would work and thus could be grouped into the same shipment of electronics with the motor to save on shipping costs.

MOTOR PULLEYS AND BELTS

The choice of a motor with a 0.25 in. shaft has made selecting a timing pulley much easier. Previous research had shown that small timing pulleys and belts were available from McMaster-Carr. From their catalog, the group filtered out all 0.25 in. inner diameter pulleys. Belt width was kept at 0.25 in. to ensure it could be securely fastened to the rail carriages and that it would be moving. Finally, the smallest pitch diameter that was left available was selected. This would help keep the resolution small. The result was part number 1375K39 which was more costly at $8.86 each for the design requirement of six total pulleys.

With the pulley determined, finding the corresponding belt was straightforward. With the plotter modeled in a CAD model, it was easy to check a belt length against the model. A belt that would match the pulley was of the MXL series. It had to be 0.25 in. wide. McMaster had a 38.6 in. long belt which worked in the model. The 1679K63 belt at $2.71 each was selected.

PEN SOLENOID

The solenoid for the pen was found at Gateway Electronics in St. Louis, Missouri. It was part of an indexing mechanism which was separated from the solenoid. It has a rotary action which was intended for a four-bar mechanism, but was changed to actuate the pen directly. The cost was under $5.

OTHER COMPONENTS

The remaining mechanical components were purchased from Lowes. The group took a trip there with the design in mind searching out what components were feasible. The solid model provided dimensions to aid component selection, but some things had to be determined by what was available.

The electrical components were primarily purchased from All Electronics in two shipments to consolidate shipping charges. Replacement and overlooked components were ordered later from Digi-Key. Components to construct the cooling system were purchased from the Electronics Exchange in Bridgeton, Missouri.

MECHANICAL DEVELOPMENT

With the major components of the plotter already selected, the physical interaction of these components needed to be identified. It was decided that virtual assembly of the components would be the best way to further define what specific sizes and components would be needed. With the assistance of the igus website, the components that were ordered through their company were downloaded in IGES format.

VIRTUAL MODELING

The remaining components of the system were measured using a caliper and a ruler. These dimensions were then used to create SolidWorks parts files. SolidWorks was chosen because it was readily available to each member of the team. The stepper motors, pulleys, belts, and rails were created in SolidWorks. Figures 7.11–7.13 are solid illustrations of a few of the parts created in SolidWorks.

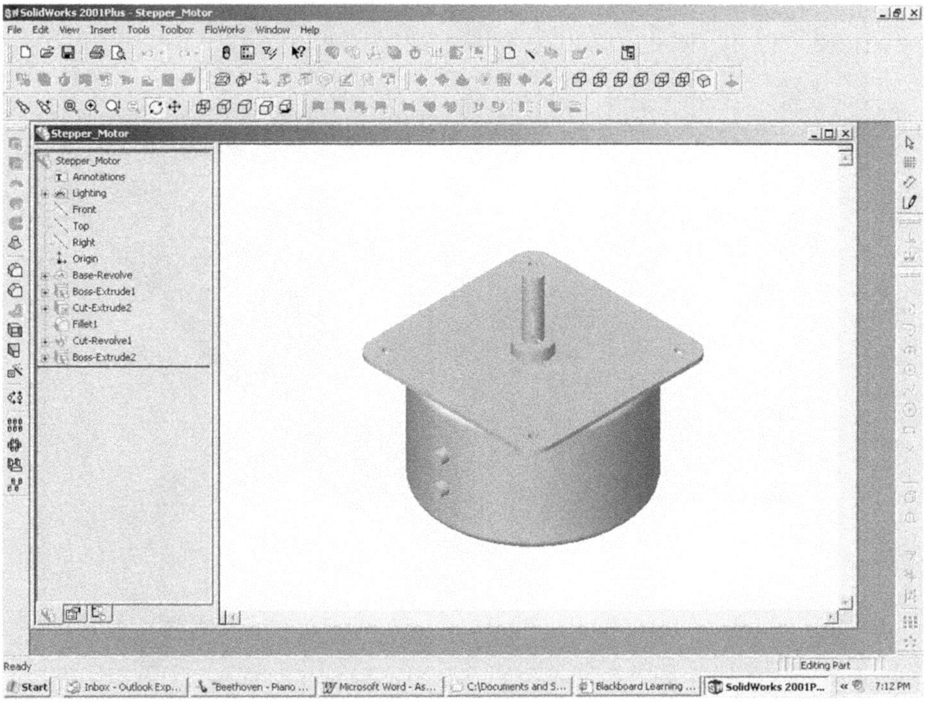

Figure 7.11 Stepper motor CAD file.

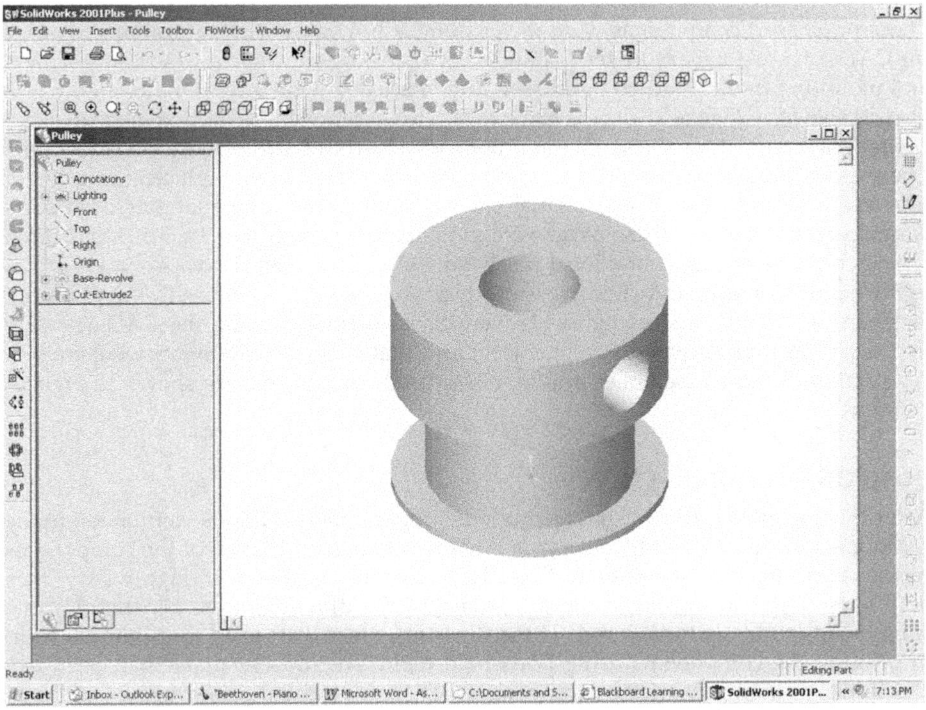

Figure 7.12 Pulley from McMaster-Carr CAD File

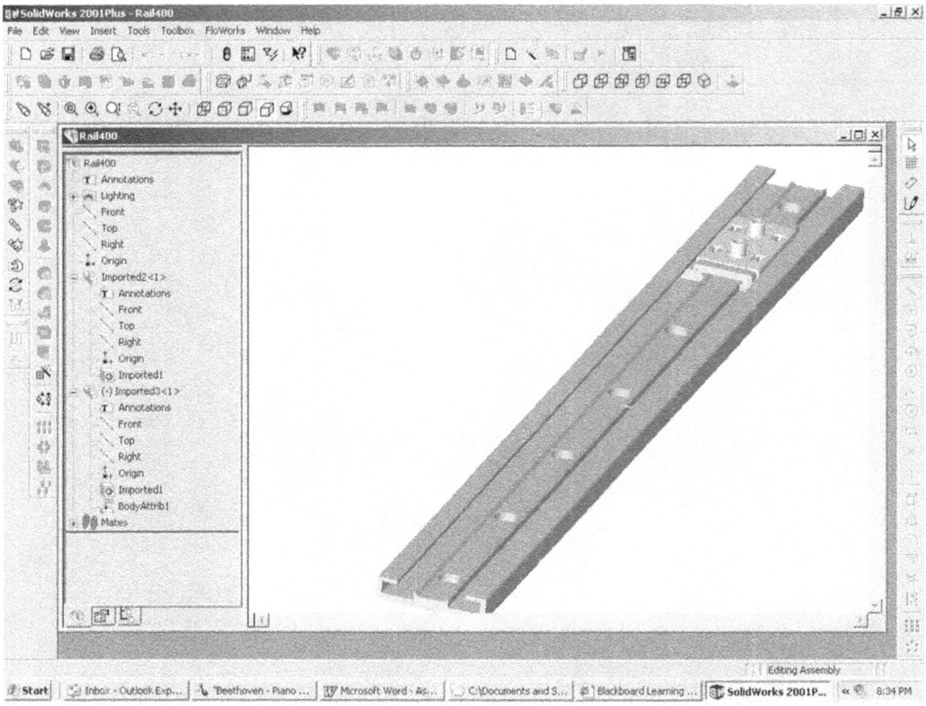

Figure 7.13 igus rail CAD file.

VIRTUAL ASSEMBLY

These purchased components were then assembled in various configurations to allow the CAD software to measure the distances between edges and surfaces so that fasteners and mounting parts could be purchased and fabricated. One important assembly was the one shown in Figure 7.14 of the three rails. This assembly was needed to ensure that clearances were left for the screws and bolts which would attach these parts together. These assemblies were also used to ensure that the parts had enough clearance to not interfere with each other. Once the major dimensions were found, manufactured parts were designed. Some of these parts were manufactured out of 1 in. by 3 in. stock birch which could be found at any local hardware store. This material was chosen for its lightweight, strength, and rigidity. These birch components along with the platform, fasteners, and purchased components were then assembled using the CAD software to confirm the measurements were correct and that proper components were chosen. A selection of these assemblies and manufactured components are shown in Figures 7.15–7.18.

COMPONENT FABRICATION

After the assemblies all fit together properly and any modifications were made to the components, a full assembly was made. This was to ensure that all of the components would work together at the same time. This assembly is shown in Figure 7.19. Once both team members approved the full assembly, the drawings were generated of the components to be manufactured. These drawings were then used to guide the team in creating the components using household tools. The tools available had been factors in designing the fabricated components. The tools created limits on how accurate a measurement could be made and the material size that could be cut or handled. This had to be taken into consideration. One of the drawings used is shown in Figure 7.20.

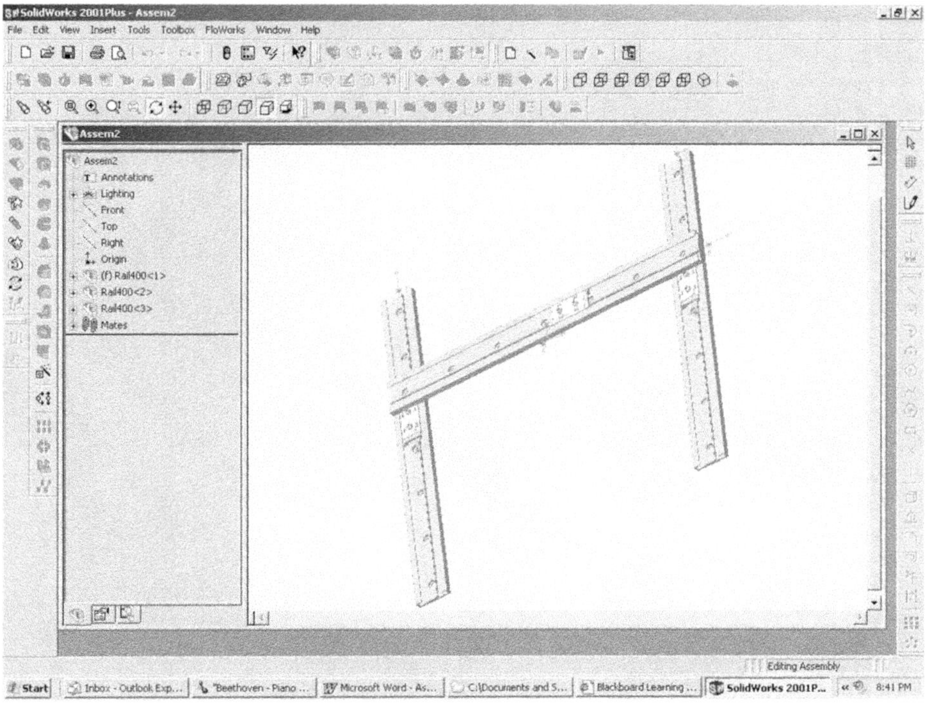

Figure 7.14 Three rails together.

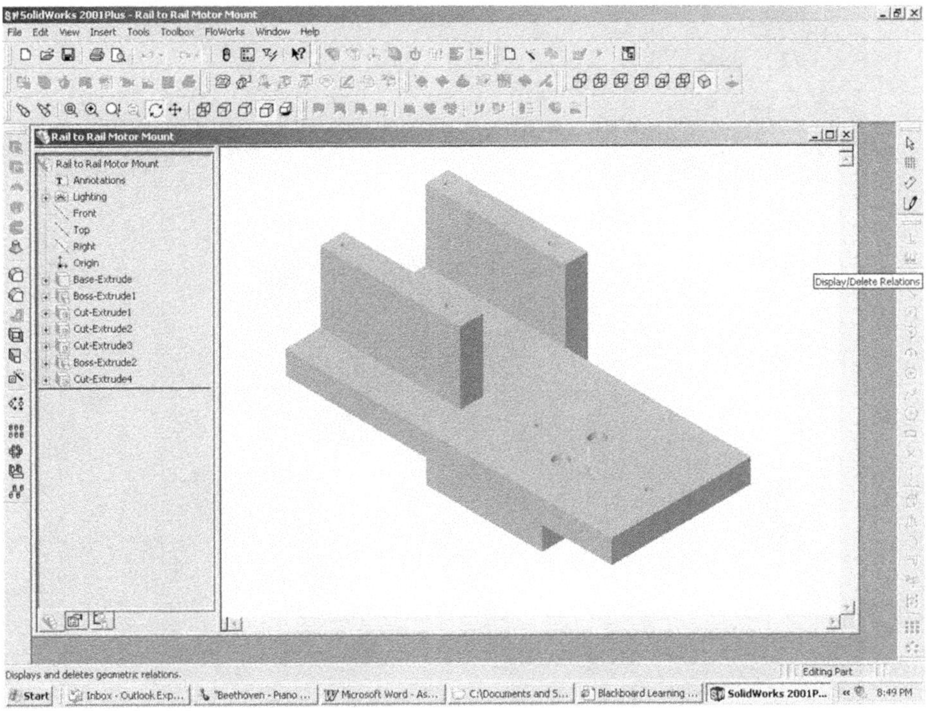

Figure 7.15 Rail to rail with motor mount.

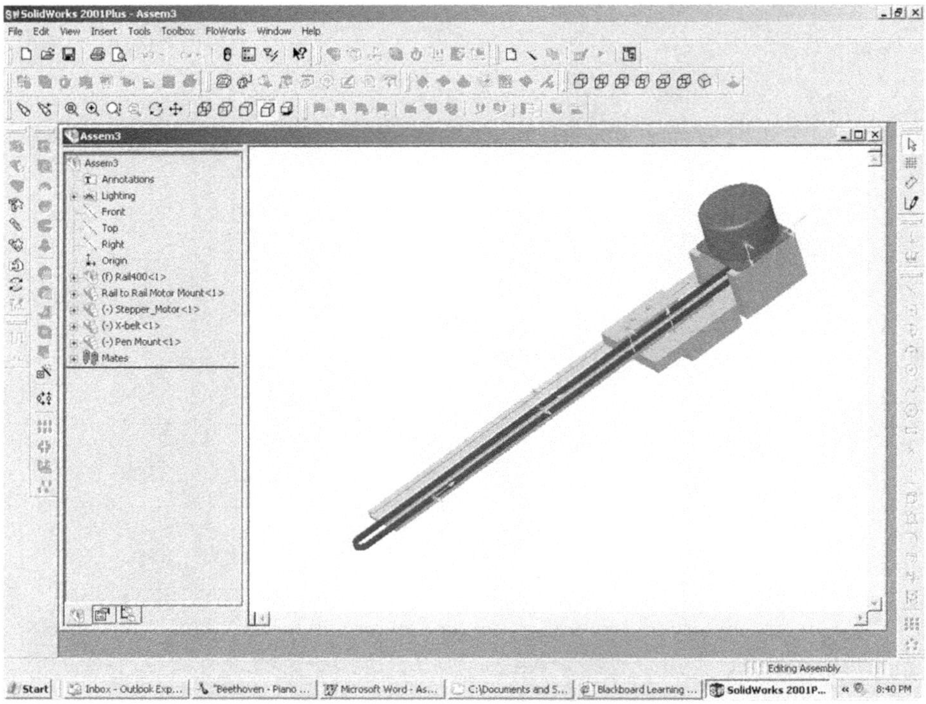

Figure 7.16 Rail, motor mount, motor, and belt.

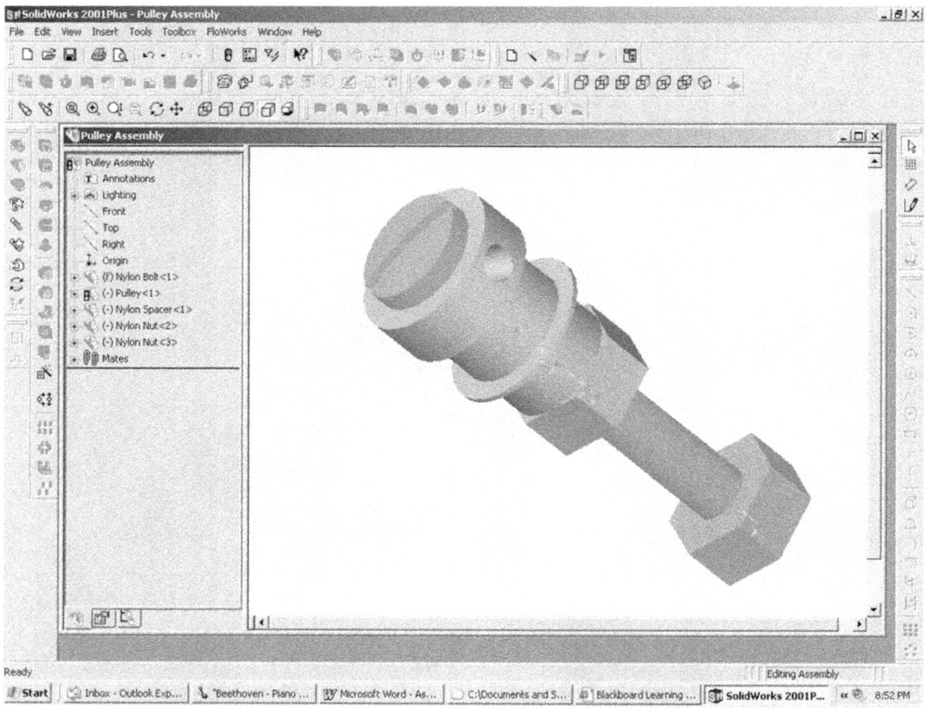

Figure 7.17 Pulley with nylon bolt, nuts, and spacer.

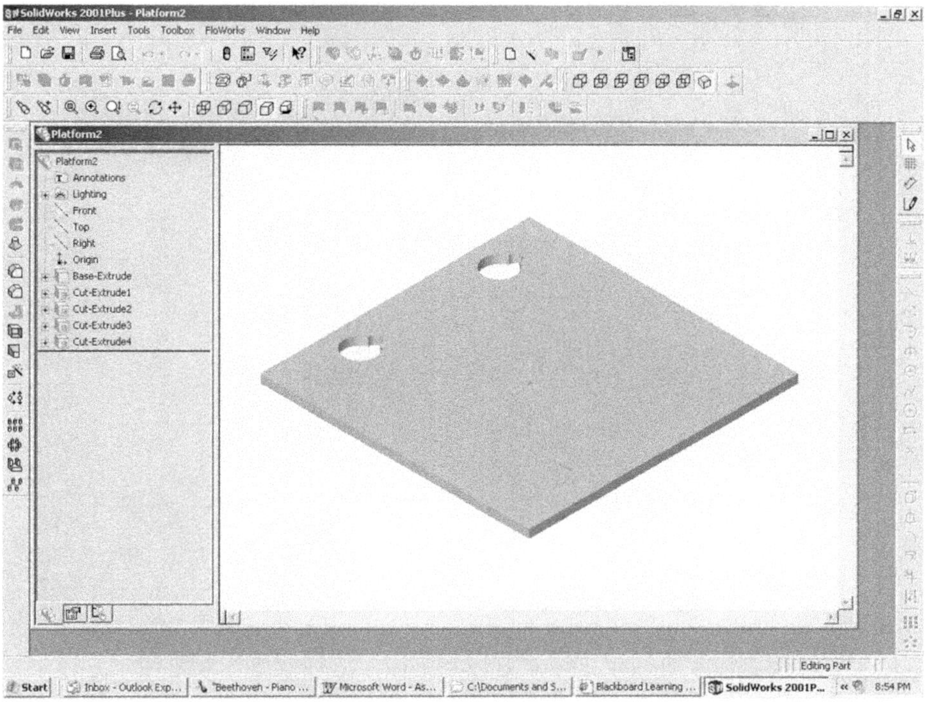

Figure 7.18 The platform.

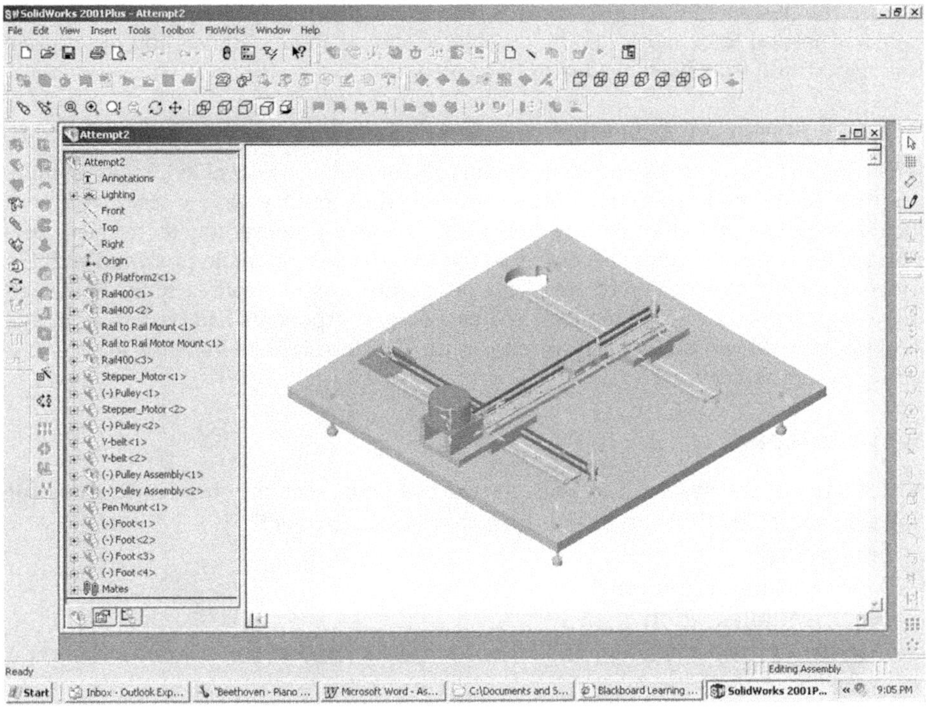

Figure 7.19 The full assembly.

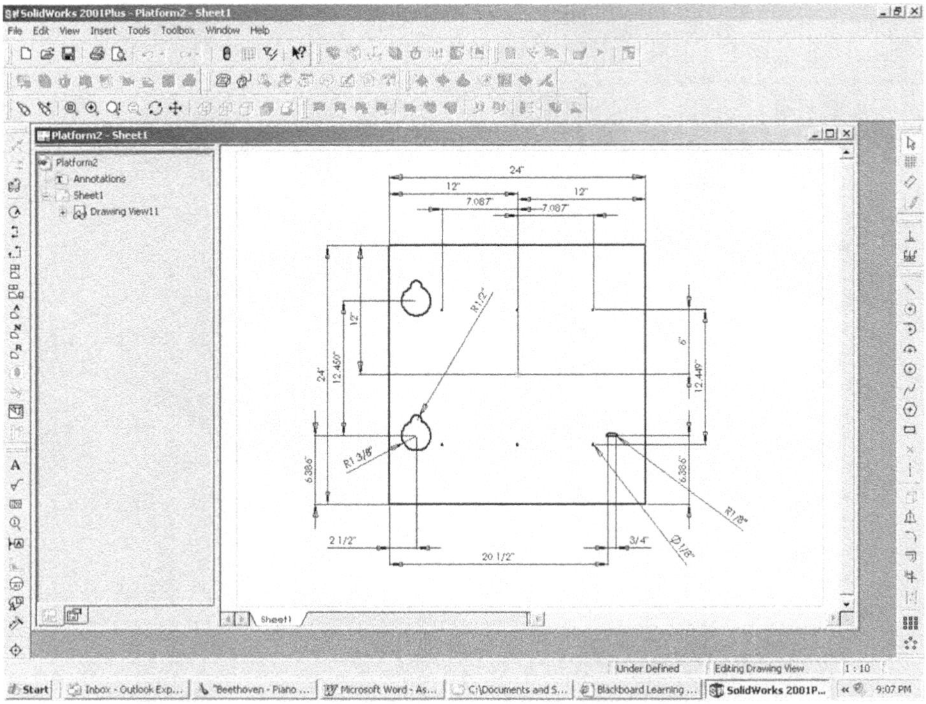

Figure 7.20 Platform drawing used for fabrication.

This particular drawing is not very clear as to what the dimensions are referring to, but because the designers were also the manufacturers, this drawing was sufficient. Had the parts needed to be fabricated by an outside source, more detailed and descriptive drawings would have been needed to be created.

PLOTTER PHYSICAL ASSEMBLY

After constructing the necessary components, the plotter was assembled. This assembly was then ready for testing and integration with the circuitry. A few tests were conducted to ensure that all of the components functioned properly together. These tests showed that it was definitely necessary to use two stepper motors to push and pull the bridge assembly along the platform. This possibility was realized early in the design phase, and the platform, rail mounts, and necessary components had been designed to accommodate this possibility if it were to occur. This foresight prevented a lot of disassembly and modification.

Review problems

1. Which area of the system integration process possesses the most amounts of uncontrollable variables that can affect the outcome of a project?
   A. Scheduling
   B. Mechanical integration
   C. Electrical integration
2. Choose the appropriate steps in "How to make sure that the purchased components will work together."
   A. We need to have good system integration, such as mechanical integration, electrical integration, and scheduling.

B. We need to have an analysis of task requirements, such as functionality, safety, and cost.
C. We can use a CAD package to model the system. This will reduce the cycle time for the design–analyze–evaluate–modify the iterative process.
D. We need to have a good schedule and follow it. This can help us figure out which components, which we ordered, can be used. If the ordered part is wrong, we then will have time to solve this problem.
E. We can search the components on some common search engines.

7.4 Tolerance analysis

> By concentrating on precision, one arrives at technique, but by concentrating on technique one does not arrive at precision.
>
> **Bruno Walter**

This section helps answer the following questions:

- How to conduct tolerance analysis so that parts can fit together?
- What are the commonly used mechanical fits? Are there any standards on these?
- What are the definitions of these mechanical fits? What are their applications?
- What is a clearance fit? What is an interference fit? What is a transition fit?
- How to assign tolerances to a specific fit?

In order to successfully integrate various components together, assembly tolerance stack analysis is very critical. This section and the rest of this chapter summarize the related techniques, such as tolerance analysis (this section), tolerance stack analysis (Section 7.5), assembly stacks (Section 7.6), process capability (Section 7.7), and statistical tolerance analysis (Section 7.8). Rather than a comprehensive reference of tolerance analysis and control, this chapter acts as a brief summary of this important subject in assembly prototyping.

It is impossible to make parts to an exact size. The tolerance required will depend on the function of the part and the particular feature being dimensioned. Therefore, the range of permissible sizes, or tolerance, must be specified for all dimensions on a drawing. A successful design creates a product assembly which contains parts that fit together well, performs desired functions efficiently, and maximizes tolerances to create a cost-effective manufactured part. It is critical to specify a tolerance as assigning a tolerance is just like selecting a particular manufacturing process to implement the specification, and thus, it is equivalent to assigning the cost to this operation. Therefore, the appropriate tolerance is decided by the following:

a. The functional importance involved (the required mission)
b. The economics of manufacture (avoid spending unnecessary time and money)

Tolerance is also called the language of assemblies. Tolerance analysis can be used to increase understanding of part function, to discover and resolve problems on paper rather than in prototype or production, and to make intelligent design decisions.

In the expression 1 ± 0.1, the limits are the maximum and minimum sizes. In this case, they are 1.1 and 0.9. The tolerance is the permissible amount of variation, or

$$1.1 - 0.9 = 0.2 \quad \text{or} \quad 1 \pm 0.1$$

Functionally important dimensions can be divided into two categories: (1) when the dimension concerned relates to a type of "fit" between two mating parts, e.g., a loose fit or a tight fit, and these can be obtained from the appropriate charts [e.g., ISO17], and (2) when the dimension is subject to a specified assembly requirement, the limits will result from a calculation.

Table 7.5 shows the ANSI (American National Standards Institute) fits standard. The eight classes are given as follows:

Table 7.5 ANSI fits standard

ANSI recommended allowances and tolerances					
Class of fit	Description	Clearance	Interference	Hole tolerance	Shaft tolerance
1	Loose	$0.0025\sqrt[3]{d^2}$		$+0.0025\sqrt[3]{d}$	$-0.0025\sqrt[3]{d}$
2	Free	$0.0014\sqrt[3]{d^2}$		$+0.0013\sqrt[3]{d}$	$-0.0013\sqrt[3]{d}$
3	Medium	$0.0009\sqrt[3]{d^2}$		$+0.0008\sqrt[3]{d}$	$-0.0008\sqrt[3]{d}$
4	Snug	0		$+0.0006\sqrt[3]{d}$	$-0.0004\sqrt[3]{d}$
5	Wringing		0	$+0.0006\sqrt[3]{d}$	$+0.0004\sqrt[3]{d}$
6	Tight		0.0025d	$+0.0006\sqrt[3]{d}$	$+0.0006\sqrt[3]{d}$
7	Medium force		0.0005d	$+0.0006\sqrt[3]{d}$	$+0.0006\sqrt[3]{d}$
8	Shrink		0.001d	$+0.0006\sqrt[3]{d}$	$+0.0006\sqrt[3]{d}$

Class 1: For fits where accuracy is not essential
Class 2: For fits where accuracy is not essential
Class 3: For sliding fits and for the more accurate machine tool and automotive parts
Class 4: Higher precision is required
Class 5: Very high precision required
Class 6: Drive fits for gears, pulleys, rocker arms, etc.
Class 7: For fastening the shafts to the locomotive and railroad car wheels; also for crank disks and for the armature of motors and generators. It is the tightest fit that is recommended when the part with a hole is of cast iron.
Class 8: The hole is of steel

Clearance fit is when the largest shaft is always smaller than the smallest hole, interference fit is when the sizes of the shaft are always larger than the sizes of the hole, and transition fit is when the largest shaft is larger than the largest hole, but the smallest shaft may fit in the largest hole. Table 7.5 is the standard to define how various mechanical components fit together.

The following are two peg and hole examples to illustrate the concept. For hole basis, the minimum size of the hole is equal to the basic size of the fit. The hole system is commonly used because holes are more difficult to produce to a given size and are more difficult to inspect. For example, if the nominal size of a fit is 1/2 in., then the minimum size of the hole in the system is 0.500 in. The calculations are summarized as follows:

$$\text{Clearance} = \text{Hole} - \text{Shaft} \tag{7.1}$$

$$\text{Clearance}_{max} = \text{Hole}_{max} - \text{Shaft}_{min} \tag{7.2}$$

$$\text{Clearance}_{min} = \text{Hole}_{min} - \text{Shaft}_{max} \tag{7.3}$$

$$\text{Allowance} = \text{Clearance}_{min} = \text{Hole}_{min} - \text{Shaft}_{max} \tag{7.4}$$

If both $clearance_{max}$ and $clearance_{min}$ are greater than zero, it is a clearance fit.
If both $clearance_{max}$ and $clearance_{min}$ are smaller than zero, it is an interference fit.
If $clearance_{max}$ is greater than zero and $clearance_{min}$ is smaller than zero, it is a transition fit.

Example 7.6

A 0.8 in. nominal diameter journal/shaft is to have a Class 3 fit. Within what size tolerances should the parts be manufactured? Use the basic hole system.

Given
Diameter of the journal/shaft = 0.8 in.
According to ANSI fits standard, from Table 7.5,

$$\text{Clearance} = 0.0009\sqrt[3]{0.8^2} = 0.0008\ \text{in.}$$

$$\text{Tolerance} = \pm 0.0008\sqrt[3]{0.8} = 0.0007\ \text{in.}$$

Hole:

Maximum dimension = 0.8 + 0.0007 = 0.8007 in.
Minimum dimension = 0.8 in.

Shaft:

Maximum dimension = 0.8 − 0.0008 = 0.7992 in.
Minimum dimension = 0.7992 − 0.0007 = 0.7985 in.

The above calculation was based on ANSI standard. ISO 286 also implements 20 grades of accuracy to satisfy the requirements of different industries.

- IT01, IT0, IT1, IT2, IT3, IT4, IT5, and IT6 are for production of gauges and instruments
- IT5, IT6, IT7, IT8, IT9, IT10, IT11, and IT12 are for precision and general industry
- IT11, IT12, IT13, IT14, IT15, and IT16 are for semi-finished products
- IT16, IT17, and IT18 are for structural engineering

Table 7.6 shows the ISO tolerance band. This table indicates the total tolerance to be applied to either a shaft or a hole diameter. The range of sizes covered is divided into a number of groups spanning from 0 to 315 mm. For each group of sizes, as many as 20 different tolerance grades are specified, labeled IT01, IT0, IT1, IT2, and so on up to IT18. The tolerance is a function of nominal size. Table 7.7 shows the typical machine capabilities for the associated IT grades. Table 7.8 shows the ISO shaft limit nearest zero. This table indicates the deviation from basic (nominal) or the way in which the tolerance is applied to the shaft diameter. Table 7.9 shows the ISO hole nearest dimension to zero and works in essentially the same way except the zones are specified as "A" through "ZC" (Table 7.9 only shows up to V8). Throughout the system, hole features are identified with uppercase characters, whereas shaft features are identified with lowercase characters.

As an example to use these tables, for a hole diameter 110 mm H11, hole diameter is 110 mm, and hole deviation is "H" with tolerance level IT11. From Table 7.9, one can find when hole diameter is 110 mm, deviation = 0 for "H," and from Table 7.6, IT11 for 110 mm nominal size, tolerance = 0.220 mm. The nominal hole size is always the minimum. Therefore, the nearest dimension to zero = 110 + 0 = 110.000 mm, and furthest from zero = 110 mm + (0 + T = 0.220) = 110.220 mm. Therefore, the resulting limits are between 110.000 mm and 110.220 mm.

Table 7.6 ISO tolerance band in micrometers (m^{-6})

	Nominal sizes (mm)										
Over	1	3	6	10	18	30	50	80	120	180	250
Inc.	3	6	10	18	30	50	80	120	180	250	315
IT grade											
1	0.8	1	1	1.2	1.5	1.5	2	2.5	3.5	4.5	6
2	1.2	1.5	1.5	2	2.5	2.5	3	4	5	7	8
3	2	2.5	2.5	3	4	4	5	6	8	10	12
4	3	4	4	5	6	7	8	10	12	14	16
5	4	5	6	8	9	11	13	15	18	20	23
6	6	8	9	11	13	16	19	22	25	29	32
7	10	12	15	18	21	25	30	35	40	46	52
8	14	18	22	27	33	39	46	54	63	72	81
9	25	30	36	43	52	62	74	87	100	115	130
10	40	48	58	70	84	100	1200	140	160	185	210
11	60	75	90	110	130	160	190	220	250	290	320
12	100	120	150	180	210	250	300	350	400	460	520
13	140	180	220	270	330	390	460	540	630	720	810
14	250	300	360	430	520	620	740	870	1,000	1,150	1,300

Table 7.7 Typical manufacturing capabilities for various IT tolerance levels

IT grade	2	3	4	5	6	7	8	9	10	11	12	13	14	15	16
Lapping	■	■	■	■											
Honing		■	■	■											
Superfinishing			■	■	■										
Cylindrical grinding			■	■	■	■									
Diamond turning			■	■	■	■									
Plan grinding				■	■	■	■	■							
Broaching				■	■	■	■	■							
Reaming				■	■	■	■	■							
Boring, turning					■	■	■	■	■	■	■				
Sawing								■	■	■					
Milling								■	■	■	■	■			
Planing, shaping									■	■	■	■	■		
Extruding									■	■	■	■			
Cold rolling, drawing									■	■	■	■	■		
Drilling										■	■	■	■		
Die casting											■	■	■	■	
Forging												■	■	■	■
Sand casting													■	■	■
Hot rolling, flame cutting													■	■	■

Similarly, for shaft 110 mm e9, shaft diameter is 110 mm, and shaft deviation is "e" with tolerance level IT9. From Table 7.8, one can find when shaft diameter is 110 mm, deviation = 0.072 mm for "e," and from Table 7.6, IT9 for 110 mm nominal size, toleranc = 0.087 mm. The nominal shaft size is always the maximum, so the nearest dimension to

Table 7.8 ISO shaft limit nearest zero (fundamental deviation)

	Nominal sizes (mm)																			
Over	1	3	6	10	14	18	24	30	40	50	65	80	100	120	140	160	180	200	225	250
Inc.	3	6	10	14	18	24	30	40	50	65	80	100	120	140	160	180	200	225	250	280
Grade	**All limits below with − sign**																			
a	270	270	280	290	290	300	300	310	320	340	360	380	410	460	520	580	660	740	820	920
b	140	140	150	150	150	160	160	170	180	190	200	220	240	260	280	310	340	380	420	480
c	60	70	80	95	95	110	110	120	130	140	150	170	180	200	210	230	240	260	280	300
d	20	30	40	50	50	65	65	80	80	100	100	120	120	145	145	145	170	170	170	190
e	14	20	25	32	32	40	40	50	50	60	60	72	72	85	85	85	100	100	100	110
f	6	10	13	16	16	20	20	25	25	30	30	36	36	43	43	43	50	50	50	56
g	2	4	5	6	6	7	7	9	9	10	10	12	12	14	14	14	15	15	15	17
h	0	0	0	0	0	0	0	0	0	0	0	0	0	0	0	0		0	0	0
j(5 and 6)	2	2	2	3	3	4	4	5	5	17	7	9	9	11	11	11	13	13	13	16
j7	4	4	5	6	6	8	8	10	10	12	12	15	15	18	18	18	21	21	21	26
js	±0.5T																			
Grade	**All limits below with + sign**																			
k(4–7)	0	1	1	1	1	2	2	2	2	2	2	3	3	3	3	3	4	4	4	4
k from 8	0	0	0	0	0	0	0	0	0	0	0	0	0	0	0	0	0	0	0	0
m	2	4	6	7	7	8	8	9	9	11	11	13	13	15	15	15	17	17	17	20
n	4	8	10	12	12	15	15	17	17	20	20	23	23	27	27	27	31	31	31	34
p	6	12	15	18	18	22	22	26	26	32	32	37	37	43	43	43	50	50	50	56
r	10	15	19	23	23	28	28	34	34	41	43	51	54	63	65	68	77	80	84	94
s	14	19	23	28	28	35	35	43	43	53	59	71	79	92	100	108	122	130	140	158
t	–	–	–	–	–	–	41	48	54	66	75	91	104	122	134	146	166	180	196	218
u	18	23	28	33	33	41	48	60	70	87	102	124	144	170	190	210	236	258	284	315
v	–	–	–	–	39	47	55	68	81	102	120	146	172	202	228	252	284	310	340	385
x	20	28	34	40	45	54	64	80	97	122	146	178	210	248	280	310	350	385	425	475
y	–	–	–	–	–	63	75	94	114	144	174	214	254	300	340	380	425	470	520	580
z	26	35	42	50	60	73	88	112	136	172	210	258	310	365	415	465	520	575	640	710

Note: Deviation in micrometers = m^{-6}.

Table 7.9 ISO hole nearest dim to zero (fundamental deviation)

	Nominal sizes (mm)																			
Over	1	3	6	10	14	18	24	30	40	50	65	80	100	120	140	160	180	200	225	250
Inc.	3	6	10	14	18	24	30	40	50	65	80	100	120	140	160	180	200	225	250	280
Grade	**All limits below with+sign**																			
A	270	270	280	290	290	300	300	310	320	340	360	380	410	460	520	580	660	740	820	920
B	140	140	150	150	150	160	160	170	180	190	200	220	240	260	280	310	340	380	420	480
C	60	70	80	95	95	110	110	120	130	140	150	170	180	200	210	230	240	260	280	300
D	20	30	40	50	50	65	65	80	80	100	100	120	120	145	145	145	170	170	170	190
E	14	20	25	32	32	40	40	50	50	60	60	72	72	85	85	85	100	100	100	110
F	6	10	13	16	16	20	20	25	25	30	30	36	36	43	43	43	50	50	50	56
G	2	4	5	6	6	7	7	9	9	10	10	12	12	14	14	14	15	15	15	17
H	0	0	0	0	0	0	0	0	0	0	0	0	0	0	0	0	0	0	0	0
J6	2	5	5	6	6	8	8	10	10	13	13	16	16	18	18	18	22	22	22	25
J7	4	6	8	10	10	12	12	14	14	18	18	22	22	26	26	26	30	30	30	36
J8	6	10	12	15	15	20	20	24	24	28	28	34	34	41	41	41	47	47	47	55
Js	±0.5T																			
K5	0	0	1	2	2	1	1	2	2	3	3	2	2	3	3	3	2	2	2	3
K6	0	2	2	2	2	2	2	3	3	4	4	4	4	4	4	4	5	5	5	5
K7	0	3	5	6	6	6	6	7	7	9	9	10	10	12	12	12	13	13	13	16
K8	0	5	6	8	8	10	10	12	12	14	14	16	16	16	20	20	22	22	22	25
Grade	**All limits below with−sign**																			
M6	2	1	3	4	4	4	4	4	4	5	5	6	6	8	8	8	8	8	8	9
M7	2	0	0	0	0	0	0	0	0	0	0	0	0	0	0	0	0	0	0	0
M8	−2	+2	+1	+2	+2	+4	+4	+5	+5	+5	+5	+6	+6	+8	+8	+8	+9	+9	+9	+9

Note: Deviation in micrometers = m^{-6}.

zero = 110 − 0.072 = 109.928 mm. Furthest from zero = 110 mm − (0.072 + T = 0.072 + 0.087) = 109.841 mm. Therefore, the resulting limits for the shaft are between 109.928 and 109.841 mm.

As an example, given a shaft of "basic size" 13 mm, a deviation zone of "d" (−50 μm), and a tolerance grade of IT8 (27 μm), it ends up with a shaft size/tolerance designated as "13 d8" mm. which will measure 12.950 − 12.923 mm in diameter, the deviation of −50 μm having been applied to the basic size and then the tolerance of 27 μm applied negatively (because this is a shaft; it would be applied positively for a hole). This shaft has a total tolerance of 27 μm, an "upper deviation" (abbreviated as ES) of −50 μm, and a "lower deviation" (abbreviated as EI) of −77 μm.

A hole of basic size 13 D8 will have a diameter of 13,050–13,077 mm, an ES of 77 μm, and an EI of 50 μm. Table 7.10 shows some preferred ISO example limits and fits using the hole basis system. Again a fit may be hole basis or shaft basis. This would indicate that one side of the hole (or shaft) is dimensioned at its nominal or basic size. In other words, the fundamental deviation is zero. The "h" and "H" deviation zones are the basis zones.

Table 7.10 Some preferred ISO example limits and fits using the hole basis system

Description	Hole	Shaft
Loose running	H11	c11
Free running	H9	d9
Loose running	H11	c11
Easy running—good quality easy to do	H8	f8
Sliding	H7	g6
Close clearance—spigots and locations	H8	f7
Location/clearance	H7	h6
Location—slight interference	H7	k6
Location/transition	H7	n6
Location/interference—press fit which can be separated	H7	p6
Medium drive	H7	s6
Force	H7	u6

Example 7.7

A hole/shaft fit is designated as "22 H11/g9". Find the hole and shaft limits.

This is a clearance, hole basis fit. The steps to find the limits are listed in the following table:

Feature	Specification
Hole designation	22 H11
Total tolerance	130 μm
Upper deviation	130 μm
Lower deviation	0 μm
Limit s of size (LMC/MMC)	22.130–22.000 mm
Shaft designation	22 g9
Total tolerance	52 μm
Upper deviation	−7 μm
Lower deviation	−59 μm
Limits of size (LMC/MMC)	21.993–21.941 mm

Comment: This fit has a maximum clearance of 189 μm and a minimum clearance of 7 μm.

Example 7.8

Use the data from Example 7.6 to calculate the limits using a H8/f7 fit. Given

Basic hole diameter = 0.8 in. = 20.32 mm
Tolerance = 0.033 mm
Deviation = 0 mm
Nearest zero = 20.32 + 0 = 20.32 mm
Furthest zero = 20.32 + (0 + 0.033) = 20.353 mm
Therefore, limits are between 20.32 and 20.353 mm.
Basic shaft diameter = 0.8 in. = 20.32 mm
Tolerance = 0.021 mm
Deviation = 0.02 mm
Nearest zero = 20.32 − 0.02 = 20.3 mm
Furthest zero = 20.32 − (0.02 + 0.021) = 20.279 mm
Therefore, limits are between 20.3 and 20.279 mm.

Review problems

1. Determine the limits of a shaft and hole assembly.
   The basic hole size is 2 in. and needs Class 3 medium fit.
2. Consider the previous problem, but calculate the limits for a Class 6 fit.
3. On a Ferris wheel, we have a 3.5 in. running journal that is to be pressure lubricated. The fit selected for this application is Class 4. Determine the tolerances required.
4. Calculate the required tolerance limits for an H6/g5 fit with a nominal diameter of 2.25 in.

7.5 Tolerance stack analysis

Design is the method of putting form and content together. Design, just as art, has multiple definitions; there is no single definition. Design can be art. Design can be aesthetics.

Design is so simple, that's why it is so complicated.

Paul Rand

This section helps answer the following questions:

- What are tolerance stacks?
- Why is tolerance stack analysis important?
- How does one conduct tolerance stack analysis?

To make sure that two parts can be assembled properly, in addition to the CAD assembly drawing, tolerance analysis is needed. Most CAD systems have the functionality to allow the user to add tolerance information to dimensions. This allows drawings to reflect design or manufacturing intent, but since the tolerances are nothing more than an additional text note applied to the dimension, no additional functionality can be gained from the tolerance value. It would be beneficial if these tolerances can be used for analyzing tolerance buildup problems. Since these tolerances are inherent in the defining dimensions of the part, they can be used in a much more intelligent way for analyzing the parts performance in an assembly as will be shown in the following scenario.

Tolerance stack-ups or tolerance chains are the buildup of tolerances or accumulation of tolerances, and either addition or subtraction of length dimensions is always characterized by adding tolerances to the individual lengths. One needs to have the basic capability to calculate and analyze a tolerance chain to determine tolerance or interface as the result of assembling components together. Clearance is the amount of space between two part surfaces, and interference is the condition where two part surfaces are trying to occupy the same surface. As shown in Figure 7.21, "virtual condition" (VC) determines whether or not parts will assemble properly. It is, in effect, "worst condition." VC of external feature-of-size is the maximum material condition (MMC) of the part plus any associated geometric tolerances. VC of internal feature-of-size is the MMC of the part minus any associated geometric tolerances. The worst case is when the hole is minimum (VC = 9.9), and the peg is maximum (VC = 9.8). Therefore, the clearance is 9.9 − 9.8 = 0.1 mm (positive). If this number is negative, then this represents the maximum interference.

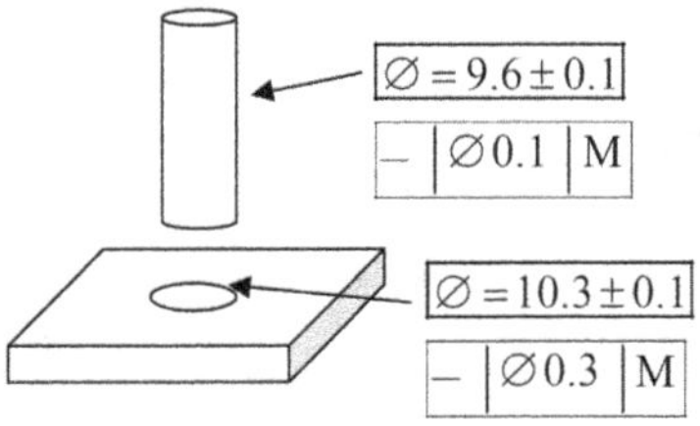

Figure 7.21 An example of the hole-and-peg assembly.

Tolerance stack is a calculation which determines the maximum and minimum distance between two features on a part or in an assembly. Stacks help us ensure good product design and reduce product costs. Stacks can be linear or radial. Radial stacks often involve diametrical dimensions; think of it as linear but in a different direction. Basic stack steps are given as follows:

1. Identify the problem: Identify the stack objective, list the conditions under which the stack is being calculated, and label the start point, end point, and direction of the stack. The conditions can include the temperature of the parts, whether the parts meet print specifications, whether the stack includes or excludes wear, the amount of deflection on the parts, and any unusual conditions present in the stack.
2. Choose the desired answer: Write down the design goal before solving the problem. "What is the extreme maximum (or minimum) answer that would be acceptable and allow the product to function as intended?"
3. Identify the stack path: A stack path is a series or chain of distances (part dimensions) from the start point of the stack to the end point of the stack. This chain of distances must consist only of known distances—which are dimensions on the drawing or a value calculated from dimensions on the drawing and must be the shortest possible chain of distances from the start point to the end point, and must be continuous—each distance must begin where the previous distance ends.
4. Perform the math: Transfer the distances onto the stack form, add each column of numbers, check the subtotals, and evaluate the answer.

Example 7.9

The objective of the case study for the part shown in Figure 7.22 is to find the maximum and minimum distance between the shoulder and the left side of the groove.

As shown in Figure 7.23, to label the start point, end point, and direction of the stack, the start point is always one of the part features named, and the endpoint is the other.

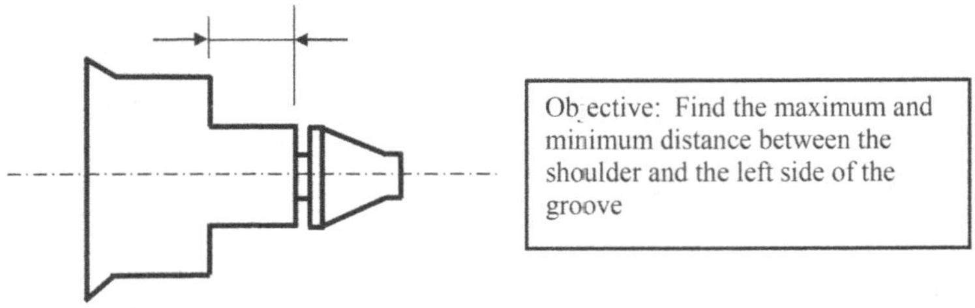

Figure 7.22 An example to define stack objective.

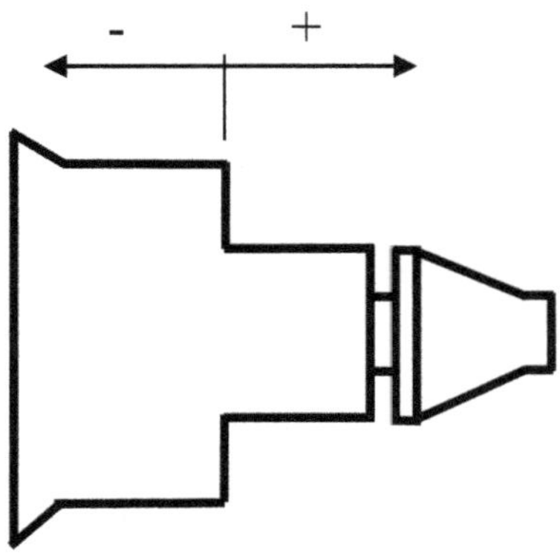

Figure 7.23 Label the part point, end point, and direction of the stack.

A stack indicator is added at the start point of the stack. A stack indicator is a pair of opposing arrows with+and−assigned to each arrow. The arrow pointing toward the end point of the stack is +, and the arrow pointing away from the end point is −. Linear stacks start on the left and radial on the bottom.

The design goal is, "what is the extreme maximum (or minimum) answer that would be acceptable and allow the product to function as intended?" As shown in Figure 7.24, stack path is a series or chain of distances (part dimensions) from the start point of the stack to the endpoint of the stack. This chain of distances must consist only of known distances—which are dimensions on the drawing or a value calculated from

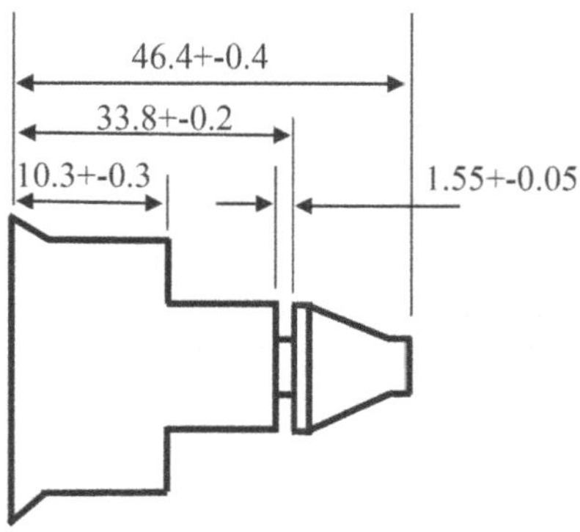

Figure 7.24 A series or chain of distances (part dimensions) from the start point of the stack to the end point of the stack.

dimensions on the drawing, must be the shortest possible chain of distances from the start point to the end point, and must also be continuous—each distance must begin where the previous distance ends.

As shown in Figure 7.25, to identify the stack path, one first locates the shortest continuous chain of distances, as defined by dimensions or values calculated from dimensions, from the start point to the end point, marks each of these distances on the sketch with a line, places a dot where the distance begins, and an arrow where it ends, and labels each distance with a code letter: A, B, C, and so on.

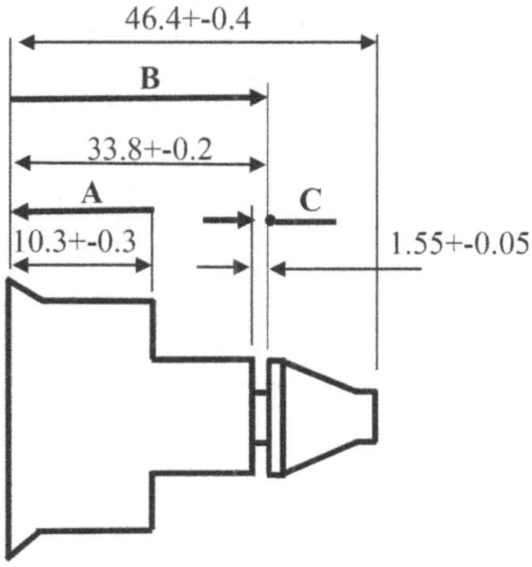

Figure 7.25 Identify the stack path.

To perform the computation, one first transfers the distances onto the stack form, adds numbers of each column, checks the subtotals, and evaluates the answer as shown in Table 7.11.

Table 7.11 Perform computation to find tolerances

Distance	Part number	Description from/to	±	Maximum (+ maximum – minimum)	±	Minimum (– maximum + minimum)	Tolerance
A	Frm-2	RS shoulder/ LS part	–	10	–	10.6	0.6
B	Frm-2	LS part/RS of groove	+	34	+	33.6	0.4
C	Frm-2	RS of groove/ LS of groove	–	1.5	–	1.6	0.1
		Subtotals	+	22.5	+	21.4	1.1

Review problems

1. For the shaft assembly example shown in Figure 7.26, find the maximum and minimum of the distance X.
2. The company is developing a new product line as shown in Figure 7.27. The designer needs to know what sized bushings the company can purchase from its supplier to make this assembly valid. The supplier makes bushings these nominal sizes: 18, 20, 22, and 24. Choose the bushing that best suits the assembly.

3. Find the maximum and minimum material between the hole and edge of the rubber spacer considering run out as shown in Figure 7.28.

Distance	Maximum	Minimum	Tolerance
A			
B			
C			
P			

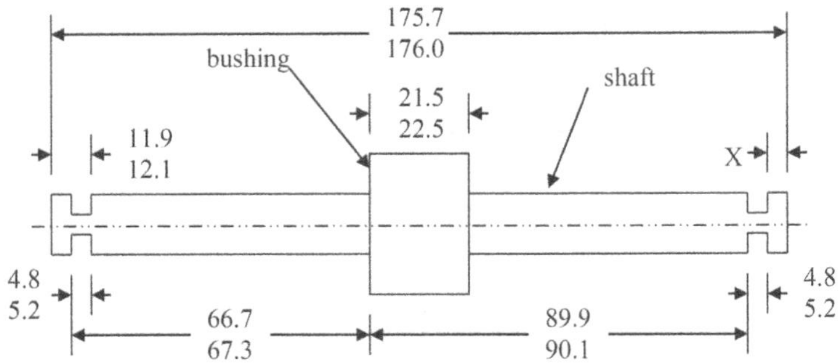

Figure 7.26 Shaft assembly (cross-sectional view 1).

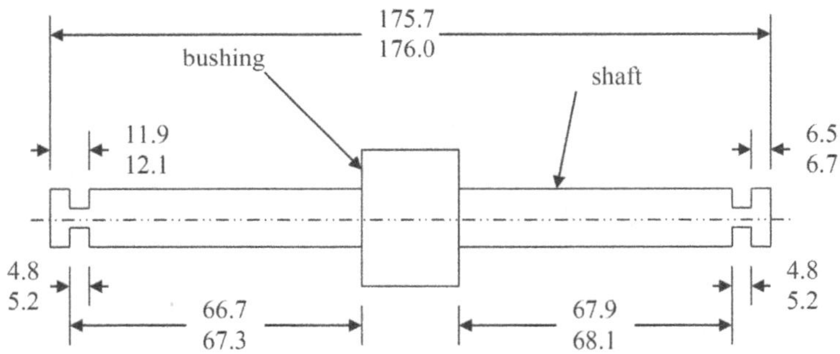

Figure 7.27 Shaft assembly (cross-sectional view 2).

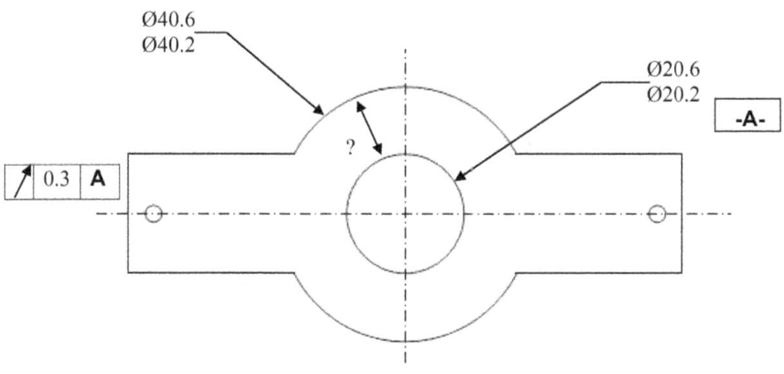

Figure 7.28 To find the maximum and minimum material between the hole and edge of the rubber spacer.

4. Calculate the maximum and minimum values of P for the drawing in Figure 7.29.

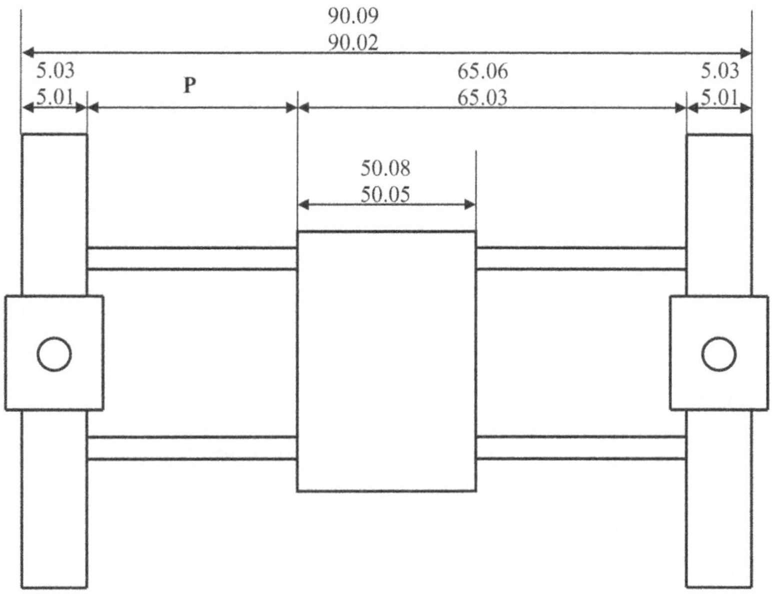

Figure 7.29 To find the maximum and minimum value of P (1).

5. Calculate the maximum and minimum values of P for the drawing in Figure 7.30.

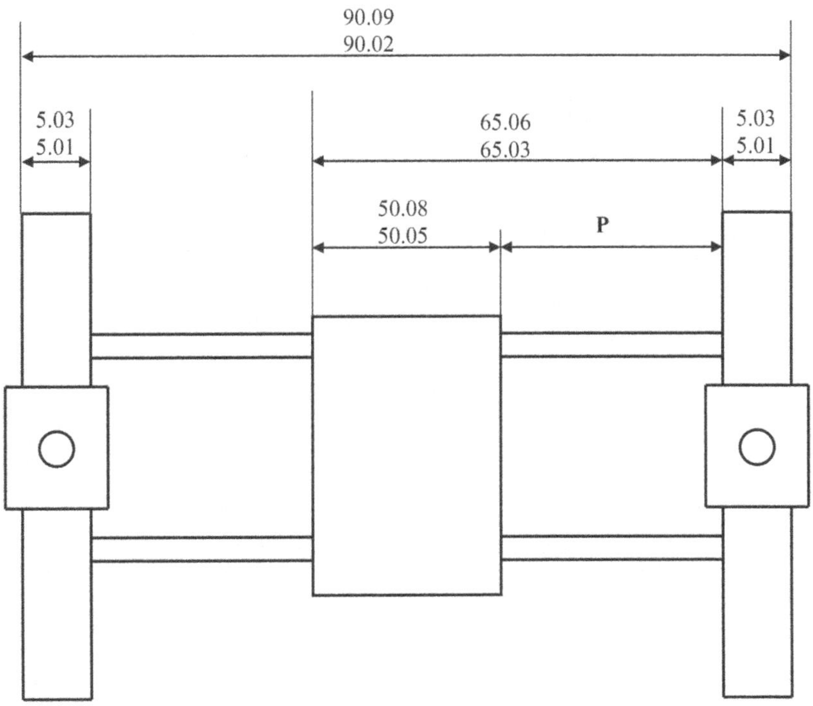

Figure 7.30 To find the maximum and minimum value of P (2).

6. Calculate the maximum and minimum values of P for the drawing in Figure 7.31.

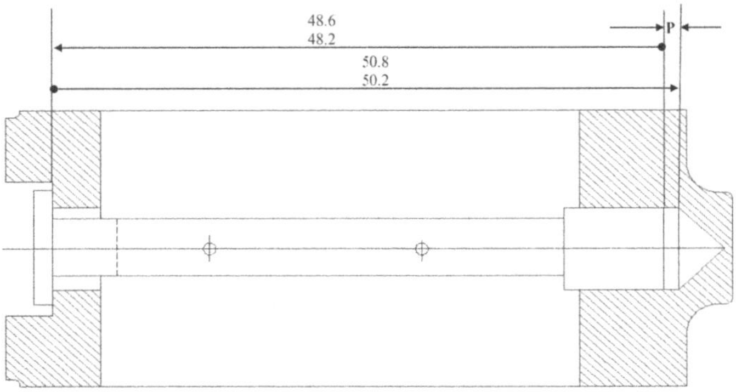

Figure 7.31 To find the maximum and minimum value of P (3).

7. Calculate the maximum and minimum values of P for the rigid coupling in Figure 7.32 (part stacks using position).

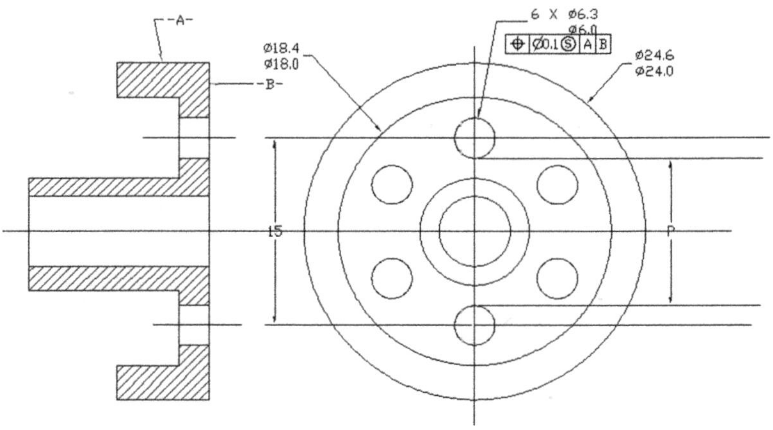

Figure 7.32 To find the maximum and minimum value of P (4).

8. Calculate the maximum and minimum distances of P of rigid coupling considering bonus as shown in Figure 7.33.

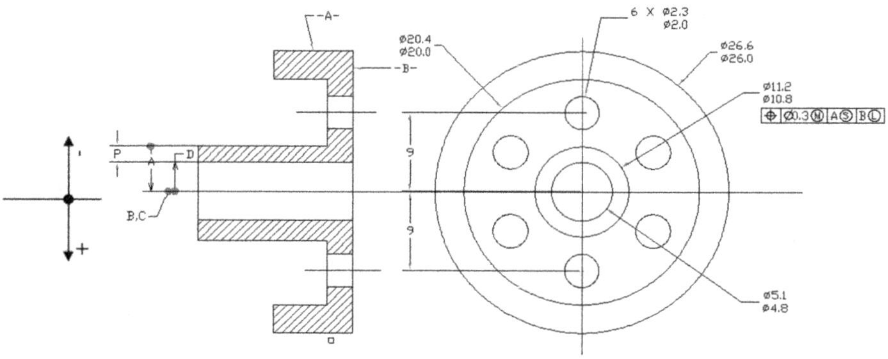

Figure 7.33 To find the maximum and minimum value of P (5).

7.6 Assembly stacks

> Providing, meaning to a mass of unrelated needs, ideas, words and pictures - it is the designer's job to select and fit this material together and make it interesting.
>
> **Paul Rand**

This section helps answer the following questions:

- How to analyze the tolerance stacks of an assembly?
- What is the procedure of assigning tolerances for components in an assembly?

When two or more parts are assembled together, assembly stack analysis is needed in order to determine whether these parts can be successfully put together. In assembly stack problems, generally there are two situations, assembly with clearances, and without clearances between assembly components. For the assembly stack problems without clearances, it is very straightforward, and the calculation is similar to that in the previous section when treating all components that are "attached" to each other. The following example is used to illustrate this procedure.

Example 7.10

As shown in Figure 7.34, this is an assembly stack problem without clearances with ring, base, and cap assembled. Find the maximum and minimum of the distance X.

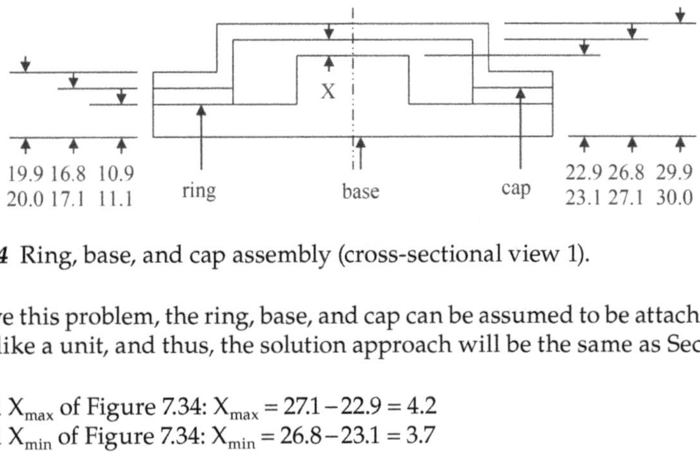

Figure 7.34 Ring, base, and cap assembly (cross-sectional view 1).

To solve this problem, the ring, base, and cap can be assumed to be attaching to each other just like a unit, and thus, the solution approach will be the same as Section 7.5.

1. Find X_{max} of Figure 7.34: $X_{max} = 27.1 - 22.9 = 4.2$
2. Find X_{min} of Figure 7.34: $X_{min} = 26.8 - 23.1 = 3.7$

To find assembly stacks with clearances, one can follow the four basic stack steps in the previous section with one additional action:

1. Identify the problem: Identify the stack objective, list the conditions under which the stack is being calculated, and label the start point, end point, and direction of the stack.
2. Choose the desired answer: Write down the design goal. "What is the extreme maximum (or minimum) answer that would be acceptable and allow the product to function as intended?"
3. Identify the stack path: A stack path is a series or chain of distances (part dimensions) from the start point of the stack to the endpoint of the stack. This chain of distances must consist only of known distances—which are dimensions on

the drawing or a value calculated from dimensions on the drawing, must be the shortest possible chain of distances from the start point to the end point, and must be continuous—each distance must begin where the previous distance ends.
4. Perform the math: Transfer the distances onto the stack form, add each column of numbers, check the subtotals, and evaluate the answer.

The additional action is when identifying the problem: first, decide if a maximum distance or a minimum distance is the objective. Then, arrange the parts of the assembly accordingly. If looking for a maximum distance, "move the parts" into the position that gives the extreme maximum condition. Do this by making a sketch that shows the parts in the position that produces the extreme maximum condition. The following example is used to illustrate this procedure.

Example 7.11

The following assembly shows a shaft and bushing assembly. The bushing can move along the shaft with clearance P. Calculate the maximum and minimum value of P for the following drawing.

Figure 7.35 shows the parts "moved" as far apart as they will go. It is necessary to "ground" them together until all of the parts in the assembly cannot move any further, and put the clearance "P" at each place that a part is touching an adjacent part as shown in Figure 7.36.

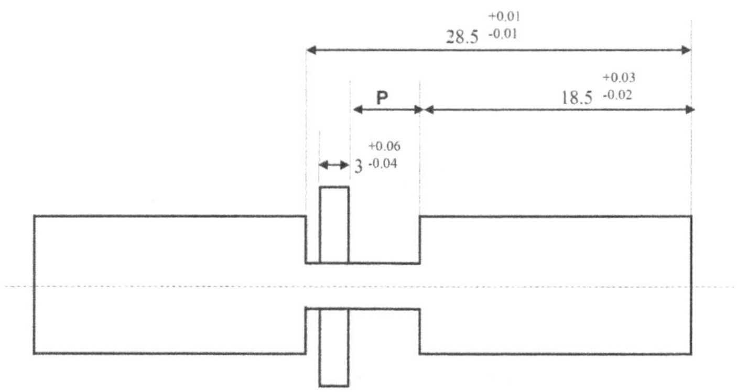

Figure 7.35 Shaft and bushing assembly (cross-sectional view).

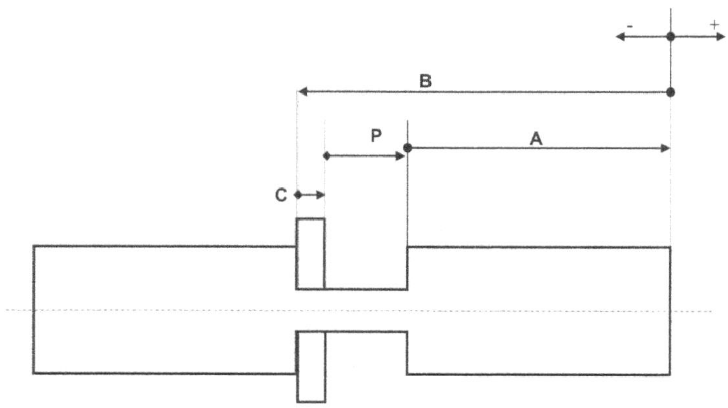

Figure 7.36 Shaft and bushing assembly. (They are "grounded" together so that all of the parts in the assembly cannot move.)

As shown in Figure 7.36, when calculating the clearance, the part is moved into the position that gives the extreme condition. The rest of the tasks can just follow the previous concept discussed in the previous section. The results are shown in Table 7.12.

Table 7.12 Finding the clearance P

Distance		Maximum		Minimum	Tolerance
A	+	18.53	+	18.48	0.05
B	−	28.49	−	28.51	0.02
C	+	3.06	+	2.96	0.1
P	+	6.9	+	7.07	0.17

Review problems

1. As shown in Figure 7.37 below, this is an assembly stack problem without clearances with ring, base, and cap assembled. Find the maximum and minimum of the distance X.

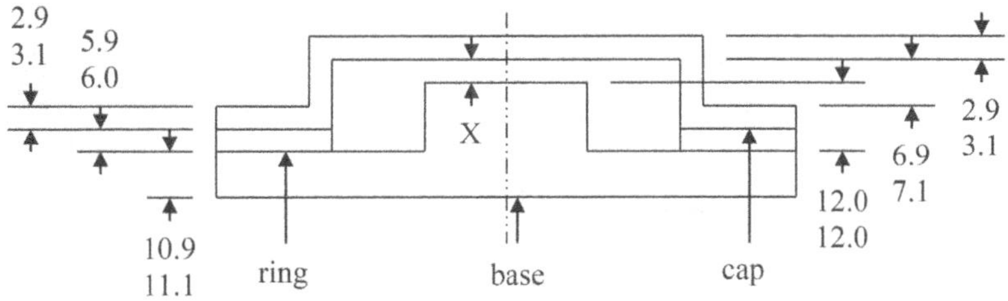

Figure 7.37 Sample assembly (cross-sectional view 1).

2. The production workers on the shop floor are having trouble with a certain assembly as shown in Figure 7.38. They say that sometimes, the pieces of this assembly do not fit correctly because of interference. Most of the time, it goes together fine though. Go back to the drawing below and prove/disprove that the assembly should go together 100% of the time as long as the parts are made to specifications. Briefly, explain your answer.

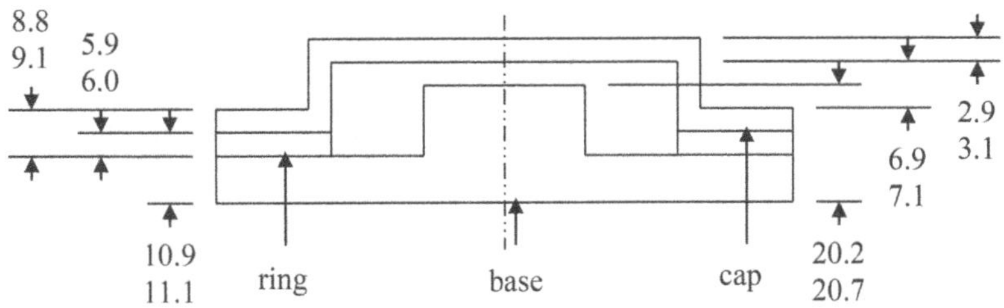

Figure 7.38 Sample assembly (cross-sectional view 2).

3. Sometimes, we would like to achieve a specific assembly tolerance requirement. For example, as shown in Figure 7.39, an assembly consists of a block which is stationary, a stud which is to be pressed fit in the block, and a pulley for a running fit on the stud. In other words, it is free to revolve on the stud. The assembly requirement is that the pulley is to have an axial degree of movement of 0.002 to 0.006 in. Assuming the pulley width is 1 in., we would like to find the width of the stud.

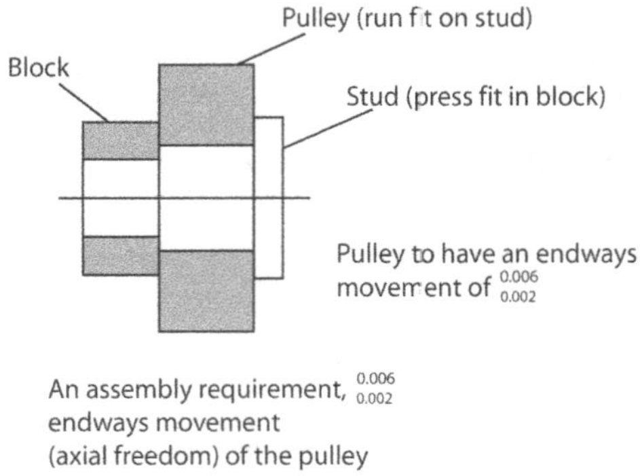

Figure 7.39 An example problem for stud–pulley assembly.

4. Consider the previous problem, but the stud needs to be pressed fit into the bearing with width of 0.25 in.; the stud width is 1.25 in. as shown in Figure 7.40.

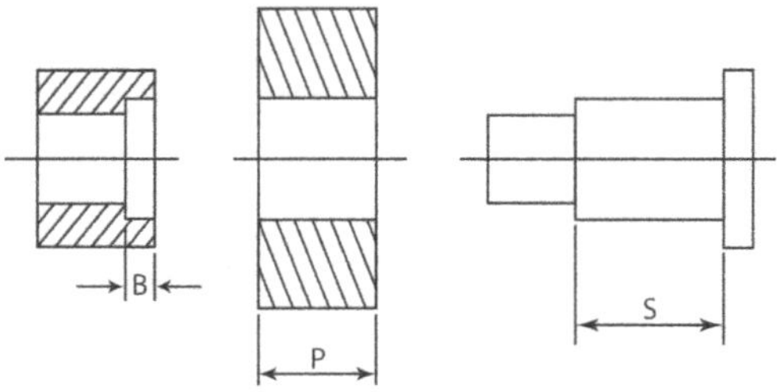

Figure 7.40 Problem for stud–pulley-bearing assembly.

5. As shown in Figure 7.41, find the extreme maximum distance.

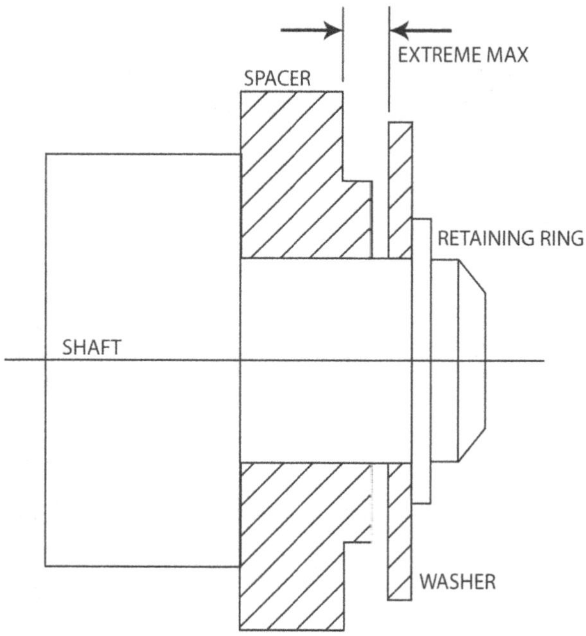

Figure 7.41 An example problem for stud–pulley-bearing assembly.

7.7 Process capability

Fast is fine, but accuracy is everything.

Xenophon

This section helps answer the following questions:

- What is process capability?
- Why is process capability important in tolerance analysis?
- What are the capability indices? How can they be used to represent a process capability?

A process which can produce a part that meets the tolerance requirement once does not mean that it can meet the tolerance requirement every time, as it is statistical in nature. For example, the diameter of a shaft in the design is 1 in. but when the machinist machined the shaft, the resulting diameter could be 1.001 or 0.999 in. If a machinist were to machine multiple shafts of the same diameter, they are likely to receive a series of dimensions from 0.999 to 1.001 in. These ranges are called tolerances. The tolerance can change due to various process conditions and parameters. Process capability describes the capability of specific manufacturing equipment or processes to meet certain tolerance requirements. Therefore, process capability can be used to help determine which machine(s) or process(es) to use to manufacture a part or a set of parts. As various processes can hold only certain tolerances, the cost of production increases geometrically for incremental tightening of tolerances as multiple processes may need to be performed for tighter tolerances. Therefore, there is a need to choose a cost-effective manufacturing process to achieve product specifications. Process capability is the repeatability and consistency of a manufacturing process, and thus should be considered when assigning tolerances in the product or prototyping design. It is necessary to make sure that a chosen machine will have the capability to meet the required tolerance limits. Process capability compares the output of an in-control process to the specification limits by using capability indices. As a capable process is one where almost all the measurements fall inside the specification limits, the process capability index uses both the process variability and the process specifications to determine whether the process is capable. Capability indices can place the distribution of the process in relation to the product specification limits.

As shown in Figure 7.42, process capability index C_p can be expressed as

$$C_p = (\text{UCL} - \text{LCL})/6\sigma \tag{7.5}$$

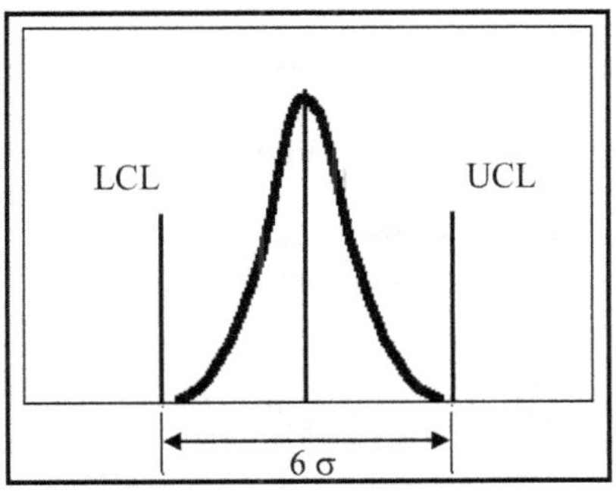

Figure 7.42 Process distribution when process capability index $C_p = 1$ and when the process is centered.

where

UCL = upper control limit

LCL = lower control limit

σ = estimated standard deviation of the process

$C_p < 1$ means the process variation exceeds specification, and a significant number of defects are being made

$C_p = 1$ means that the process is just meeting specifications. Defects will be made and more if the process is not centered

$C_p > 1$ means that the process variation is less than the specification; however, defects might be made if the process is not centered on the target value.

Therefore, it is necessary to address how well the process average, μ, is centered to the target value. C_p is often referred to as process "potential," while C_{pk} measures not only the process variation with respect to allowable specifications, but also considers the location of the process average.

$$C_{pk} = \min\left[(\mu - LCL)/3\sigma, (UCL - \mu)/3\sigma\right] \tag{7.6}$$

If $C_{pk} > 1$, then the process variation is less than the specification.

Many companies may typically start with 1.33 for supplier qualification and have an expected goal of 2.0. If the process is near normal and in statistical control, C_{pk} can be used to estimate the expected percentage of the defective material.

Example 7.12

A welding machine process has an upper control limit of 10.00 in. stroke and lower control limit of 1.000 in. stroke. After several hundred samples, the mean is determined to be 6.371 in.

1. The company usually adheres to $\sigma = 1.33$. What is this machine's C_p level under these conditions? What does the result mean?

$$C_p = (UCL - LCL)/6\sigma = (10.00 - 1.000)/(6 \times 1.33) = 1.128$$

Since $C_p > 1$, then the process variation is in control, but defects could be made if the process is not centered.

2. What is the machine's C_{pk} value? What does the result mean?

$$C_{pk} = \min\left[(\mu - LCL)/3\sigma, (UCL - \mu)/3\sigma\right]$$

$$C_{pk} = \min\left[(6.371 - 1.000)/(3 \times 1.33), (10.00 - 6.371)/(3 \times 1.33)\right] = 0.9095$$

Since $C_{pk} < 1$, the process is off center enough to cause more variations than normal.

3. What happens to the C_p value if the company wants to adhere to $\sigma = 2$? What does the result mean?

$$C_p = (UCL - LCL)/6\sigma = (10.00 - 1.000)/(6 \times 2) = 0.75$$

Since $C_p < 1$, then the process variation is out of control, and many defects are being produced.

4. What would be the machine's new C_{pk} value?

$$C_{pk} = \min\left[(\mu - LCL)/3\sigma, (UCL - \mu)/3\sigma\right]$$

$$C_{pk} = \min\left[(6.371 - 1.000)/(3 \times 2), (10.00 - 6.371)/(3 \times 2)\right] = 0.6048$$

Example 7.13

For a certain process, UCL = 20 and LCL = 8. The observed process average, $\mu = 16$, and the standard deviation, $\sigma = 2$. What is the C_p level of this process? What is the C_{pk} value of this process? How can the process be improved?

$$C_p = (UCL - LCL)/6\sigma = (20 - 8)/(6 \times 2) = 1.0$$

$$C_{pk} = \min\left[(\mu - LCL)/3\sigma, (UCL - \mu)/3\sigma\right]$$

$$= \min\left[(16 - 8)/6, (20 - 16)/6\right] = 0.6667$$

In order to make $C_{pk} > 1$, one can either reduce σ or center the process.

$$\text{If } \sigma = 2, \mu = 14, C_{pk} = \min\left[(14 - 8)/6, (20 - 14)/6\right] = 1.0$$

$$\text{If } \sigma = 1, \mu = 16, C_{pk} = \min\left[(16 - 8)/3, (20 - 16)/3\right] = 1.333 > 1$$

So far, the assumption was based on all available data such as σ, UCL, and LCL. However, how can these numbers be established? Statistical process control (SPC), a method for achieving quality control in manufacturing processes, can be used to obtain these data. It employs control charts to detect whether the process observed is under control. SPC involves using statistical techniques to measure and analyze the variation in processes. Some sample control charts as shown in Figures 7.43 and 7.44 are often used.

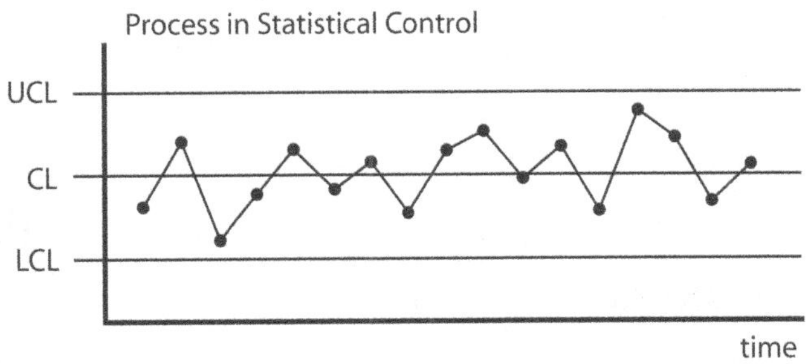

Figure 7.43 Process in statistical control.

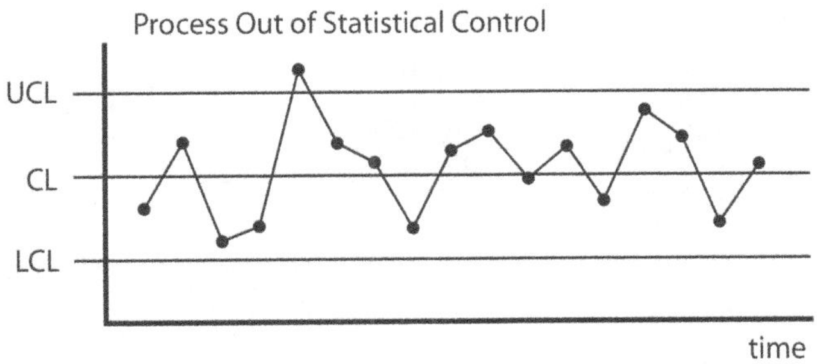

Figure 7.44 Process out of statistical control.

The following steps can be used to conduct an SPC:

Step 1: Collect samples from 25 or more subgroups of consecutively produced units.
Step 2: Plot the results on the appropriate control chart (e.g., C-chart).

To find the C-chart, it is necessary to find the mean:

$$\bar{c} = \frac{1}{k}\sum_{i=1}^{k} c(i) \tag{7.7}$$

where
k = number of inspection units.
c(i) = number of nonconformities in the ith sample, and in this case

$$\sigma = \sqrt{\bar{c}} \tag{7.8}$$

$$UCL = \mu + 3\sigma \tag{7.9}$$

$$LCL = \mu - 3\sigma \tag{7.10}$$

Step 3: If all groups are in statistical control, use the control limits (UCL and LCL).
Step 4: Observe these for some time until satisfied. UCL and LCL are then the limits.
Step 5: Otherwise, identify the special cause of variation and take action to eliminate it.

Example 7.14

Table 7.13 shows drill error per every 100 samples. Establish the C-chart.

Table 7.13 Drill error per 100 drills

Tune	1	2	3	4	5	6	7	8	9	10	11	12	13	14	15	16	17	18	19	20	21	22	23	24	25
Drill error	6	5	8	1	6	2	5	5	4	3	3	2	0	2	5	1	7	7	1	5	5	8	8	4	5

Mean c = 4.32

$UCL = 4.32 + (3 \times 2.0785) = 10.56$

$LCL = 4.32 - (3 \times 2.0785) = -1.92$ and thus set to 0; when $LCL < 0$, set $LCL = 0$

(negative has no physical meaning)

Figure 7.45 shows the data chart of Table 7.13.

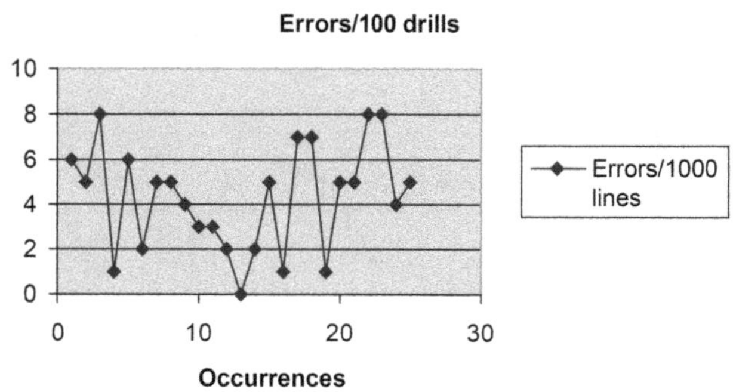

Figure 7.45 Data chart for Table 7.13.

According to the definition of "in control," the process should meet four criteria:

1. No sample points outside of process limits
2. Most points near average
3. Nearly equal number of points above and below average
4. Points are randomly distributed

Some typical manufacturing capabilities for various tolerance levels are listed in Table 7.14.

Table 7.14 Some typical surface finishes of various manufacturing processes

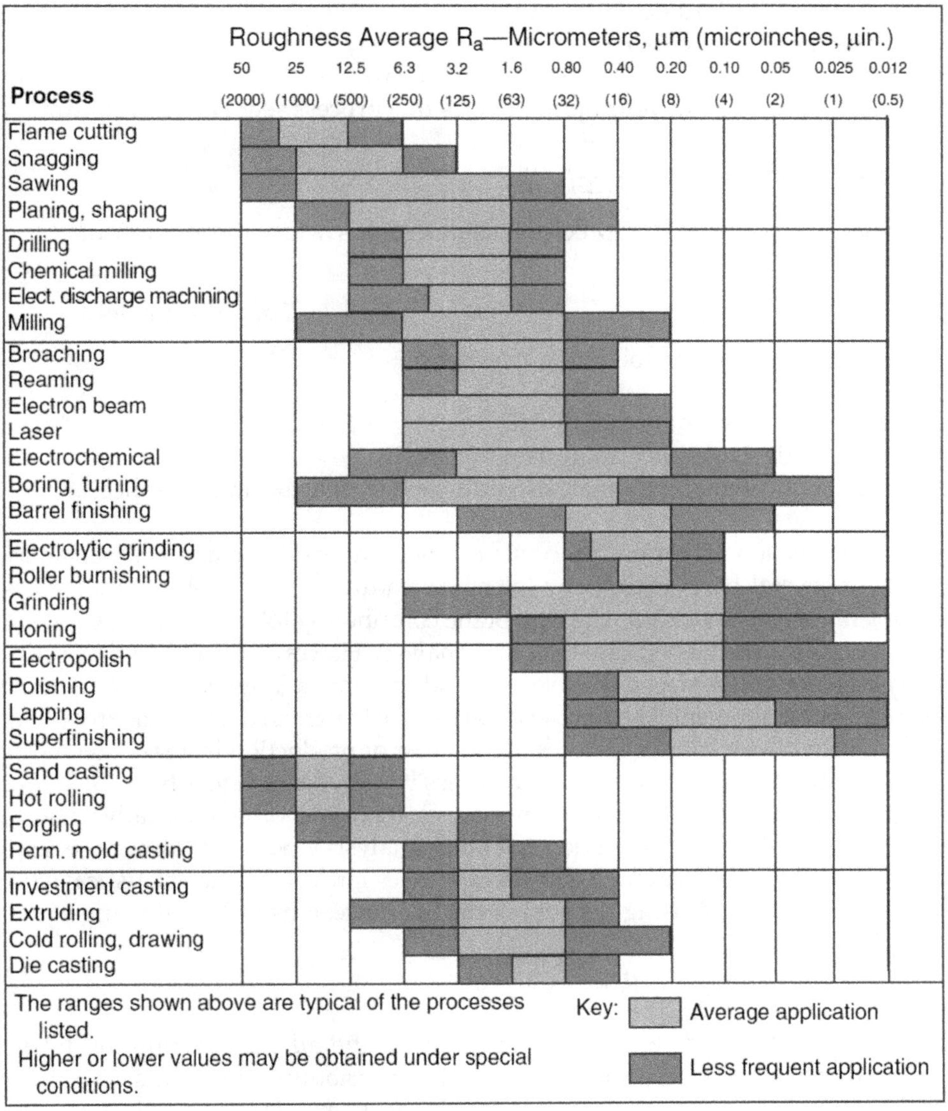

Review problems

You work at a customer service center where calls come in about peoples' computers. There have been several complaints about the response times of the employees. You perform a study where you monitor how long customers have to wait before being helped.

	Number of minutes on hold																								
Trial	1	2	3	4	5	6	7	8	9	10	11	12	13	14	15	16	17	18	19	20	21	22	23	24	25
Minutes	5	5	9	10	6	6	6	7	10	12	9	5	11	9	9	13	4	8	9	11	2	3	4	7	5

Below are the results:

1. Calculate the mean of the process.
2. Plot your results on a chart.
3. Calculate the standard deviation of the process.
4. Calculate the UCL and LCL.
5. In order to place the UCL and LCL on the chart you plotted in step 2, the process must look in control. In your opinion, does this process seem to be in control?

7.8 *Statistical tolerance analysis*

Be precise. A lack of precision is dangerous when the margin of error is small.

Donald Rumsfeld

This section helps answer the following questions:

- What is statistical tolerance analysis?
- Why is statistical tolerance analysis performed?
- How is statistical tolerance analysis performed in distributing tolerances?

As discussed in Section 7.7, various processes can hold different tolerances. When an assembly consists of several parts with many tolerance features and assembly complexities, the resulting tolerance chain adds up. The sum of the contributing tolerances as discussed in the previous sections is called worst case tolerance analysis. However, often the worst case tolerance analysis is not practical as these tolerances add up into a large number that would not meet the process requirement. The only solution may be to tighten the tolerance requirements for all the contributing tolerance features. As the cost of production increases geometrically for incremental tightening of tolerances, sometimes it is not possible to achieve economically.

In addition to the worst case tolerance analysis, there is another alternative, that is, statistical tolerance analysis. The statistical tolerance analysis would not change the resulting tolerance sum of the tolerance chain. However, it takes into account the statistical nature of the process, and instead of using the sum of the tolerance, it uses the following equation:

$$\sigma_{sum} = \sqrt{\sigma_1{}^2 + \sigma_2{}^2 + \cdots + \sigma_n{}^2}$$

where σ_{sum} is the standard deviation of the sum of the all the contributing tolerances ($i = 1 \ldots n$). By using this equation, the more the contributing tolerances, the larger the statistical sum is going to be. Therefore, statically, the sum is larger and thus has a better chance to meet the process requirements. However, the statistical fact remains: for 3σ condition, 2.7 parts out of 1,000 will fail based on normal distribution. In other words, when the process is controlled within 3σ, the resulting sum is σ_{sum} which yields a larger resulting sum. However, for every 1,000 parts, there will be two to three parts that may fall outside the prediction and may result in a failure.

Statistical tolerance analysis is conducted to specify appropriate tolerances about the nominal values established in parameter design to achieve a balance between setting wide tolerances to facilitate manufacture and minimizing tolerances to optimize product performance. The design guidance for total assembly can be summarized below:

1. Datum surface: When many parts are fitted together, the buildup of maximum or minimum from datum surfaces should be within the capability of the automatic workhead.
2. Location points: The product should have locating points to facilitate automated assembly.

Automation engineers should evaluate the control of their processes to determine whether the processes really qualify. Process standard deviations should be known or estimated and then compared with specified tolerances.

Example 7.15

A robot gripper can pick up parts within 2±0.010 in. dimensional specification. The process standard deviation σ = 0.005 in. Is the process ready for this robot application? If the mean, μ, of the part dimension is 2 in. and the variation is normally distributed, what % of the parts will the robot be unable to pick up?

$$\mu = \frac{\sum_{i=1}^{n} x_i}{n} \qquad \sigma = \sqrt{\frac{\sum_{i=1}^{n}(x_i - \bar{x})^2}{n-1}}$$

$$2.010 - 1.990 = 0.020$$

$$0.020/0.005 = 4$$

That is, four times the process deviation

Upper limit: 2.000+0.010 = 2.000±0.010/0.005 σ = 2.000+2 σ

Lower limit: 2.000−0.010 = 2.000−2σ

From areas of the probability table as shown in Figure 7.46, the following can be obtained: 100−95.44 = 4.56% of the parts will be missed.

Quality and automation go hand in hand. The classic 3σ limits have been challenged 100%−99.74% = 0.26%., that is, 3 out of 1,000 parts will fail. Many big companies are shooting for 6σ limits, that is, 100−99.9999998 = 0.0000002%.

Example 7.16

As shown in Figure 7.47, it is necessary to determine the assembly tolerances. The process variations are (all dimensions in inches): $\sigma_1 = 0.01$ for part 1, $\sigma_2 = 0.02$ for part 2, and $\sigma_3 = 0.01$ for part 3.

1. Find the worst case condition of the tolerance of the assembly.
2. Find the assembly variation σ.

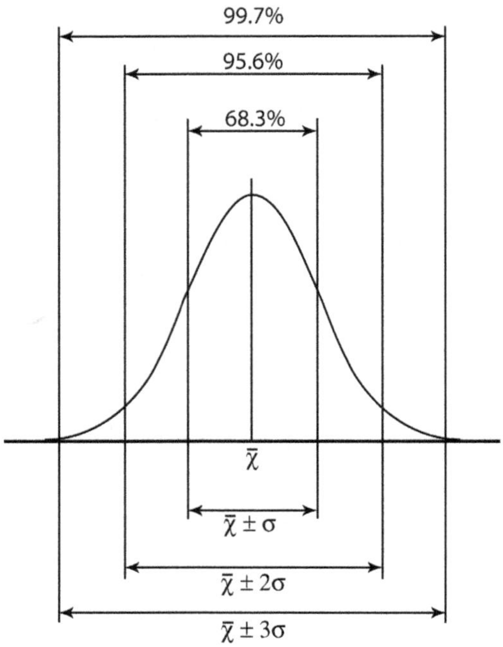

The Gaussian or normal distribution of a process variable.

Areas of Probability Density of the Normally Distributed Random Variable

Region	Area (%)
$\mu \pm \sigma$	68.26
$\mu \pm 2\sigma$	95.44
$\mu \pm 3\sigma$	99.74
$\mu \pm 4\sigma$	99.994
$\mu \pm 5\sigma$	99.99994
$\mu \pm 6\sigma$	99.9999998

Figure 7.46 Areas of the probability density of the normally distributed random variable.

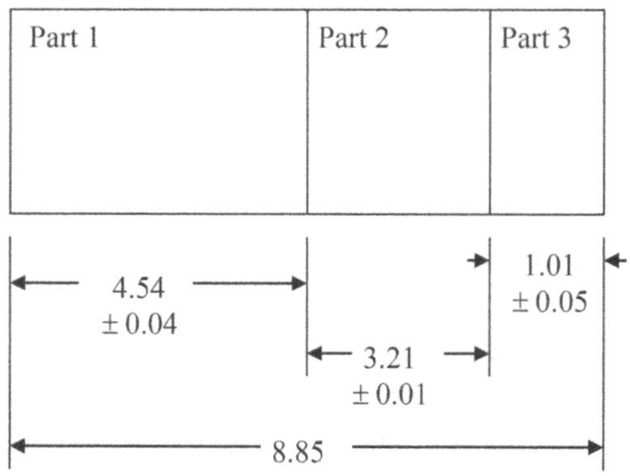

Figure 7.47 Sample assembly (top view 1).

3. All assemblies will fall within what range if the assembly's natural tolerance is set to
   a. $\pm\sigma = ?$
   b. $\pm 3\sigma = ?$
   c. $\pm 6\sigma = ?$

Solution:

1. $0.04 + 0.01 + 0.05 = 0.10$ in.
2. $\sigma = \sqrt{0.01^2 + 0.02^2 + 0.01^2} = 0.02$ in.

3. (a) $\pm\sigma = 0.02$ in.

   (b) $\pm 3\sigma = 0.06$ in.

   (c) $\pm 6\sigma = 0.12$ in.

As shown in Figure 7.48, for a three-part assembly,

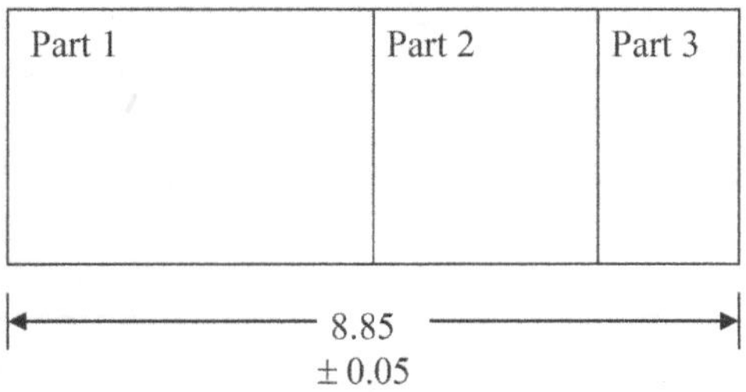

Figure 7.48 Sample assembly (top view 2).

1. Calculate σ of the assembly if its natural tolerance is 4σ.
2. Calculate the individual tolerances assuming each part had an equal σ value.

Solution:

1. Calculate σ of the assembly if its natural tolerance is 4σ.
   a. $4\sigma = 0.05$

      $\sigma = 0.0125$ in.
2. Calculate the individual tolerances assuming each part had an equal σ value.
   b. $\sigma(\text{assembly}) = (3\sigma^2)0.5$

      $0.0125^2 = 3\sigma^2$

      $\sigma(\text{individual}) = 0.007$

      $4\sigma = 0.029$ in., thus ± 0.029 in.

Review problems

1. If a process is centered upon the mean and under statistical control,
   A$\pm\sigma$ = ____ % of valid parts
   B$\pm 2\sigma$ = ____ % of valid parts
   C$\pm 3\sigma$ = ____ % of valid parts
   D$\pm 4\sigma$ = ____% of valid parts
   E$\pm 5\sigma$ =____ % of valid parts
   F$\pm 6\sigma$ = ____% of valid parts
2. Why is statistical tolerance analysis important to a company?
3. The robot arm can pick up parts measuring 5±0.005 in. dimensional specification. The normally distributed process that creates these parts has a μ = 5 in. and σ = 0.005 in. If you were working for this company, prove why you would or would not want this machine on your assembly line. Briefly explain your answer.
4. As shown in Figure 7.49, determine the total assembly tolerance. Statistical control is normally distributed, and the process capability is ±4σ.
5. Reverse case: see Figure 7.50 to determine part tolerances.

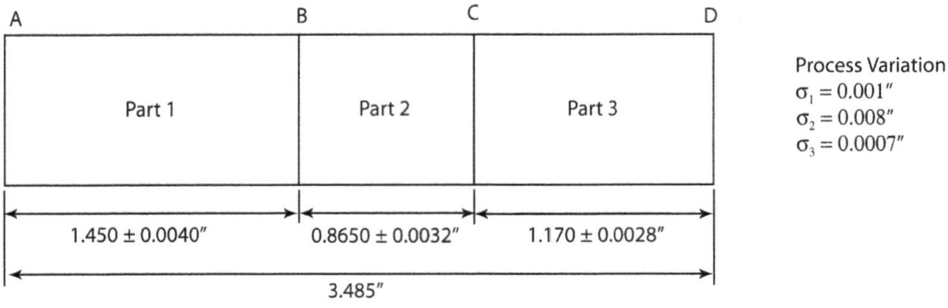

Figure 7.49 To determine assembly tolerances.

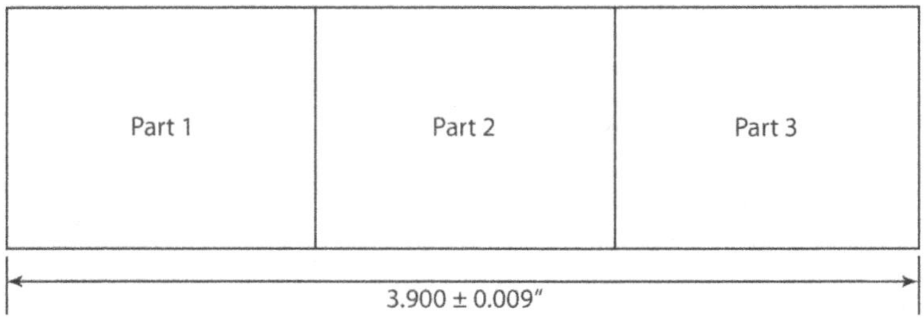

Figure 7.50 To determine part tolerances.

6. Why is statistical analysis control performed?
7. Does the distribution of a process as shown in Figure 7.51 indicate that the process is valid for SPC?

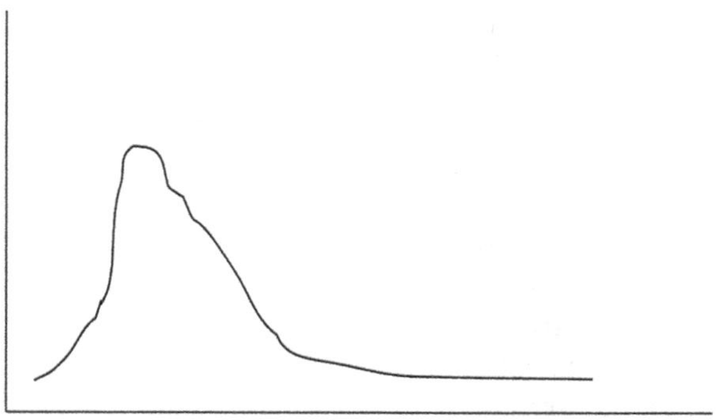

Figure 7.51 The distribution of a process.

7.9 Case study: conceptual design of a chamber cover

7.9.1 Problem description

The main purpose for the design of the argon chamber is to allow argon gas into the chamber and let oxygen out. It should not allow oxygen to enter into the chamber and leak out argon. The chamber's top cover should fit to the CNC spindle and laser hub, and also sustain the temperature of the laser's working environment. The conceptual design of the chamber needs to be developed.

7.9.2 Requirement definition

The design requirements of the chamber design are summarized in Figure 7.52. This process identifies the critical design parameters.

TECHNICAL REQUIREMENTS

CUSTOMER REQUIREMENTS (HIGH ○ 5; MEDIUM □ 3; LOW Δ 1)	OVERALL WEIGHTING	OXYGEN FREE 500-1000PPM ALLOWABLE	HOT LASER/LASER PARTICLES DO NOT BURN ARGON CHAMBER	DOES NOT RESTRICT THE MOVEMENT	SHOULD BE COMPACT AS MUCH AS POSSIBLE	SHOULD GO IN DEFINED SCHEDULE	SHOULD LOOK GOOD WHEN NOT IN USE
COST-EFFECTIVE (Saving of argon, replaceable if expensive)	5	○	□	□	○	□	
MUST BE COMPLETED IN ONE SEMESTER	3	□	Δ	Δ	Δ	○	
TITANIUM DOES NOT MIX WITH AIR	3	○			□		
AESTHETIC DESIGN	1						○
ABSOLUTE IMPORTANCE		49	18	18	31	30	5

Figure 7.52 Matrix of the house of quality for the chamber design.

7.9.3 Component identification and design

Figure 7.53 shows the argon gas chamber component design based on the functional efficiency technique. This design process results in the manufacture of the hardtop dome, the customization of the plastic bag, the purchase of rubber and adhesive sealant, and a fire-resistant coating (Table 7.15).

For the evaluation of various concepts, the digital logic approach was used as shown in Tables 7.16 and 7.17.

Ratings produced by the digital logic approach are used to generate the final dome concept as shown in Figure 7.54. The layout of the dome cover in the final chamber is shown in Figure 7.55.

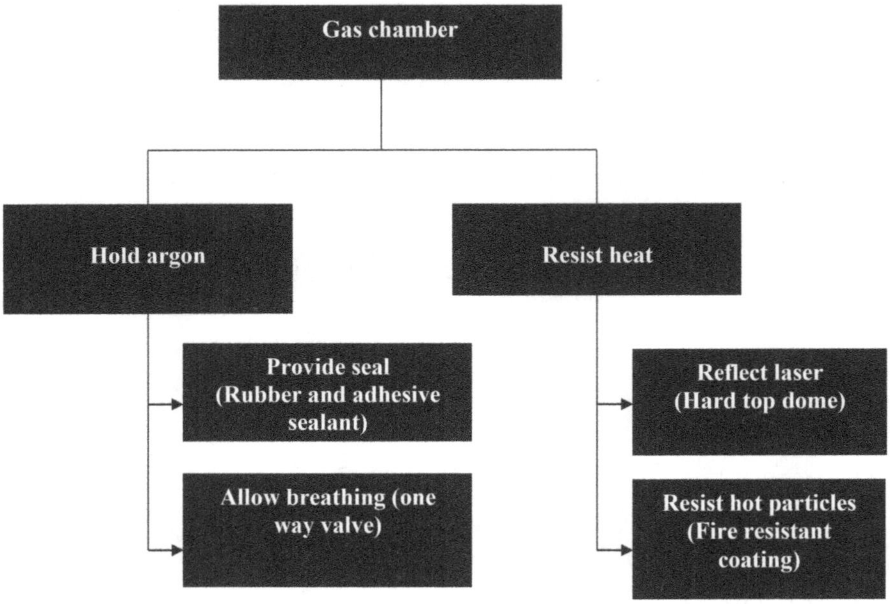

Figure 7.53 The argon gas chamber component design based on the functional efficiency technique.

Table 7.15 Chamber design objectives

S. No.	Design objectives	Representation
1	No leakage factor	A
2	Argon saving factor	B
3	Tool changing	C
4	Setting time effectiveness	D
5	Ease in handling (managing of wires)	E
6	Ease in manufacturing	F
7	Ease in assembly	G
8	Cost effectiveness	H
9	Resistance to burn	I

Table 7.16 The digital logic representation of the chamber design

	A	B	C	D	E	F	G	H	I	Total
A	—	1	1	1	1	1	1	1	0	7
B	0	—	0	1	0	1	1	0	0	3
C	0	1	—	1	0	1	1	1	0	5
D	1	1	1	—	0	1	0	0	1	5
E	0	0	0	1	—	1	1	0	0	3
F	0	0	0	1	0	—	0	0	0	1
G	0	0	0	1	0	0	—	0	0	1
H	0	0	0	1	0	1	1	—	0	3
I	0	1	1	1	1	1	1	1	—	7

Table 7.17 Finding the wall thickness between the left side of hole and surface, "C"

Distance		Maximum		Minimum	Tolerance
A	+	4	+	4	0
B	–	5.15462/2 = 2.57731	–	5.15938/2 = 2.57969	0.00238
C	+	1.42269	–	1.42031	0.00238

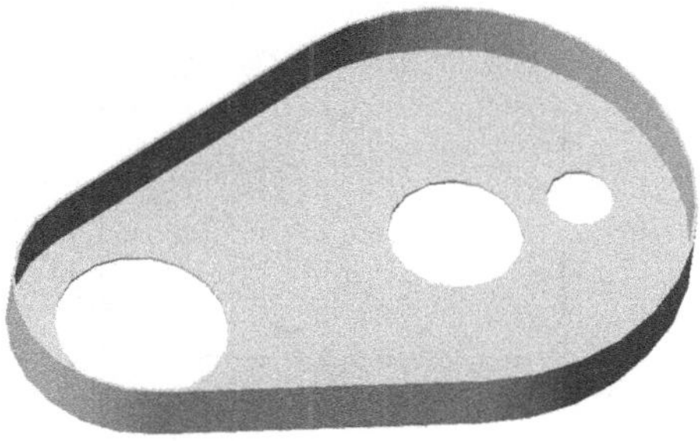

Figure 7.54 Final dome concept.

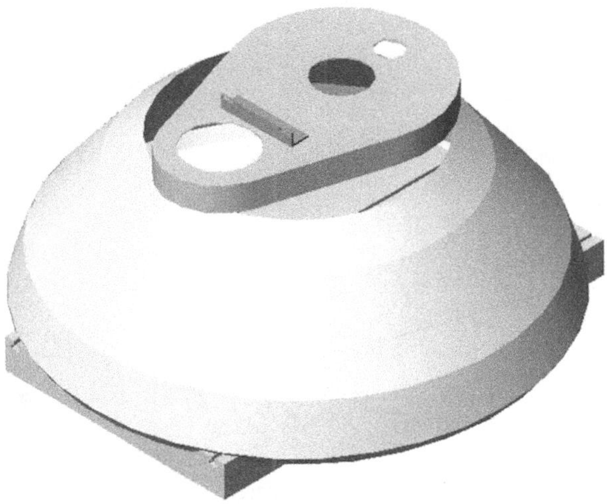

Figure 7.55 The conceptual design of the dome (top) and the chamber.

7.9.4 *Tolerance analysis*

In this design, only the dome is identified as the critical tolerance feature. The holes in the dome must fit in the spindle and laser hub. The following two steps were considered:

a. Fitting with spindle and laser hub: In order to fit the dome with spindle and laser hub, for the dimensions of a 5.157 in. diameter hole to fit in the spindle and a 4 in. diameter hole to fit in the laser hub as shown in Figure 7.56, a Class 3 free fit (close sliding fit) was selected from the ANSI fit standard chart. This avoids leakage, but if there is any gap between the hole and the spindle or laser hub, a rubber gasket can be applied between respective gaps.

b. When the dimension is subject to a specified assembly requirement, the limits will result from a calculation:

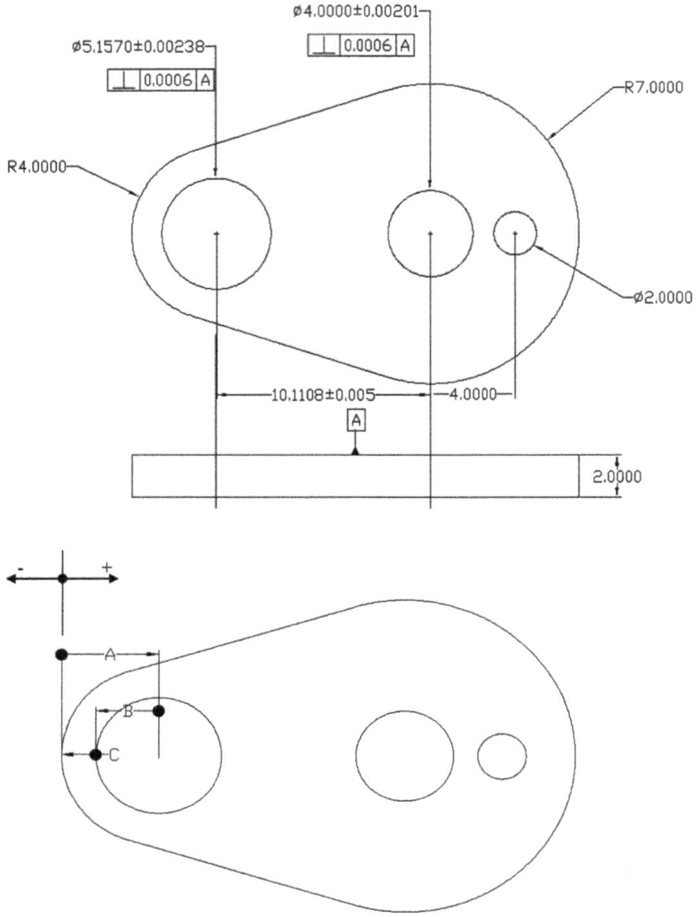

Figure 7.56 Dome after tolerance analysis.

For the 7.2 in. diameter hole to fit in the spindle

$$\text{From Table 7.5, the allowance(clearance)} = 0.0009\sqrt[3]{(d^2)}$$

$$= 0.0009\sqrt[3]{(5.517^2)} = 0.00268\text{ in.}$$

$$\text{The tolerance} = \pm 0.0008\sqrt[3]{(5.517^2)} = \pm 0.00238\text{ in.}$$

For a shaft basis system, because the spindle and the laser hub are already there:

Hole +
Maximum dimension = 5.157 + 0.00238 = 5.159 in.
Minimum dimension = 5.157 − 0.00238 = 5.155 in.
Similarly for the 4 in. diameter hole to fit in the laser hub Allowance = 0.00267 in.
Tolerance = ±0.00201 in.

Hole
Maximum dimension 4.002 in.
Minimum dimension 3.998 in.

To fit the dome smoothly into the spindle and laser hub, perpendicularity tolerance to the datum feature "A" as shown in the figure is critical. The wall thickness between the left side of the hole and surface "C" was not known and thus needs to be found by using tolerance chain analysis as shown in Figure 7.56. The result is shown in Table 7.17.

7.9.5 A focused prototype

To check the critical tolerances and dimensions of the CAD model, a focused prototype made of cardboard as shown in Figure 7.57 for the hardtop dome was developed. The figure shows the prototype installed on the CNC mill to check for fit with the CNC spindle and the laser hub. It was found to be very helpful to use a focused prototype to validate the critical dimensions of the dome before installation as more details were found to help in the procedure in installing the dome.

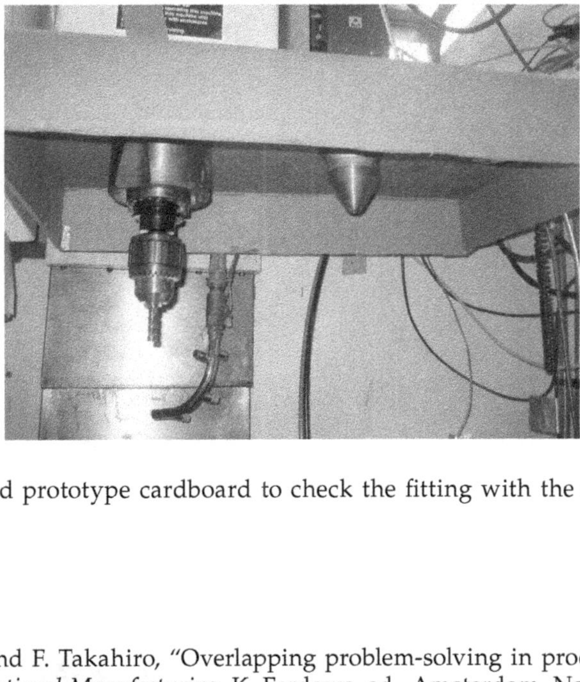

Figure 7.57 A focused prototype cardboard to check the fitting with the CNC spindle and the laser hub.

References

[Clark89]. Clark, K. and F. Takahiro, "Overlapping problem-solving in product development," in *Managing International Manufacturing*, K. Ferdows, ed., Amsterdam, North Holland, 1989. See also Chapter 5 of Womack, Jones, and Roos, The Machine That Changed the World, New York: Harper Perennial, 1991.

[ISO17]. Dimensional and geometrical product specifications and verification, Retrieved on September 1, 2017 from www.iso.org/committee/54924/x/catalogue.

chapter eight

Prototyping of automated systems

Many of the recent products involve some level of automated devices, such as most electronic devices, appliances, and even toys. Also, almost all of the current products are manufactured using automated processes, that is, automation. Automation is a way of increasing productivity by incorporating tasks and performing them with automatic systems. Automation in a factory can take many forms and can be used in many areas of a production facility. Therefore, it is critical to have the capability to quickly prototype or process a product with automated devices. The basic elements of an automated system consist of the machine instructions or codes and the machines themselves. The machine instructions are the software that is encoded to tell the machines how to perform their specific function. The machines are the system of components that carry out the tasks as instructed. These machines can perform the same tasks over and over, or they can be programmed to perform multiple jobs.

There are four basic elements in an automated system. The first element includes the actuators that conduct the movement of the system. These are the workhorses of the system. They provide the movement to drive the system and can obtain that energy from different sources, with electricity, pneumatics, or hydraulics being the most popular. The second element includes the sensors that monitor the environment for action. They measure certain quantities like temperature, stress, light, or even just an on/off condition in the instance of a switch. This can give indications of the running conditions of the system, a step in a process, an error condition, or be used to validate a safe operating condition. The third element involves the controller. It controls the motion and operation of the system components based on inputs from sensors and the internal programmed instructions. A controller would be thought of as "the brain" of the system. It makes decisions and performs the corresponding action. The final element is the mechanical system or mechanism. This category consists of the other hardware included in the autonomous system. These mechanisms or components are used with actuators and drives to perform specific tasks. These basic elements and their selections for prototyping purpose are summarized in this chapter.

8.1 Actuators

This section helps answer the following questions:

- What is to be considered when choosing an actuator for automation?
- What are some examples of actuators and how can they be applied to assist in environments that are not fully automated?
- What are the specific applications of each type of actuator?
- What are the advantages and disadvantages of each type of actuator?
- Is there a rule of thumb to choose actuators?

An actuator is a device that performs a mechanical action in response to an input signal. There are three general types of actuators to choose from when creating an automated process: electric, hydraulic, and pneumatic. A major advantage of using these types of actuators is that they are used so much that they are considered reliable in creating an automated process. Each type of actuator has its advantages and disadvantages for each possible application. The key to properly selecting an actuator is to understand the advantages and limitations of each class, and within that class, the subclasses of actuators. For example, when a motor is required to rotate a mechanism, at an accuracy of 0.28, it would allow the use of several different types of motors. An added requirement to this operation that the motor be able to rotate at 2,000 rpm eliminates the selection of the stepper motor subclass. The need to index with precision eliminates the standard electric motor. The subclass of servomotors would be best suited to this operation. The specific motor that is chosen would be dependent on loads, acceleration rates, torque requirements, etc. Such restrictions and selections exist for all major and minor classes of actuators, and being able to select the appropriate actuator requires the understanding of these traits and characteristics.

8.1.1 Types of actuators

A solenoid is an electromechanical device used to convert electrical energy into linear mechanical work to push or pull a ferrous plunger against a nonferrous load. A typical solenoid is shown in Figure 8.1. It can generate quick, linear motion. The small force (several ounces) is generated by magnetic force. Solenoids are found in many conventional machines, such as coin-operated machines, vending machines, coin-operated arcade games, or change machines; they all work on the same principle. They are also found in many electromechanical devices, such as copiers. One can also use a solenoid to switch an electrical circuit on and off; this is called a (mechanical) relay. When used as relays, small solenoid allows a low-power circuit to move a switch controlling the current in a higher-power circuit. The resulting linear work produced could either pull or push in configuration.

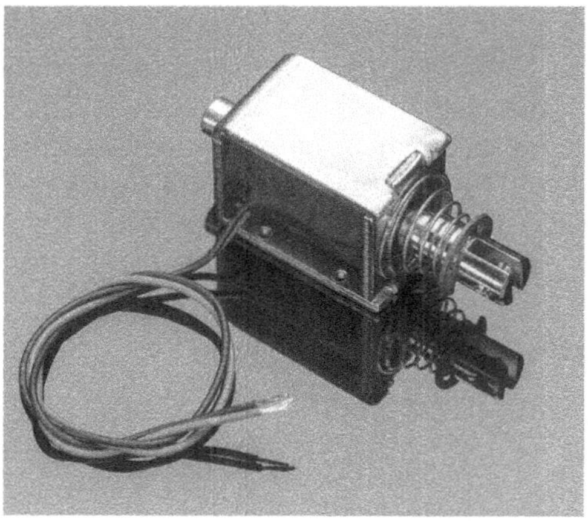

Figure 8.1 A typical solenoid.

Some solenoids can indirectly move heavier loads, such as the spool in a solenoid-actuated pneumatic or hydraulic valve. The larger the coil, the longer it takes to actuate. Solenoids can be made to act more quickly to move light loads, such as pins in a dot-matrix printer, if larger power supplies are used to drive them. These solenoids must be built to withstand the high temperatures they generate. Larger and slower solenoids are available to move heavier loads, such as a diverter in a mail sorting transport. Solenoids, because they are electromagnets, do not exert the same force over their whole stroke—some solenoids provide more force at the end of their stroke, whereas others provide more force at the start of their stroke.

Pneumatic cylinders as shown in Figure 8.2 are generally chosen for low-load applications in places where piping can convey air pressure around a building. As the typical shop air is about 80–100 psi, the force exerted from an air cylinder is the cross section of cylinder times the air pressure. It is commonly used for applications in which the required force is less than 200 lbs. The advantages of pneumatic cylinders include their small size and weight, low cost (around $20 to ~$50), and good reliability. The disadvantages include the need of an air supply, possible water leakage, delay time (when the line is over 2 ft), and uneven motion.

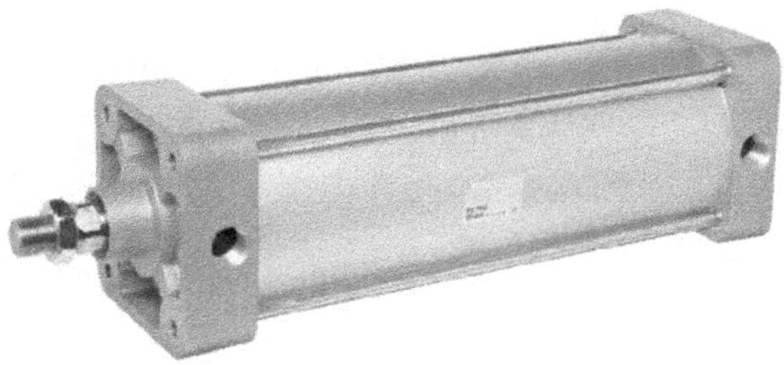

Figure 8.2 An air cylinder with a connecting joint.

Hydraulic cylinders, however, can support much larger loads than pneumatic cylinders. These actuators are generally used for linear motion. The advantages of hydraulic cylinders include the capability to exert extremely high force, excellent damping, smooth motion, and less compressibility and delay. The disadvantages include high cost and possible leakage.

Note that although cylinders are mainly used for linear motion, there are options for rotary actuators. Two types of rotary actuators are often used: the vane rotary actuator and rack and pinion gear to translate linear motion to rotary motion.

Some applications using cylinders are shown in Figure 8.3. Note that these devices can be used to replace a robot for some simple motion in a very cost-effective way.

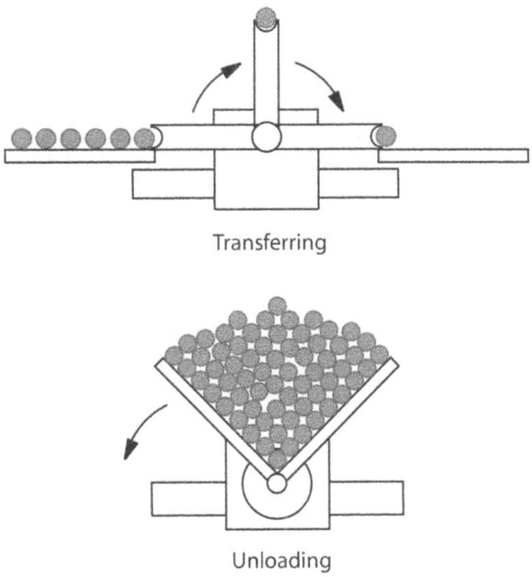

Figure 8.3 Some simple motions can be achieved using cost-effective cylinders.

8.1.2 Drives

Drives or motors are devices that provide continuous movements and usually rotational motion. As shown in Figure 8.4, a stepping motor rotates through a given angle for every electrical pulse from the controller. The advantages of a stepping motor include the facts that motor and control boards are small in size and position control is excellent. The disadvantages of the stepping motor are that torque control is limited, the motion is not as smooth at low rpm, and calibration is often needed. Figure 8.5 shows a schematic diagram when using a pair of rollers to advance plastic film using friction. As the film is printed and will be cut precisely into labels right after the roller advances, a stepping motor may be considered as a candidate to drive the rotor. One issue is that after some usage, the film may be out of position for cutting due to slippage of the film on the rollers, and thus, recalibration may be needed.

A DC motor can be controlled by a timer or an indicator switch as a cost-effective way to form an automated system. The advantages of this setup include smoother motion (than

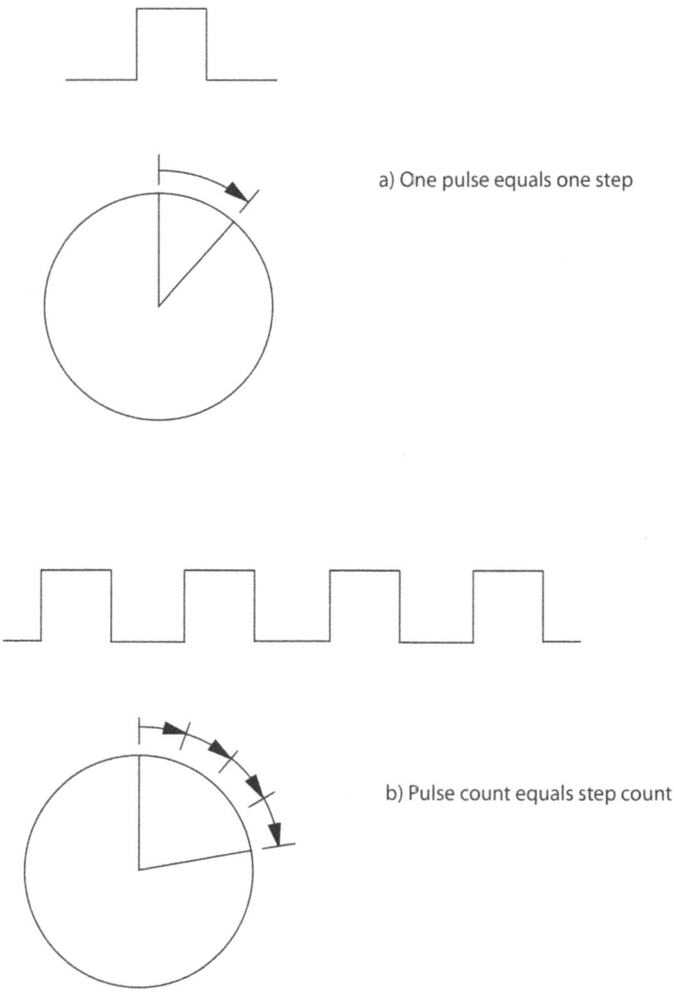

Figure 8.4 A stepping motor rotates through a given angle for every electrical pulse from the controller.

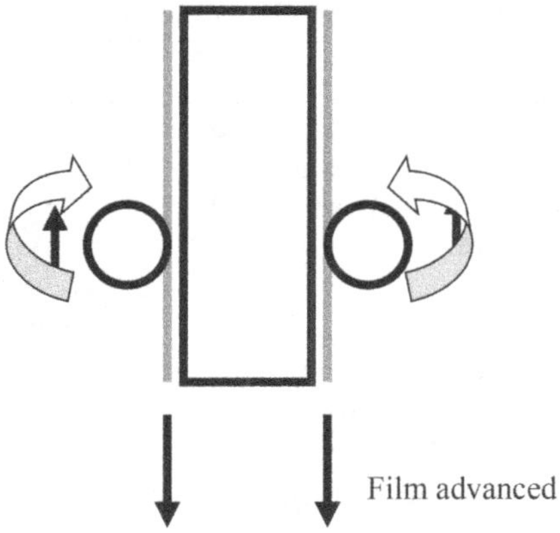

Figure 8.5 An example to use a pair of rollers to advance plastic film using friction.

stepping motor), the possibility of constant torque/velocity, and low installation cost. The disadvantages include poor position control, the need for a controller for constant velocity/torque control, and the need for a separate DC power supply.

A servomotor as shown in Figure 8.6 can have complete control of motor position, velocity, and acceleration. The advantages of a servomotor include excellent position, velocity, and acceleration control, and they are flexible and programmable. The disadvantages include higher cost; they are bulky compared to DC and stepping motors, and the need for special power and wiring.

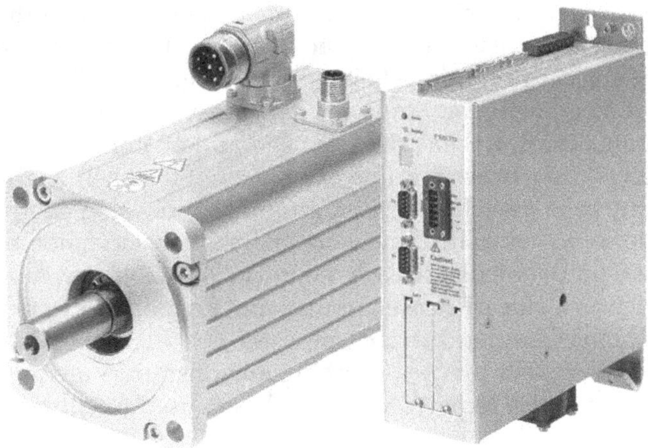

Figure 8.6 An example of servomotors.

Table 8.1 summarizes the comparisons between servomotors and stepper motors. Stepper motors can be found almost anywhere, such as printers and plotters, disc drives in computers, robots, machine tools, and CD players, whereas servomotors are used in more specific applications, such as robots and synchronization of motion in automation.

There are other types of actuators. For example, a piezoelectric actuator makes use of the phenomenon in which forces applied to a segment of material lead to the appearance of an electrical charge on the surface of the segment. Therefore, a piezoelectric actuator

Table 8.1 Comparisons between servomotor and stepper motor

	Stepper motor	Servomotor
Resolution and accuracy	Typical stepper motors can produce 200 full steps per revolution. They do not necessarily achieve the desired location, especially under load.	The positional feedback is used to correct any discrepancy between a desired and an actual position. Typical encoders produce 2,000–4,000 pulses per revolution.
Reliability and maintenance	They experience little or no wear, and are virtually maintenance free.	Brush-type servomotors require a change of brushes.
Speed and torque	Not as smooth at lower speed. Poor torque characteristics at higher speeds (over 2,000 rpm).	Can produce speeds and powers two to four times that of similarly sized steppers.
Cost	Less expensive for smaller motors and standard steps (instead of microsteps).	10%–30% more expensive than similar stepper systems. Brushless servomotor systems tend to be even more expensive.

can be very simple in shape and can be used in micromachines as shown in Figure 8.7. A piezoelectric element is placed in between an electromagnetic positioning element and a weight. Initially, the electromagnetic positioning element is fixed to the ground with electromagnetic force. When the voltage is applied to the piezoelectric element in the middle, the piezoelectric element will expand and thus the weight will move to the right side. When the weight travels to the right, the electromagnetic force will then be released so that the entire object will move to the right due to inertia. This process can repeat quickly, and thus, the entire object can continue to move to the right side. Therefore, the piezoelectric element can act as an actuator. Many micro/nanosystems use this type of principle to make a micro/nanoactuator. Some devices used as actuators are complete automated systems in themselves. Robots are a prime example. Off-the-shelf X–Y tables are used as components in many systems where a reprogrammable position control is desired. The X–Y table is just two servo-controlled linear actuators (often servomotors and ball screws) with their linear axes perpendicular to each other. The X–Y table may move the table under a stationary tool such as a glue dispenser, or may move a tool over a stationary worktable. Programmable X–Y tables are used to control water-jet cutters that are used to cut shoe leather, plastics, and floor mats for cars. Clothing manufacturers are improving efficiency by using powerful X–Y tables to move knives or lasers used to cut materials. Numerical

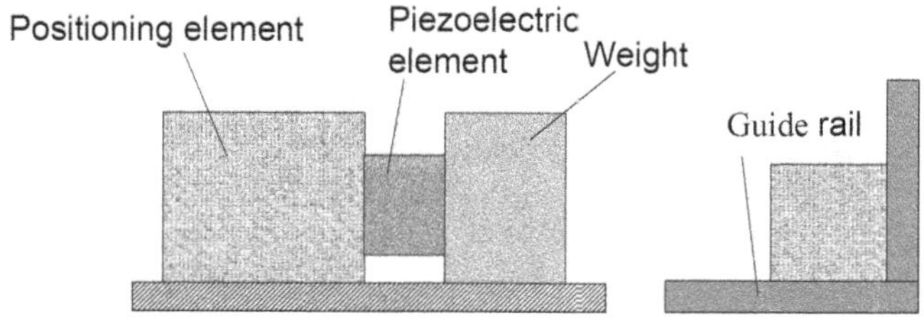

Figure 8.7 A piezoelectric element is placed in between an electromagnetic positioning element and a weight.

control (NC) equipment can cut metal under the control of a built-in digital controller instead of being controlled by a human operator.

8.1.3 When to choose an actuator

Actuators should be chosen early in the design of the automated system, but after the motion of the system has been determined. By deciding what tasks and motions will be performed first, a correct type of actuator can be chosen. If high forces are needed, hydraulics would be a good choice. Faster applications could use solenoids. The advantages and disadvantages can be weighed against the functions that need to be accomplished. After the actuators are chosen, the feedback necessary to control the motion can be determined, the appropriate sensors chosen, and the controller appropriate for the actuators selected. The design of the mechanisms will depend on the type of actuator selected. One would not want to end up requiring a unique-sized actuator for a job and not be able to obtain it.

The commonly used actuators include solenoids, pneumatic motors, hydraulic motors, stepper motors, DC motors and servomotors. The selection of a particular actuator depends on many issues. Some commonly considered factors are summarized as follows:

1. Cost: Solenoids and relays are generally cheaper than pneumatic and hydraulic systems.
2. Function: If a job requires high loads, solenoids and relays may not be appropriate. For heavy duty jobs, hydraulic cylinders may be a preferable option.
3. Space: Solenoids and relays usually come in small sizes. Motors can be small if the torque requirement is not high; they can generally be accommodated and fixed in a compact system.
4. Accuracy: Stepping motors are cost-effective in terms of position control, whereas servomotors can be controlled very precisely in terms of position, velocity, and acceleration. Pneumatic cylinders, because of the compressibility of air, do not provide a smooth and consistent movement.
5. Power source: For either pneumatic or hydraulic cylinders, a power source to drive these is required. These power sources such as air compressors or hydraulic pumps are very bulky and require a lot of piping to eventually connect to the cylinders. However, if the power source is available, such as shop air (which is often available in many factories), all one needs is the pneumatic cylinder. On the other hand, solenoids, relays, and most motors require electrical energy, which can be provided easily and requires marginal maintenance.
6. Related environment: For example, if the job requires a fire or hazard-proof system, the electrical systems need to be frequently inspected to avoid any sparks or fire hazard.

8.1.3.1 Base/manifold-mount solenoid control valves

One can use a single pressure source for multiple valves as shown in Figure 8.8. These electrically operated valves are mounted to a base or manifold, plus have much lower power consumption than other valves with similar flow rates. For multiple control valves, the multiple-station base can accommodate several stations as shown in Figure 8.9. They can be conveniently connected to a computer or programmable logic controller (PLC) to provide electronic control of the on/off valves for pneumatic or hydraulic cylinders.

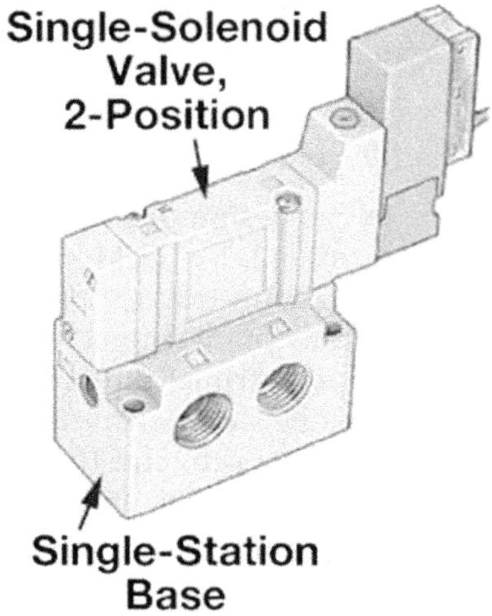

Figure 8.8 An example of a single station base/manifold-mount solenoid control valve.

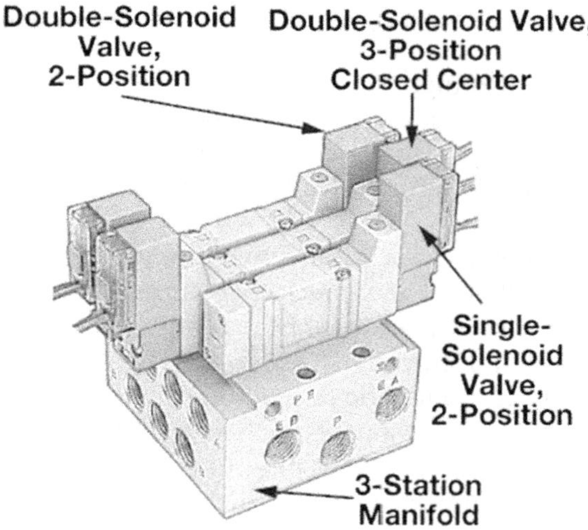

Figure 8.9 An example of multiple-station base/manifold-mount solenoid control valves.

Review problems

1. Which actuators listed below can effectively control the angular displacement of rotation?
   A. Servomotors
   B. Hydraulic motors
   C. DC motors
   D. Stepper motors
2. What are the advantages and disadvantages of using solenoids?
3. What is the difference between a DC motor and a servomotor?
4. What is a simple relay and what is its main purpose?
5. Why must stepper motors be monitored and recalibrated when required?

6. List three advantages of pneumatic systems over hydraulic systems.
7. What is the difference between DC motors and stepping (stepper) motors?
8. What are the practical use, advantage, and disadvantage of a solenoid actuator?
9. Figure 8.10 shows a double-action pneumatic cylinder. Double-action means the pneumatic cylinders can be moved both inward and outward by compressed air, as compared to single-action pneumatic cylinders, which can only be moved pneumatically in one direction. The return movement is caused by a spring. There are two valves required. When the cylinder moves outward and inward, respectively, which valve is open and which valve is closed?

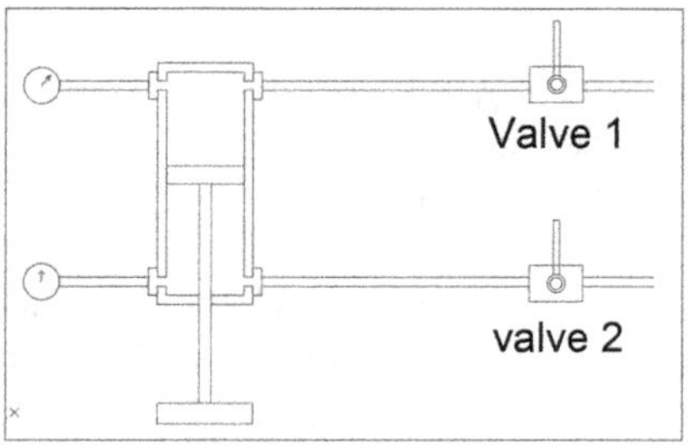

Figure 8.10 A double-action pneumatic cylinder.

10. An engineer needs a small and cheap motor with excellent position control to realize rotational movements for an industrial application. Given that the control of torque, speed, and smoothness of motion are not considered, which of the following motors would be the best choice?
    a. DC motor
    b. Pneumatic motor
    c. Solenoids
    d. Stepper motor
    e. Servomotor
11. Match each type of actuator and drive to the appropriate characteristics.

1. Solenoids	a. Use piping compressed air through a manufacturing plant
2. Relays	b. Rotate through a given angle for every electrical pulse from controller
3. Cylinders	c. Complete control of motor position, velocity, and acceleration
4. Stepper motors	d. Controlled by timers or indicator switches
5. DC motors	e. An electromechanical device to convert electrical energy into linear mechanical work
6. Servomotors	f. It is an electromagnetic device for remote control

12. A 3D NC machine of a ceramic slurry extrusion prototyping project needs three motors to drive the three axes. The motions of the axes need to be very precise in velocity and position of the trajectory. The motors will be mounted on the end of the slides. Which kind of actuators will be the best to be used?

8.2 Sensors

> Equipped with his five senses, man explores the universe around him and calls the adventure Science.
>
> **Edwin Powell Hubble**

This section helps answer the following questions:

- What are the common characteristics one should be aware of that help determine the quality and capabilities of a sensor?
- Is there an organized approach for the sensor selection and to get knowledge about the latest advances in sensor technologies?
- What are the optimum operating environmental conditions for each type of sensor?
- Why are mechanical switches still being used?
- What are the examples of sensor applications?

A sensor is a device to sense system environment. It helps the controller detect the changing conditions in a working environment. In other words, sensors are devices that provide an interface between electronic equipment and the physical world. They help to see, hear, smell, taste, and touch the physical world by converting input objective physical or chemical signals into electrical signals.

There are several ways to parameterize sensors. A transfer function expresses the functional relationship between a physical input signal and an electrical output signal. The sensitivity is the ratio between a small change in an electrical signal to a small change in a physical signal. Accuracy is the largest expected error between actual and ideal output signals. The resolution is the minimum detectable signal fluctuation. Hysteresis is the width of the expected error in terms of the measured quantity. As shown in Figure 8.11, for example, hysteresis is the percent of the received light in a photosensor. For instance, when a part has occupied 10% of the light beam of a photosensor, it is considered as "on" or the object is considered as "in existence." When a part has occupied 5% in the beam, it is considered as "off" or the object is considered as "not in existence."

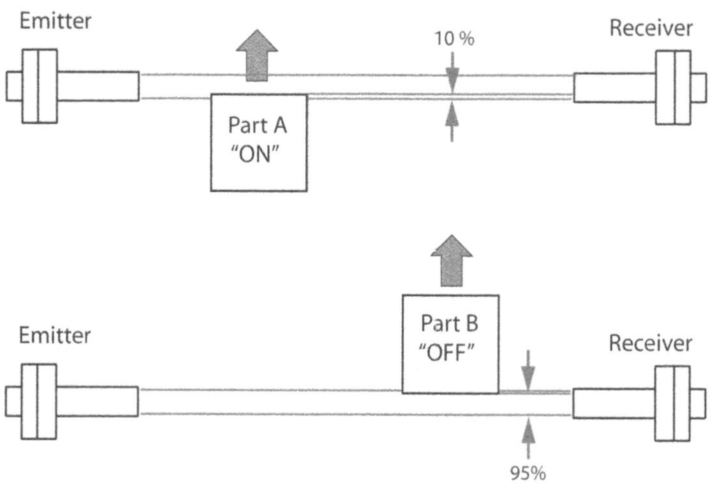

Figure 8.11 Hysteresis is the percent of the received light in a photosensor.

On the basis of the sensed property, there are several types of sensors, such as thermal signal sensors, mechanical signal sensors, chemical signal sensors, magnetic signal sensors, radiant signal sensors, and electrical signal sensors, as shown in Table 8.2. Thermal signal sensors are used to sense temperature or heat flow. For example, to sense the temperature in a controlled chamber, one needs a thermal sensor with a range –508°F to 2,208°F, and accuracy of +0.348°F. The cost is around $180. A thermocouple can detect the temperature difference of the environment instead of absolute temperature. It can sense a rise or drop in temperature. Thermocouples are a strand of two different metallic insulated wires connected together at one end and to a voltage sensing meter at the other. Thermocouples work as a device to detect temperature because when heating the junction of a circuit formed by two dissimilar conductors, a voltage is created. By measuring this voltage created, a temperature can be derived. The higher the temperature, the higher the voltage created and vice versa. Different combinations of wire pairs are used for different temperature ranges, and the size of the wires helps to determine the maximum operating temperature. In addition, different insulation materials are used for different applications (i.e., corrosive environments).

Table 8.2 Classification of sensors based on sensed property

Sensor type	Applications
Thermal signal	Temperature, energy, flux, flow, conductivity, dispersion
Mechanical signal	Linear displacement, velocity, acceleration, density, fluid How, strain, torque, force, angular velocity, acceleration, pressure, humidity, etc.
Chemical signal	Gas detection: ammonia, carbon dioxide, oxygen, nitrogen, etc.
Magnetic signal	Direction of magnetic field, strength of magnetic field
Radiant signal	Radiation detection: alpha, gamma, beta, photonic emissions, among others; infrared or ultraviolet light, visible light
Electrical signal	Conductivity, impedance, resistance, capacitance, power, charge, frequency, voltage, etc.

Mechanical sensors measure force, position, velocity, etc., whereas chemical sensors can detect concentration, reaction rate, etc. Magnetic sensors can sense magnetization, whereas radiant sensors detect electromagnetic waves and may quantify wavelength, intensity, etc.

Figure 8.12 shows an example of an oxygen sensor and analyzer. It is a kind of electrochemical sensor and can measure oxygen content from 0% to 30% with a resolution of 0.1%. It can be a critical sensor for safety. In case the oxygen level is below a certain threshold value, it can trigger an internal buzzer. Most cars produced after 1980 have an oxygen sensor. The sensor is part of the emissions control system and feeds data to the engine management computer. The goal of the sensor is to help the engine run as efficiently as possible and also to produce as few emissions as possible. If there is less air than the perfect air-to-fuel ratio, then there will be fuel left over after combustion. This is called a rich mixture. Rich mixtures are bad because the unburned fuel creates pollution. If there is more air than this perfect ratio, then there is excess oxygen. This is called a lean mixture. A lean mixture tends to produce more nitrogen oxide pollutants, and in some cases, it can cause poor performance and even engine damage. The oxygen sensor is positioned in the exhaust pipe and can detect rich and lean mixtures.

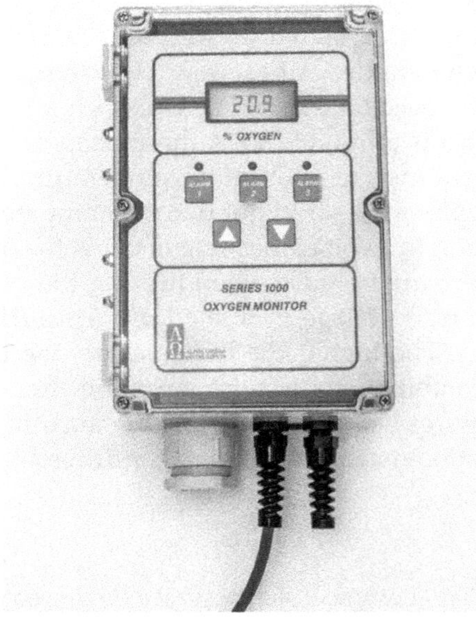

Figure 8.12 An example of an electrochemical sensor for oxygen content.

8.2.1 Sensor classification based on sensor technology

On the basis of the technology used, there can be several types of sensors, such as acoustic, inductive, magnetic, electromechanical, piezoresistance, piezoelectric, and pressure. For example, sonar sensors can detect how far something is without touching it. Inductive sensors change inductance with different applied magnetic field strengths. Varying inductance has been used to create very sensitive magnetic sensors that are low in cost and power. Electromagnetic sensors use an electromagnetic field to detect objects. One application is for indicating the level of fluid. Piezoresistance sensors are devices which exhibit a change in resistance when strained. It separates two regions with different pressures on either side as shown in Figure 8.13. Pressure sensors can detect the physical force that arises due to pressure. This kind of sensor transforms pressure into an electrical quantity. Normally, a diaphragm construction is used with strain gauges either bonded to or diffused into it, acting as resistive elements. Under the pressure-induced strain, the resistive values change.

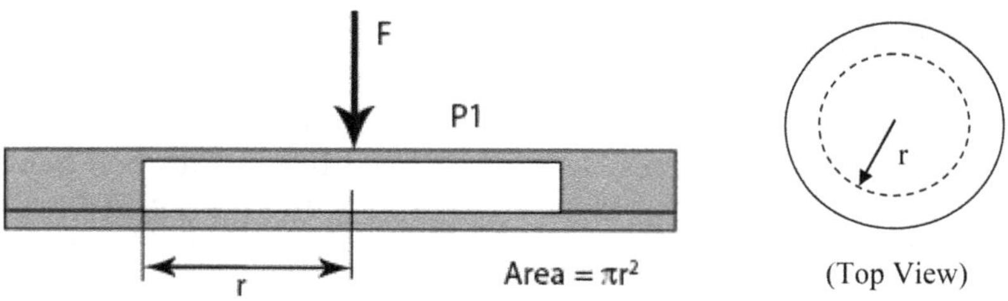

Figure 8.13 Schematics of a pressure sensor.

8.2.1.1 Manual switches

In automation, binary sensors or switches are commonly used to detect the presence of an object, including manual switches, limit switches, proximity switches, photoelectric switches, infrared switches, etc. Manual switches are activated by human response. It plays a critical role in the world of automation as it serves as an opportunity to interrupt the process for safety. For example, an emergency stop button allows the operator to overwrite all automated processes for safety reasons. Limit switches are mechanically activated. Due to the robustness, they are often used to detect the presence of a critical, safety-related object as shown in Figure 8.14. Although most of the switches are now electronics-related products such as photosensors, mechanical switches are still very useful in many areas due to ruggedness, ease of installation, and reliability. For example, in some cases, one would like to ensure a physical contact to proceed with a subsequent action. Photosensors, for instance, may be interrupted due to ambient light or environment and produce a fault signal. A mechanical switch can specify certain range of motion or force to send a signal to the control mechanism.

Figure 8.14 Some limit switches, including push-button type and rotating type.

8.2.1.2 Proximity switch

A proximity switch, as shown in Figure 8.15, can detect the presence of an object using inductive, magnetic, or capacitive properties. Three common types include electromagnetic, sonar, and Hall effect. No physical contact is required. They are useful for humid and dusty environments. For example, within a bottle washing machine or metal polishing machine, photosensors can become ineffective, and the proximity switch is a more appropriate choice. Figure 8.16 shows an application for using a proximity switch to detect the liquid level.

8.2.1.3 Photosensor

Photosensors are used for the detection of light. The commonly seen photosensors are photocells used to detect the presence of light radiating naturally from some object in the process. For example, the device sold in the supermarket to turn a light on at dusk and off

Figure 8.15 Some example of proximity switches.

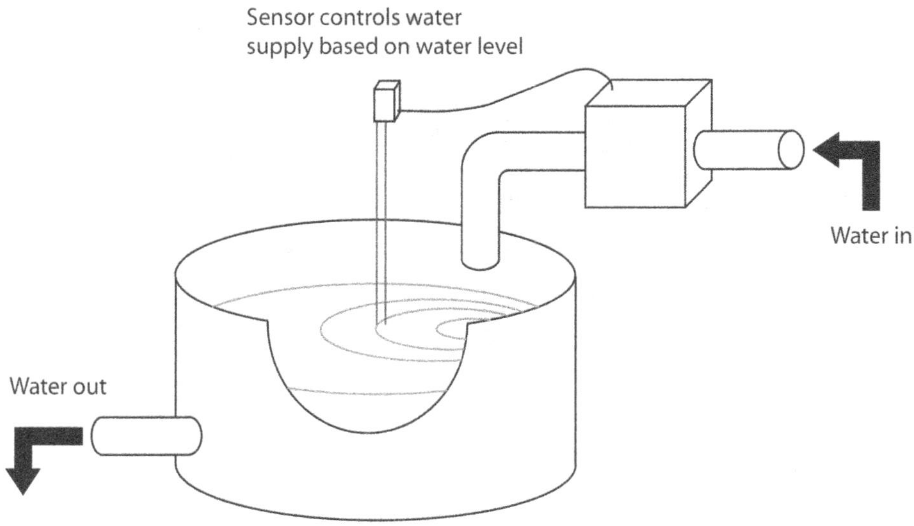

Figure 8.16 Use of a proximity switch to detect the liquid level.

at dawn is a photosensor. Many of the industrial photosensors use an artificial light source instead of the sun. Photosensors are on/off switches that change state as a result of light radiation. They are an alternative to proximity sensors when the sensing distance is longer or when the item to be detected is nonmetallic. On the basis of reflecting surfaces, reflectors are of three types, namely, diffusive reflectors, specular reflectors, and retroreflectors, as shown in Figure 8.17. Different photosensors will have different reflector requirements. A diffusive surface can be any flat surface as long as it can reflect light. A specular reflector is a very smooth surface such as a mirror, and thus, only when the light is aiming from a particular direction, can the sensor detect it. The light source and receiver are placed at a

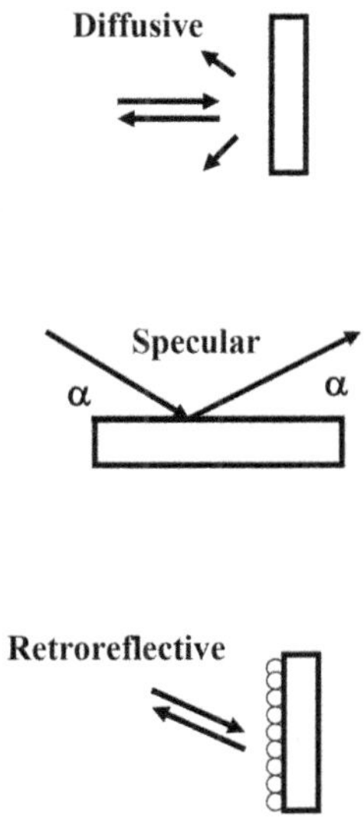

Figure 8.17 Three types of reflecting surfaces.

particular angle. This can selectively activate the sensor for the light source from a specific direction. A retroreflector is able to reflect the light in the same direction from which it comes. Because of this characteristic, it is always used to make safety signs such as the rear lights of bicycles and vehicles and the safety uniforms worn by people working at night. Figure 8.18 shows an example of using a retroreflector.

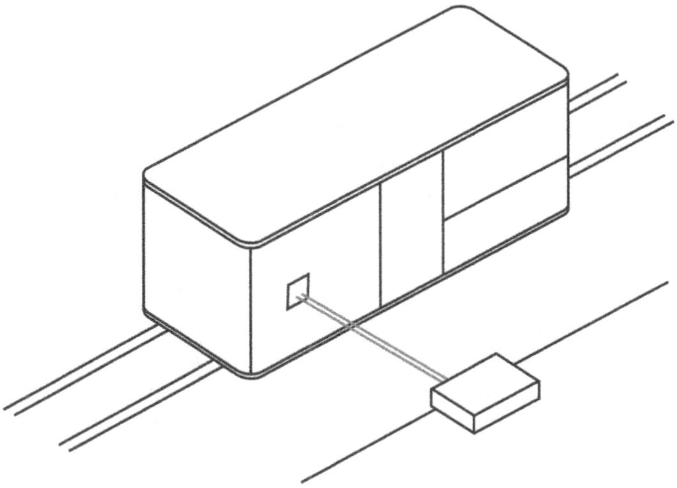

Figure 8.18 A retroreflector is placed on a cart to stop the automated guided vehicle in the desired location.

8.2.1.4 Fiber optics sensor

One useful feature is to use fiber optics in combination with the photosensors to bend light around corners or to achieve more accurate results. Figure 8.19 shows the application of using fiber optics to stop the boat in a precise location for robotic painting.

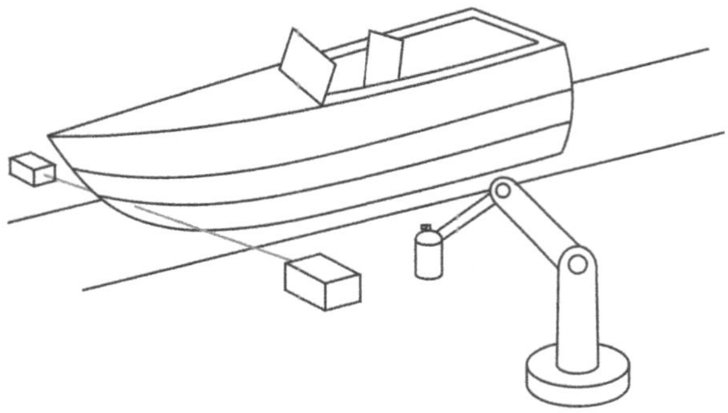

Figure 8.19 Using fiber optics to stop the boat in a precise location for robotic painting.

8.2.1.5 *Infrared sensor*

An infrared sensor can detect heat, which is infrared radiation. It can be used to detect the hot objects. One example of this application is in a small assembly environment for the laser welding of a part that is difficult to observe from outside. As a welded component is hot, it can be revealed using an infrared sensor just like an X-ray. One can trace whether two welded parts were successfully bound together with an infrared sensor.

8.2.2 *Sensor selection*

The process for selecting a sensor for a particular application in an automation system includes the following steps. The first step is to determine where a sensor may be required or more specifically, "What do I want to sense?" "Where am I sending the signal?" "What form of output is required?" and "What signal conditioning will be required?" This involves identifying the functions of the system and identifying the areas of focus for the system. The areas of focus include the workpiece itself and the components of the automated system. Sensors are selected for automation depending upon the application and environment, including inspection, indexing part train, finding parts (like for a robot), edge-finding, heat detection, proximity detection, fluid flow measurement, shaft position, force and torque, velocity, safety (light curtain and chemical detection), and pressure. The next step is to determine if a sensor should be used in a particular situation. If a particular operation in an automated system is critical, a sensor should be implemented to monitor the system.

Once a decision has been made to incorporate a sensor, several factors may be considered before a specific type of sensor can be chosen. The first factor to consider is the power source for the sensor. Power sources can be AC or DC with a range of voltages. Typically, power for the sensors should come from a stable source. The next factor is the properties of the item or feature that should be detected by the sensor. The physical properties of the item—size, material, color, resolution, cost, reliability, environment and background, the range of detection, etc.—will determine the types of sensors that may be used. The last factor to be considered is the environmental conditions in which the sensor will operate. These factors will help to identify a sensor type or types that will be able to give an accurate, reliable, and cost-effective performance in the system.

A flowchart is summarized in Figure 8.20. The steps are given as follows:

1. Determine the objective and constraints. For example, an objective could be to know that plastic pellets are ready to be injected into a mold. Constraints could be the operating environment, accuracy, maintenance, etc.

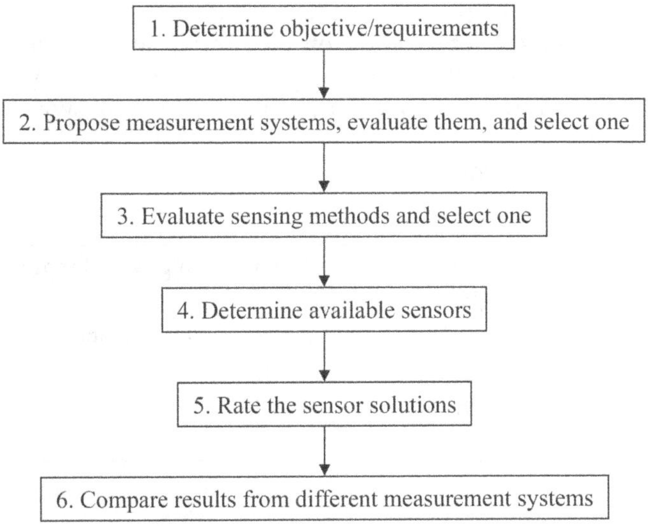

Figure 8.20 The flowchart to summarize the sensor selection process.

2. Determine measurement systems that will meet the requirements and evaluate them. For example, at a specific temperature, the plastic may be ready to be injected. The sensor will measure the temperature of the plastic. The sensor could also measure the viscosity of the plastic. The designer evaluates these measurement systems and chooses what appears to be the best.
3. Determine the sensing method (ways that the temperature can be measured). For example, a thermocouple could measure temperature with electrical resistance or an infrared sensor could measure the heat radiation. The designer evaluates these against the requirements of the system.
4. Select sensors from a commercial catalog or consider a custom-fabricated sensor. Search for a sensor that will meet the requirements for the constraints placed on the system.
5. Most likely, there is more than one sensor solution. The solutions must be compared and evaluated on whether they meet the objective.
6. If the objective is not met and if time allows, another measurement system could be chosen and taken through steps 3–5.

Review problems

1. Match different types of sensors with their functions.

Sensors	Functions
1. Thermal signal sensor	a. Detects concentration, reaction rate, etc.
2. Mechanical signal sensor	b. Detects electromagnetic waves and may quantify wavelength, intensity, etc.
3. Chemical signal sensor	c. Senses magnetization
4. Magnetic signal sensor	d. Detects the temperature difference
5. Radiant signal sensor	e. Senses current, voltage, and other electrical signals
6. Electrical signal sensor	f. Measures force, position, velocity, etc.

2. Explain the working principles for pressure sensors.
3. Explain the characteristics of the retroreflector and its application.
4. Manual switches are normally used for emergency stops. Which is more suited for a backup to the manual switch as an emergency stop, the limit switch or the proximity? Why?
5. Match the following parameters of a sensor.

1. Sensitivity	a. Functional relationship between physical input signal and electrical output signal
2. Accuracy	b. Largest expected error between actual and ideal output signals
3. Resolution	c. Ratio between a small change in electrical signal and a small change in physical signal
4. Transfer function	d. Minimum detectable signal fluctuation

6. On the basis of the technology used, match the following sensors.

1. Sonar switches	a. Detect a physical force that arises due to pressure
2. Pressure sensors	b. Sense a load or added weight
3. Load cells	c. Exhibit a change in resistance when it is strained
4. Piezoresistance sensors	d. Detect how far something is without touching it

7. Explain how to detect whether or not an object is at a station along an assembly line using three different sensors.

8.3 Controllers and analyzers

Our senses don't deceive us: our judgment does

Johann Wolfgang von Goethe

This section helps answer the following questions:

- What are the alternatives to PLC?
- What is the reason to select a PLC over a computer? Is it strictly because the computer has more functions than are possibly required and therefore cost more for the given task? Or are there other parameters that the PLC offers as an advantage over the computer?
- How does a PLC interface with an automated system?
- Can parallel ports and serial ports be used in the computer for automation?
- How could parallel ports be used for automation?

In order to control a process, a controller is needed. A controller can be a simple analyzer such as a counter, timer, or bar code reader; a PLC; or a computer. In industry, a PLC is commonly used. The main function of an analyzer or controller is to receive process inputs, register and analyze inputs, and output responses to the process (display results or actuate output continuously). A counter/timer could be used to measure time, frequency, period, duty cycle, phase angle, event counting, time interval, and pulse width. Specifically, times are for time-delay relays between two or more processes, to countdown time for an operation to be completed, and to count-up time that an operation has run over. Counters can be used to count the number of parts available for assembly, parts produced in a machining cell, and testing cycles. Barcode readers can collect data rapidly and accurately, and encode text information by electronic readers. Barcode readers can also be used for optical stock-taking purposes. These analyzers are very cost-effective and can be stand-alone.

8.3.1 PLC control

The primary function of a PLC is to control automated processes in a system. A PLC is designed to track a large number of inputs and issue a large number of commands in a very short amount of time (milliseconds). A PLC uses a symbolic programming language that makes it simple for the programmer to modify the way that the outputs are sent based on the input signals. In other words, it can be programmed with a form of programming known as ladder logic. Ladder logic is a program that is basically structured like a flowchart. The program uses four different types of registers in order to function. For example, the X registers are used as inputs, the Y registers are used as outputs, the D registers are data that can form integers, hex, and real numbers, and the R registers are the internal relays. X's and Y's are pointers to the actual terminal strip connectors on the PLC.

A PLC is a digital operating electronic apparatus. It uses a programmable memory for the internal storage of instructions for implementing specific functions. This cyclic controller receives input, analyzes, and outputs continuously as shown in Figure 8.21. It is cheaper than relays in a larger system of 100 or more relays, but a higher cost per relay function in a small system.

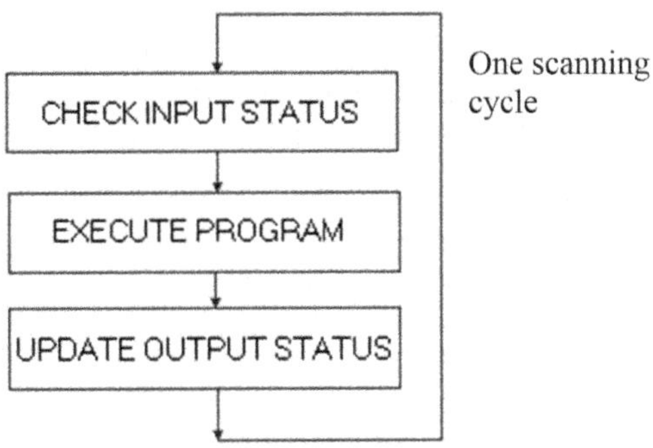

Figure 8.21 PLC receives input, analyzes, and outputs continuously.

One can use a PLC to control a conveyer belt as shown in Figure 8.22. On this conveyer belt is a part that must move back and forth from one end to the other continuously. To move the conveyor belt, one could use a DC motor. To change the direction of rotation, one would use a relay switch attached to the input power on the DC motor. In its normal state, the input switch would allow power to the DC motor to run clockwise. When the switch is thrown, it will allow reversed wires to power the DC motor to run counterclockwise.

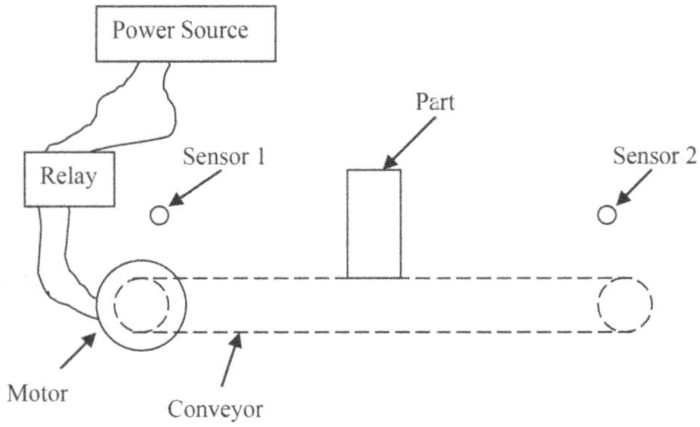

Figure 8.22 Pictorial example.

To sense whether the part is at one end or the other, one would use optical sensors (a type of photosensor) with reflective targets. When the optical sensor no longer "sees" its light from the reflective target, one will know that the part is at the end.

Here is a set of if/then statements for example:

If sensor 1 is blocked and sensor 2 is not blocked, then turn on the motor.
If sensor 1 is not blocked and sensor 2 is blocked, then turn on the motor.
If sensor 1 and sensor 2 are not blocked, and the motor is on, keep the motor on.
If sensor 2 is blocked and sensor 1 is not blocked, then turn on the relay.
If the relay is on and sensor 1 is not blocked, keep the relay on.

Assume the sensors are normally open when not blocked. Also assume when the part is at sensor 2's location, the relay must be turned on to reverse the rotation of the motor. The ladder logic diagram is shown in Figure 8.23. PLCs are selected by the following factors [Jack05]:

1. What is the cost of the PLC? Will the system be installed from scratch or are there existing products already installed that the rest of the system will need to be compatible with? As certain PLC products will not be compatible with others, make sure

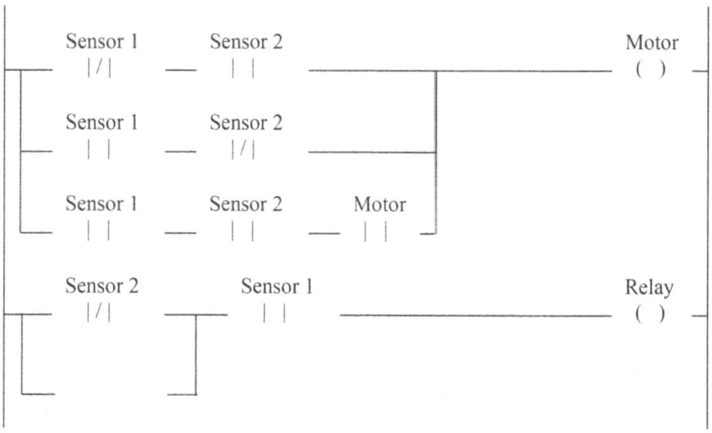

Figure 8.23 Ladder logic program.

the existing products are compatible with any PLC products. The plant may already be using a certain controller type that makes training and communication between systems easier. If the system is large, should it be controlled by a single controller or multiple smaller PLCs?
2. Are there specific environmental issues that will affect the applications such as temperature, dust, vibration, and codes specific to the facility? Certain environments may affect the operation of a PLC. For example, a typical PLC is operating at room temperature. If an application will be in an extreme environment, one needs to find a PLC to meet those specifications.
3. How many discrete and analog devices will be needed? Which types (AC, DC, etc.) are needed? Does the PLC need a capability of including more I/O and being expanded or changed when the process changes? The number and type of devices the system will include is directly linked to the amount of I/O that will be needed. The PLC model should support the I/O requirements and have modules that support the signal types. Some spare I/O ports will be useful for future expansion.
4. Will the application require high-speed counting or positioning? What about a real-time clock or other specialty feature? What is the required scan time of the system? Specialty functions are not necessarily available using standard I/O modules. Plan ahead and determine whether or not the application will require any special features. Larger programs and faster processes require shorter scan times, which result in a higher cost controller.
5. How much memory will the system require? How many devices the system will have determines data memory. How large is the program, and what types of instructions will the program include? The amount of memory is dictated by the size of the ladder logic program (ladder elements usually take only a few bytes).
6. Will the system require only local I/O, or both local and remote I/O locations? If subsystems are needed at long distances from the CPU, the PLC should support remote I/O. It is needed to determine if the remote distances and speeds supported by the PLC will be adequate for the application.
7. Will the system be communicating to other networks or systems?
8. Does the application require only traditional programming instructions, or are special instructions necessary? Certain PLCs may not support every type of instruction. A PLC should support all instructions for a specific application. For example, built-in proportional–integral–derivative (PID) control functions are convenient to perform closed-loop process control.
9. The PLC vendor should provide adequate documentation, training, and support for installation, setup, and troubleshooting throughout the life of the PLC.

The PLC selection steps are listed in Figure 8.24.

8.3.2 Computer control

Computer control is another approach to achieve automation in a project. Computer control could be at various levels: (1) hardware level, such as to turn on the disk motor; (2) basic input–output system (BIOS) level, which consists of a collection of programs that control the hardware, such as to save data to a single sector on disk; (3) disk operating system (DOS) operating system level, such as to save a file to the disk; and (4) application level, such as to replace a file on the disk in WordPerfect with an equivalent file in Word. One

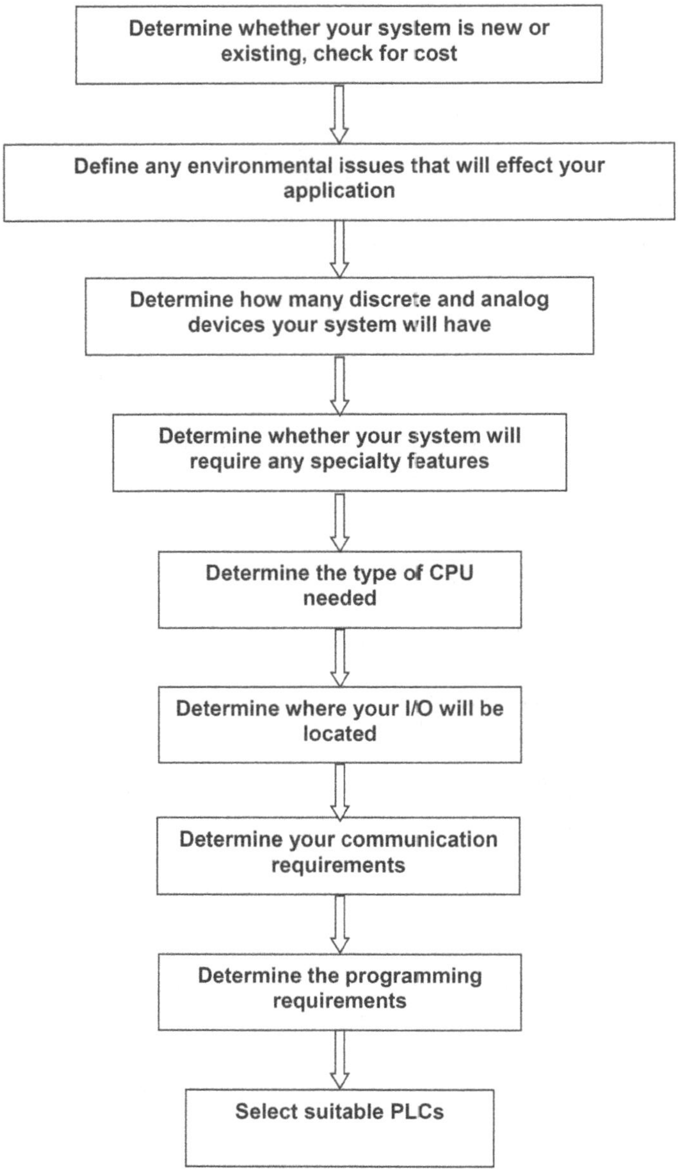

Figure 8.24 Flow diagram of PLC selection steps.

could often use lower levels such as hardware and BIOS to control external devices. For example, at the hardware level, one could use a command:

```
OUT 5F4, 3
```

to send a value of "3" (in hexadecimal or hex) through port number 5F4 (in hex).

Parallel ports, or printer ports, are the commonly used ports for interfacing computer projects. It is found commonly on the back of the PC as a D-type 25-pin female connector. The voltage can range from 0.5 to 2.4 V DC and the current ranges from 2.6 to 24 mA. These data are critical to whether they have sufficient power to be used to drive motors and actuators. A parallel port, labeled as LPT1, LPT2, or LPT3, as shown in Figure 8.25 is a 25-pin sub-D female port, and facilitates bidirectional communication, which includes a feedback system from the machine. A serial port, labeled as COM1, COM2, COM3, and COM4, facilitates single-way communication between a computer and a machine. A serial port is male in gender, either nine pins or 25 pins, slightly D-shaped. It may be used for a mouse, external modem, scanner, an internal or external infrared device, or other linking software, and is marked as COM, SER, or with a symbol representing 10101.

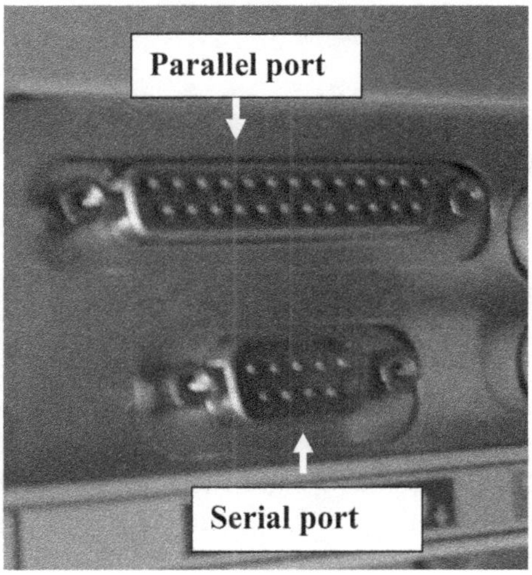

Figure 8.25 A parallel port and a serial port in the back of the computer.

The schematic of the parallel port is shown in Figure 8.26. The port is composed of four control lines, five status lines, and eight data lines. The control ports send commands from the computer to the printer during a printing operation, for example, to let the printer know about the incoming data. The status ports are for the printer to report to the computer about the status of the printer, such as data received and busy. The eight DATA ports are always active and carry the information to be printed and also special printer codes to set the printer in different modes like italics. Data is sent in the form of eight bits, "all at a time" unlike the serial port where bits are transmitted "one by one." The parallel port

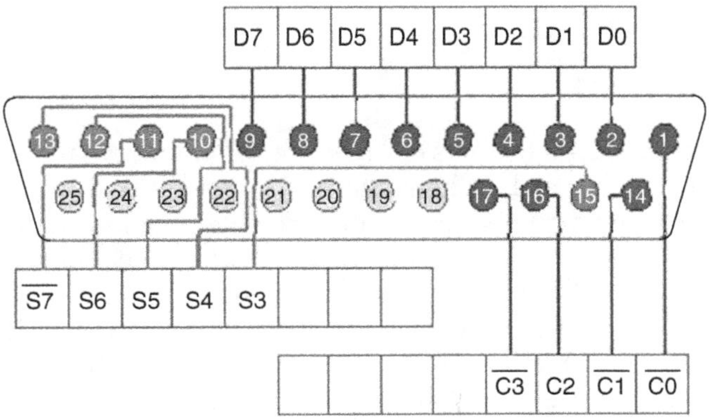

Figure 8.26 A parallel port consists of D-DATA port, S-Status port, and C-Control port.

has three commonly used base addresses. For example, 378(H)-37f(H) is the usual address for LPT1, and 278(H)-27f(H) is the usual address for LPT2. If the DATA port is at address 0×0378, the corresponding status port is at 0×0379, and the control port is at 0×037a.

One can use a printer port to control machines. Status and control pins are incomplete, and there are a lot of bits with an inverter between the bits and the output connector pin. The DATA port is free and can be used cleanly. Depending on the language used, there are commands that can be used to access the ports. For example, to transfer the port to an output port using a "C" command, one can use

```
outportb (port_address, data)
```

where "port_address" is the address of the destination, such as 378(Hex); "data" can be a byte of data or a variable of type char.

In BASIC language, to send data to the address, one can use

```
OUT 888, 0
```

where

888=378(H) is port_address

0=data value

As the eight digit DATA port is used as a combined representation, the output from each of the pins or digits is shown in Table 8.3. For example,

Table 8.3 Binary digits output

Input	Digit 8	Digit 7	Digit 6	Digit 5	Digit 4	Digit 3	Digit 2	Digit 1
0	0	0	0	0	0	0	0	0
1	0	0	0	0	0	0	0	1
2	0	0	0	0	0	0	1	0
3	0	0	0	0	0	0	1	1
4	0	0	0	0	0	1	0	0
5	0	0	0	0	0	1	0	1
6	0	0	0	0	0	1	1	0
7	0	0	0	0	0	1	1	1
...								

```
outportb (378, data)
```

When data is equal to 0, all the pins or digits are "0" or low, but when data is equal to 1, pin 1 or digit 1 will be "1" or high. When data is "7," pins 1, 2, and 3 will be high. One can thus use this logic to input a number to set the corresponding pins high.

As there are eight pins, one can control up to eight external devices (motors, relays, solenoids, etc.). To control more than eight devices, one can either use multiple parallel ports or use some or all of the control lines. One can also use a data demultiplexer (chip 74154), which converts four binary digits into four binary weighted digits to provide 16 outputs as shown in Figure 8.27. The demultiplexer is the inverse of the multiplexer, in that it takes a single data input and n address inputs. It has 2n outputs. The address input determines the data output having the same value as the data input. The other data outputs will

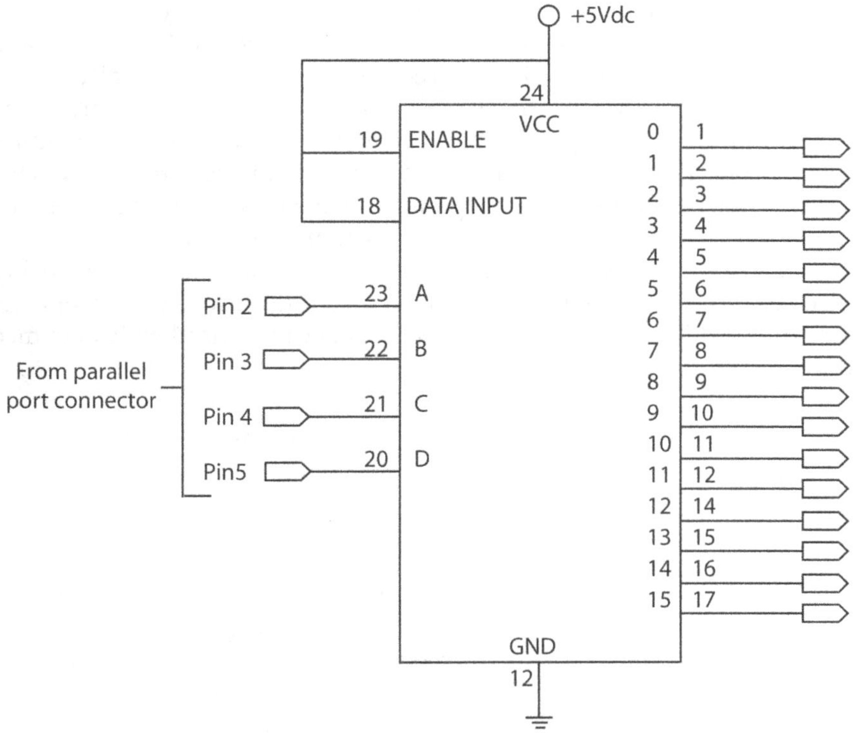

Figure 8.27 A Demultiplexer Chip 74154 provides 16 outputs. The left side could be the four digits from the parallel port.

have the value 0. A multiplexer is a combinatorial circuit that is given a certain number (usually a power of two) data inputs, let us say 2n, and n address inputs used as a binary number to select one of the data inputs. The multiplexer has a single output, which has the same value as the selected data input. Similarly, one can use the five status lines to send data back into the computer. In other words, sensors can send the status from these lines to the computer. For example, to read data from an input variable (C):

```
data_variable = inportb(port_address)
byte is transferred into the variable ''data_variable.''
In BASIC language:
Y = INP(888)
```

where

Y will store the decimal value of the port.
888 represents the decimal address of the port.

Also for multiple sensors, chip 74150 multiplexer can be used.

One could also use a breakout box to simplify troubleshooting in parallel port connections. It consists of a few LEDs, resistors, jumpers mounted on a printed circuit board (PCB), with a 25-pin male sub-D connector so that one can just connect it to the parallel port for testing. An I/O buffer can be provided by chip 74367 hex buffer driver to protect the parallel port from damage.

To control the power to a device either in the on/off mode or continuous mode, one can connect the output of the parallel ports to transistor switches. A transistor allows it to function as an amplifier or a switch. This is accomplished by using a small amount of electricity to control a gate on a much larger supply of electricity, much like turning a valve to control a supply of water. As the parallel port can only source and sink a few milliamperes and can mostly work using transistor–transistor logic (TTL) at 5 V, the outputs have to be buffered before it can be connected to any peripheral device that needs switches with high current and voltage. Stepper motors need switches at high frequencies.

As shown in Figure 8.28, one can turn a motor, light, or solenoid on/off by using a relay. One can also select different voltage levels to control motor speed or reverse the direction of motor rotation. In addition to relays, transistors can be used instead of mechanical relays to provide more durable service.

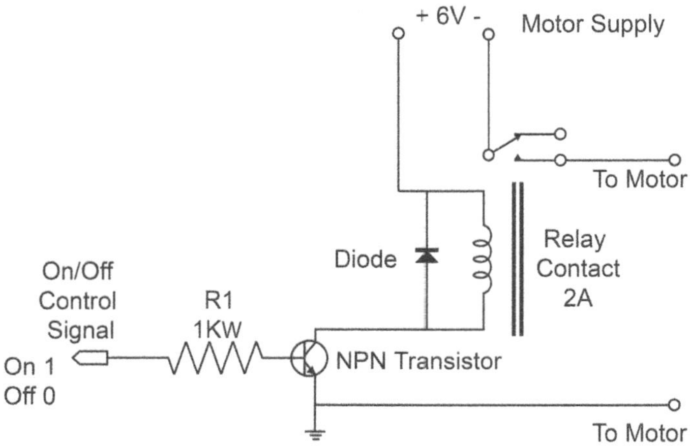

Figure 8.28 The parallel port output is connected to a relay to control the power to a device.

To use a parallel port to control a stepper motor, in general, at least two binary digits are needed to control the stepper motor: they individually control the "step" and "direction." For example, the step line of the driver is connected to the first output bit of the parallel port, and the direction line is connected to the following bit. When the software outputs the decimal number 2 (00000010) to the parallel port, then the driver step line receives a 0, and the direction line of the driver receives a 1. If the number 3 is pumped out, then the output will be 00000011. Consequentially, the motor shaft will turn one step whenever the motor driver receives a pulse. The output is then changed again to 2. When nothing happens, the motion is paused. After that, the output is set to 3, and the motor will turn one step again. If the number 1 (0000001) is then pumped out, the shaft will turn one step in the opposite direction. It turns in the opposite direction because the second digit is now a zero. The direction line is now low. The computer pumps out the number 3 (00000011) over and over again. In between these, it is pumping out the number 2 and then await or delay for a moment. Hence, 3 then 2, pause then again 3, and so on. The stepper motor will continuously rotate. The delay will set the speed of the stepper motor.

Example 8.1: Maneuvering a toy car using a PC parallel port

This project integrates the CAD/CAM program developed in C programming language with a PC and parallel port to maneuver a toy car. When a PC sends data to a car or a printer or any other device using a parallel port, it sends eight bits of data (which is equal to one byte) at a time. These eight bits are transmitted parallel to each other. This is quite the opposite to what happens in a serial port where the same eight bits are transmitted serially in a single row through a serial port. The following are the general conventions used:

1. An electrical "high" on a pin or line is high, the value of which can be in the range of +2.4 to +5 V.
2. An electrical "low" is low, whose value can be in the range of 0 to +0.8 V.
3. A data high is designated as a 1; a data low is designated as a 0.
4. The connection between data and pins is said to be direct if a data 1 is associated with an electrical high. It is said to be inverted if data 1 is associated with low.
5. An overall connection (data to TTL to data) is considered direct if outputting a 1 produces a 1 on input at the other end, or inverted if outputting a 1 produces a 0 on the other end.
6. The prefix "–" (or a line drawn over the name) implies that the signal is "active low"; that is, that when the signal is in its active state when electrically low. The "+" prefix means "active high," just the opposite.

The DATA ports are labeled from D0 to D7 with D0 being the least significant bit and D7 the most significant. All of the data out bits are direct and are not inverted. The control out bits are labeled C0–C3 (which go to pins) and C4 (for IRQ enable). The status bits are labeled from S3 to S7 (corresponding to data and CPU bit positions). Generally, suffixes C0 to C3 and S3 to S7 are designated with a "+" or "–" to tell whether or not the bits are inverted as compared to the output or input pin with which it is associated.

The design consisted of a voltage supply and relay switches. A diagram of the new circuit is provided below in Figure 8.29. The new circuit would provide all the functions needed.

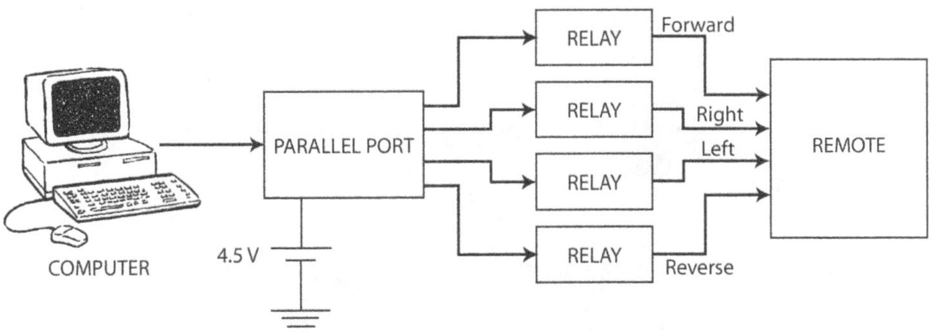

Figure 8.29 Diagram of the revised circuit.

The idea of this circuit is that the relay required 6 V to close the circuit in the remote. There is 4.5 V across each coil in each relay when the parallel port data lines are at low signal "0."

When a data line became active or changed to high signal "1," the data line was raised at a potential of 2.15 V. The data lines were wired in series with the 4.5 V source. The total voltage across the corresponding relay was then at 6.15 V, which is enough to activate the coil and latch the switch. This was a very simple solution. Any combination of relays could be activated allowing the car to turn and go forward at the same time or other combinations that are reasonable. Apparently, the car could not go forward and reverse at the same time. This may cause damage to the car.

Wires were soldered to the contacts of these manual switches and connected to the relays. The manual switches were left in working condition for backup. A simple diagram of the wiring of the remote is below in Figure 8.30. Each relay is connected to one of the functions of the remote (forward, reverse, right, or left). The remote was provided standard with two dual manual switches. The radio frequency of the remote control was 49 MHz. The range of the remote was 65 ft or 20 m. The circuitry of the remote operated from a 9 V battery.

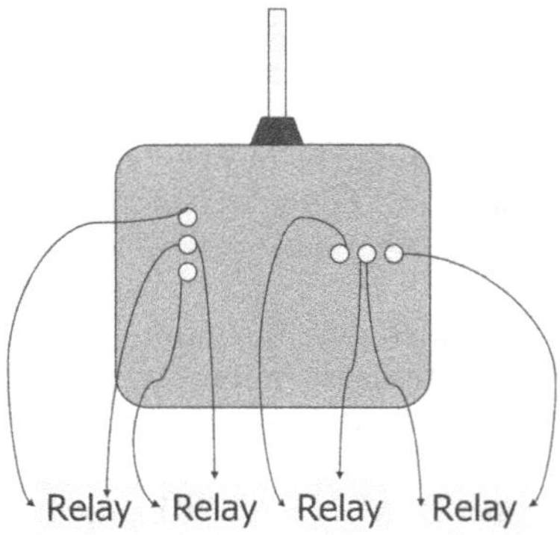

Figure 8.30 Simple wiring diagram of the remote control.

The circuit was built and tested. It was a success and performed as it was designed to perform.

A picture of the actual circuit with the remote and car is shown in Figure 8.31.

Figure 8.31 Picture of the setup of the circuit with computer and remote.

C programming language has been used for controlling the data transfer through the parallel port. The C program involves identifying the address of the DATA port and activating the required data line alone, for example, if one needs to activate the least significant bit, then the concerned bit alone should be modified without affecting the other bits. DATA 0×03bc is the address of the DATA port of the parallel port, and the address is in the Hexadecimal format. "Data" is the variable name that holds the value that can be assigned to the parallel port. "Outportb" is the command used for this purpose, whose syntax will be "outportb"(DATA, data). "Data" can be set to a certain value and reset to 0 at any time. Once the data is set to some nonzero value (i.e., 1), the car moves in the direction according to the circuitry and comes to the stationary position if reset (i.e., set to 0). "Delay" command is used to prolong an action; if a data line is set to "high" or "1," the delay is the length of time in which the data line stays in high, etc. A sample C program to control the motion of the car is listed below:

```
#include<stdio.h>
#include<dos.h>
#define DATA 0 x 03bc f the DATA port
int data;
void main(void)
{
int n;
outportb(DATA,data);
delay(25);
data = 0 x 0c;
outportb(DATA,data);
delay(25);
data = 0 x 01;
outportb(DATA,data);
delay(300);
data = 0 x 02;
outportb(DATA,data);
delay(250);
data = 0 x 00;
outportb(DATA,data);
delay(30);
data = 0 x 04;
outportb(DATA,data);
delay(250);
data = 0 x 00;
outportb(DATA,data);
delay(30);
data = 0 x 08;
outportb(DATA,data);
delay(250);}
```

The only concerned data bit is set or reset, using the command output_bit (bit number, bit value). The project was completed, and the desired results were obtained. As it only controls four motors, the motion was not as smooth as expected. For example, if a motor is stopped, the motion of the car is not, due to inertia. However, this project demonstrated the concept of using a parallel port to control four motors of an external device.

Example 8.2: Assembly and automation of a 2D plotter (courtesy of Jake Strait and Joe Dersch)

The objective of this project is to automate a 2D plotter using a parallel port. The project would use the data lines on the parallel port as digital 5 V signals. To ensure successful control of the port, a parallel port tester was constructed. It was designed so that the 5 V signal would pass through an LED and light it when that data line was activated. Figure 8.32 shows the circuit for the tester. The eight data lines are on pins 2 through 9. Current runs from a pin, through the LED and a resistor load, to ground, which is available on pins 18 through 25.

For the plotter, a stepper motor was selected. The EDE1200, a unipolar driver chip, was selected to operate the phases of the motor. Figure 8.33 shows digital logic I/O of the chip. The chip is capable of two modes of operation: STEP and RUN. The RUN mode does not count steps and runs the motor in the indicated direction at a speed indicated by the speed control. The STEP mode is what the group used. In the STEP mode, a logic

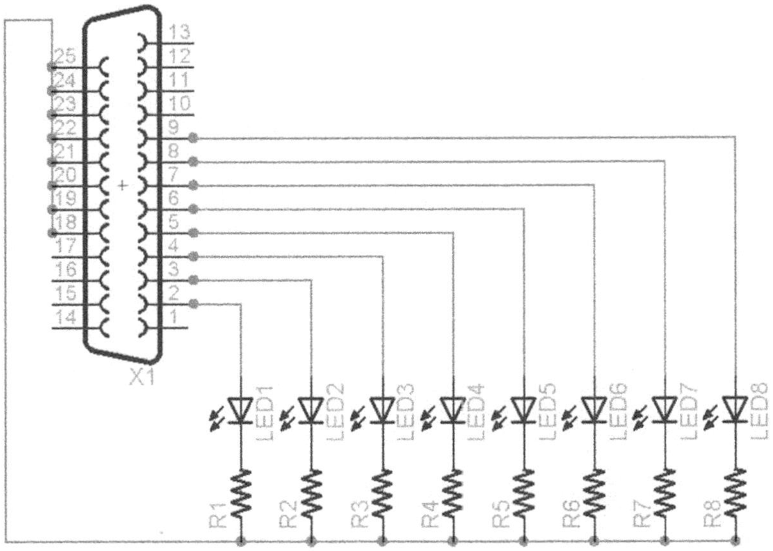

Figure 8.32 Parallel port tester schematic

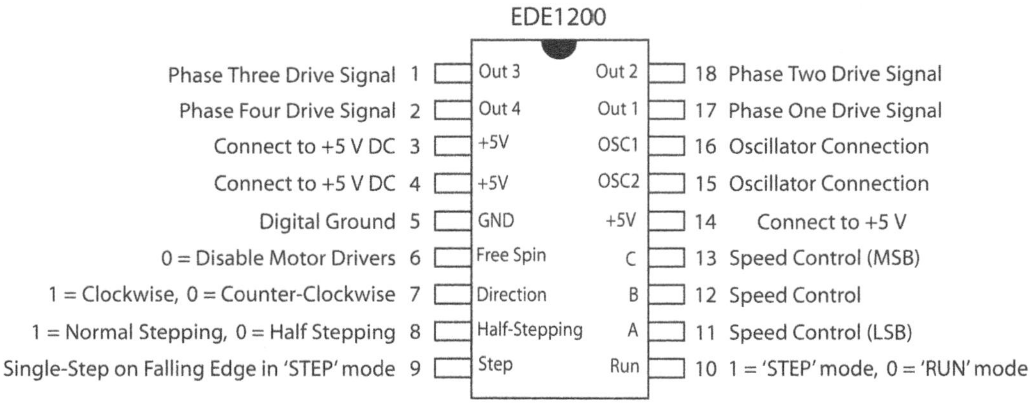

Figure 8.33 Pin I/O on the EDE1200.

transition from high to low causes the chip to step the motor by changing the phase signals. A data line from the parallel port is connected to this pin. Pins 7 and 8 were also connected to data lines so that the program could select the direction and the half-stepping or full-stepping mode. A resonator crystal must be connected between pins 15 and 16 for operation in the step mode. The pulsing of the oscillator cues the integrated circuit (IC) to check for a change in the status of the pins. Initially, this was omitted on the first prototype board and then corrected.

In order to turn the motor, each phase must be turned on in order. To find this firing order, 5 V was connected to the common leads of the motor, which were distinguished by their twisted color pair. Wires were then grounded to determine their influence. Black was grounded, which caused the center rotor to turn to face that phase winding. While holding black to ground, blue was grounded, which caused a slight clockwise movement. With black still grounded, red was grounded, and a counterclockwise movement was the result. Grounding black and white caused a larger random movement in either direction. From this series of tests, the firing order had been determined to be black, blue, white, red, and then repeated. Physically, the winding orientation looks like Figure 8.34 and demonstrates how that firing order causes the rotor to follow the energized winding clockwise.

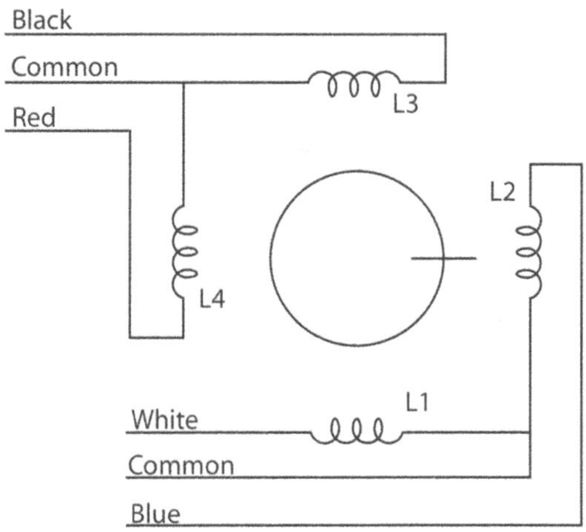

Figure 8.34 The stepper motor windings and corresponding wire colors.

Before constructing a board that would control three stepper motors and a solenoid, it is necessary to prove it could control a single stepper motor. After consulting the EDE1200 datasheet, the circuit in Figure 8.35 was constructed.

A 5 V power supply provided the voltage to drive the stepper motors as well as the source voltage for the IC chip. The grounds of the parallel port and supply were tied together. Otherwise, it was possible to get a voltage discrepancy between the two grounds. Pins on the chip were tied to ground or 5 V as required. The parts of the chip one wanted to control were connected to the data lines of the parallel port.

The IC cannot source enough current to operate the stepper motors alone. Instead, the phase drive signals coming from the chip operate the transistors. The transistors allow current to flow through the winding when current from the IC is delivered to the base of the transistor. The diodes in the circuit allow inductive current from the windings to feedback and diminish rather than induce problems.

The mechanical portion had been completed by this time, and this board was used to drive the main axis. The single motor did not appear to have enough power by itself to drive the pen and bridge mechanism along the rail pair of the main axis. Two motors were needed to overcome friction. The second motor of the main axis was installed while the final board was produced.

The final board was produced by applying the circuit design of the prototype board to two ICs. One IC would control the motor of the secondary axis, and the other would control two motors wired in parallel on the main axis. The solenoid of the pen was wired through a transistor that the parallel port would control directly. Figure 8.36 shows the schematic of this final circuit.

The addition of the second motor did not help drive the main axis any harder. The motor appeared to be weaker when wired in parallel. Upon closer inspection, it was determined that the same current was flowing to the single motor as was flowing to the motor pair. The current was expected to double for the motor pair. Something appeared to be limiting the current to the stepper motors. The specification on the transistor said the base junction was capable of receiving 0.4 A. The 1 kV resistor suggested by the EDE1200 spec was only delivering 5 mA to the base of the transistor and restricting the

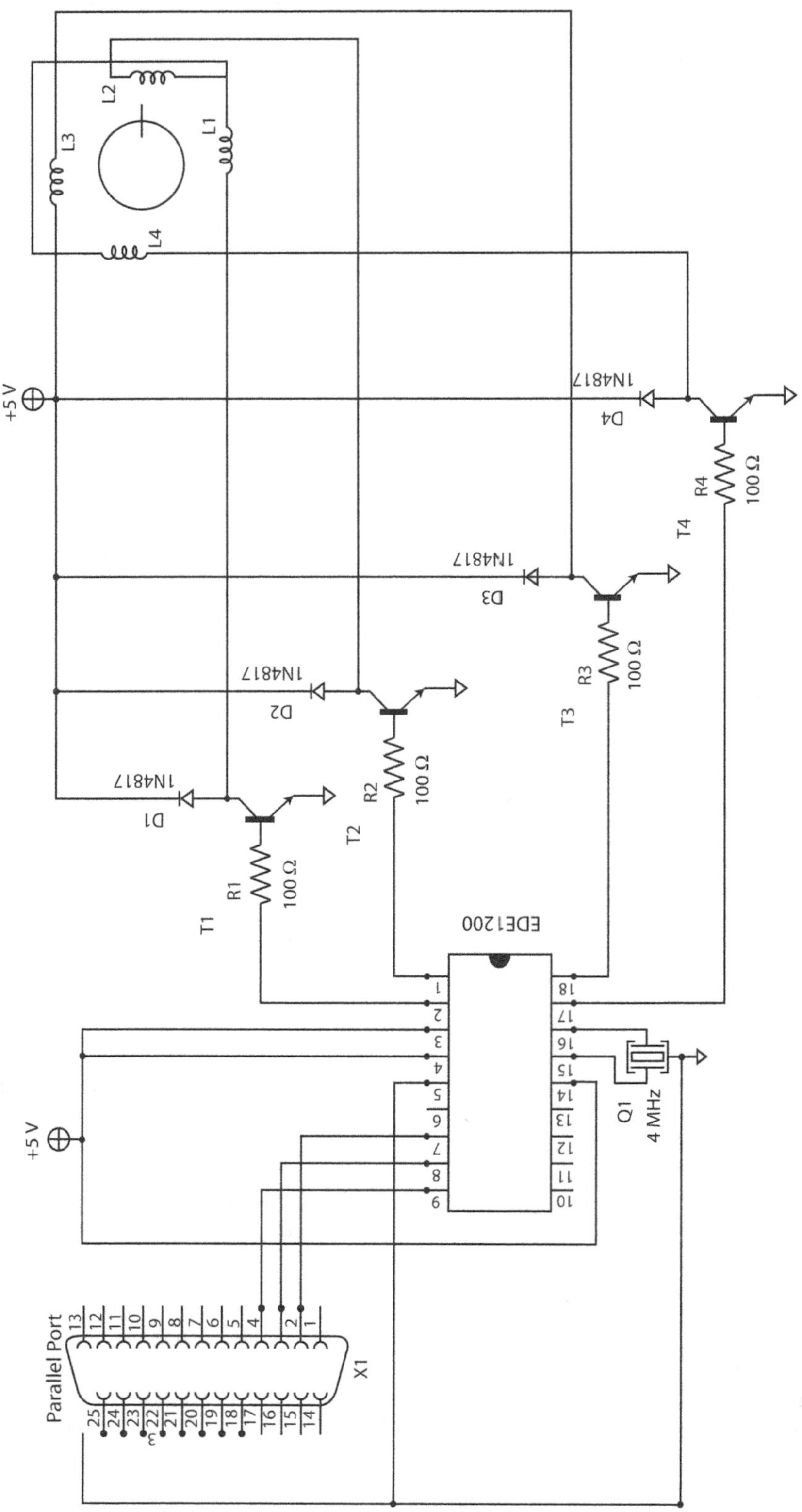

Figure 8.35 The circuit of the prototype board to drive one stepper motor.

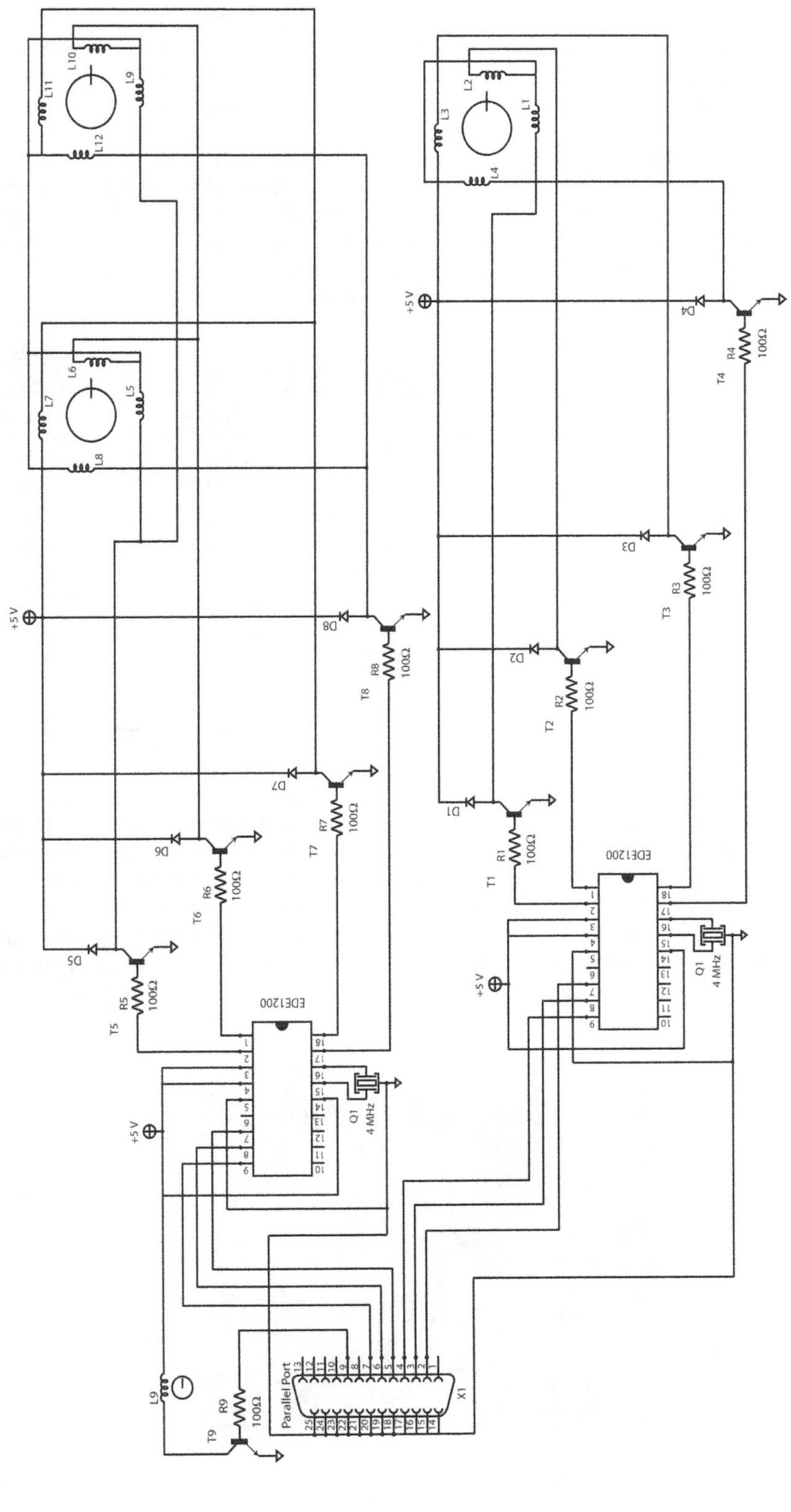

Figure 8.36 The schematic of the final circuit controlling three motors and one solenoid.

current flow through the collector–emitter junction of the transistor. The 1 kV resistor was replaced by a 10 V resistor and motor power increased dramatically.

With the increased current flow, the transistors operated at a higher temperature, which could inadvertently burn someone. To reduce the temperature, large heat sinks were attached to the transistors, and a computer fan was rigged to blow air across the board.

Initially, it was thought that the motor driver chip fired each phase in order so that it would half step the motor by activating signal 1, signal 2, signal 3, and signal 4 in that order. After plotting lines were started, it appeared that the motors were skipping a step and then taking a large step. It was discovered that the IC operated each phase in the order of Table 8.4. After determining this, the wires of the motors were wired in the correct firing order to the corresponding phase signal.

Table 8.4 Motor driver phase output signals (1 is on, 0 is off)

	Phase 1	Phase 2	Phase 3	Phase 4
Step progression	**1**	0	0	0
⇓	**1**	0	0	**1**
	0	0	0	**1**
	0	**1**	0	**1**
	0	**1**	0	0
	0	**1**	**1**	0
	0	0	**1**	0
	1	0	**1**	0
Phase wired to:	Red	Blue	White	Black

The completed parallel port tester is shown in Figures 8.37 and 8.38. The tester was valuable in confirming that the program was controlling the parallel port. In one instance, it was useful to determine the port number of a separate PCI parallel port card. In another instance, the team was running Qbasic under Windows rather than DOS and discovered that Windows was not allowing the program to control the port as the lights on the tester were always lit. If the tester was to be made again, the male pins would be replaced with a female receptacle so that a parallel port cable can be used. With this tester, having access to the back of the computer was required at all times.

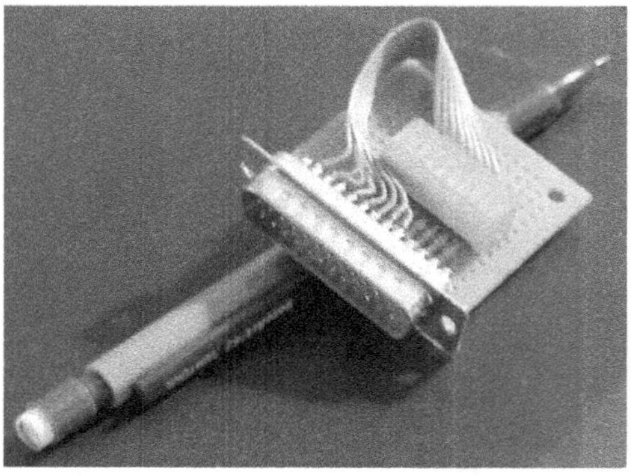

Figure 8.37 Parallel port tester.

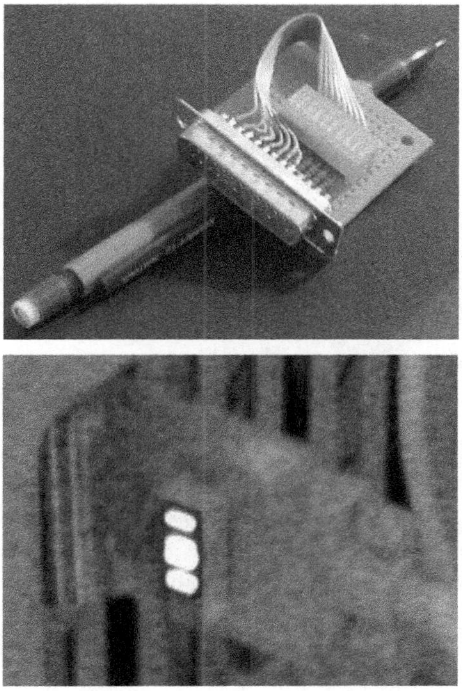

Figure 8.38 Lighting LEDs with the parallel port.

The cannibalized prototype board is shown in Figure 8.39. It was helpful in determining the resonator requirement. It was also good practice for the final board. The final board had better component placement and wiring after the learning experience provided by this board.

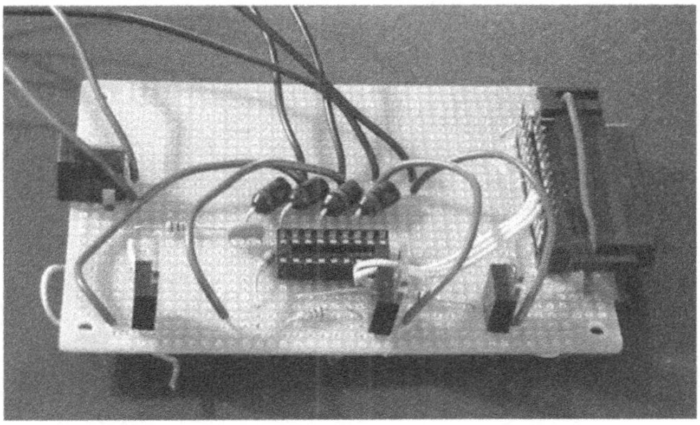

Figure 8.39 Cannibalized prototype board.

The final circuit board is shown in Figure 8.40. It functioned flawlessly as expected. The lead wires to the motor and solenoid have gone through a series of reconnections, which ruined the connectors that were used. Better wire identification could have helped prevent that, but no additional colors of wire were purchased. It would not have been an issue except for the amount of rewiring that was required to get the motor phases correct.

Figure 8.40 Final control board with the power supply unplugged.

While half stepping, the motor did increase resolution, which was not desirable. Full-stepping mode operates two phases all the time as shown in Table 8.5. This produces a stronger pull on the rotor and helps improve accuracy. To improve accuracy without sacrificing resolution, one should purchase stepper motors with more steps per revolution. This project chose motors with 48 steps per revolution—mostly for budgetary reasons.

Table 8.5 Full-step mode operation

	Phase 1	Phase 2	Phase 3	Phase 4
Step progression	**1**	0	0	**1**
⇓	0	**1**	0	**1**
	0	**1**	**1**	0
	1	0	**1**	0

Figure 8.41 shows a completed plot by the machine. There is room for improvement with the resolution of the steps and the tolerances in the pen mechanism, which would improve the quality of the plot. Nevertheless, this project met the goals and objectives put forth at the beginning of the semester.

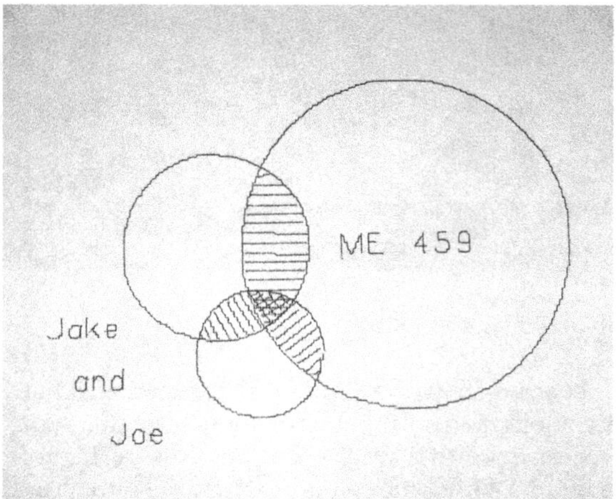

Figure 8.41 Completed drawing plot.

Review problems

1. Which is true about parallel ports?
   - A. The port is composed of five control lines, eight status lines, and five data lines.
   - B. The port is commonly used on the back of a PC as a D-type 25-pin male connector.
   - C. The voltage of the port is often greater than 5 V DC.
   - D. A printer port is a parallel port.
   - E. COM1, COM2, COM3, and COM4 are all parallel ports.
2. Explain the difference between the serial port and the parallel port.
3. Match the analyzers with their applications.

Analyzers	Applications
1. Timer	a. Decodes text information rapidly and accurately
2. Counter	b. Count number of parts
3. Bar code reader	c. It is cost-effective to be used in large system of 100 or more relays
4. Programmable logic controller	d. To count down/up lime for an operation to be completed

4. Describe the three tasks or processes performed by analyzers.
5. Match the operation and the corresponding control level:

1. Replace a file on the disk in WordPerfect equivalent file in Word	a. Hardware level
2. Save a file to the disk	b. BIOS level
3. Save data to a single sector on disk	c. DOS operating system level
4. Turn on the disk motor	d. Application level

6. Convert the following decimal values into hexadecimal ones: 1000(D). And convert the following hexadecimal values into decimal ones: B35F(H).
7. Convert the HEX number 70B.C9 to a decimal number.
8. Match the following:

1. 00001000 (binary)	a. 032 (octal)
2. 014 (octal)	b. 8 (hexadecimal)
3. 1A (hexadecimal)	c. C (hexadecimal)
4. 360 (octal)	d. 003 (octal)
5. 00000011 (binary)	e. 11110000 (binary)
	f. 00001010 (binary)

9. Which of the following in Figure 8.42 are described correctly?
   - A. 1 and 3 are normally open, 2 and 4 are normally closed.
   - B. 1 and 2 are normally open, 3 and 4 are normally closed.

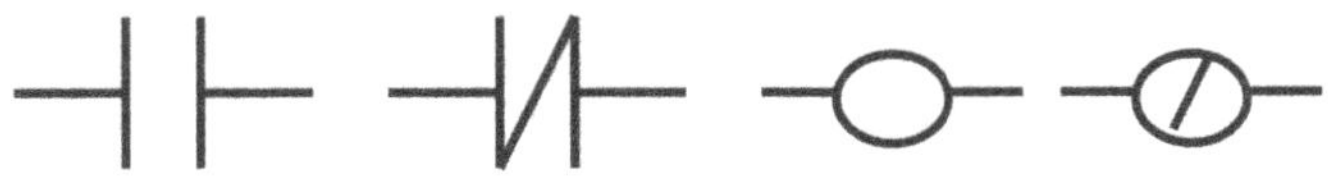

Figure 8.42 PLC symbols.

C. 1, 2, and 3 are normally open, 4 is normally closed.
D. 1 is normally open, 2, 3, and 4 are normally closed.

10. How to use PLC to realize the basic logic unit (draw the ladder diagrams)? Suppose A and B are two inputs and C is an output:

No.	Basic logic unit	Ladder diagram
1	**AND:** If both **A** and **B** are on, **C** is on. In all other conditions, **C** is off.	
2	**OR:** If **A** or **B** is on or **A** and **B** are both on, **C** is on. In all other conditions, **C** is off.	
3	**Exclusive-OR:** If only one of the two inputs is on, **C** is on. In all other conditions, **C** is off.	
4	**NAND:** If both **A** and **B** are on, **C** is off. In all other conditions, **C** is on.	
5	**NOR:** If **A** or **B** is on or **A** and **B** are both on, **C** is off. In all other conditions, **C** is on.	

11. Which of the following computer peripherals can often be connected to a computer using parallel ports? And which of the following computer peripherals can often be connected to a computer using serial ports?
    A. Printers
    B. Scanners
    C. Modems
    D. External hard drives
    E. Network adaptors
    F. PDAs (personal digital assistant)
    G. Digital cameras
12. Which of the following is wrong regarding serial ports?
    A. A single-way communication between a computer and a machine
    B. Male in gender, either 9 or 25 pins, slightly D-shaped
    C. May be used for a mouse, external modem, scanner, an internal or external infrared device, or other linking software
    D. Marked as LPT, or with a symbol representing 10101
13. Explain how to control more than eight devices with a parallel port and compute how many devices can be controlled with four and six inputs by using a demultiplexer.
14. Below are some descriptions about the port address. Which of the following is/are wrong?

  A. The parallel port has three commonly used base addresses.
  B. The computer can exchange information with controlled devices through the address.
  C. Demultiplexers are limited to two data signals.
  D. Printer port consists of three port addresses: data, status, and control port.
15. How do two to four demultiplexers enable work?
16. Which of the following is/are true?
  A. DATA port has two-way communication between the computer and machine.
  B. Using a parallel port, bits are transmitted one at a time.
  C. Data strobe is used in a parallel port to tell the printer that data is coming.
  D. There are only three types of parallel ports used: LPT1, LPT2, and LPT3.
  E. BASIC language uses a hexadecimal system.
17. Can the parallel port typically be used to control a device directly? Explain your answer.
18. If a data demultiplexer takes n binary weighted inputs, how many outputs can it provide?
  A. 4n
  B. 2n
  C. 2n
  D. 4n
  E. 8n

8.4 Mechanisms

> Man, alone, has the power to transform his thoughts into physical reality; man, alone, can dream and make his dreams come true.
>
> **Napoleon Hill**

This section helps answer the following questions:

- What types of mechanisms are used in automation?
- Are there any mechanisms that can be found in all forms of automation or are the mechanisms used dictated by the situation?
- What are the criteria for selecting a particular type of mechanism for an automated system? Are any particular types of mechanisms more reliable or prone to wear than others?
- Is there a systematic way of classifying various mechanisms for automation?

The definition of mechanism here according to Merriam-Webster is a piece of machinery or mechanical operation or action. Mechanisms for unique designs often seem to be a collection of off-the-shelf commercial items, whether it is an assortment of gears and links, or computing hardware. Mechanisms cover a broad area that ultimately provides the action or enables the action to take place for the automation sequence. The focus of this section is on some useful mechanisms specifically used for automated systems.

8.4.1 Mechanisms in automation

Mechanisms are used to convert between one type of motion and another [Erdman01]. Any machine can be looked on as a group of interconnected mechanisms, which convert one type of motion to a variety of other motions. In some automation projects, linkages and mechanisms outperform motors and are cheaper in cost. A distinct advantage of linkages over motors would be synchronization. As shown in Figure 8.43, the basic mechanisms can be classified into 2D and 3D, and there are lower-pair and higher-pair mechanisms [Yeh93, Huang94]. Some motions are linear, rotary, reciprocating, and oscillating. Then, there are modifiers, where the motions can be intermittent, constant, irregular, and regular.

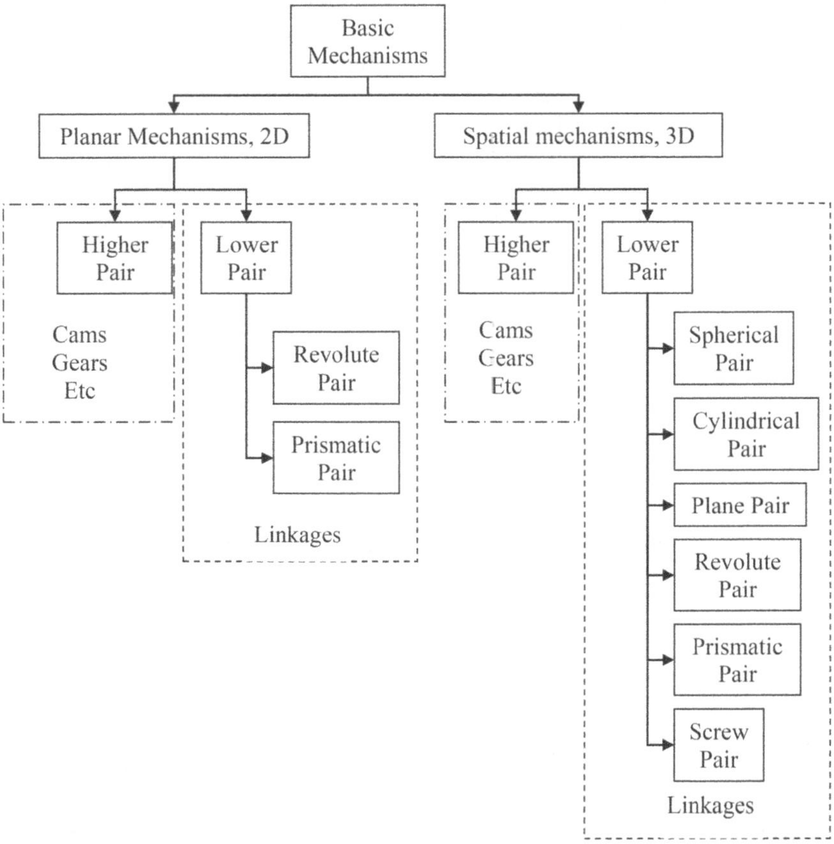

Figure 8.43 Classifications of mechanisms.

These changes may be able to convert rotary motion to straight line motion or to convert reciprocal (back and forth) motion to intermittent motion. They may also transform a fixed type of motion, for example by magnifying a linear motion or by slowing down a rotary motion.

To select a mechanism for automation, one must select one in parallel with an actuator. Mechanisms can transform motion like making rotary motion into linear motion, linear into the rotary, transfer rotary motion, and create oscillatory, intermittent, and irregular motion. One would select a mechanism based on the type of actuator to be used and the type of motion created. A function tree, as shown in Figure 8.44, can be used to organize the functions required in a prototype. One summarizes the required functions according to the objectives of the design. In this example, the designer selected three functions: cutting, holding, and transmitting. The corresponding devices, that is, cutting device, holding

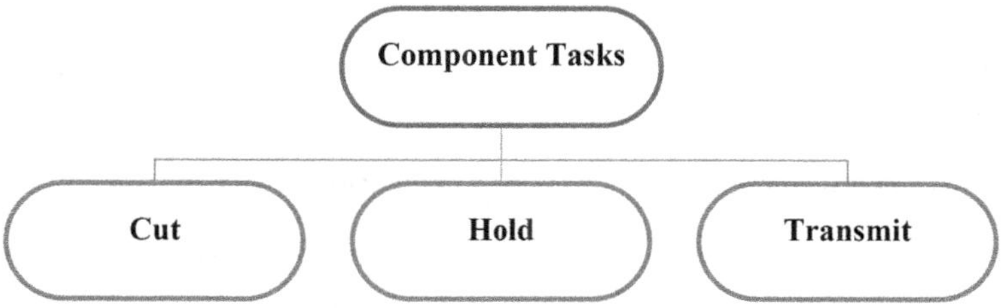

Figure 8.44 An example of a function tree which shows the function of a designed component.

device, and transmitting device will be needed. These devices, although independent, could also be closely related to each other.

Once the function is identified, one needs to design a way to implement the function. Figure 8.45 shows an example of a function tree to define a transmit function. Note that to implement the "transmit" function, if a fixture is proposed, a "hold" function is required. This design process may need several iterations to optimize the configuration of various devices. The function tree of the linear output motion is shown in Figure 8.46.

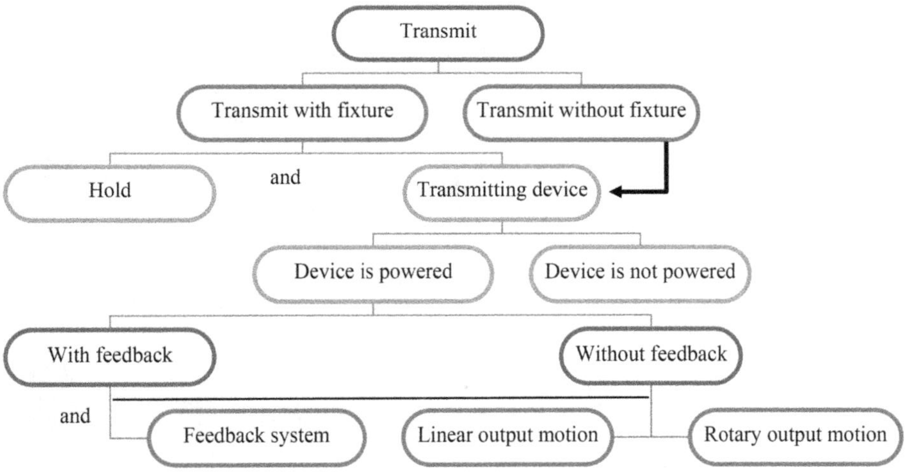

Figure 8.45 An example of "transmit" function.

Besides performance consideration, very often, the factors that dominate a mechanical design are cost and space constraints. The space constraints include transmitting direction and transmitting distance. One can define forward transfer when the vector of the axis of the input motion is the same as the vector of the axis of the output motion. The backward transfer is similar except that the purpose is to provide a returned motion. For example, a spring may be an option to return the solenoid back to the original position. The parallel transfer is when the vector of the axis of the input motion is parallel to the vector of the axis of the output motion. The angular transfer is when the vector of the axis of the input motion has an angle with the vector of the axis of the output motion.

Four basic types of transformations are specified: linear–linear, linear–rotary, rotary–linear, and rotary–rotary transformations. Figure 8.47 shows one of the subfunction trees of these transformations. The designer may select the input driver and

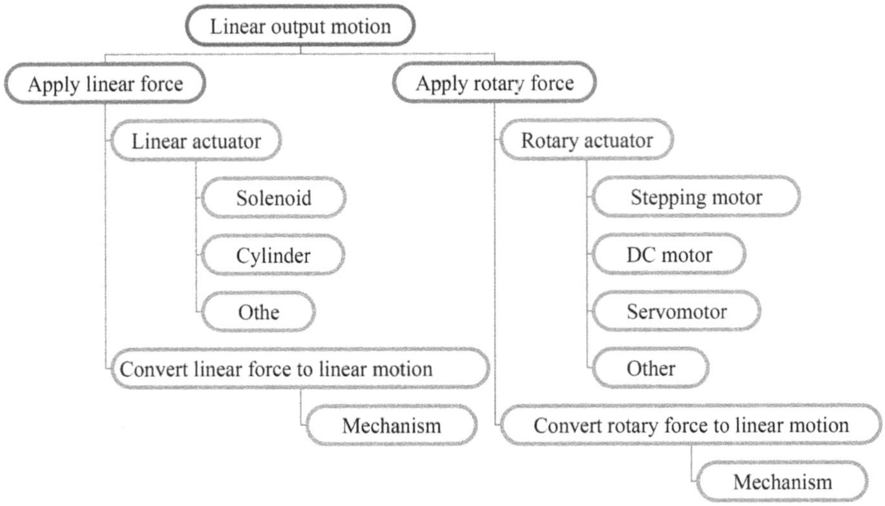

Figure 8.46 An example to produce function "linear output motion."

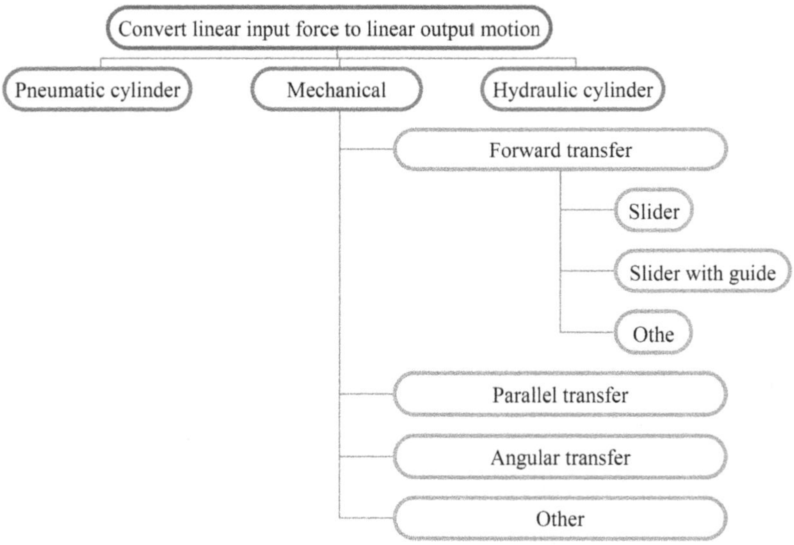

Figure 8.47 An example to produce function "convert linear input force to linear output motion."

output motion according to the design objectives and then follow these trees to search for the corresponding devices. The complexity of these devices may be expanded by selecting more steps of transformation. For example, instead of using a linear input power to linear output motion directly, the designer may use a linear–rotary transformation as the first step and use a rotary–linear transformation as the second step according to the space constraints. Therefore, the process is converted from a direct linear–linear transformation to a two-step linear–rotary and rotary–linear transformation. Table 8.6 shows some commonly used mechanisms and mechanical components for linear–linear transformation.

Table 8.6 Some commonly used mechanisms and mechanical components for linear–linear transformation

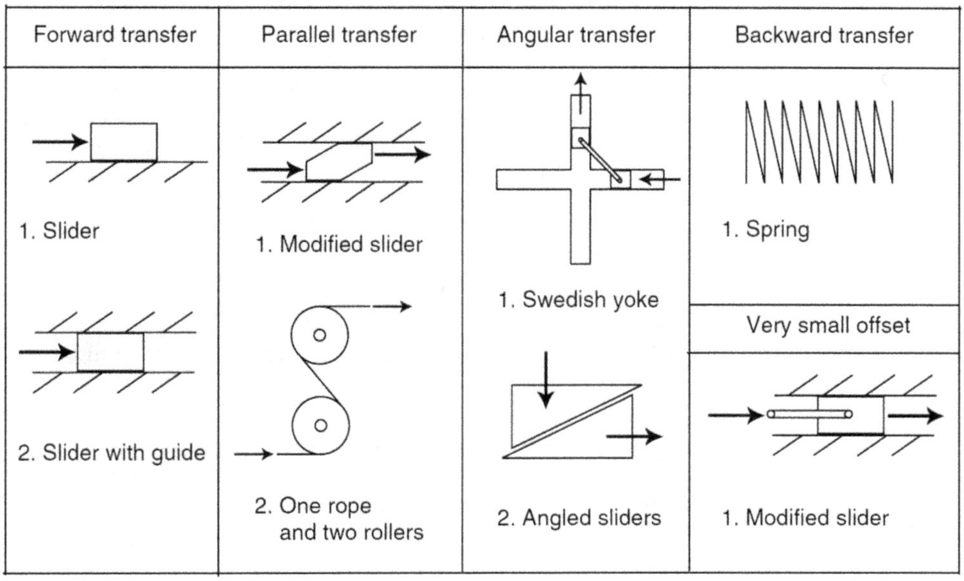

Table 8.7 shows a summary of the most commonly used power conversion options. Table 8.8 shows the same commonly used components for linear–rotary or rotary–linear transformation [RoyMech05] and Table 8.9 shows the rotary–rotary transformation.

Table 8.7 A summary of most commonly used power conversion option

Spatial directions	Linear input to linear output	Rotary input to rotary output
Forward (or advancing) transfer	1. Slider—for short stroke length 2. Slider with guide—long stroke	1. Shaft—for short or normal transfer 2. Shaft and bearings—for long or precise transfer
Parallel transfer	1. Modified slider—for short stroke 2. Slider with more than one rope and two rollers—for long stroke	1. Gear pair—strong, precise, and short but costly 2. Pulleys and Belt—for long transfer cheaper than gear pair 3. Four-bar linkage—cheapest but needs more designs
Angular transfer	1. Swedish yoke—for right angle 2. Angled slider—for any angle	1. Gear pair—precise and strong
Backward (or reversing) transfer	1. Spring—for low cost and small space 2. Using other power source—costly	1. Gear pair—simple, precise, and strong 2. Others—more parts, more complex
Very small offset	1. Modified slider—allow more tolerance	1. Coupling—to protect the driver, especially in high speed rotation

(Continued)

Table 8.7 (Continued) A summary of most commonly used power conversion option

Spatial directions	Linear input to linear output	Rotary input to rotary output
Others (more than one cycle)	1. Linear to rotary first, than rotary to linear last 2. Linear to linear first, than linear or rotary to linear last	1. Rotary to linear first, than linear or rotary to linear last 2. Rotary to rotary first, than rotary or linear to rotary last

Spatial directions	Rotary input to linear output	Linear input to rotary output
Forward (or advancing) transfer	1. Power screw set with guide—best for small space	(same as the left column)
Parallel transfer	1. Power screw set with guide—best for small space	(same as the left column)
Angular transfer	1. Pack and Pinion—for small space 2. Roller pair—cheaper and strong but not very precise 3. Slider–crank—for back and forth motion 4. Radio turning mechanism—for long or cycle motion 5. Cam mechanism—short stroke but strong and looks simple	(same as the left column)
Backward (or reversing) transfer	1. Power screw set with guide—best for small space	(same as the left column)
Very small offset	1. Lubricant—for low cost and small space 2. Others—allow more tolerance	(same as the left column)
Others (more than one cycle)	1. Rotary to rotary first, than rotary to linear last 2. Rotary to linear first, than linear or rotary to linear last	1. Linear to linear first, than linear or rotary to rotary last 2. Linear to rotary first, than linear or rotary to rotary last

Linear motion is the most basic of all motions. Linear motion is measured in speed, and direction. If a linear motion is used as a starting point of a system, actuators such as solenoids and cylinders are commonly used. Rotary motion is motion in a circle. If it is used as the starting point of a system, motors such as DC/AC motors, stepper motors, and servomotors are used. Rotary motion is measured in either angular velocity, the number of degrees turned in a given time, or in revolutions per minute (rpm). The direction of the turn, either clockwise or counterclockwise, is also part of the measurement of rotary motion. The strength of rotary motion is known as the torque or turning force. Reciprocating motion is back-and-forth motion. Reciprocating motion is measured by the distance between the two extremes of motion and by its cycle time. If it is used as the starting point of a system, cylinders or solenoids can be used, just like linear motion. Oscillation is the back-and-forth motion about a pivot point. It is measured in terms of both the angle of amplitude and the cycle time for one complete cycle. Oscillation tends to be an ending point for a mechanism rather than the starting point. However, some mechanisms are available to convert or transform oscillations.

Intermittent motion is a motion that starts and stops regularly. For example, in assembly stations, a table can be moved by a Geneva mechanism one station at a time and then

Table 8.8 Some commonly used mechanisms and mechanical components for linear–rotary or rotary–linear transformations

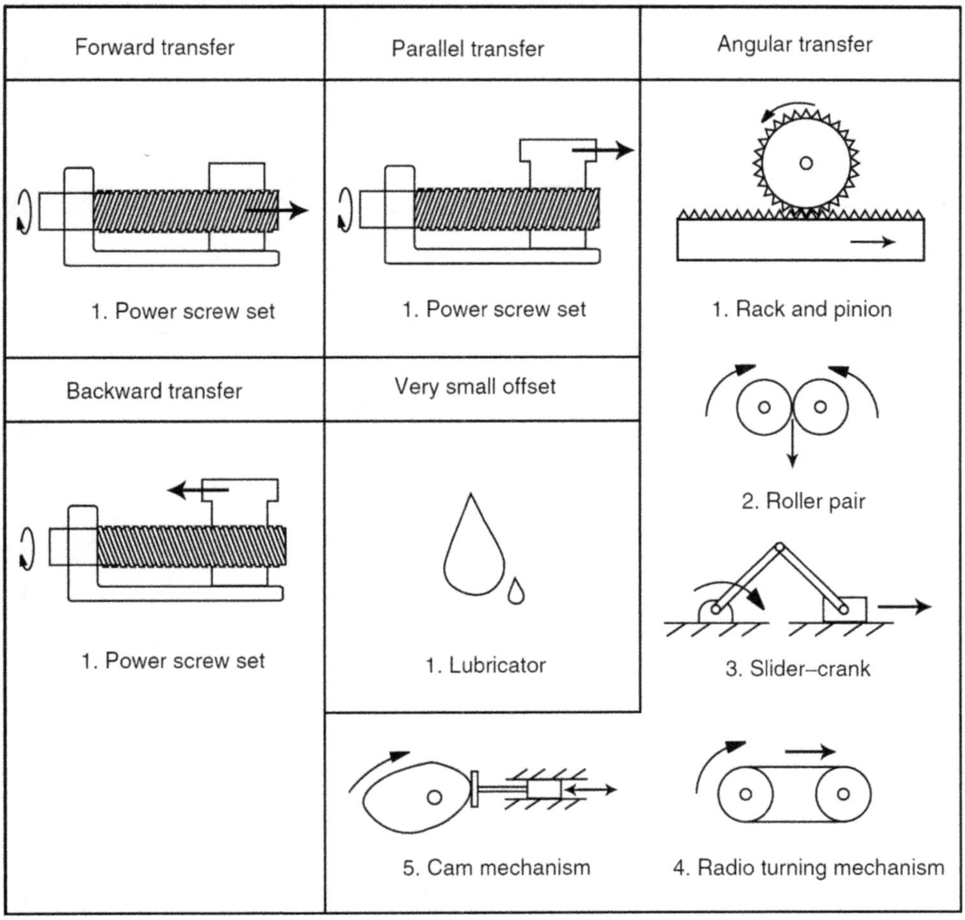

held stationary while the part is being assembled. Intermittent motion is usually the end result of a mechanism rather than the starting point for conversion. Irregular motion is a motion that has no obvious pattern to its movement. Irregular motion is usually created using a cam or series of cams. Irregular motion is not often used as the starting point for a mechanism. The conversion of various motions is summarized in Table 8.10.

8.4.2 Applications and selection of mechanisms

There are many other ways to convert motions with mechanisms. Some of the commonly used mechanisms and mechanical components are further illustrated here.

8.4.2.1 Linear or reciprocating input, linear output

Sometimes with linear input and output, one can gain a mechanical advantage. With the use of the inclined plane as shown in Figure 8.48, a given load can be overcome with a smaller force.

Using an inclined plane requires a smaller force exerted through a greater distance to do a certain amount of work. Letting F represent the force required to raise a given weight

Table 8.9 Some commonly used mechanisms and mechanical components for rotary–rotary transformation

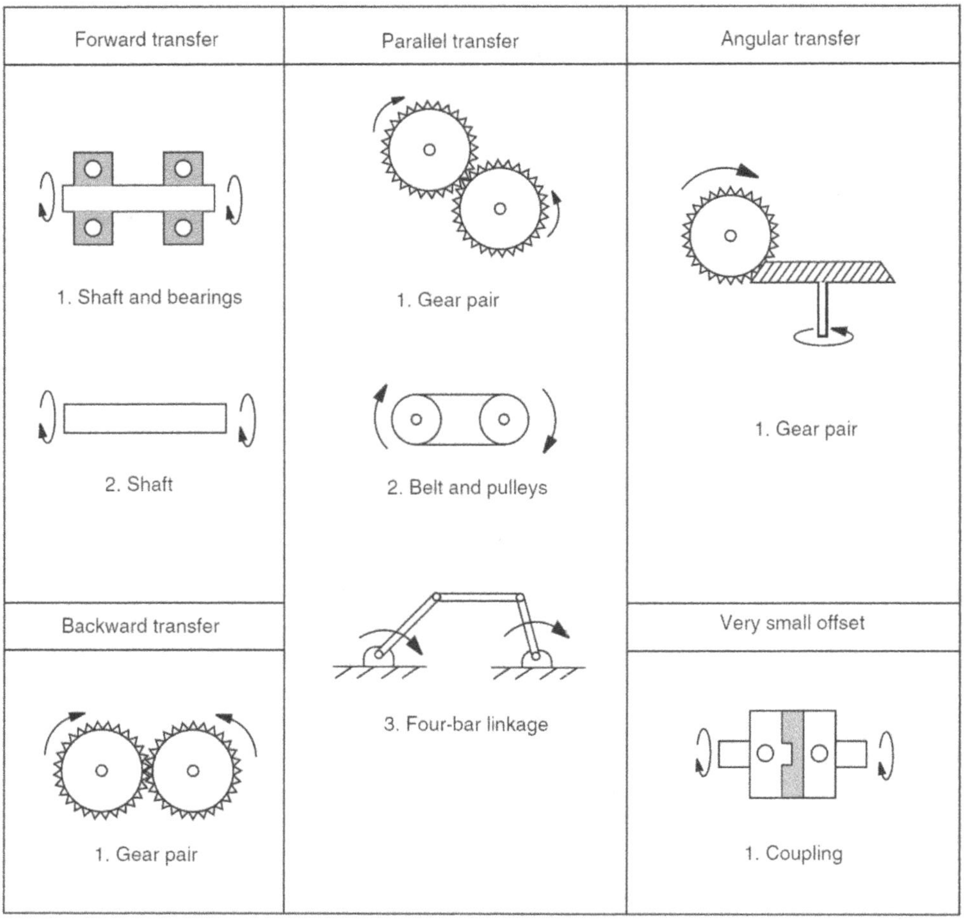

on the inclined plane with inclined slope length l and height h, and W the weight to be raised, the following proportion can be obtained:

$$F/W = h/l \tag{8.1}$$

One of the most common applications of the principle of the inclined plane is in the screw jack as shown in Figure 8.49, which is used to overcome a heavy pressure or raise a heavy weight of W by a much smaller force F applied at the handle. Note this is rotary input and linear motion output. R represents the length of the handle and P the pitch of the screw, or the distance advanced in one complete turn. The following relationship can hold:

$$F(2\pi R) = WP \tag{8.2}$$

$$F = WP/(2\pi R) \tag{8.3}$$

Table 8.10 Conversion between various motions

Conversion from	Conversion to: Linear motion	Rotary motion	Reciprocating motion	Oscillation	Intermittent motion	Irregular motion
Linear motion	Lever, inclined plane					
Rotary motion		Gear sets, belts and pulleys, chain, etc.	Piston, geared mechanism, Cardan gear	Linkages	Geneva	Cam, linkages
Reciprocating motion		Piston, scotch yoke mechanism, etc.	Lever, inclined plane, linkages, etc.	Rack and pinion, piston, scotch yoke mechanism, linkages, etc.	Ratchet	Cam, linkages
Oscillation		Crank–rocker	Cam, crank–rocker	Linkages	Ratchet	Cam

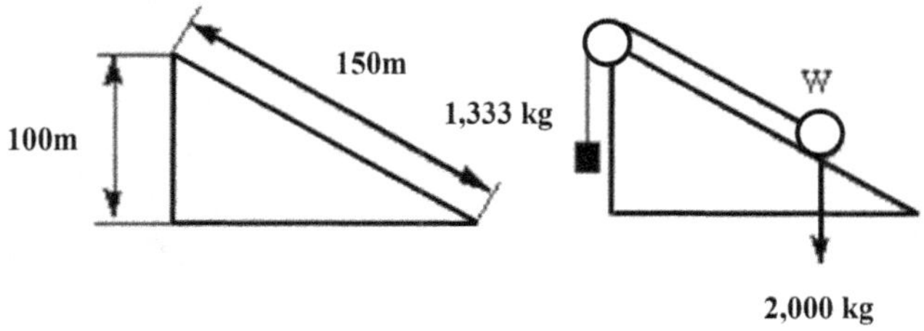

Figure 8.48 Inclined plane.

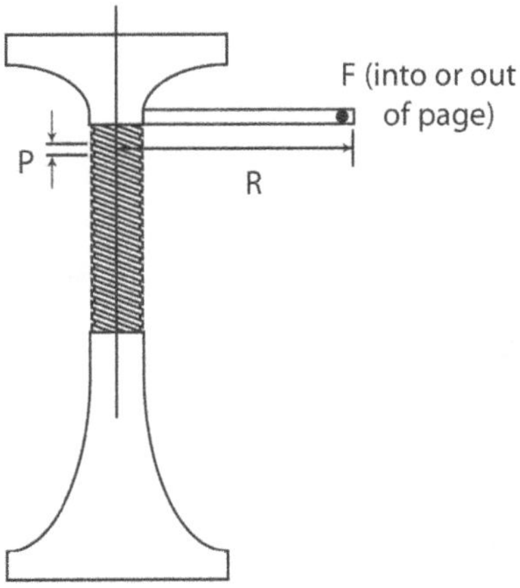

Figure 8.49 The screw jack.

Although levers, as shown in Figure 8.50, are not exactly linear, they are essential mechanisms to change the amount, the strength, and the direction of movement. The position of the force and the load are interchangeable, and by moving them to different points on the lever, different effects can be produced.

Figure 8.50 A lever can be used to magnify force or velocity.

8.4.2.2 Rotary input, rotary output

The commonly used mechanisms for rotary input and rotary output are gear sets. Gears are also used to change speed in rotational movement. As shown in Figure 8.51 [Zhang07], they can be used to increase or decrease output torque from a motor by using a reduction gearbox. They can also be used to increase or reduce the output speed for a particular application. Sometimes, offset output shaft position can be achieved using a gear set as shown in Figure 8.51b, when the input is A and output is C. Gear sets can also be used to generate or synchronize multiple outputs as shown in Figure 8.51b, when the input is B and outputs are A and C. Depending on the gear set design, one can also use a gear set to provide multiple outputs at the same speed. Sometimes, a gear set is used to change the direction of the output as shown in Figure 8.51a. The diameters of the gears can be designed to output the proper speed.

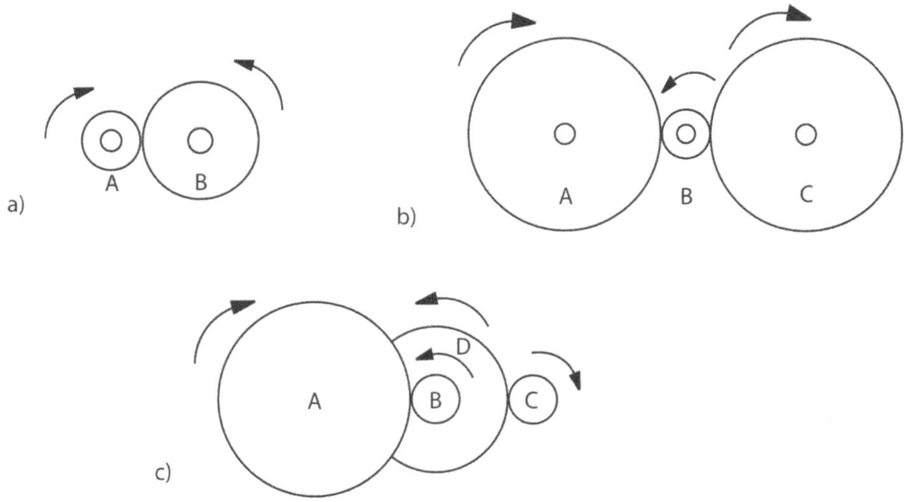

Figure 8.51 Gear sets can be used for several different tasks.

Belts and pulleys, as shown in Figure 8.52, can also be used for a similar purpose, except that the precision may not be as good unless timing belts are used, but the belts and pulleys can provide much wider offset than a gear set. They can also be used to increase output torque, increase/reduce speed, offset output position, and generate or synchronize multiple outputs.

Chains can also be used to connect gears. They work in a similar way to pulleys but with a positive drive rather than a reliance on friction. Gears that are connected by a chain turn in the same direction, unlike gears that mesh against each other.

Figure 8.52 Belts and pulleys system can be used for many rotary input and rotary output applications.

8.4.2.3 Rotary input, reciprocating output

Reciprocating motion can be converted into linear motion by pistons as shown in Figure 8.53, oscillation by linkages in Figures 8.54 and 8.55, and intermittent motion by ratchets as shown in Figure 8.69. This piston is used to convert rotary motion into reciprocating motion and vice versa. Notice how the speed of the piston changes. The piston starts from one end and increases its speed. It reaches maximum speed in the middle of its travel and then gradually slows down until it reaches the end of its travel. The oscillation mechanism and the Scotch yoke mechanism function like the piston or crank–slider mechanism.

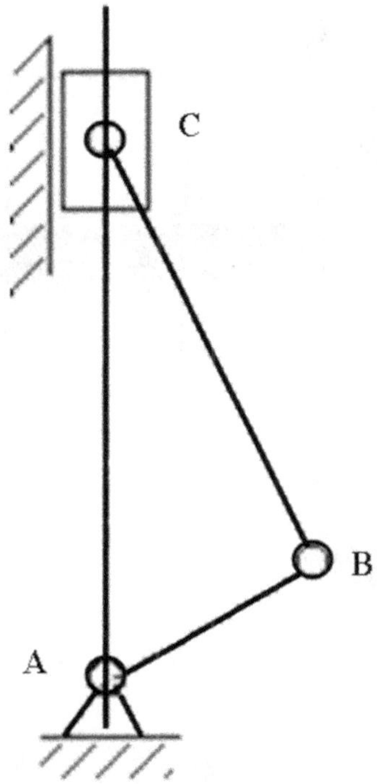

Figure 8.53 A crank–slider mechanism.

The rack and pinion, as shown in Figure 8.56, can also be used to convert energy between rotary and linear motions. A good example is the auto-steering system, which rotates a rack and pinion. As the gear turns, it slides the rack either to the right or left, depending on which way to turn the wheel. A rack and pinion can convert motion from rotary to linear and from linear to rotary. A rack and pinion is commonly used in the

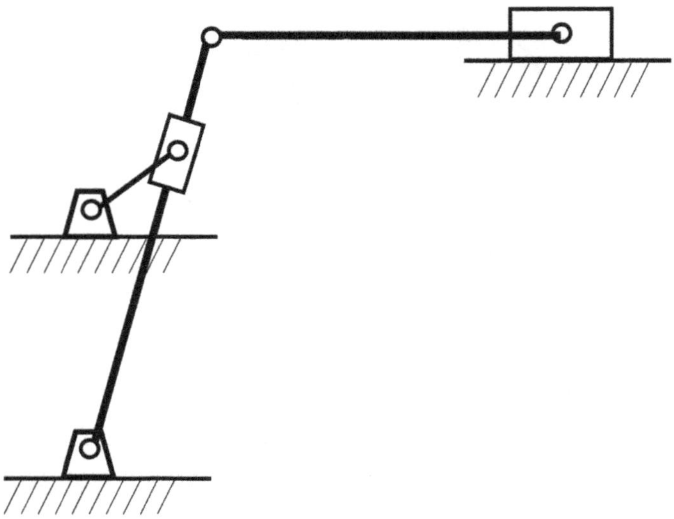

Figure 8.54 An oscillation mechanism.

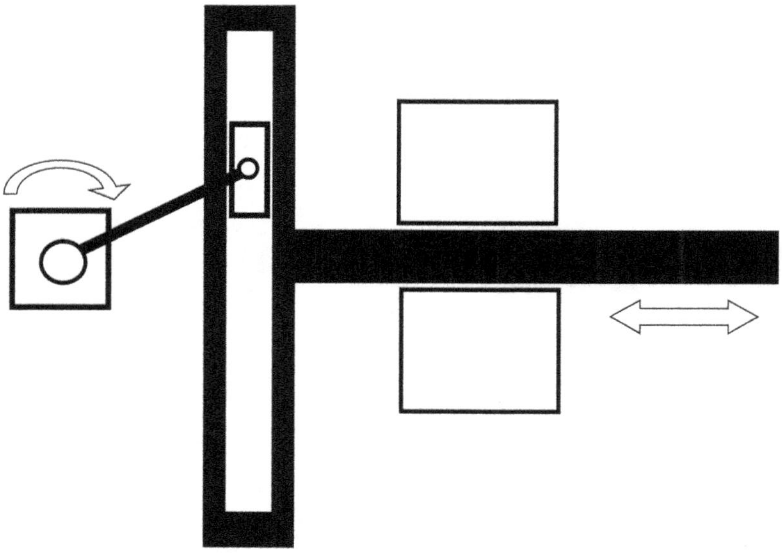

Figure 8.55 Scotch yoke mechanism.

Figure 8.56 A rack and pinion system can convert between linear and rotary motion.

steering system of cars to convert the rotary motion of the steering wheel to the side-to-side motion in the wheels. Rack and pinion gears give a positive motion, especially compared to the friction of a wheel driving on tarmac. For example, in the rack and pinion railway, a central rack between the two rails engages with a pinion on the engine allowing the train to be pulled up very steep slopes.

8.4.2.4 Rotary input, intermittent output

Intermittent motion is often used as an indexing motion to convert a rotating (or oscillatory) motion to a series of step movements of the output link or shaft. It is useful for counters, and machine tool feeds. In automation, a pair of rotating members may be used so that, for continuous rotation of the driver, the follower will alternately roll with the driver and remain stationary for assembly or manufacturing. One arrangement is known as intermittent gearing as shown in Figure 8.57. The applications include counting mechanisms, motion-picture machines, and feed mechanisms.

Figure 8.57 Intermittent gearing.

Another example is the Geneva mechanism as shown in Figure 8.58. In this case, the follower, B, makes one-fourth of a turn for one turn of the driver, A. The drive pin on the lower wheel engages with the slots on the Geneva wheel and makes it turn just enough so that it is in position when the pin comes around again. The Geneva mechanism is used in cinema film projectors to step the film on one frame at a time. It is also widely used in assembly and manufacturing in automation. The minimum number of stations is three, and the maximum could be about eight depending on the slot diameter.

Another example is shown in Figure 8.59. A walking beam can be used to index a linear type of automated line. In this example, it is driven by a constant rotary crank to operate the beam. In this example, the radius of rotation of the drive pins and the length of engagement of the machine member (drive arm) that contact the article to move it. The

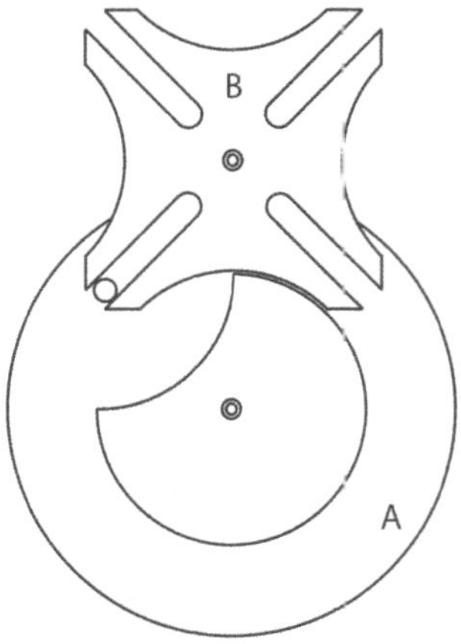

Figure 8.58 A Geneva mechanism with four stations.

Figure 8.59 A walking beam with a rotary input and an intermittent output.

radius determines the maximum displacement (the diameter of rotation), and the length of engagement can be adjusted to reduce the contact time of the drive arm and shorten the displacement.

8.4.2.5 Rotary input, irregular output

A cam mechanism as shown in Figure 8.60 usually consists of two moving elements, the cam, and the follower, mounted on a fixed frame. A cam may be defined as a machine element having a curved outline or a curved groove, which, by its oscillation or rotation motion, gives a predetermined specified motion to another element called the follower. The cam has a very important function in the operation of many classes of machines, especially those of the automatic type, such as printing presses, shoe machinery, textile machinery, gear cutting machines, and screw machines. A cam can be used to convert rotary motion into designed motion. The motion created can be simple and regular, or complex and irregular. A cam is driven by the circular motion, it turns, and the cam follower traces the surface of the cam transmitting its motion to the required mechanism.

8.4.2.6 Reciprocating input, rotary output

Some mechanisms that can convert the rotary input into linear output can also convert the linear input into rotary output, such as the piston as shown in Figure 8.53. A good example

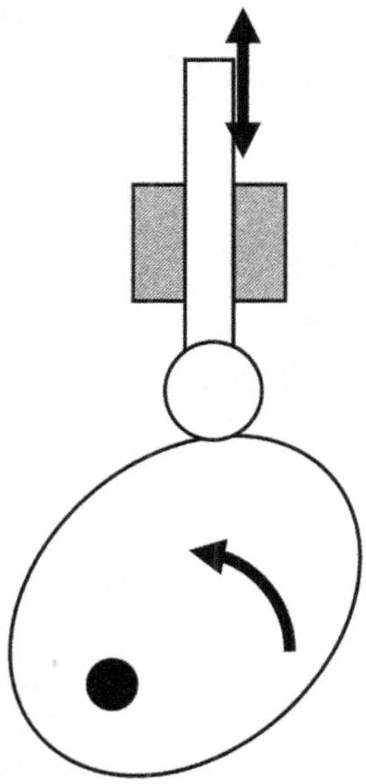

Figure 8.60 A cam mechanism with rotary input.

is the engine piston assembly. In this type of setup, the firing order and certain arrangement of the reciprocating input force are important to ensure that the rotation can be in a specific direction and no binding position will occur for a successful operation.

8.4.2.7 Reciprocating input, oscillation output

Many mechanisms that can convert the rotary input into linear output can also convert the linear input into oscillation output, such as a rack and pinions, and a piston. Rocker–rocker four-bar mechanisms (shortest link is the link opposite to the ground link) as shown in Figure 8.61 can also facilitate this transfer. Linkages are an essential part of many mechanisms. They can be used to change direction, alter speed, and change the timing of moving parts.

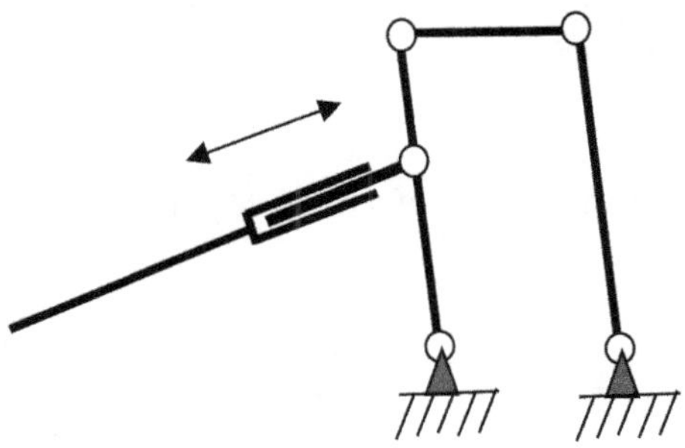

Figure 8.61 A rocker–rocker four-bar linkage can be used to transfer a linear motion into oscillation or rocking motion.

A four-bar linkage comprises four bar-shaped links and four turning pairs as shown in Figure 8.62. The link opposite the frame is called the coupler link, and the links which are hinged to the frame are called side-links. A link free to rotate through 3,608 with

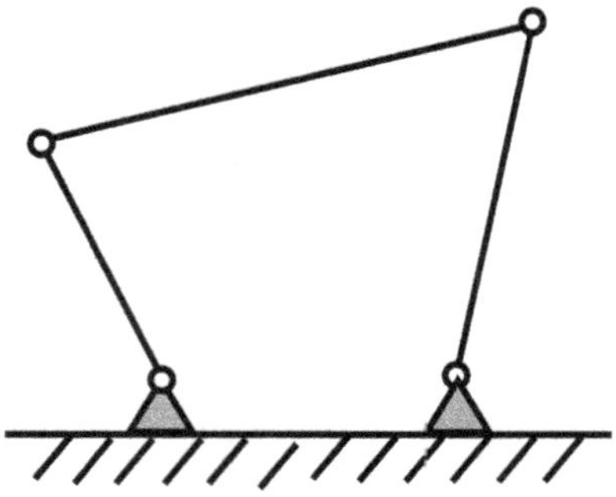

Figure 8.62 Four-bar linkage.

respect to a second link will be said to revolve relative to the second link. Some important concepts in link mechanisms are as follows:

- Crank: A side-link that revolves relative to the frame is called a crank.
- Rocker: Any link that does not revolve is called a rocker.
- Crank–rocker mechanism: In a four-bar linkage, if the shorter side-link revolves and the other one rocks (i.e., oscillates), it is called a crank–rocker mechanism.
- Double-crank mechanism: In a four-bar linkage, if both of the side-links revolve, it is called a double-crank mechanism.
- Double-rocker mechanism: In a four-bar linkage, if both of the side-links rock, it is called a double-rocker mechanism.

A linkage is implemented using various links and joints. A revolute joint, or pin joint, can be implemented using bearing joints as shown in Figure 8.63. Some ball joints are shown in Figure 8.64, and some slider joints are shown in Figure 8.65.

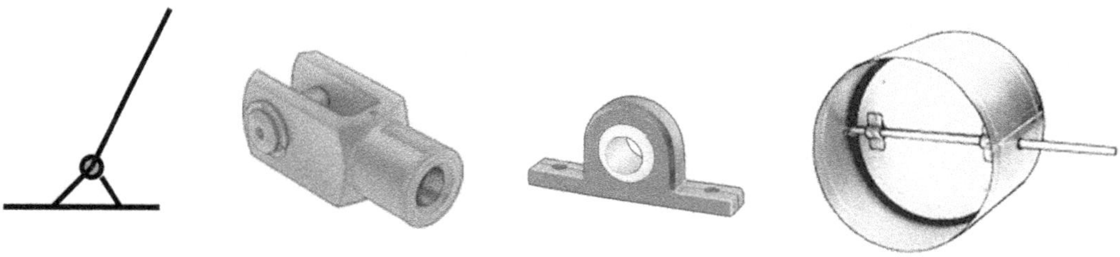

Figure 8.63 Some possible implementations of a revolute joint.

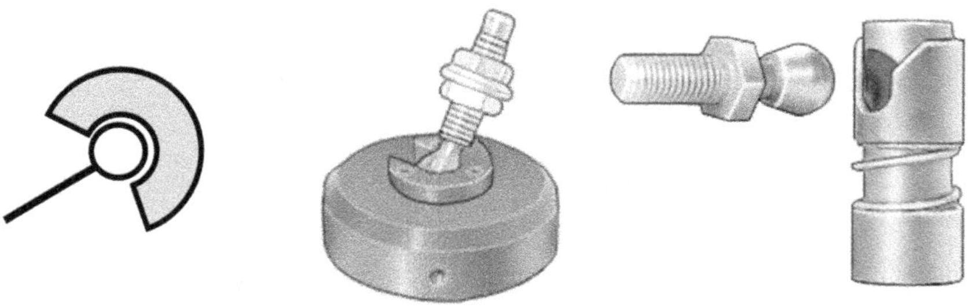

Figure 8.64 Some possible implementations of a ball joint.

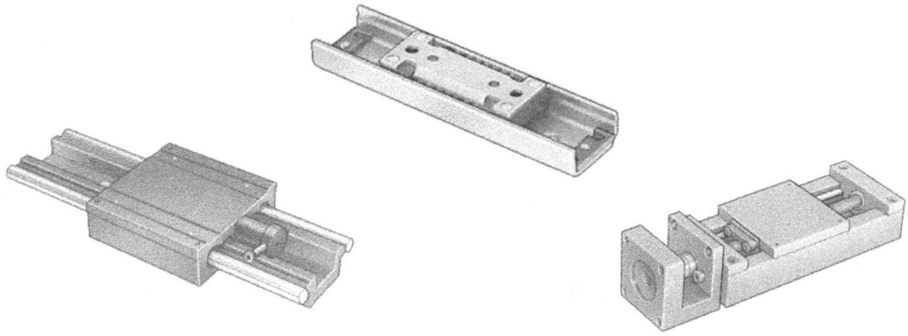

Figure 8.65 Some possible implementations of a slider joint.

8.4.2.8 Reciprocating input, intermittent output

A good example for reciprocating input and intermittent output is a walking beam as shown in Figure 8.66.

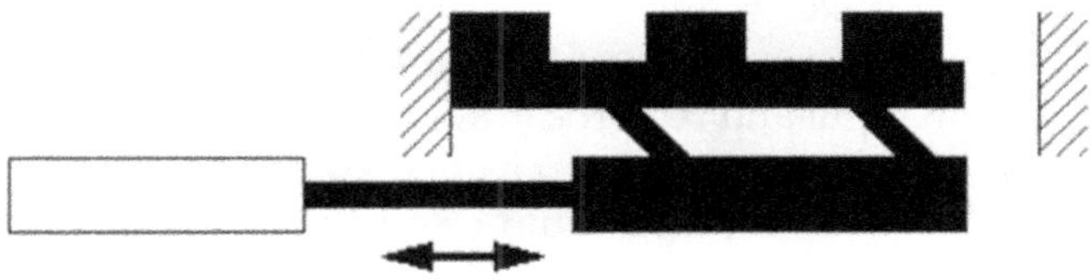

Figure 8.66 A walking beam with reciprocating input and intermittent output is commonly used in industrial automation.

8.4.2.9 Reciprocating input, irregular output

Cam motion can also be generated from a reciprocating input as shown in Figure 8.67.

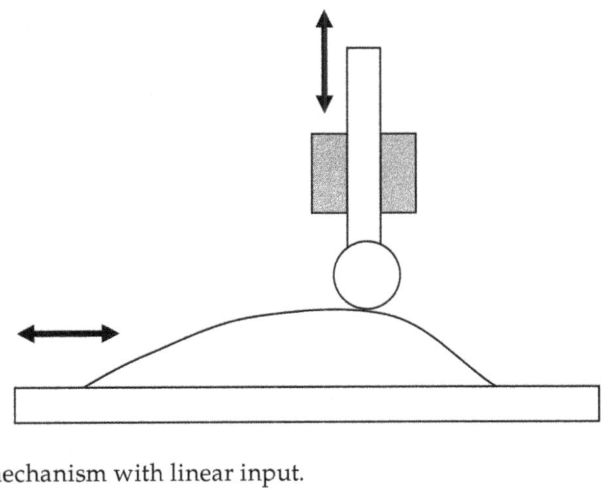

Figure 8.67 A cam mechanism with linear input.

8.4.2.10 Oscillation input, rotary output

One possible linkage to facilitate an oscillation input and produce a rotary output is the crank–rocker four-bar linkage as shown in Figure 8.68. A crank–rocker four-base linkage is when either link adjacent to the grounded link is the shortest. Similar to the reciprocating input, rotary output mechanism, particular firing order, or arrangement may be needed to ensure specific rotational direction and smoother motion.

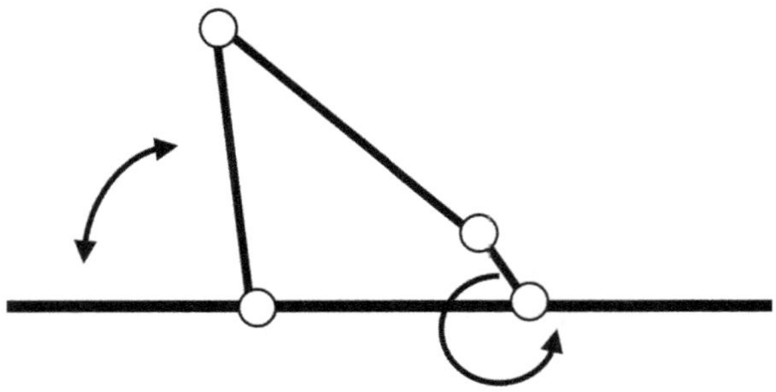

Figure 8.68 A crank–rocker four-bar linkage can be used to transfer an oscillation motion into rotational motion.

8.4.2.11 *Oscillation input, reciprocating output*

The piston, oscillation mechanism, or Scotch yoke mechanism as shown in Figure 8.55 can also be used for oscillation input and reciprocating output transfer. A cam-follower system as shown in Figure 8.60 can also be used to transfer an oscillation motion into the reciprocating output. Instead of a rotary input, it could be an oscillation input.

8.4.2.12 *Oscillation input, intermittent output*

A wheel provided with suitably shaped teeth, receiving an intermittent circular motion from an oscillating or reciprocating member, is called a ratchet. A ratchet can be used to move a toothed wheel one tooth at a time. A simple form of ratchet mechanism is shown in Figure 8.69. A is the ratchet wheel, and B is an oscillating lever carrying the driving pawl C. A supplementary pawl at D prevents backward motion of the wheel. Ratchets are also used to ensure that motion only occurs in one direction, useful for winding gear that must not be allowed to drop. Ratchets are also used in the free-wheel mechanism of a bicycle.

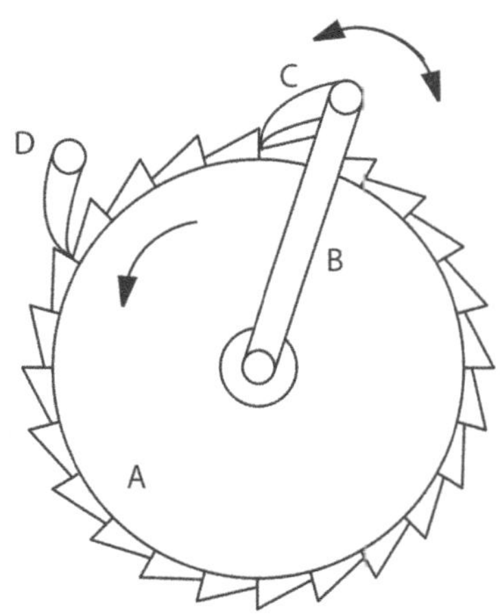

Figure 8.69 A ratchet mechanism.

A special form of a ratchet is the overrunning clutch. It is a type of free-wheel mechanism. Figure 8.70 illustrates a simplified model. As the driver delivers torque to the driven member, the rollers or balls are wedged into the tapered recesses. This gives the positive drive. Should the driven member attempt to drive the driver in the directions shown, the rollers or balls become free, and no torque is transmitted.

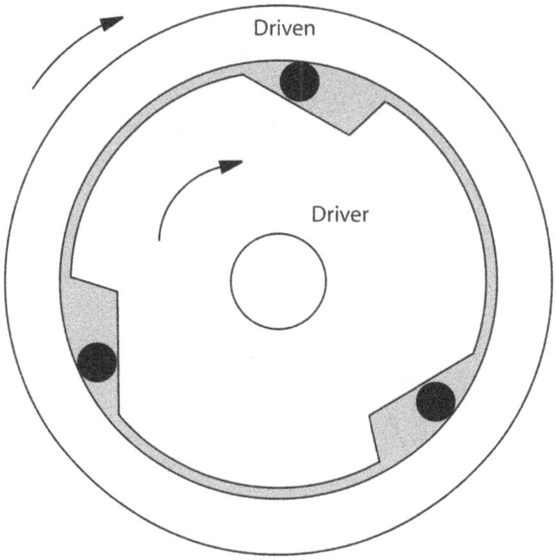

Figure 8.70 Overrunning clutch.

8.4.2.13 *Oscillation input, irregular output*

Similar to Figure 8.60, but instead of rotary input, one can use oscillation input to drive a cam to generate irregular output motion.

8.4.2.14 *Rotary input, linear output*

A commonly used mechanism for rotary input and linear output is the fine adjustment mechanism as shown in Figure 8.71.

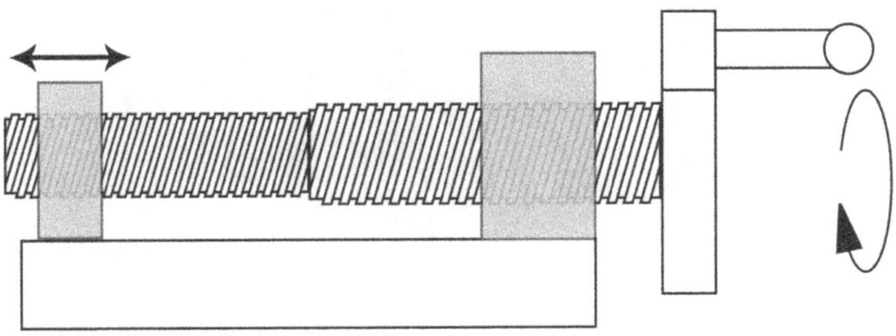

Figure 8.71 A fine adjustment mechanism can provide rotary input and linear output motion.

8.4.2.15 *Other complex motions*

Linkages have many different functions, which can be classified according to the primary goal of the mechanism: (1) function generation—the relative motion between the links connected to the frame, (2) path generation—the path of a tracer point, and (3) motion generation—the motion of the coupler link. Examples of linkage mechanism include the straight line mechanism as shown in Figure 8.72 [Erdman01].

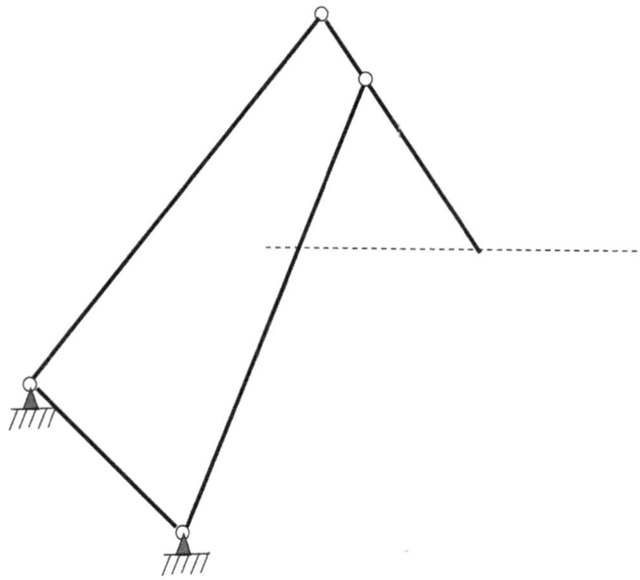

Figure 8.72 A four-bar mechanism to produce a straight line motion.

8.4.2.16 Universal joint mechanisms

A typical universal joint, as shown in Figure 8.73, can be used to connect two spinning shafts that are not aligned. To choose a universal joint, one needs to consider the angle of the shafts, speed, and torque the joint will need to transmit. Generally, the more the shafts are offset, the less speed and torque the joints will be able to transmit.

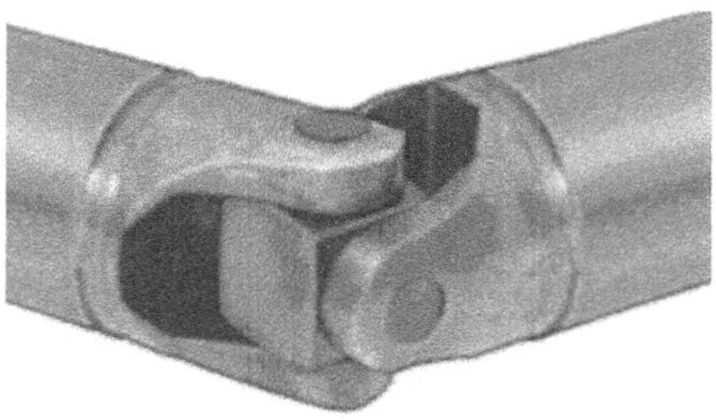

Figure 8.73 A universal joint.

8.4.2.17 Wedges and stopping

A basic wedge as shown in Figure 8.74 [Vranish06] can be used in many applications, such as a doorstop. It has been used as a machining tool, a fixture, or a tool stop. It could also have many forms such as the flex wedge as shown in Figure 8.75.

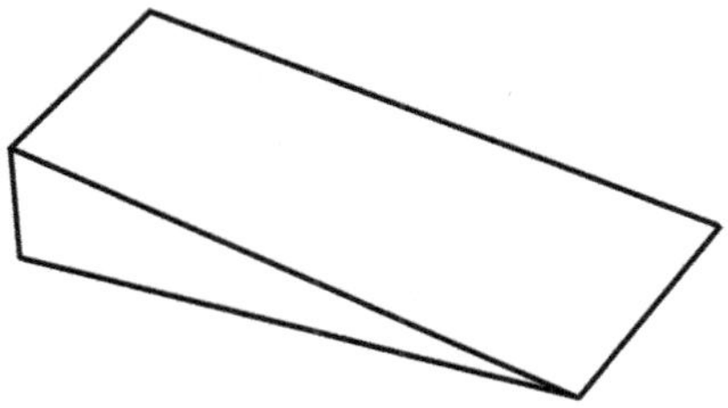

Figure 8.74 Basic form of a wedge.

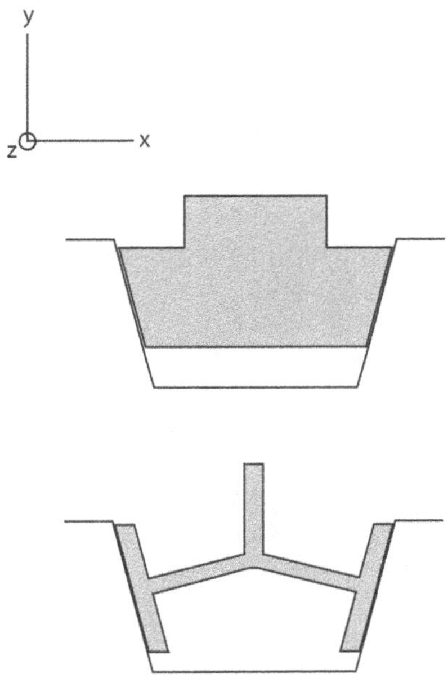

Figure 8.75 A flex wedge can be used as a brake or clutch to generate large contact forces between the wedge and groove surfaces in order to generate large friction forces to resist relative motion in the z-direction.

Review problems

1. Match the mechanical linkages and their functions.

Mechanical linkages	Functions
1. Inclined plane	a. Rotation⇒ intermittent rotation
2. Gear	b. Linear motion⇒ linear motion
3. Cam	c. Rotation⇒ reciprocating motion
4. Geneva	d. Rotation⇒ rotation

2. Explain the difference between gears and belts and pulleys.
3. An indexing table driven by a Geneva mechanism has eight stations and a driver speed of 20 rpm. Calculate
   a. Index time
   b. Dwell time
   c. Ideal production rate per hour

4. A Geneva mechanism has four stations and a driver speed of 10 rpm. Calculate the index time, dwell time, and the ideal production rate for one 8 h work shift.
5. Find the force required to perform the function as shown in Figure 8.76.

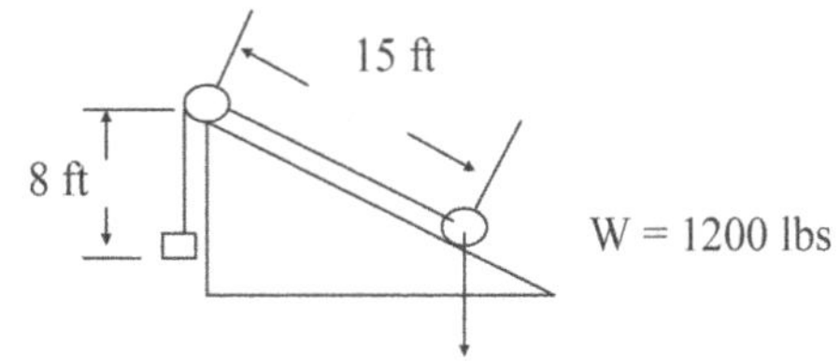

Figure 8.76 Find the minimum vertical force needed to carry the weight W.

6. The driver pulley has a diameter D1 of 180 mm and rotates at 1,000 rpm. If it was required to drive another pulley that must rotate at 250 rpm, what must be the diameter D2 of the driven pulley? What is the velocity ratio of the pulley system?
7. To design a capsule filling machine, its production rate is 900 capsules per minute. Its dosage assembly had six stations and used a Geneva mechanism for indexing. Each Geneva mechanism station advances a die that allows the filling of seven capsules simultaneously. One die has seven holes for holding capsules. So during each cycle time, seven capsules can be filled. Compute the driver speed for this assembly.
8. What are the degrees of freedom of the joints shown in Figure 8.77?

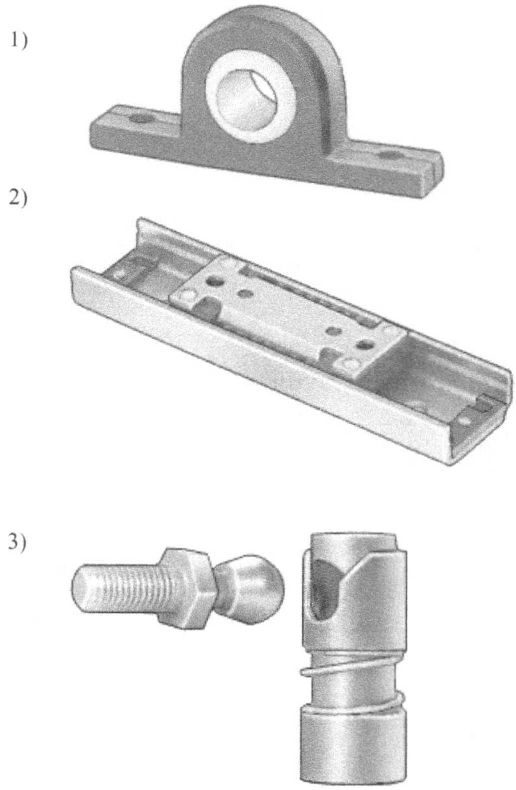

Figure 8.77 Find the degrees of freedom of various joints.

9. Give the value of force F in order to lift the bin as shown in Figure 8.78.

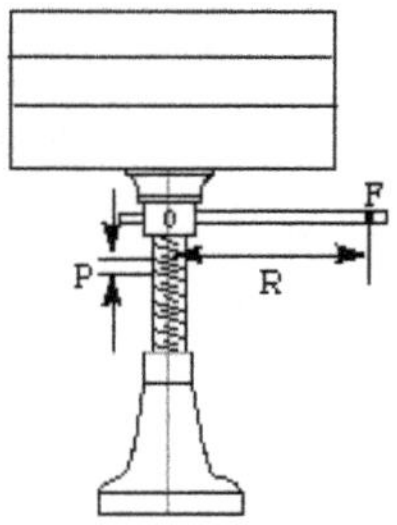

Given: R=1 m, W= 1000 pound, P= 2π mm

Figure 8.78 Find force F in order to lift the bin.

10. Design a pulley system (frictionless) where a motor that can pull 25 lbs has to lift 100 lbs.
11. Your boss tells you that the assembly line must have a production rate of almost 65,000 pieces per hour. Find the needed motor at Grainger.com with the driver speed capable of driving an indexing table driven by a Geneva mechanism that has eight stations along this line. What is the cycle time with the picked motor?

References

[Erdman01]. Erdman, A.G., G.N. Sandor, and S. Kota, *Mechanism Design: Analysis and Synthesis*, 4th edition, Prentice Hall, Upper Saddle River, NJ, May 15, 2001.

[Jack05]. Hugh, J., Selecting a PLC, 31 August, 2001, Retrieved on January 19, 2005 from http://claymore.engineer.gvsu.edu/~jackh/eod/mechtron/mechtron-395.html.

[Huang94]. Huang, C.-P., An advisory system for the conceptual design of mechanical components, M.S., M.E., University of Missouri-Rolla, February 1994.

[RoyMech05]. RoyMech, Mechanisms, Retrieved on January 27, 2005 from www.roymech.co.uk/Useful_Tables/Cams_Springs/Mechanism.html.

[Vranish06]. Vranish, J.M., Retrieved on February 5, 2006 from www.nasatech.com/Briefs/Oct01/GSC14006.html.

[Yeh93]. Yeh, J.H., The preliminary design of mechanical drive and transmission components, M.S., M.E., University of Missouri-Rolla, November 1993.

[Zhang07]. Zhang, Y., S. Finger, and S. Behrens, Introduction to mechanisms, Retrieved on January 22, 2007 from www.cs.cmu.edu/People/rapidproto/mechanisms/tablecontents.html.

chapter nine

Using prototypes for product assessment

Prototyping is a high-value development activity. The reason for prototyping is to find out something one does not know and not to demonstrate something already known. Often, people lost in the prototyping activities forget the fact that the major purpose, or sometimes the only purpose, of developing a prototype is to evaluate a new concept or an unknown. By paying attention to the prototyping results and assessment, one can greatly improve the product performance. The purpose of prototype assessment is to conduct an evaluation to (1) provide answers to design questions, (2) improve the prototype or design, (3) review the usefulness of evaluation provided by the prototype, and (4) transform evaluation results into design or revision recommendations. Therefore, before conducting an assessment, the objective of the prototyping activities should be clearly defined, the prototyping methods should be consistent with the objective, and the testing or measurement should be able to achieve the original prototype objective. Therefore, the following structure can be followed to set up the prototype assessment plan:

1. Review prototype objective
   Is the prototype objective to learn more about customer requirements, to show proof of concept to senior management, to validate system specifications, to resolve fuzziness in early stages of design, to explore solutions to specific usability or design problems, to integrate subsystems, to manage change requests, to pretrain users or to create a marketing demo, to improve customer acceptance, to assure product quality, or to enforce milestones, etc.?
2. Identify the key issues
   Prepare a list of the critical uncertainties or questions one still needs to address to determine the viability of one's design or process. For each one, specify an associated plan of action, such as analysis, mockups, interviews, experiments, etc.
3. Review the level of approximation of the prototype
   Has the simplest prototype been adapted? What is the level of approximation selected? Is it an analytical or physical prototype? Is it a focused or comprehensive prototype?
4. Review the experimental plan
   Explain how and why the prototype was constructed and tested. What are the expected experimental values? What is the prototype testing protocol? What measurements are to be performed? What are the ways to analyze the resulting data?
5. Review the schedule for procurement, construction, and testing
   Is the prototype project on schedule? Is the cost under control? Should the project scopes be modified?
6. Complete test and measurement preparation
   Layout a test and measurement plan. Can evidence be provided that the prototype testing meets the objective of the prototype? Can the test and measurement yield the expected experimental value? Is any special measurement equipment such as video recorders or cameras needed? Is it necessary to have any observers or testers perform

the test or measurement? Which variable or variables are likely to be important? Which variables are noise variables and which are the design variables? Which variables are not controllable? What are the prototype objectives? What are the constraints? What are the required personnel, tools, and test facilities? What are the deliverables?

7. Analyze the measurements

   Analysis usually consists of standard deviations, data summary, themes or trends, and key representative examples. Use plots or tables to summarize the data. The design of experiments (DOE) is detailed in the next session.

8. Perform the assessment and report including a proposal for improvements.

   Could the results show insights that resulted from the prototyping process and offer recommendations for further development of the prototype or design concept? Has the prototype made effective use of available time? Discuss the quality of the prototype in meeting the prototyping objective. What parts of the prototype work well and what still needs improvement?

   The report needs to summarize the key insights gained from the evaluation in terms of the key strengths and weaknesses of the prototype, the relative importance of these strengths and weaknesses learned, and how the view of the situation has changed from prior to the evaluation. Have the prototype objectives been met? Have the questions been answered? Can the prototype be improved? What is the next step?

9. Reflect on the design process

   There is nothing worse than having a successful prototype demonstration and then having to start again from scratch to build something that is affordable and serves some useful purpose. One should build a prototype with an eye toward making a smooth transition to beyond the prototype stage by learning from the current prototype. The possible outcomes of the current prototyping could be the following: (1) design validated as is; (2) minor adjustments needed, an overall approach validated; (3) concept still worth investigating but serious problems identified; and (4) design approach not validated. It is necessary to decide whether the prototyping result has changed in the design perspective. What has been learned about the design process and over the course of the entire project? Did it work? A brainstorming session may be needed. Was the design changed based on the current prototyping result? What were the biggest surprises—the things learned that would not have been predicted based on the experience and intuition? Did the prototyping methods for evaluation and prototyping get at what was being looked for? In hindsight, would a different prototyping approach have been better? What was the most, and least, valuable among the activities that were tried, either generally or specifically? Having gone through this course, how might the next prototype be approached differently? Could the prototypes be used to optimize the design? The engineering optimization is discussed in Chapter 10.

9.1 Introduction to DOE

> If your result needs a statistician then you should design a better experiment.
>
> **Ernest Rutherford**

This section helps answer the following questions.

- What is DOE?
- How to design an experiment?

- What are the main components of Taguchi's philosophy?
- What is Taguchi's loss function? What is the philosophy of the loss function?
- What is "sum of squares (SS)"? What is "variance"? What is "confidence interval"?
- What is standard deviation? What does it mean?

DOE is a technique used to lay out an experimental (investigation, studies, survey, tests, etc.) plan in the most logical, economical, and statistical way. Very often when one needs to conduct a test for a prototype, using DOE will be more systematic. One can potentially benefit from it when one wants to determine the most desirable design of the product, best combination of parameters for the process, most robust recipe for the formulation, permanent solution for some of the production problems, most critical validation or durability test condition, and most effective survey or data collection plan, etc.

9.1.1 Design of experiments

Traditionally, engineers tend to change only one variable of an experiment at a time. This chapter introduces the Taguchi's technique or quality engineering by design. The strength of the Taguchi's technique is that the engineer can change many variables at the same time and still retain control of the experiment. The Taguchi's method, a DOE, is a system of cost-driven quality engineering principles that emphasize the robust application of engineering strategies rather than advanced statistical techniques. The Taguchi's method allows a company to rapidly and accurately acquire information to design and produce low cost, reliable products, and processes. It is, however, not without weaknesses. The main ones can be found in the use of linear graphs for assigning factors and interactions, using signal-to-noise (S/N) ratio as a performance indicator of robustness, etc. [Ranjit01, Pignatiello91]. As there are many DOE technologies, one can evaluate the pros and cons of each method for a specific application before adopting it.

The Taguchi's method of experimentation provides an orderly way to collect, analyze, and interpret data to satisfy the objectives of the study. By using this method, in the DOE, one can obtain the maximum amount of information for the amount of experimentation used. For example, one can optimize product and process designs, study the effects of multiple factors, that is, variables, parameters, and ingredients, on the performance, and solve production problems by objectively laying out the investigative experiments. One can use this method to study the influence of individual factors on the performance and determine which factors have more or less influence. One can also use it to find out which factor should have a tighter tolerance and which tolerance should be relaxed. The information from the experiment will tell how to allocate quality assurance resources based on the objective data. It will indicate whether a supplier's part causes problems or not (analysis of variance [ANOVA] data), and how to combine different factors in their proper settings to get the best results. It is accomplished by the efficient use of experimental runs to the combinations of variables studied. This section summarizes Taguchi's DOE methods and some of the basic terms used, including loss function, orthogonal array (OA), and ANOVA.

9.1.2 Standard deviation

An average (or mean) of a set of data can tell us only where the center of a data set exists. Very often, one needs the second piece of information: "how much variation there is in the data." The standard deviation is a statistic that tells how tightly all the various examples are clustered around the mean in a set of data. For example, the number of students absent in a class is listed in Table 9.1. How does one use a simple term to describe this?

Table 9.1 Standard deviation example: number of students absent

	Day 1	Day 2	Day 3	Day 4	Day 5
Absent	3	5	7	6	9

$$\text{The average or mean: } \bar{x} = \frac{\sum x}{n} = (3+5+7+6+9)/5 = 6 \tag{9.1}$$

It tells us the "center" of the data set, but does not tell us about the variations. The range of these data sets the difference between the maximum and the minimum and yields $9 - 3 = 6$, or 6 ± 3. It tells how extreme a data set reaches, but does not tell about the density of the data point. Some people thus tried to find the deviation from the mean as shown in Table 9.2. As one hopes to use one number to represent these deviations, one way to do this is to add all deviations together, that is, $(-3) + (-1) + (1) + (0) + (3) = 0$. This does not help much as all the cases will end up with zero when adding all deviations together.

Table 9.2 Standard deviation example: deviations from the mean

	Day 1	Day 2	Day 3	Day 4	Day 5
Deviation	$3-6=-3$	$5-6=-1$	$7-6=1$	$6-6=0$	$9-6=3$

To avoid summing to zero, one can find the total deviation squared or SS, that is,

$$(-3)^2 + (-1)^2 + (1)^2 + (0)^2 + (3)^2 = 9+1+1+0+9 = 20$$

One can thus define mean squared deviation (MSD) as total deviation squared/number of samples = 20/5 = 4

One can also define V, or variance, as total deviation squared/(number of samples − 1) = 20/4 = 5

where (number of samples − 1) is also called the number of degrees of freedom.

Define

$$V = \frac{\sum_{i=1}^{n}\left(x - \frac{\sum x}{n}\right)^2}{n-1} = \frac{\sum_{i=1}^{n} x^2 - \frac{\sum x^2}{n}}{n-1} \tag{9.2}$$

$$\text{Or } V = \left[(9+25+36+49+81) - (3+5+6+7+9)^2/5\right]/(5-1) = (200-180)/4 = 5$$

As one may be interested in the original unit, one can then define standard deviation, or σ, as

$$\sigma = \sqrt{V} = \sqrt{\frac{\sum_{i=1}^{n} x^2 - \frac{\sum x^2}{n}}{n-1}} \tag{9.3}$$

Or in this case, $\sigma = 5^{1/2} = 2.236$. One can also define the confidence interval to relate the standard deviation. The *confidence interval* is an interval in which a measurement or trial falls corresponding to a given probability. Table 9.3 shows the confidence interval for various standard deviations.

Table 9.3 Confidence interval for various standard deviations

Range	Confidence interval
σ	0.6826895
2σ	0.9544997
3σ	0.9973002
4σ	0.9999366
5σ	0.9999994

In this particular example, one standard deviation is 6 ± 2.236, or [3.764···8.236]. In Table 9.1, only the values 5, 7, and 6 lie within this range, or 60% of them. The confidence interval for the first standard deviation is 0.683 or 68.3% meaning that 68.3% of the values should lie within the first standard deviation. Similarly, for two standard deviations, it is $6 \pm 2.236 \times 2$, or [1.528···10.472]. In this case, all (or 100%) of the values in Table 9.1 lie within this range. The confidence interval for two standard deviations is 0.954 or 95.4%, which also predicts the interval in data set well. With more data points, the difference between the predicted confidence interval and actual interval should be closer.

9.1.3 *Loss function*

Taguchi's philosophy of product quality is to build quality in product design and to measure quality by deviation from the target. The conventional way of thinking is that a high-quality product will cost more, but Taguchi's philosophy states that a high-quality product will cost less to society, which includes the production company, consumers, repairing cost, and maintenance. Moreover, the quality loss, based on Taguchi's concept, is the monetary loss incurred by the manufacturer due to a loss in product sales, and due to a loss in customer satisfaction as the product performance deviates from its desired or target value. The loss function measures quality. The loss function establishes a financial measure of the user dissatisfaction with a product's performance as it deviates from a target value. Thus, both average performance and variation are critical measures of quality. In other words, the Taguchi's loss function is a way to show how each defective part produced results in a loss for the company. The loss function can be defined as follows:

$$\text{Loss at a point: } L(x) = k(x - t)^2 \tag{9.4}$$

where

$L(x)$ = loss due to the deviation of the product specification from the desired target specification
k = quality loss coefficient
x = measured performance value of a product
t = target value

$$\text{Average loss of a sample set: } L = k\left[\sigma^2 + (p_m - t)^2\right] \quad (9.5)$$

where

σ = standard deviation of the sample
p_m = process mean

$$\text{Total loss} = (\text{Average loss}) \times (\text{number of samples}) \quad (9.6)$$

The quality loss function is a continuous function that is defined in terms of the deviation of a design parameter from an ideal or target value, as in Figure 9.1. This is a quadratic curve to represent a customer's dissatisfaction with a product's performance. The quality loss function penalizes the deviation of a parameter from the specified value that contributes to deteriorating the performance of the product, resulting in a loss to the customer.

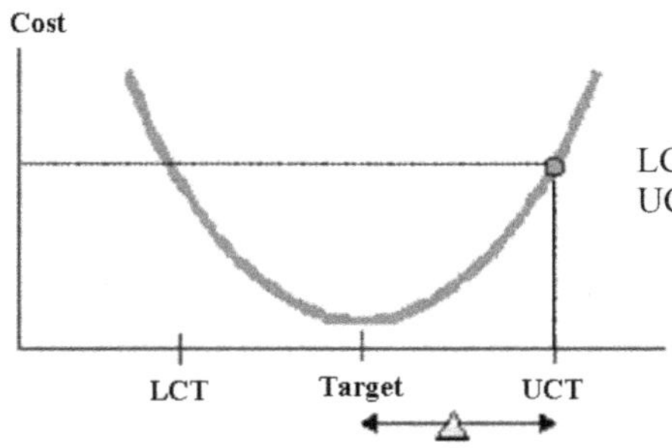

Figure 9.1 The loss function.

Example 9.1

Assume that a machine has a warranty cost of \$50 for each part, and typically when the part's dimension specification deviates ±2 in. from the target, customers often requested to have it replaced. Based on this information, find the loss function of this part.

$$L(x) = \$50 = k(x-t)^2 = k(\pm 2)^2 = 4k$$

$$k = \$12.5/\text{in.}$$

Therefore, the loss function $L(x) = 12.5(x-t)^2$

Example 9.2

Using the same conditions Example 9.1, assume that a machine has a rework cost of \$2 for each part when it is reworked in-house. Based on this information, find the appropriate tolerance level to rework the part in-house.

$$L(x) = 12.5(x-t)^2 = 2;\ (x-t)^2 = 0.16;\ (x-t) = \pm 0.4\,\text{in.}$$

Example 9.3

A group of engineers need to study the brightness of automobile headlights as shown in Figure 9.2. On the one hand, they need to be as bright as possible to give maximum visibility and range for the driver, but on the other hand, they should not be too bright for oncoming traffic according to DOT (US Department of Transportation) guidelines. Therefore, considering brightness as a safety factor, there will be more accidents if there is not enough illumination and there will be more accidents if there is too much illumination blinding oncoming traffic. It is known that to replace the out of range light will cost \$5 to the manufacturer in-house and the warranty cost is \$30 if replaced at the dealer's place. The current typical tolerance when the consumers complained was when the light is out of the range of 1,200 ± 200 lumens. Use Taguchi's loss function to design the light tolerance.

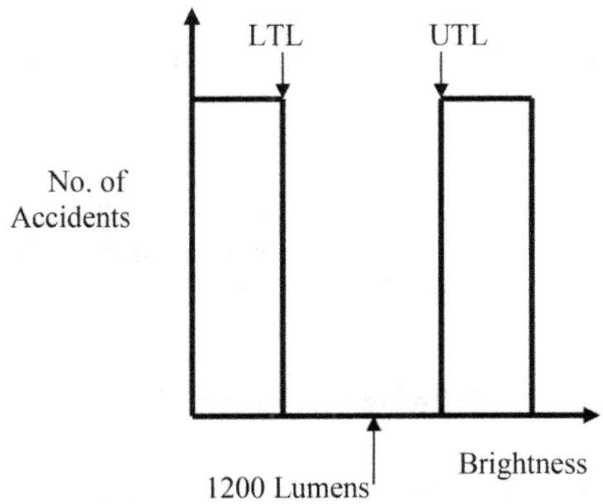

Figure 9.2 Automobile headlights with an average of 1,200 on high beam. The current typical tolerance when the consumers complained was when the light was out of the range of 1,200 ± 200 lumens.

Answer:

As shown in Figure 9.3, it is known:

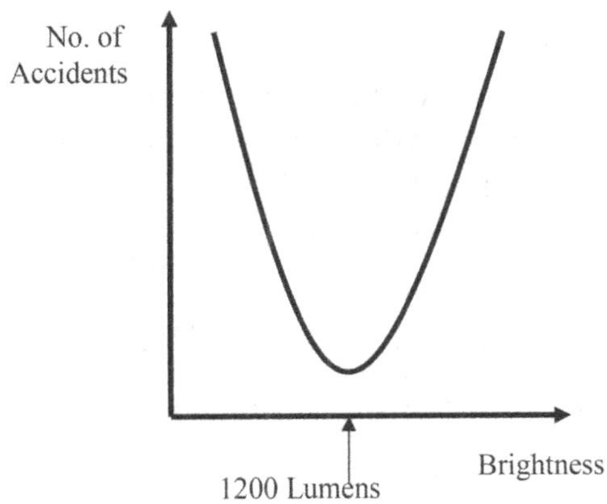

Figure 9.3 Taguchi's loss function.

$$LM(\text{loss to manufacturer}) = \$5$$

$$LW(\text{warranty cost}) = \$30$$

$$\text{Tolerance } 1200 \pm 200$$

$$L(x) = k(x - t)^2$$

$$30 = k(200)^2$$

$$k = (\text{cost to correct})/\text{tolerance}^2 = 0.00075$$

Therefore, the loss function is $L(x) = 0.00075(x - T)^2$

$$5 = 0.00075\,\delta^2$$

$$\delta = 81.6 \text{ lumens}$$

Tolerance should be set at 1,200 ± 81.6 lumens for in-house repair.

Example 9.4

A machine produces a part that has a hole measuring 0.2 ± 0.050 in. Failure costs per part are \$10. From a batch of 1,000 parts, the average of the hole is measured at 0.201, and the standard deviation is about 0.02. Find the loss of the total 1,000 parts produced.

Find k, using $L(x) = k(x - t)^2$

$$\$10 = k(\pm 0.05)^2 ; k = 4{,}000$$

$$\text{The average loss equation}: L = k\left(\sigma^2 + (p_m - t)^2\right)$$

$$L = 4{,}000\left(0.02^2 + (0.201 - 0.2)^2\right) = 2.0$$

So the average loss per part in this set is \$2.

For the loss of the total 1,000 parts produced

$$= L(\text{number of samples}) = \$2(1{,}000) = \$2{,}000$$

Review problems

1. What is DOE?
2. Why use Taguchi's method for DOE?
3. A student is doing investment casting with an ice pattern. After casting, the width of the metal casting was measured. Calculate the average and standard deviation for the measurements that are provided in the table below.

	Width				
Sample no.	W_1	W_2	W_3	W_4	Average
37	1.83	1.58	1.76	2.02	
40	1.95	1.80	1.76	1.84	
41	1.71	1.85	1.81	1.69	
42	1.81	1.74	1.75	1.75	
43	1.69	1.92	1.72	1.67	

4. What are the advantages of the Taguchi's approach?
   a. Find out which factor should have tighter tolerance.
   b. Tell how to allocate quality assurance resources based on the objective data.
   c. Indicate how to combine different factors into their proper settings to get the best results.
   d. In general, it requires fewer experiments than traditional approaches.
5. What are the main philosophies of product quality?
   - Building quality in product design.
   - Measuring quality by deviation from target, not by rejection.
   - Conventional: high-quality product costs more.
   - Taguchi: a high-quality product will cost less to society.
6. A medical company produces a part that has a hole measuring 0.5 ± 0.050 in. The tooling used to make the hole is worn and needs replacing, but management does not feel it is necessary since it still makes "good parts." All parts pass QC, but several parts have been rejected by assembly. Failure costs per part are $0.45. Using the loss function, explain why it may be to the benefit of the company and customer to replace or sharpen the tool more frequently. Use the data below:

   Measured values:

   0.459 | 0.478 | 0.495 | 0.501 | 0.511 | 0.527
   0.462 | 0.483 | 0.495 | 0.501 | 0.516 | 0.532
   0.467 | 0.489 | 0.495 | 0.502 | 0.521 | 0.532
   0.474 | 0.491 | 0.498 | 0.505 | 0.524 | 0.533

7. In laser deposition experiments, there are three kinds of particles typically being used: Titanium, H13, and Steel 4140 powder particles, below are the sizes of 25 particles measured on the SEM:

	Size (μm)		
No.	Titanium	H13	Steel 4140
1	114.089	123.455	101.088
2	154.513	119.424	120.85
3	175.545	224.689	136.767
4	154.513	167.303	147.528
5	114.866	107.77	106.887
6	121.273	140.008	75.581
7	160.239	97.243	108.111
8	99.951	104.734	141.728
9	73.832	152.264	76.446
10	130.806	145.693	83.375
11	44.5	129.781	134.463
12	106.001	129.524	138.361
13	140.246	133.166	148.084
14	106.001	130.806	135.276
15	151.017	114.186	121.483
16	127.009	162.445	139.112

(Continued)

	Size (μm)		
No.	Titanium	H13	Steel 4140
17	147.363	144.697	98.339
18	146.91	129.781	67.941
19	128.749	138.731	121.438
20	152.41	101.278	93.481
21	145.693	154.728	103.129
22	99.169	154.728	88.355
23	115.156	194.486	92.36
24	104.734	81.154	131.87
25	112.121	110.119	147.044

What are the average size and standard deviation?

8. The quantity of the television sets a factory manufactured in a week is listed in the following table. Find the mean, variance, and standard deviation.

	Monday	Tuesday	Wednesday	Thursday	Friday
Production	1,000	1,250	980	1,030	950

9. A domestic refrigerator needs to operate within a set range of temperatures. If the temperature is too high, food will spoil by rotting and if the temperature is too low, food will freeze. To work most efficiently, one's refrigerator should be set between 388F and 408F. The freezer temperature should be set at 08F. Use Taguchi's loss function to design the tolerance for the temperature control system. It is known that to replace an out of range control system will cost \$4 to the manufacturer and the warranty cost is \$50. The current typical tolerance when the consumers complained was when the refrigerator was out of the range of 39° ± 4°.
10. It is determined that a scraped SLS (selective laser sintering) build cost \$1,250 in material for your company. It is known that a scraped build is recorded if the tolerance variation exceeds ±0.030 in. dimensionally. Using the Taguchi's quadratic loss function, determine what loss in dollars is incurred from ±0.018 in. The target would be 0.
11. You work at a manufacturing company that stamps sheet metal covers for a motor. The diameter of the cover is 3.50 in. with a design tolerance of ±0.05 in. When this part is out of specification, it leads to motor failures which result in a warranty claim cost of \$6 per motor. The stamping tool is worn and produces parts on average of 3.48 in. with a standard deviation of 0.03 in. A replacement progressive die will cost \$20,000 and should last for 6 years. The monthly production of motors is 5,000. If you replace the progressive die, how many months will it take before the payback?

9.2 Orthogonal arrays

The ability to simplify means to eliminate the unnecessary so that the necessary may speak.

Hans Hofmann

This section helps answer the following questions.

- What is an OA? Why is it useful?
- What is a factor? What is a level?
- What is the procedure of the Taguchi's DOE method?
- How is a Taguchi's DOE method OA formulated?

Taguchi used a partial OA that is a set of well balanced (minimum) experiments in which all parameters of interest are varied over a specified range, and thus gives much reduced "variance" for the experiment with "optimum settings" of control parameters. OA can be used to optimize product and process designs, study the effects of multiple factors (i.e., variables, parameters, and ingredients) on the performance, and solve production problems by objectively laying out the investigative experiments. It can study the influence of individual factors on the performance to find out which factor should have a tighter tolerance, to tell how to allocate quality assurance resources based on the objective data, and to indicate how to combine different factors in their proper settings to get the best results.

9.2.1 *What is OA?*

In OA, factors are design parameters that influence the performance and the input that can be controlled. Levels are values that a factor assumes when used in the experiment. The number of factors and their levels determines the choice of an OA. Tables 9.4–9.7 show L4, L8, L9, and L18 OAs, respectively. Both L4 and L8 arrays are for two-level experiments, and L9 is for three-level experiments. L18 is for two levels for one factor, and three levels and up to seven factors. The various types of Taguchi's tables are listed in Table 9.8.

Table 9.4 L4 orthogonal array: two levels, up to three factors

	Column number		
Run	1	2	3
1	1	1	1
2	1	2	2
3	2	1	2
4	2	2	1

Table 9.5 L8 orthogonal array: two levels, up to seven factors

	Column number						
Run	1	2	3	4	5	6	7
1	1	1	1	1	1	1	1
2	1	1	1	2	2	2	2
3	1	2	2	1	1	2	2
4	1	2	2	2	2	1	1
5	2	1	2	1	2	1	2
6	2	1	2	2	1	2	1
7	2	2	1	1	2	2	1
8	2	2	1	2	1	1	2

Table 9.6 L9 orthogonal array: three levels, up to four factors

Run	Column number 1	2	3	4	Result
	Factors A	B	C	D	
1	1	1	1	1	539
2	1	2	2	2	229
3	1	3	3	3	230
4	2	1	2	3	623
5	2	2	3	1	479
6	2	3	1	2	198
7	3	1	3	2	576
8	3	2	1	3	583
9	3	3	2	1	342

Table 9.7 L18 orthogonal array: $2^1 \times 3^7$: two levels for one factor and three levels up to seven factors

Experiment number	1	2	3	4	5	6	7	8
1	1	1	1	1	1	1	1	1
2	1	1	2	2	2	2	2	2
3	1	1	3	3	3	3	3	3
4	1	2	1	1	2	2	3	3
5	1	2	2	2	3	3	1	1
6	1	2	3	3	1	1	2	2
7	1	3	1	2	1	3	2	3
8	1	3	2	3	2	1	3	1
9	1	3	3	1	3	2	1	2
10	2	1	1	3	3	2	2	1
11	2	1	2	1	1	3	3	2
12	2	1	3	2	2	1	1	3
13	2	2	1	2	3	1	3	2
14	2	2	2	3	1	2	1	3
15	2	2	3	1	2	3	2	1
16	2	3	1	3	2	3	1	2
17	2	3	2	1	3	1	2	3
18	2	3	3	2	1	2	3	1

Table 9.8 Commonly used orthogonal arrays

Level	Orthogonal arrays
Two-level (fractional factorial) arrays	$L_4(2^3)$, $L_8(2^7)$, $L_{12}(2^{11})$ $L_{16}(2^{15})$, $L_{32}(2^{31})$, $L_{64}(2^{63})$
Three-level arrays	$L_9(3^4)$, $L_{27}(3^{13})$, $L_{81}(3^{40})$
Four-level arrays	$L_{16}(4^5)$, $L_{64}(4^{21})$
Five-level arrays	L_{25} (5^6)
Mixed-level arrays	$L_{18}(2^1 \times 3^7)$, $L_{32}(2^1 \times 4^9)$, $L_{50}(2^1 \times 5^{11})$ $L_{36}(2^{11} \times 3^{12})$, $L_{36}(2^3 \times 3^{13})$, $L_{54}(2^1 \times 3^{25})$

9.2.2 *Taguchi's DOE procedure*

The sequence of procedures for the Taguchi's method is given as follows:

1. Identifying the control factors and the noise factors and evaluating the interactions.
2. Choosing the levels for the factors: In order to locate maxima or minima, one usually uses two levels, and three levels where necessary, to check the linearity in the variation.
3. Selecting an appropriate OA: An OA "A" is defined as an "N × k" array with a number of levels "s". The control and noise factors, their levels, and the study of their interactions help to decide the appropriate OA.
4. Assigning the factors and interactions to columns of the OA: One column is allocated to measure the factor interactions and other factors and errors that were not taken into consideration.
5. Conducting the experiments: Three samples are used for each experiment for repeatability.
6. Analyzing the data and determining the optimal levels: This is an important part of the whole process and helps in obtaining the optimized values depending on the output one chooses to measure. In the analysis of data, S/N ratios are used in controlling the target as well as reducing the target's variation. ANOVA is used in this process to calculate the percentage contribution of the control factors associated with the conclusions.
7. Predicting the responses for the optimal levels: A predictive model is used to estimate the optimal response for various weightings of the overall evaluation criterion (OEC), which is a kind of objective function.
8. Running the F-tests and estimating the confidence intervals for the responses: The F-test shows the significance of the control parameters on the process and the confidence interval that is used to estimate the response values with a fixed level of confidence.
9. Conducting the confirmation experiment: This is conducted to validate the accuracy of the experiment. If the predicted values that one obtains after optimization match the results of the confirmation experiment, then the experiment is valid.

Example 9.5: Solve the following problem by using the Taguchi's approach

The three main parameters that affect the density of laser deposition are laser power, powder feed rate, and tool travel speed. Based on the experiments shown in Table 9.9, find the most influencing parameter. This is an L4 array with two levels for each factor.

Table 9.9 Design of experiments example using an L4 orthogonal array

	Factor			
Run	A	B	C	Density
1	1	1	1	5.5
2	1	2	2	5.8
3	2	1	2	6.0
4	2	2	1	5.3

Factor A: Laser power: 1,000 W: 1; 800 W: 2
Factor B: Powder feed rate: 1 g/min: 1; 0.8 g/min: 2
Factor C: Tool travel speed: 0.5 in./s: 1; 0.3 in./s: 2

One can find the sum of each corresponding level:

$$A_1 = 5.5 + 5.8 = 11.3$$
$$A_2 = 6.0 + 5.3 = 11.3$$
$$B_1 = 5.5 + 6.0 = 11.5$$
$$B_2 = 5.8 + 5.3 = 11.1$$
$$C_1 = 5.5 + 5.3 = 10.8$$
$$C_2 = 5.8 + 6.0 = 11.8$$

The analysis of the means (ANOM) is listed in Table 9.10.

Table 9.10 Design of experiments result using an L4 orthogonal array

	Factor		
Level	A	B	C
1	11.3	11.5	10.8
2	11.3	11.5	11.8
Range	0	0.4	1

Therefore, factor C (tool travel speed) has the greatest effect as it causes the process to vary and thus has the maximum range. In this case, factor A (laser power) has little effect on the outcomes.

Example 9.6: Solve the following problem by using the Taguchi's approach

The laser scribed linewidth is mainly dependent on laser specifications, beam delivery system, and the interaction mechanism between the laser beam and the workpiece material. To produce linewidth of 6 mm, the factors to be studied are the pulse repetition rate, table speed, and attenuator setting on linewidth. Table 9.11 summarizes the levels assigned to the various factors.

Table 9.11 Level assignment for the various factors

		Levels		
Factor symbol	Factor name	1	2	3
Q	Q-switch frequency (kHz)	10	20	30
S	Table speed (mm/min)	1,000	2,000	3,000
A	Attenuator setting	20	30	40

An L9 OA is employed to provide the minimum number of degrees of freedom required for the experiment. The experimental layout and raw data are shown in Table 9.12. Each experimental trial is performed with two repetitions. Complete randomization is utilized for the entire experiment.

Table 9.12 Experimental layout and raw data

	Factors			Linewidth (μm)		
Trial no.	Q	S	A	Replicate 1	Replicate 2	Average
1	1	1	1	11	12	11.5
2	1	2	2	10	11	10.5
3	1	3	3	9	11	10
4	2	1	2	10	10	10
5	2	2	3	6	8	7
6	2	3	1	9	8	8.5
7	3	1	3	6	6	6
8	3	2	1	8	10	9
9	3	3	2	8	8	8

$$Q_1 = 11.5 + 10.5 + 10 = 32$$
$$Q_2 = 10 + 7 + 8.5 = 25.5$$
$$Q_3 = 6 + 9 + 8 = 23$$
$$S_1 = 11.5 + 10 + 6 = 27.5$$
$$S_2 = 10.5 + 7 + 9 = 26.5$$
$$S_3 = 10 + 8.5 + 8 = 26.5$$
$$A_1 = 11.5 + 8.5 + 9 = 29$$
$$A_2 = 10.5 + 10 + 8 = 28.5$$
$$A_3 = 10 + 7 + 6 = 23$$

The ANOM is summarized in Table 9.13. Factor Q (Q-switch frequency) has the greatest effect, and factor S (table speed) has the least effect.

Table 9.13 Linewidth range for various factors

Factors	Q	S	A
Level 1	32	27.5	29
Level 2	25.5	26.5	28.5
Level 3	23	26.5	23
Range	9	1	6

OPTIMIZATION OF CONTROL FACTORS

Factor Q: It could be observed that factor Q at level 3 produced a mean response closer to the target value. Therefore, this level is selected as the optimized level.

Factor S: Level 2 or 3 of factor S can be chosen as they generate the mean response closer to the target value, and is more robust.

Factor A: Level 3 produced a mean response closer to the target value. Therefore, this level is selected as the optimized level.

Based on the above analysis, the recommended parameter settings are therefore chosen to be Q_3, S_1 (or S_2), and A_3.

Example 9.7

The test of an electronic device prototype yields the following performance scores. The scores should be as close to 100 points as possible. As shown in Table 9.14, factors A–F are the components of this prototype. The two levels of each factor represent the low and high tolerance levels. Three experiments under three slightly different conditions were conducted as Y_1, Y_2, and Y_3. Find the most critical factor that affects the performance scores. How could this critical factor be improved?

Table 9.14 L8 array for testing of electronic device prototype

A	B	C	D	E	F	G	Y_1	Y_2	Y_3
1	1	1	1	1	1	1	85	84	84
1	1	1	2	2	2	2	84	82	89
1	2	2	1	1	2	2	80	82	84
1	2	2	2	2	1	1	85	89	91
2	1	2	1	2	1	2	93	91	90
2	1	2	2	1	2	1	91	90	95
2	2	1	1	2	2	1	92	85	86
2	2	1	2	1	1	2	91	92	85

Answer:

The ANOM is summarized in Table 9.15:

Table 9.15 ANOM for testing of electronic device prototype

	A	B	C	D	E	F	G
Level 1	1,019	1,061	1,042	1,036	1,043	1,060	1,040
Level 2	1,081	1,039	1,061	1,064	1,057	1,050	1,060
Range	62	22	19	28	14	10	20

The calculation, for example, in A_1 is

$$85 + 84 + 84 + 84 + 82 + 89 + 80 + 82 + 84 + 85 + 89 + 91 = 1019.$$

From the table, one knows that factor A has the largest range and thus has the greatest influence on the result. The tolerance of the associated component should be tightened. Also as level 2 can yield more scores, one should pick A_2 (level 2) as the optimal factor. One can also study the robustness of factor A in Table 9.16. It can be seen that in both cases, level 2 is fairly robust as the range is not large.

Table 9.16 Level cross-checked against various environmental conditions

Factor A	Y_1	Y_2	Y_3	Range	%
Level 1	334	337	348	14	4.12
Level 2	367	358	356	11	3.05

The calculation, for example, in A_1 at Y_1 is

$$85 + 84 + 80 + 85 = 334$$

Review problems

1. Study the example of a diesel engine control calibration for a six-cylinder 9.0 L common-rail diesel engine with variable geometry turbo (VGT) and cooled EGR as provided below. It is applied in an off-road application with a very narrow engine speed range from 1,600 to 2,200 rpm. The aim of the case study is to produce optimal start of injection (SOI), base fuel, VGT, EGR, and fuel schedules as a function of torque and rpm as shown in the table below. It involves models for torque, NOx, peak pressure, equivalence ratio, exhaust temperature, VGT speed, and EGR mass fraction. The optimization setup in CAGE (Calibration Generation) tool is based on an eight-mode off-road emission test. An L8 OA array is used as shown in the following table. Find the most critical factors that affect the performance.

	Variable	Symbol	Min	Max
A	Engine speed (rpm)	RPM	1600	2200
B	Fuel quantity per injection (mg)	BFM	20	200
C	Fuel pressure (Pa)	P	90	160
D	EGR valve position (mm)	EGR	0.5	5
E	VTG rack position (%)	VTG	0.2	0.9

A	B	C	D	E	F	G	Y_1
1	1	1	1	1	1	1	9
1	1	1	2	2	2	2	9
1	2	2	1	1	2	2	6
1	2	2	2	2	1	1	8
2	1	2	1	2	1	2	4
2	1	2	2	1	2	1	5
2	2	1	1	2	2	1	8
2	2	1	2	1	1	2	8

2. In the cake-baking process, there are three aspects of the process that are analyzed by a designed experiment. An L8 OA array is used as shown in the following table. Find the critical parameter to bake a good cake.

	Factors	Levels	Responses
A	Oven	Temp	Taste
B	Sugar	Cup	Color
C	Flour	Cup	Consistency
D	Eggs	Eggs	

	Factors	High	Low
A	Oven	350°C	300°C
B	Sugar	Two cups	One cup
C	Flour	Five cups	Three cups
D	Eggs	Eight eggs	Three eggs

A	B	C	D	E	F	G	Y_1
1	1	1	1	1	1	1	8
1	1	1	2	2	2	2	7
1	2	2	1	1	2	2	6
1	2	2	2	2	1	1	8
2	1	2	1	2	1	2	5
2	1	2	2	1	2	1	2
2	2	1	1	2	2	1	2
2	2	1	2	1	1	2	1

3. The test of an electronic device prototype yields the performance scores. The scores should be as close to 100 points as possible. As shown in the following table, factors A–F are the components of this prototype. The two levels of each factor represent the low and high tolerance levels. Three experiments under three slightly different conditions were conducted as Y_1, Y_2, and Y_3. Find the most critical factor that affects the performance scores. How could this critical factor be improved?

A	B	C	D	E	F	G	Y_1	Y_2	Y_3
1	1	1	1	1	1	1	7	8	8
1	1	1	2	2	2	2	8	7	8
1	2	2	1	1	2	2	7	8	8
1	2	2	2	2	1	1	8	8	9
2	1	2	1	2	1	2	9	9	9
2	1	2	2	1	2	1	9	9	9
2	2	1	1	2	2	1	9	8	8
2	2	1	2	1	1	2	9	9	8

4. What is the smallest OA that can be chosen for factors that have three levels?
5. What is OA? Why is it useful?

9.3 *Analysis of variance*

Do not put your faith in what statistics say until you have carefully considered what they do not say.

William W. Watt

This section helps answer the following questions.

- What is ANOVA? Why is it useful?
- What is F-distribution? Why and how does one calculate F-ratio?
- What are one-way ANOVA and two-way ANOVA?
- What is an interaction? How can one find the interaction effect?
- What is pooling? How can one pool a factor?
- What is the percent influence?
- What is the S/N ratio?

The purpose of ANOVA is to test for significant differences between means [StatSoft07]. In other words, ANOVA is a technique that can be used to test the hypothesis that the means between two or more groups are equal, under the assumption that the sampled populations are normally distributed. In addition, it also assumes that the variance is equal and error is independent and identically normally distributed as well. If this is violated, it has an effect on the power of the F-test statistic. However, it is fairly robust to mild departures from all three assumptions. If there is a significant difference, the corresponding factor has an influence on the process. In statistics, ANOVA is a collection of statistical models and their associated procedures, which compare means by splitting the overall observed variance into different parts. The fundamental technique is a partitioning of the total SS into components related to the effects in the model used. ANOVA, sometimes called an F-test, is closely related to the t-test. The major difference is that, where the t-test measures the difference between the means of two groups, an ANOVA tests the difference between the means of two or more groups. The advantage of using ANOVA rather than multiple t-tests is that it reduces the probability of error and is more efficient for more complicated cases.

The F-distribution is a right-skewed distribution used most commonly in ANOVA. The F-test or F-ratio is an overall test of the null hypothesis that group means on the dependent variable do not differ. The logic of the F-test is that the larger the ratio of between-groups variance (a measure of effect) to within-groups variance (a measure of noise), the less likely that the null hypothesis is true. Once the F-ratio is obtained, one can transfer it to various distributions, such as P-value (probability value) or confidence intervals, with the degrees of freedoms of the numerator and denominator. The P-value of a statistical hypothesis test is the probability of getting a value of the test statistic as extreme as or more extreme than that observed by chance alone, if the null hypothesis is true. It is equal to the significance level of the test for which one would only just reject the null hypothesis. The P-value is compared with the actual significance level of the test, and, if it is smaller, the result is significant. That is, if the null hypothesis were to be rejected at the 5% significance level, this would be reported as $p<0.05$. There are tables available for the transformation of F-distribution to P-value. Such distributions are listed in Appendix A. The right tail area is given in the name of the table, for example, as $a = 0.05$. The F-distribution transformation is also available using Excel as discussed in Section 9.4.

9.3.1 One-way ANOVA

A one-way ANOVA tests the differences between groups that are only classified by one independent variable [Fowlkes98]. The following example is used to illustrate the terms and the ANOVA procedure.

Example 9.8

Suppose that one is interested in the study habits of our students. Table 9.17 shows the test scores of studying alone or in a group based on three observations. One can find the means of the two separate habits, 74 and 86, and the overall mean of 80.

The SS is simply used to measure the deviation of the control factor effects on the average response:

$$SS = \sum_{i=1}^{N} \left(Y_i - \bar{Y}\right)^2 \tag{9.7}$$

Table 9.17 An example to show sum of squares of two study habits

	Study habits	
	Alone	In a group
Observation 1	78	88
Observation 2	74	86
Observation 3	70	84
Mean	74	86
Sum of squares (SS)	32	8
Overall mean	80	
Total sum of squares (Total SS)	256	

where
Y_i = observation
$\bar{Y}$ = mean of the observations

Therefore,

$$SS_{Alone} = (78-74)^2 + (74-74)^2 + (70-74)^2 = 32$$

$$SS_{In\ a\ group} = (88-86)^2 + (86-86)^2 + (84-86)^2 = 8$$

$$\text{Total SS} = (78-80)^2 + (74-80)^2 + (70-80)^2 + (88-80)^2 + (86-80)^2 + (84-80)^2 = 256$$

The variance computed using the SS based on the within-group variability, or the individual mean (i.e., 74 and 86), yields a smaller estimate of variance than computing the variance based on the total variability, or the total SS based on the overall mean (i.e., 80). This means that the results of the two contributing factors (study alone or study in a group) are quite different, and thus, the confidence level should be higher than normal, which is 1. From this example, one can see how the SS can be used to check whether there are significant differences between factors. As the variances within groups are relatively small compared to the total variability, the variability must be from between the two factors. In other words, the observations have resulted in data that show good confidence that the contributing factors are quite different.

To perform an ANOVA on the above data, the following result, as shown in Table 9.18, can be obtained:

Table 9.18 One-way ANOVA results of Example 9.3.1

	SS	df	F
Effect, or between factor SS or SS_F	216	1	21.6
Error, or error SS or within group SS_e	40	4	

One-way ANOVA was calculated based on the following procedure:

Step 1: Compute total SS or SS_T and its components, SS_F and SS_e

$$\begin{aligned} SS_T &= \sum_{i=1}^{N}\left(Y_i - \bar{Y}\right)^2 = \sum_{i=1}^{N}\left(Y_i{}^2 - 2\bar{Y} + \bar{Y}^2\right) = \sum_{i=1}^{N} Y_i{}^2 - 2\bar{Y}\sum_{i=1}^{N} Y_i + \sum_{i=1}^{N}\bar{Y}^2 \\ &= \sum_{i=1}^{N} Y_i{}^2 - 2\bar{Y}T + N\bar{Y}^2 = \sum_{i=1}^{N} Y_i{}^2 - \frac{T^2}{N} \end{aligned} \tag{9.8}$$

where $T = \text{total} = N\bar{Y}$

Note that for general expression, the total sums of squares,

$$SS_T = \sum_{i=1}^{a}\sum_{j=1}^{n}\left(Y_{ij} - \bar{Y}_{..}\right)^2 = \sum_{i=1}^{a}\sum_{j=1}^{n}\left[\left(\bar{Y}_{i.} - \bar{Y}_{..}\right) + \left(Y_{ij} - \bar{Y}_{i.}\right)\right]^2$$
$$= n\sum_{i=1}^{a}\left(\bar{Y}_{i.} - \bar{Y}_{..}\right)^2 + \sum_{i=1}^{a}\sum_{j=1}^{n}\left(Y_{ij} - \bar{Y}_{i.}\right)^2 + 2\sum_{i=1}^{a}\sum_{j=1}^{n}\left(\bar{Y}_{i.} - \bar{Y}_{..}\right)\left(Y_{ij} - \bar{Y}_{i.}\right) \tag{9.9}$$
$$= n\sum_{i=1}^{a}\left(\bar{Y}_{i.} - \bar{Y}_{..}\right)^2 + \sum_{i=1}^{a}\sum_{j=1}^{n}\left(Y_{ij} - \bar{Y}_{i.}\right)^2 = SS_F + SS_e$$

where

n = number of factors

a = number of observations

$$Y_{i.} = \sum_{j=1}^{n} Y_{ij} \quad \bar{Y}_{i.} = \frac{Y_{..}}{n}, \quad i = 1,2,\ldots,a, \quad j = 1,2,\ldots,n \tag{9.10}$$

$$Y_{..} = \sum_{i=1}^{a}\sum_{j=1}^{n} Y_{ijk}, \quad \bar{Y}_{..} = \frac{Y_{..}}{an} \tag{9.11}$$

In Example 9.8, $SS_T = 256$

$$SS_F = 3\left[(74 - 80)^2 + (86 - 80)^2\right] = 216$$

$$SS_e = (78 - 74)^2 + (74 - 74)^2 + (70 - 74)^2 + (88 - 86)^2 + (86 - 86)^2 + (84 - 86)^2 = 40$$

As one can see, in Table 9.18 the total SS (256) was partitioned into the SS due to within-group variability (32 + 8 = 40) and variability due to differences between means (256 – 40 = 216).

Step 2: Find the mean squares. The SS needs to be divided by the number of degrees of freedom (df) to yield the average or mean squares, MS, for a fair comparison with the other factors. This is similar to calculating the variance in that the SS needs to be divided by (n – 1). In other words,

$$MS_{Factor} = \frac{SS_F}{a - 1}. \tag{9.12}$$

$$MS_e = \frac{SS_e}{N - a}. \tag{9.13}$$

In Example 9.8

$$MS_F = 216/(2 - 1) = 216$$

$$MS_e = 40/(6 - 2) = 10$$

Step 3: Find F-ratio, F. F-ratio = $MS_F/MS_e = 216/10 = 21.6 \gg 1$. Therefore, the error mean squares are much smaller when compared to the between-factor mean squares. This means that good confidence in the experiment is confirmed.

9.3.2 *Two-way ANOVA*

A factorial ANOVA can examine data for multiple independent variables. For example, a two-way ANOVA can measure the difference among factors simultaneously. One can use more than two independent variables in an ANOVA (e.g., three-way, four-way). A factorial ANOVA can show whether there are significant main effects of the independent variables and whether there are significant interaction effects between independent variables in a set of data. Interaction effects occur when the impact of one independent variable depends on the level of the second independent variable. Computation can be done on statistical software.

For various factors and levels, the total SS,

$$SS_T = \sum_{i=1}^{a}\sum_{j=1}^{b}\sum_{k=1}^{n}\left(Y_{ijk} - \bar{Y}_{...}\right)^2$$

$$= \sum_{i=1}^{a}\sum_{j=1}^{b}\sum_{k=1}^{n}\left[\left(\bar{Y}_{i..} - \bar{Y}_{...}\right) + \left(\bar{Y}_{.j.} - \bar{Y}_{...}\right) + \left(\bar{Y}_{ij.} - \bar{Y}_{i..} - \bar{Y}_{.j.} + \bar{Y}_{...}\right) + \left(Y_{ijk} - \bar{Y}_{ij.}\right)\right]^2$$

$$= bn\sum_{i=1}^{a}\left(\bar{Y}_{i..} - \bar{Y}_{...}\right)^2 + an\sum_{j=1}^{b}\left(\bar{Y}_{.j.} - \bar{Y}_{...}\right)^2 + n\sum_{i=1}^{a}\sum_{j=1}^{b}\left(\bar{Y}_{ij.} - \bar{Y}_{i..} - \bar{Y}_{.j.} + \bar{Y}_{...}\right)^2 \tag{9.14}$$

$$+ \sum_{i=1}^{a}\sum_{j=1}^{b}\sum_{k=1}^{n}\left(Y_{ijk} - \bar{Y}_{ij.}\right)^2$$

$$= SS_A + SS_B + SS_{AB} + SS_e = \sum_{i=1}^{a}\sum_{j=1}^{b}\sum_{k=1}^{n} Y_{ijk}^2 - \frac{Y_{...}^2}{abn}$$

where

a and b = numbers of levels for factors A and B, respectively

n = number of factors

$$Y_{i..} = \sum_{j=1}^{b}\sum_{k=1}^{n} Y_{ijk}, \quad \bar{Y}_{i..} = \frac{Y_{i..}}{bn}, \quad i = 1,2,\ldots,a \tag{9.15}$$

$$Y_{.j.} = \sum_{i=1}^{a}\sum_{k=1}^{n} Y_{ijk}, \quad \bar{Y}_{.j.} = \frac{Y_{.j.}}{an}, \quad j = 1,2,\ldots,b \tag{9.16}$$

$$Y_{ij.} = \sum_{k=1}^{n} Y_{ijk}\bar{Y}_{ij.} = \frac{Y_{...}}{n}, \quad i = 1,2,\ldots,a, \quad j = 1,2,\ldots,b \tag{9.17}$$

$$Y_{...} = \sum_{i=1}^{a}\sum_{j=1}^{b}\sum_{k=1}^{n} Y_{ijk}, \quad \bar{Y}_{...} = \frac{Y_{...}}{abn}. \tag{9.18}$$

Again, SS_T can be separated into several components. SS_A is the SS for factor A, SS_B is the SS for factor B, SS_e is the error SS, and SS_{AB} is the SS for interference between A and B, if any. From this equation, one can find the expressions of all the terms. For example,

$$SS_A = bn\sum_{i=1}^{a}\left(\bar{Y}_{i..} - \bar{Y}_{...}\right)^2 = bn\sum_{i=1}^{a}\left(\bar{Y}_{i..}{}^2 - 2\bar{Y}_{i..}\bar{Y}_{...} + \bar{Y}_{i...}{}^2\right)$$

$$= bn\sum_{i=1}^{a}\left(\bar{Y}_{i..}{}^2\right) + bn\sum_{i=1}^{a}(-2\bar{Y}_{i..}\bar{Y}_{...}) + abn\,\bar{Y}_{i...}{}^2$$

$$= bn\sum_{i=1}^{a}\left(\frac{Y_{i..}}{bn}\right)^2 - 2bn\bar{Y}_{...}\sum_{i=1}^{a}(\bar{Y}_{i..}) + abn\,\bar{Y}_{i...}{}^2 \tag{9.19}$$

$$= \sum_{i=1}^{a}\frac{\bar{Y}_{i..}{}^2}{bn} - 2bn\bar{Y}_{...}(a\bar{Y}_{...}) + abn\bar{Y}_{i..}{}^2$$

$$= \sum_{i=1}^{a}\frac{\bar{Y}_{i..}{}^2}{bn} - abn\bar{Y}_{...}^2 = \sum_{i=1}^{a}\frac{\bar{Y}_{i..}{}^2}{bn} - \frac{\bar{Y}_{...}{}^2}{abn}$$

$$SS_B = \sum_{j=1}^{b}\frac{Y_{.j.}{}^2}{an} - \frac{Y_{...}{}^2}{abn}. \tag{9.20}$$

SS_{AB}, the interaction between A and B, can be calculated from the following equation:

$$SS_{AB} = \sum_{i=1}^{a}\sum_{j=1}^{b}\frac{Y_{ij.}{}^2}{n} - \frac{Y_{...}{}^2}{abn} - SS_A - SS_B \tag{9.21}$$

The mean squares can be computed as

$$MS_A = \frac{SS_A}{a-1}. \tag{9.22}$$

$$MS_B = \frac{SS_B}{b-1}. \tag{9.23}$$

$$MS_e = \frac{SS_e}{ab(n-1)} \tag{9.24}$$

$$MS_{AB} = \frac{SS_{AB}}{(a-1)(b-1)} \tag{9.25}$$

The F-ratios are

$$F_A = \frac{MS_A}{MS_e} \tag{9.26}$$

$$F_B = \frac{MS_B}{MS_e} \quad (9.27)$$

$$F_{AB} = \frac{MS_{AB}}{MS_e} \quad (9.28)$$

Example 9.9

Table 9.19 shows the laser deposition quality experiment based on table velocity and laser power. Find SS_T (total SS), SS_V (SS for velocity), SS_P (SS for power), SS_e (error SS), and F-ratios.

Table 9.19 Data for laser deposition quality experiment

Power	V = 0.5	V = 0.4	V = 0.3
250	12	20	41
300	15	25	46
350	19	28	53
400	15	22	40

In this case, a = 3, b = 4, and n = 1 for one sample and no replication.

$$SS_T = (12)^2 + (20)^2 + (41)^2 + (15)^2 + \cdots - 336^2/12 = 2046$$

$$SS_V = bn\sum_{i=1}^{a}\left(\bar{Y}_{i..} - \bar{Y}_{...}\right)^2 = 4\left[(15.25-28)^2 + (23.75-28)^2 + (45-28)^2\right] = 1878.5$$

$$SS_P = an\sum_{i=1}^{b}\left(\bar{Y}_{.j.} - \bar{Y}_{...}\right)^2 = 3\left[(24.33-28)^2 + (28.67-28)^2 + (33.33-28)^2 + (25.67-28)^2\right]$$

$$= 143.33$$

Another way to calculate SS_V and SS_P is

$$SS_V = \sum_{i=1}^{a}\frac{Y_{i..}^{\ 2}}{bn} - \frac{Y_{...}^{\ 2}}{abn} = \left(\frac{61^2 + 95^2 + 180^2}{4}\right) - \left(\frac{336^2}{12}\right) = 1878.5$$

$$SS_P = \sum_{j=1}^{b}\frac{Y_{.j.}^{\ 2}}{an} - \frac{Y_{...}^{\ 2}}{abn} = \left(\frac{73^2 + 86^2 + 100^2 + 77^2}{3}\right) - \left(\frac{336^2}{12}\right) = 143.33$$

SS_{VP} will not be calculated as this is a randomized complete block problem (n = 1).

$$SS_e = SS_T - SS_V - SS_P = 2{,}046 - 1878.5 - 143.33 = 24.17$$

$$MS_V = SS_V/(a-1) = 1{,}878.5/2 = 924.25$$

$$MS_P = SS_P/(b-1) = 143.33/3 = 47.78$$

$$MS_e = SS_e/\left[(abn-1) - (a-1) - (b-1)\right] = 24.17/6 = 4.03$$

$$F_V = MS_V/MS_e = 924.25/4.03 = 229$$

$$F_P = MS_P/MS_e = 47.78/4.03 = 11.86$$

9.3.3 Three-way ANOVA

Similar to the previous section, the equations for a three-way ANOVA is summarized in Table 9.20.

Table 9.20 Three-way ANOVA

Source of variation	Sum of squares	Degrees of freedom
A	$SS_A = \sum_{i=1}^{a} \frac{Y_{i...}^2}{bcn} \quad \frac{Y_{....}^2}{abcn}$	a–1
B	$SS_B = \sum_{j=1}^{b} \frac{Y_{j...}^2}{acn} \quad \frac{Y_{....}^2}{abcn}$	b–1
C	$SS_C = \sum_{i=1}^{a} \frac{Y_{..K.}^2}{abn} \quad \frac{Y_{....}^2}{abcn}$	c–1
AB	$SS_{AB} = \sum_{i=1}^{a} \sum_{j=1}^{b} \frac{Y_{ij..}^2}{cn} - \frac{Y_{....}^2}{abcn} - SS_A - SS_B$	(a–1) (b–1)
AC	$SS_{AC} = \sum_{j=1}^{a} \sum_{k=1}^{c} \frac{Y_{i.k.}^2}{bn} - \frac{Y_{....}^2}{abcn} - SS_A - SS_C$	(a–1) (c–1)
BC	$SS_{BC} = \sum_{j=1}^{b} \sum_{k=1}^{c} \frac{Y_{.jk.}^2}{an} - \frac{Y_{....}^2}{abcn} - SS_B - SS_C$	(b–1) (c–1)
ABC	$SS_{ABC} = \sum_{i=1}^{a} \sum_{j=1}^{b} \sum_{k=1}^{c} \frac{Y_{ijk.}^2}{n} - \frac{Y_{....}^2}{abcn} - SS_A - SS_B - SS_C - SS_{AB} - SS_{AC} - SS_{BC} - SS_{AC}$	(a–1)(b–1) (c–1)
Error	$SS_e = SS_T - SS_A - SS_B - SS_C - SS_{AB} - SS_{AC} - SS_{BC} - SS_{ABC}$	abc(n–1)
Total	$SS_T = \sum_{i=1}^{a} \sum_{j=1}^{b} \sum_{k=1}^{c} \sum_{l=1}^{n} Y_{ijkl}^2 - \frac{Y_{....}^2}{abcn}$	abcn–1

9.3.4 Interaction effects

There is an interaction between two factors if the effect of one factor depends on the levels of the second factor. Imagine that one has a sample of study habit results for students studying in a group or not. To measure how well the students work on the test, the means of this study are as shown in Table 9.21.

Table 9.21 Study habit example to show the interaction effect

	Study habit	
	Alone	In a group
Easy test	90	60
Tough test	60	80

Can one summarize that the students' scores depend on the study habit? Can one say students test better when they have studied in a group? Or, can one say when students study in a group they tend to get higher scores? Obviously, one cannot make such conclusions, and this means that the main effects that lead to the above questions are not significant based on the data. One can only summarize that for an easy test, studying alone is likely to yield good grades, whereas studying for a tough test in a group is likely to yield better grades. In other words, study habit and test difficulty interact in their effect.

Figure 9.4 shows the line graph of the case. Interaction is indicated by nonparallel lines in a line graph. If the lines are crossed or would cross if extended, then there is an interaction. Since the lines are rarely perfectly parallel, the question is whether the different pattern of means across the subgroups is likely to occur. The F-test of the interaction can reveal the chance of interaction. In the Taguchi's method, as one often has small data sets, there is a great amount of variation in these graphs from sample to sample. If the interaction is significant, one cannot examine the main effects because the main effects do not tell the complete story. The Taguchi's method often confounds all, but assumes important interactions. This requires one to understand significant factors to the response well ahead of conducting the experiment. A superior alternative is conducting a series of iterative experiments that can be folded to determine what factors are important.

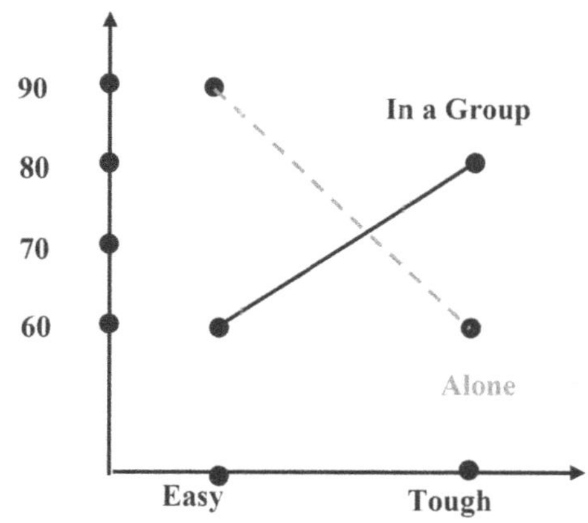

Figure 9.4 The line graph shows interaction.

Example 9.10

Table 9.22 shows the laser deposition quality experiment based on table velocity and laser power. Find SS_T, SS_V, SS_P, SS_{VP}, SS_e, and F-ratios.

Table 9.22 Laser deposition quality experiment

Power	V = 0.5	V = 0.4	V = 0.3
250	12, 13, 11, 12	20, 22, 23, 19	41, 40, 42, 41
300	15, 16, 15, 17	25, 24, 27, 23	46, 45, 47, 48
350	19, 20, 22, 20	28, 27, 29, 30	53, 52, 53, 53
400	15, 16, 15, 27	22, 23, 21, 24	40, 41, 42, 39

Answer:

$$SS_T = \sum_{i=1}^{a}\sum_{j=1}^{b}\sum_{k=1}^{n} Y_{ijk}{}^2 - \frac{Y_{...}{}^2}{abn} = (12)^2 + (13)^2 + (11)^2 + (12)^2 + (20)^2 + (22)^2 + \cdots - \frac{1,375^2}{48}$$

$$= 7,858.98$$

$$SS_V = \sum_{i=1}^{a} \frac{Y_{i..}{}^2}{bn} - \frac{Y_{...}{}^2}{abn} = \left(\frac{265^2 + 387^2 + 723^2}{16}\right) - \left(\frac{1,375^2}{48}\right) = 7,032.17$$

$$SS_P = \sum_{j=1}^{b} \frac{Y_{.j.}{}^2}{an} - \frac{Y_{...}{}^2}{abn} = \left(\frac{296^2 + 348^2 + 406^2 + 325^2}{12}\right) - \left(\frac{1375^2}{48}\right) = 543.73$$

$$SS_{VP} = \sum_{i=1}^{a}\sum_{j=1}^{b} \frac{Y_{ij.}{}^2}{n} - \frac{Y_{...}{}^2}{abn} - SS_V - SS_P = \frac{48^2 + 84^2 + 164^2 + \ldots + (162)^2}{4}$$

$$-\left(\frac{1,375^2}{48}\right) - 7,032.17 - 543.73 = 129.33\,(\text{interaction})$$

$$SS_e = SS_T - SS_V - SS_P - SS_{VP} = 7,858.98 - 7,032.17 - 543.17 - 129.33 = 153.75$$

$$MS_V = 7,032.17/(3-1) = 3,516.08$$

$$MS_P = 543.73/(4-1) = 181.24$$

$$MS_{VP} = 129.33/(3-1)(4-1) = 21.16$$

$$MS_e = 153.33/(47-2-3-6) = 153.33/36 = 4.270833$$

$$F_V = 3,516.08/4.270833 = 823\,(\text{P-value} = 6.56\text{E}-12)$$

$$F_P = 181.24/4.270833 = 42\,(\text{P-value} = 8.83066\text{E}-31)$$

$$F_{VP} = 21.56/4.270833 = 5\,(\text{P-value} = 0.00076)$$

9.3.5 *Two-way ANOVA and OAs*

The two-way ANOVA can also be applied to the OAs. SSA, a factor SS, is a measure of factor A variability. When the number of levels for factor A, or "a," is equal to 2, then

$$SS_A = \frac{A_1{}^2}{N_{A_1}} + \frac{A_2{}^2}{N_{A_2}} - CF, \quad \text{and} \quad CF = \frac{T^2}{N} \tag{9.29}$$

Also, if

$$N_{A_1} = N_{A_2} = N/2$$

it can be shown that

$$SS_A = \frac{(A_1 - A_2)^2}{N} \tag{9.30}$$

where

N_{A_1} = total number of experiments in which level 1 of factor A is present
N_{A_2} = total number of experiments in which level 2 of factor A is present

Similarly,

$$SS_B = \frac{B_1^{\,2}}{N_{B_1}} + \frac{B_2^{\,2}}{N_{B_2}} - CF, \quad \text{and} \quad CF = \frac{T^2}{N} \tag{9.31}$$

$$SS_B = \frac{(B_1 - B_2)^2}{N} \tag{9.32}$$

where

N_{B_1} = total number of experiments in which level 1 of factor B is present
N_{B_2} = total number of experiments in which level 2 of factor B is present

Note that although there are only two factors, A and B in this case, the other one, C, actually represents the interaction A × B.

$$SS_{AB} = \frac{(C_1 - C_2)^2}{N} \tag{9.33}$$

Let df be the number of degrees of freedom; and df for a factor is always equal to 1 less than the number of levels of that factor, similar to Table 9.20, which was for a three-way ANOVA. For example, for factor A, df = (a – 1), for factor B, df = (b – 1), and for interaction A × B, df = (a – 1)(b – 1). Total df = (N – 1).

Also, one can calculate the percent influence of each factor using the following equations:

$$\text{The pure sum of square } SS'_A = SS_A - (SS_e \times df_A) \tag{9.34}$$

$$\text{Percent influence } PI_A = \frac{SS'_A}{SS_T} \tag{9.35}$$

The percent influence of all factors should be equal to 100%.

Example 9.11

Conduct a two-way ANOVA with two levels using an L4 OA as shown in Table 9.23. Find SS_T, SS_A, SS_B, SS_C, SS_e, and F-ratio.

Table 9.23 L4 orthogonal array example

Run	A	B	C	Result
1	1	1	1	2
2	1	2	2	5
3	2	1	2	3
4	2	2	1	1

In this case a = 2, b = 2, c = 2, and n = 1 as there is only one column.

$$SS_A = \frac{A_1{}^2}{N_{A_1}} + \frac{A_2{}^2}{N_{A_2}} - \frac{T^2}{N} = \frac{7^2}{2} + \frac{4^2}{2} - \frac{(2+5+3+1)^2}{2\times 2} = 2.25$$

$$SS_B = \frac{B_1{}^2}{N_{B_1}} + \frac{B_2{}^2}{N_{B_2}} - \frac{T^2}{N} = \frac{5^2}{2} + \frac{6^2}{2} - \frac{(2+5+3+1)^2}{2\times 2} = 0.25$$

$$SS_C = \frac{C_1{}^2}{N_{C_1}} + \frac{C_2{}^2}{N_{C_2}} - \frac{T^2}{N} = \frac{3^2}{2} + \frac{8^2}{2} - \frac{(2+5+3+1)^2}{2\times 2} = 6.25$$

However, the degree of freedom is zero as $n - 1 = 0$, and thus, the F-ratio cannot be found in this case. This is due to the fact that only one set of data is available.

$$\bar{Y}_{...} = (2+5+3+1)/4 = 2.75$$

$$SS_T = (2-2.75)^2 + (5-2.75)^2 + (3-2.75)^2 + (1-2.75)^2 = 8.75$$

$$SS_e = SS_T - SS_A - SS_B - SS_C = 8.75 - 2.25 - 0.25 - 6.25 = 0$$

The pure SS

$$SS_A{}' = SS_A - (SS_e \times df_A) = SS_A = 2.25$$

$$SS_B{}' = SS_B - (SS_e \times df_B) = SS_B = 0.25$$

$$SS_C{}' = SS_C - (SS_e \times df_C) = SS_C = 6.25$$

Percent influence

$$PI_A = \frac{SS_A{}'}{SS_T} = \frac{2.25}{8.75} = 25.71\%$$

$$PI_B = \frac{SS_B{}'}{SS_T} = \frac{2.25}{8.75} = 2.86\%$$

$$PI_C = \frac{SS_C{}'}{SS_T} = \frac{6.25}{8.75} = 71.43\%$$

This means that factor B has the smallest influence and thus can be ignored, and this process is called pooling. If factor B is ignored and pooled, one can recalculate SS_e as 0.25.

$$SS_A{}' = SS_A - (SS_e \times df_A) = SS_A - (0.25 \times 1) = 2$$

$$SS_C{}' = SS_C - (SS_e \times df_C) = SS_C - (0.25 \times 1) = 6$$

$$PI_A = \frac{SS_A'}{SS_T} = \frac{2}{8.75} = 22.86\%$$

$$PI_C = \frac{SS_C'}{SS_T} = \frac{6}{8.75} = 68.57\%$$

$$PI_e = 100\% - 22.86\% - 68.57\% = 8.57\%$$

$$F_A = \frac{MS_A}{MS_e} = \frac{2.25}{0.25} = 9$$

$$F_C = \frac{MS_C}{MS_e} = \frac{6.25}{0.25} = 25$$

Example 9.12

Using an L8 OA as shown in Table 9.24, find SS_T, SS_A, SS_B, SS_C, SS_e, and F-ratio.

Table 9.24 L8 orthogonal array example

	A	B	C = A × B		
	1	2	3	4–7	Data
1	1	1	1		3
2	1	1	1		4
3	1	2	2		5
4	1	2	2		7
5	2	1	2		5
6	2	1	2		3
7	2	2	1		9
8	2	2	1		10

$$SS_A = \frac{(A_1 - A_2)^2}{N} = \frac{((3+4+5+7)-(5+3+9+10))^2}{8} = 8$$

$$SS_B = \frac{(B_1 - B_2)^2}{N} = \frac{((3+4+5+3)-(5+7+9+10))^2}{8} = 32$$

$$SS_{AB} = \frac{(C_1 - C_2)^2}{N} = \frac{((3+4+9+10)-(5+7+5+3))^2}{8} = 4.5$$

$$SS_T = \sum_{i=1}^{a}\sum_{j=1}^{b}\sum_{k=1}^{n} Y_{ijk}{}^2 - \frac{Y_{...}{}^2}{abn} = (3)^2 + (4)^2 + (5)^2 + (7)^2 + (5)^2 + (3)^2 + \cdots - \frac{46^2}{8} = 49.5$$

$$SS_e = SS_T - SS_A - SS_B - SS_C = 49.5 - 8 - 32 - 4.5 = 5$$

$$SS_A' = SS_A - (SS_e \times df_A) = SS_A - 5 = 3$$

The pure sum of square $SS_B' = SS_B - (SS_e \times df_B) = SS_B - 5 = 27$

$$SS_C' = SS_C - (SS_e \times df_C) = SS_C - 5 = -0.5$$

$$PI_A = \frac{SS_A'}{SS_T'} = \frac{3}{49.5} = 6.06\%$$

$$\text{Percent influence } PI_B = \frac{SS_B'}{SS_T} = \frac{27}{49.5} = 54.54\%$$

$$PI_C = \frac{SS_C'}{SS_T} = \frac{-.5}{49.5} = -0.1\%$$

The interaction A × B can be ignored. If factor A × B is ignored and pooled, one can recalculate SS_e as 0.25.

$$F_A = \frac{MS_A}{MS_e} = \frac{8/1}{9.5/5} = 4.2$$

$$F_B = \frac{MS_A}{MS_e} = \frac{27}{9.5/5} = 14.2$$

9.3.6 S/N ratios

The S/N ratio is a measurement that compares a signal's strength to any background noise present and is used to calculate the design robustness. The larger the S/N ratio, the more robust the performance. However, the S/N ratio alone does not measure how well the performance at that point minimizes the Taguchi's loss function. Taguchi has developed various types of S/N ratios as universal indices for the evaluation of quality and reliability of products and processes. The S/N ratio gives a sense of how close the design is to the optimum performance of a product or process. The S/N ratio is a utility function that is merely a transformation of the loss function. When the S/N is maximized, one has minimized the loss and has found the optimal quality that the system can deliver. As one recalls that the average quality loss per unit was given in Equation 9.5, the customer's average quality loss depends on the deviation of the mean from the target and also on the variance. The product/process/system design phase involves finding the best values or levels for the control factors. The S/N ratio is an ideal metric for that purpose. Between the mean and standard deviation, it is typically easy to adjust the mean on target, but reducing the variance is difficult. Therefore, the designer should minimize the variance first and then adjust the mean on target. Among the available control factors, most of them should be used to reduce variance. Only one or two control factors are adequate for adjusting the mean on target. Variability in the product–process function is the enemy of quality. Taguchi relates this deviation from ideal to noise, which is variability. Factors that cause variations are referred to as noise factors. By definition, noise factors are uncontrolled either from a practical or cost standpoint. The S/N ratio is defined as

$$S/N = -10 \log MSD \tag{9.36}$$

MSD is the mean square deviation and is calculated differently for nominal-the-best, smaller-the-best, and larger-the-best situations.

Most parts in mechanical fittings have dimensions that are nominal-the-best type:

$$MSD = \frac{1}{n}\sum_{i=1}^{n}\left(Y_i - \bar{Y}\right)^2 \tag{9.37}$$

For all undesirable characteristics, such as defects and costs, for which the ideal value is zero, choose the smaller-the-best type. (The minimization problem is often called least square mean deviation.):

$$\text{MSD} = \frac{1}{n}\sum_{i=1}^{n}(Y_i)^2 \tag{9.38}$$

For all desirable characteristics such as profits and efficiency, choose larger-the-best type:

$$\text{MSD} = \frac{1}{n}\sum_{i=1}^{n}\frac{1}{Y_i^2} \tag{9.39}$$

Example 9.13

A rapid prototyping machine was tested for its performance in terms of producing a quality surface. The parameter design base was an L8 array as shown in Table 9.25, and the surface quality index is the lower-the-better. Find the S/N ratio.

Table 9.25 L8 orthogonal array S/N ratio example

	A	B	C					
	1	2	3	4–7	Y_1	Y_2	Y_3	S/N
1	1	1	1		3	3	4	−10.54
2	1	1	1		4	5	2	−11.76
3	1	2	2		5	6	5	−14.57
4	1	2	2		7	8	6	−16.96
5	2	1	2		5	6	7	−15.64
6	2	1	2		3	2	1	−6.69
7	2	2	1		9	9	10	−19.41
8	2	2	1		10	10	12	−20.59

As this is a smaller-the-best problem, one needs to use Equation 9.38 to find MSD. For the first row:

$$\text{MSD} = \frac{1}{n}\sum_{i=1}^{n}(Y_i)^2 = \frac{1}{3}\left(3^2 + 3^2 + 4^2\right) = 11.33$$

$$\text{S/N} = -10\log\text{MSD} = -10(1.054) = -10.54$$

The rest of the S/N results are shown in the same table. As discussed before, S/N ratio is opposite to loss function, and thus, the process is to find a process to maximize the S/N ratio. As can be observed in this case, the smaller the test results, the better the quality, and thus, the larger the S/N ratio.

Review problems

1. Suppose that we are interested in the study habits of our students. The following table shows the test scores of studying alone or in a group based on three observations. Fill in the blanks on the table:

	Study habits	
	Alone	In a group
Observation 1	90	88
Observation 2	84	86
Observation 3	70	84
Mean	81	86
SS		
Overall mean	83.67	
Total SS		

2. The following table shows the laser deposition quality experiment based on table velocity and laser power. Find SS_T (total SS), SS_V (SS for velocity), SS_P (SS for power), SS_e (error SS), and F-ratios.

Power	V = 0.5	V = 0.4	V = 0.3
250	24	20	23
300	25	25	47
350	28	28	23
400	25	22	30

3. An L4 OA as shown in the following table. Find SS_T, SS_A, SS_B, SS_C, SS_e, and F-ratio.

Run	A	B	C	Result
1	1	1	1	2
2	1	2	2	2
3	2	1	2	3
4	2	2	1	4

4. The thickness required for a bulletproof glass window needs to be determined. The standard is given as

   For momentum range, 32–34 kg m/s thickness should be 5 mm.
   For momentum range, 34–36 kg m/s thickness should be 8 mm.
   For momentum range, 36–38 kg m/s thickness should be 10 mm.
   Momentum is calculated as the product of the mass of bullet (kg) and its velocity (m/s).

   The glass is intended to withstand a bullet from a 0.38 revolver. The variation in the mass of the bullet is 90–100 g. The variation in speed due to proximity is 350–400 m/s. To conduct the experiment, our factors are A (weight) and B (speed), and the function is $A \times B = M$ (momentum). An L4 array was used to construct the table shown below. Find the percentage influence of A and B and F-ratios.

Run	A	B	M
1	1	1	31.5
2	1	2	36
3	2	1	35
4	2	2	40

5. A rapid prototyping machine was tested for its performance in terms of producing a quality surface. The parameter design base was an L8 array as shown in the following table, and the surface quality index is the lower-the-better. Find the S/N ratio.

	A	B	C					
	1	2	3	4–7	Y_1	Y_2	Y_3	S/N
1	1	1	1		5	5	6	–14.57
2	1	1	1		6	7	4	–15.27
3	1	2	2		7	8	7	–17.32
4	1	2	2		9	10	8	–19.12
5	2	1	2		7	8	9	–18.11
6	2	1	2		5	4	3	–12.22
7	2	2	1		11	11	12	–21.09
8	2	2	1		12	12	14	–22.08

9.4 *ANOVA using Excel*

Computers are useless. They can only give you answers.

Pablo Picasso

This section helps answer the following questions.

- How to use Microsoft Excel to find ANOVA?
- Can one use Microsoft Excel to find F-table or P-value? How?

This section summarizes an Add-In tool in Microsoft Excel for ANOVA. If Data Analysis cannot be found in Tools, one can find Add-Ins under Tools. One will have to add-in the Analysis ToolPak by hand. Data Analysis will appear under Tools. Various versions of Excel may have slightly different setup procedures.

9.4.1 *Single-factor (one-way) ANOVA*

Conduct a one-way ANOVA using the following steps:

Step 1: Enter data into a spreadsheet. Let us use Example 9.8 as an example. Figure 9.5 shows the data input.

Step 2: Select Data Analysis from the Tools Menu. Choose ANOVA: Single Factor, as shown in Figure 9.6. Since the only factor is "study habit," one-way ANOVA is used. The rows are the observations.

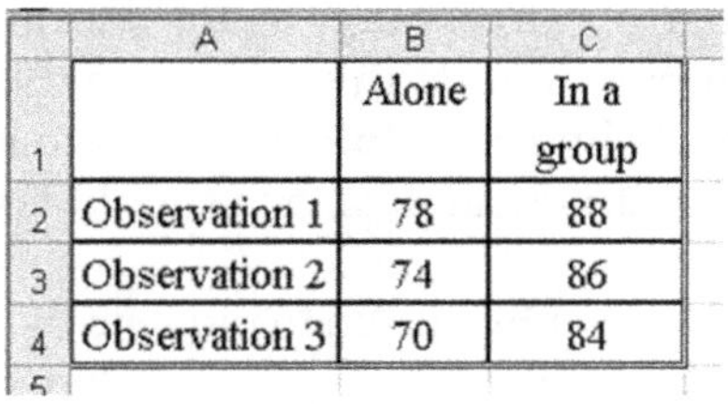

	A	B	C
1		Alone	In a group
2	Observation 1	78	88
3	Observation 2	74	86
4	Observation 3	70	84

Figure 9.5 The input data in Excel.

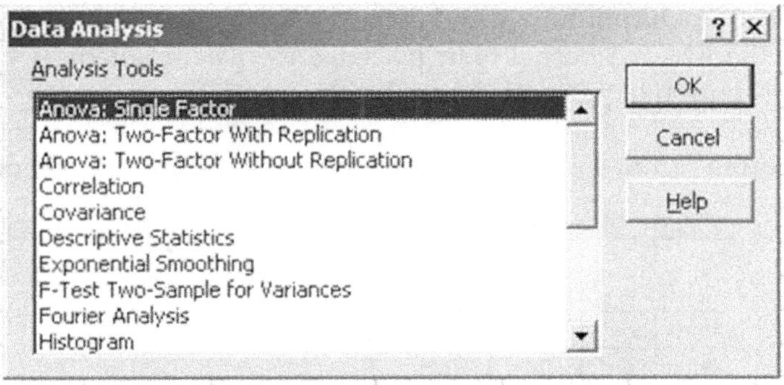

Figure 9.6 Select single factor for one-way ANOVA.

Step 3: Fill in the required information in the dialog boxes. As shown in Figure 9.7, an input dialog will appear once Step 2 is carried out. The following items will need to be specified.

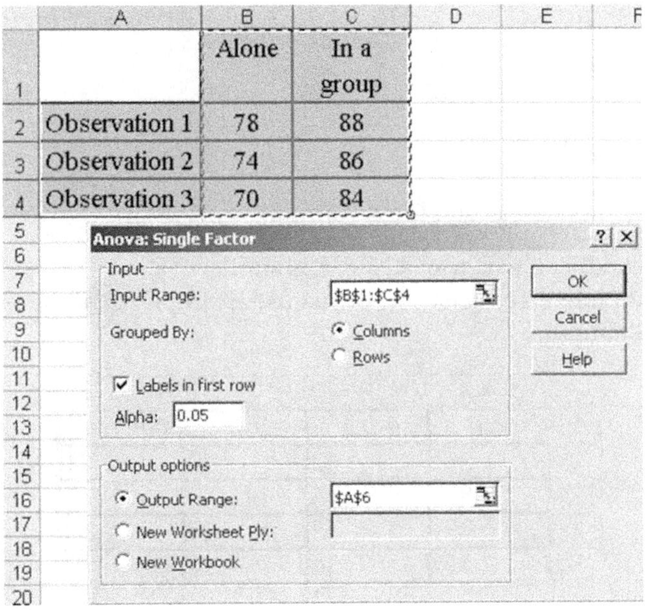

Figure 9.7 Specifying input range and grouped by to specify the data in Excel.

- Input range with the range of cells containing the data and column headings. Use the mouse to highlight the input range. In this case, it was \$B\$1:\$C\$4. Note that the range could include the major labeling cells, but one needs to let Excel know about it. If the first column contains nonnumerical data, one should not include the first column in the Input Range.
- Check the appropriate Grouped By button to indicate the data is grouped by columns.
- Since one has specified that column headings are included, check the Labels in the first row box. This tells Excel that the words in the column titles are not numerical data.
- Enter an appropriate level in the Alpha box which is used for the F-test. The default is 0.05.
- Check the button Output Range and enter the top left cell on the spreadsheet where one wishes the ANOVA output to be placed. \$A\$6 was selected.
- Press the OK button to perform the analysis.

Step 4: ANOVA output summary. One should see the output summary as shown in Figure 9.8.

	A	B	C	D	E	F	G
1		Alone	In a group				
2	Observation 1	78	88				
3	Observation 2	74	86				
4	Observation 3	70	84				
5							
6	Anova: Single Factor						
7							
8	SUMMARY						
9	*Groups*	*Count*	*Sum*	*Average*	*Variance*		
10	Alone	3	222	74	16		
11	In a group	3	258	86	4		
12							
13							
14	ANOVA						
15	*Source of Variation*	*SS*	*df*	*MS*	*F*	*P-value*	*F crit*
16	Between Groups	216	1	216	21.6	0.009679	7.70865
17	Within Groups	40	4	10			
18							
19	Total	256	5				

Figure 9.8 Output data in Excel.

9.4.2 *Two-factor (two-way) ANOVA without replication*

The procedure to conduct a two-way ANOVA is similar.

Step 1: Enter data into a spreadsheet. Let us use Example 9.9 as an example. Figure 9.9 shows the data input.

	A	B	C	D
1	Power	V = 0.5	V=0.4	V=0.3
2	250	12	20	41
3	300	15	25	46
4	350	19	28	53
5	400	15	22	40

Figure 9.9 Input data in Excel.

Step 2: Select Data Analysis from the Tools Menu. Choose ANOVA: Two-Factor without Replication, as shown in Figure 9.10. Since there are two factors in the example (laser power and table velocity), two-way ANOVA is used. As there is no replication data, Two-Factor without Replication should be selected.

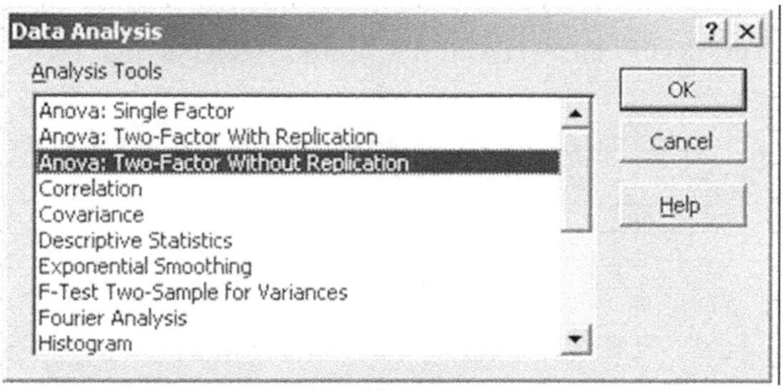

Figure 9.10 Select two-factor without replication for two-way ANOVA.

Step 3: Fill in the required information in the dialog boxes. As shown in Figure 9.11, an input dialog box will appear once Step 2 is carried out. The following items will need to be specified.

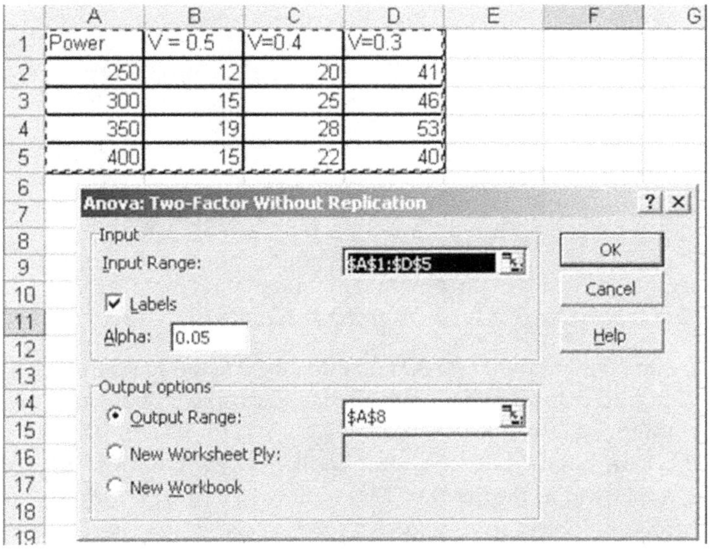

Figure 9.11 Specifying input range to specify the data in Excel.

- Input range with the range of cells containing the data and column and row headings. Use the mouse to highlight the input range. In this case, it was $A$1:$D$5. Note that the range could include the major labeling cells, but one needs to let Excel know about it. If the first column and first row contains nonnumerical data, one should check the Labels box. This tells Excel that the words in the column titles are not numerical data.

- Enter an appropriate level in the Alpha box which is used for the F-test. The default is 0.05.
- Check the button Output Range and enter the top left cell on the spreadsheet where one wishes the ANOVA output to be placed. $A$8 was selected.
- Press the OK button to perform the analysis.

Step 4: ANOVA output summary. The output summary is shown in Figure 9.12.

	A	B	C	D	E	F	G
1	Power	V = 0.5	V=0.4	V=0.3			
2	250	12	20	41			
3	300	15	25	46			
4	350	19	28	53			
5	400	15	22	40			
6							
7							
8	Anova: Two-Factor Without Replication						
9							
10	SUMMARY	Count	Sum	Average	Variance		
11	250	3	73	24.33333	224.3333		
12	300	3	86	28.66667	250.3333		
13	350	3	100	33.33333	310.3333		
14	400	3	77	25.66667	166.3333		
15							
16	V = 0.5	4	61	15.25	8.25		
17	V=0.4	4	95	23.75	12.25		
18	V=0.3	4	180	45	35.33333		
19							
20							
21	ANOVA						
22	ce of Varia	SS	df	MS	F	P-value	F crit
23	Rows	143.3333	3	47.77778	11.86207	0.006203	4.757055
24	Columns	1878.5	2	939.25	233.1931	2.05E-06	5.143249
25	Error	24.16667	6	4.027778			
26							
27	Total	2046	11				

Figure 9.12 The output data in Excel.

9.4.3 *Two-factor (two-way) ANOVA with replication*

The procedure to conduct two-way ANOVA with replication is similar.

Step 1: Enter data into a spreadsheet. Let us use Example 9.11 as an example. Figure 9.13 shows the data input.

Step 2: Select Data Analysis from the Tools Menu. Choose ANOVA: Two-Factor Replication, as shown in Figure 9.14. Since there are two factors in the example (laser power and table velocity), two-way ANOVA is used. As there are replication data, Two-Factor Replication should be selected.

Step 3: Fill in the required information in the dialog boxes. As shown in Figure 9.15, an input dialog will appear once Step 2 is carried out. The following items will need to be specified.

- Input range with the range of cells containing the data and column and row headings. Use the mouse to highlight the input range. In this case, it was $A$1:$D$17. Note that the range could include the major labeling cells, but one needs to let Excel know about it. If the first column and first row contains nonnumerical data,

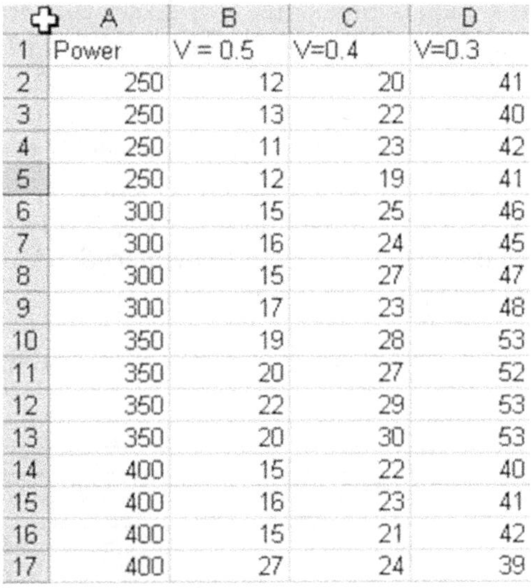

	A	B	C	D
1	Power	V = 0.5	V=0.4	V=0.3
2	250	12	20	41
3	250	13	22	40
4	250	11	23	42
5	250	12	19	41
6	300	15	25	46
7	300	16	24	45
8	300	15	27	47
9	300	17	23	48
10	350	19	28	53
11	350	20	27	52
12	350	22	29	53
13	350	20	30	53
14	400	15	22	40
15	400	16	23	41
16	400	15	21	42
17	400	27	24	39

Figure 9.13 The input data in Excel.

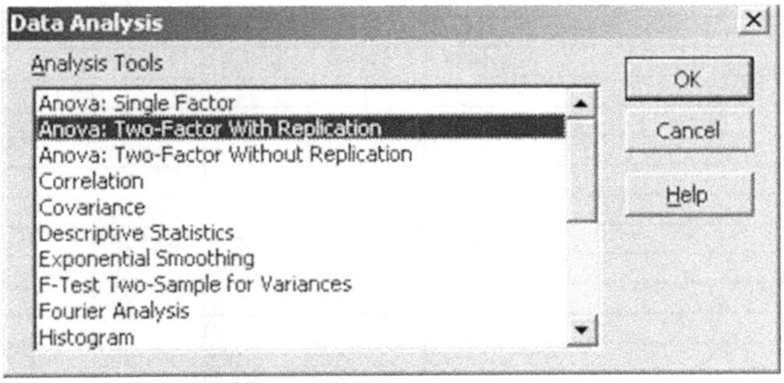

Figure 9.14 Select two-factor replication for two-way ANOVA.

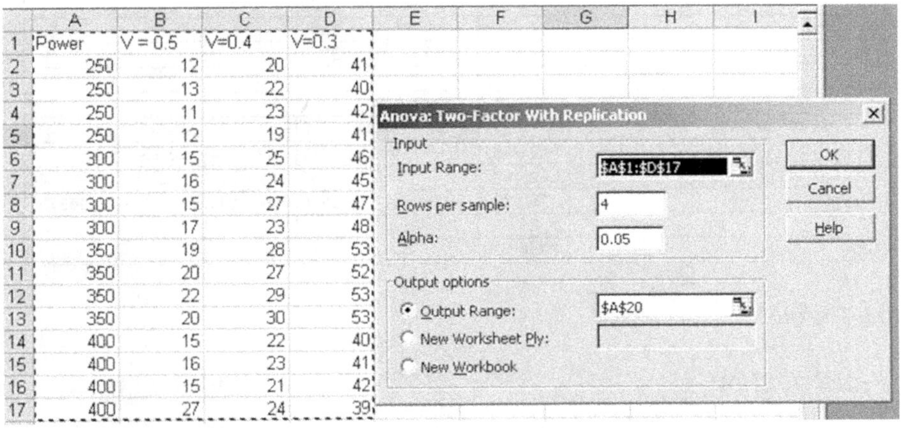

Figure 9.15 Specifying input range to specify the data in Excel.

one should check the Labels box. This tells Excel that the words in the column titles are not numerical data. As there are four sets of replications, the rows per sample are 4.

- Enter an appropriate level in the Alpha box which is used for the F-test. The default is 0.05.
- Check the button Output Range and enter the top left cell on the spreadsheet where one wishes the ANOVA output to be placed. $A$20 was selected.
- Press the OK button to perform the analysis.

Step 4: ANOVA output summary. The output summary is shown in Figure 9.16.

Anova: Two-Factor with Replication						
SUMMAF	V=0.5	V=0.4	V=0.3	Total		
250						
Count	4	4	4	12		
Sum	48	84	164	296		
Average	12	21	41	24.66667		
Variance	0.666667	3.333333	0.666667	161.5152		
300						
Count	4	4	4	12		
Sum	63	99	s	348		
Average	15.75	24.75	46.5	29		
Variance	0.916667	2.916667	1.666667	183.2727		
350						
Count	4	4	4	12		
Sum	81	114	211	406		
Average	20.25	28.5	52.75	33.83333		
Variance	1.583333	1.666667	0.25	208.5152		
400						
Count	4	4	4	12		
Sum	73	90	162	325		
Average	18.25	22.5	40.5	27.08333		
Variance	34.25	1.666667	1.666667	111.7197		
Total						
Count	16	16	16			
Sum	265	387	723			
Average	16.5625	24.185	45.1875			
Variance	17.59583	10.42917	27.09583			
ANOVA						
Variance of Variance	SS	df	MS	F	P-Value	Fcrit
Sample	543.7292	3	181.2431	42.4374	6.56E-12	2.866266
Columns	7.32.167	2	3516.083	823.278	8.83E-31	3.259446
Interaction	129.3333	6	21.55556	5.047154	0.00076	2.363751
Total	7858.979	47				

Figure 9.16 The output data in Excel.

9.4.4 F-distribution

The F-distribution to P-value transformation table can be obtained in the statistical functions of Microsoft's Excel spreadsheet package. The upper percentage points of the

F-distribution can be best tabulated as a separate table for each tail probability, alpha or α. Figure 9.17 shows a possible layout, where the formula in C3 is = FINV($A$1, C$2, $B3) = 4,052.18. In the table, V_1 = the numerator degrees of freedom and V_2 = the denominator degrees of freedom.

	A	B	C	D	E
1	0.01				
2		v1=	1	2	3
3	v_2=	1	4052.18	4999.34	5403.53
4		2	98.50	99.00	99.16
5		3	34.12	30.82	29.46
6		4	21.20	18.00	16.69
7		5	16.26	13.27	12.06
8		6	13.75	10.92	9.78
9		7	12.25	9.55	8.45
10		8	11.26	8.65	7.59
11		9	10.56	8.02	6.99
12		10	10.04	7.56	6.55

Figure 9.17 Using the Excel FINV function to generate percentage points of the F-distribution when α = 0.01.

A layout for using function FDIST as shown in Figure 9.18 could be used for most problems. For example, the formula in C3 is = FDIST($B3,C$1,C$2), which is copied throughout the table.

	A	B	C	D	E	F	G	H	I	J
1		v_1=	1	1	1	1	1	2	2	2
2		v_2=	1	2	3	4	5	1	2	3
3	F=	1	0.500	0.423	0.391	0.374	0.363	0.577	0.500	0.465
4		2	0.392	0.293	0.252	0.230	0.216	0.447	0.333	0.281
5		3	0.333	0.225	0.182	0.158	0.144	0.378	0.250	0.192
6		4	0.295	0.184	0.139	0.116	0.102	0.333	0.200	0.142
7		5	0.268	0.155	0.111	0.089	0.076	0.302	0.167	0.111
8		6	0.247	0.134	0.092	0.070	0.058	0.277	0.143	0.089
9		7	0.230	0.118	0.077	0.057	0.046	0.258	0.125	0.074
10		8	0.216	0.106	0.066	0.047	0.037	0.243	0.111	0.063
11		9	0.205	0.095	0.058	0.040	0.030	0.229	0.100	0.054
12		10	0.195	0.087	0.051	0.034	0.025	0.218	0.091	0.047
13		11	0.186	0.080	0.045	0.029	0.021	0.209	0.083	0.042
14		12	0.179	0.074	0.041	0.026	0.018	0.200	0.077	0.037
15		13	0.172	0.069	0.037	0.023	0.015	0.192	0.071	0.033
16		14	0.166	0.065	0.033	0.020	0.013	0.186	0.067	0.030
17		15	0.161	0.061	0.030	0.018	0.012	0.180	0.062	0.027
18		16	0.156	0.057	0.028	0.016	0.010	0.174	0.059	0.025
19		17	0.151	0.054	0.026	0.015	0.009	0.169	0.056	0.023
20		18	0.147	0.051	0.024	0.013	0.008	0.164	0.053	0.021

Figure 9.18 Using Excel to generate percentage points of the F-distribution.

As an example, to calculate the P-value in Figure 9.16, one can find the P-value by inputting $V_1 = 3$, $V_2 = 36$, and F = 42.4374 to obtain P-value = 0.00000000000656 as shown in Figure 9.19.

	A	B	C	D
1		v_1=	3	
2		v_2=	36	
3	F=	42.4374	0.00000000000656	
4				

Figure 9.19 Using FDIST to find the P-value.

Review problems

1. Use Excel to solve Review Problem 1 in Section 9.3.
2. Use Excel to solve Review Problem 2 in Section 9.3.
3. Use Excel to solve Review Problem 2 in Section 9.2.
4. Use Excel to solve Review Problem 3 in Section 9.2.
5. Use Excel to find P-value when $F = 12$, $V_1 = 3$, and $V_2 = 6$.
6. Use Excel to find P-value when $F = 1.5$, $V_1 = 3$, and $V_2 = 1$.

9.5 Quality characteristic

You must keep your mind on the objective, not on the obstacle.

William Randolph Hearst

This section helps answer the following questions.

- What is a quality characteristic (QC)? How do we define a QC function?
- How to combine several QC functions into one?
- What is a predictive model when conducting ANOM?

QC is any dimensional, mechanical, and physical property, and functional or appearance characteristic that can be used for measuring the unit quality of the prototype, product, or service. It is critical to be able to set up a QC function to effectively evaluate the performance in experiments. This needs to predefine the prototype objectives as discussed in the previous chapters. Why is the prototype made? What are the objectives of the prototype? What are the main features to be measured? How are they going to be measured? The selection of the QC function will greatly influence the number of experiments.

9.5.1 Overall evaluation criterion

OEC comes into the picture when there is more than one performance criterion or response in the experiment. It should be checked, and if the respective responses give different optimal results for the process, only then the OEC must be formulated or else it would be redundant to do an overall evaluation. In this case, a certain weight percentage must be allocated to each QC subjectively during the experiment planning session. The different QCs are adjusted in such a way that they give rise to either a smaller-the-better or a larger-the-better QC. The QCs must be normalized and weighted in order to give a single OEC. The nominal-the-best QC can be adapted to represent the deviation from the nominal value which would then become a smaller-the-better. The OEC is devised as shown below:

$$\text{OEC} = \left\{ \frac{QC_a - QC_{\min(a)}}{QC_{\max(a)} - QC_{\min(a)}} \times W_a\% \right\} + \left\{ \frac{QC_b - QC_{\min(b)}}{QC_{\max(b)} - QC_{\min(b)}} \times W_b\% \right\} \cdots \tag{9.40}$$

Equation 9.40 is for a larger-the-better case where QC_a is the performance value of a (otherwise known as the mean or a weighted mean), QC_b is the mean of b, QC_{max} and QC_{min} are the best, and the worst means of the respective QCs and W_x is the weight percentage allotted to the QC_x. For a smaller-the-better case (say, characteristic b), one can modify the above equation into the following form:

$$OEC = \left\{ \frac{QC_a - QC_{min(a)}}{QC_{max(a)} - QC_{min(a)}} \times W_a\% \right\} + \left\{ \left(1 - \frac{QC_{max(b)} - QC_b}{QC_{max(b)} - QC_{min(b)}} \times W_b\% \right) \right\} + \cdots \quad (9.41)$$

In Equation 9.41, characteristic a is larger-the-better, and characteristic b is smaller-the-better, whereas the OEC is formulated for the larger-the-better approach. For nominal-the-best types of QC, it must be noted that the OEC will be evaluated based on the magnitude of deviation from the nominal value. This magnitude would be categorized as smaller-the-best, which can be easily reformulated to fit larger-the-better as shown in Equation 9.41 [Ranjit01].

Example 9.14

The QCs for prototyping tests are shown below:

						Data	
	Criterion	Worst reading	Best reading	QC	Weighting (%)	1	2
A	Speed	2	8	L	25	6	4
B	Acceleration	35	90	L	25	40	50
C	Fuel	45	10	S	50	30	32

Formulate the OEC for this problem.

$$OEC = \left\{ \frac{QC_a - QC_{min(a)}}{QC_{max(a)} - QC_{min(a)}} \times W_a\% \right\} + \left\{ \left(1 - \frac{QC_{max(b)} - QC_b}{QC_{max(b)} - QC_{min(b)}} \right) \times W_b\% \right\} + \cdots$$

$$= \left\{ \frac{QC_a - 2}{8 - 2} \times 25\% \right\} + \left\{ \frac{QC_b - 35}{90 - 35} \times 25\% \right\} + \left\{ \left(1 - \frac{45 - QC_c}{45 - 10} \right) \times 50\% \right\}$$

9.5.2 *Predictive model*

The predictive model is constructed from the results of the OA analysis and is formed by using the optimum level contribution of each factor in relation to the deviation from the overall mean value for the experiment. The general formula for a predictive model is

$$Y(A,B,C,D) = \bar{Y} + \left(\overline{Y_A} - \bar{Y}\right) + \left(\overline{Y_B} - \bar{Y}\right) + \left(\overline{Y_C} - \bar{Y}\right) + \cdots \quad (9.42)$$

where

Y = QC
A, B, C, D, etc. = control factors
YA = best response value of A and so on

For example, in problem Example 9.6, the total output average is 9.94, and the resulting optimal control factors are given as follows:

Factor Q: Level 3 is chosen; $\bar{Y}_Q = 23/3 = 7.67$.
Factor S: Level 2 or 3 is chosen and $\bar{Y}_S = 26.5/3 = 8.83$.
Factor A: Level 3 and $\bar{Y}_A = 23/3 = 7.67$.

Based on the above analysis, the recommended parameter settings are therefore chosen to be

Q_3, S_1 (or S_2), and A_3, and thus, the predictive result for this combination is given as follows:

$$Y(Q,S,A) = \bar{Y} + \left(\overline{Y_Q} - \bar{Y}\right) + \left(\overline{Y_S} - \bar{Y}\right) + \left(\overline{Y_A} - \bar{Y}\right) = 8.94 + (7.67 - 8.94)$$

$$+ (8.83 - 8.94) + (7.67 - 8.94) = 6.29$$

Review problems

1. In Example 9.5, if density is considered to be the higher-the-better, what should the predictive model be for that example?
2. In Example 9.7, what should the predictive model be for that example?
3. As shown in the table, formulate OEC for this problem. Note that hardness (H) and build rate (B) are larger-the-better cases; the remaining QCs, such as porosity (P), and dendrite (D), are smaller-the-better cases.

		Average values			
Build rate ($1\,mm^3/s$)	Hardness knoop (HK)	Porosity rank	Dendrite size (μ)	SDAS (μ)	Gain size (ASTM)
0.37	562.48	2.57	22.19	1.45	2.5
0.85	580.83	1.50	68.09	1.60	5.0
0.91	591.12	4.25	44.59	2.39	6.5
1.19	610.50	2.50	26.53	2.41	9.5
1.27	576.58	3.17	60.26	1.68	7.0
3.36	596.87	3.50	58.56	3.12	8.5
3.40	612.17	3.50	59.15	2.89	5.5
4.17	633.56	4.33	61.35	2.96	9.0
3.41	580.88	4.67	52.03	2.11	6.0
2.18	587.17	1.50	43.26	2.15	10.5
0.82	649.33	1.66	22.61	2.32	4.5
2.98	569.25	3.50	49.77	1.05	3.5
1.04	580.97	3.66	43.73	2.62	5.0
4.20	577.91	2.50	10.45	0.33	5.0
1.59	562.71	3.50	26.75	1.82	6.0
2.52	555.00	3.83	7.85	2.26	5.0
1.67	608.29	4.33	37.81	0.00	8.5
2.24	628.94	4.33	54.98	2.87	2.5

9.6 An example: optimization of a prototype laser deposition process

> The only purpose of education is to teach a student how to live his life by developing his mind and equipping him to deal with reality. The training he needs is theoretical, i.e., conceptual. He has to be taught to think, to understand, to integrate, to prove. He has to be taught the essentials of the knowledge discovered in the past and he has to be equipped to acquire further knowledge by his own effort.
>
> **Ayn Rand**

The actual implementation of DOE could involve a lot of details, and the parameters and conditions could be problem dependent. The example chosen [Prakash04] further illustrates the practice of the topics discussed in the previous sessions.

9.6.1 Problem statement

In laser deposition, a laser is used to fuse metal powder onto a substrate in the form of many layers and the part is gradually fabricated to near-net-shape. This problem is to conduct DOE optimization of the laser deposition process using H13 tool steel as the deposit material and 316L stainless steel as the substrate material. Stainless steel is corrosion resistant, whereas tool steel is known for its inherent toughness. The microstructural properties such as the Knoop hardness, dendrite size, secondary dendrite arm spacing (SDAS), ASTM (American Society for Testing and Materials) grain size, the build rate, and porosity are used as multiple evaluation criteria.

As the grain size of H13 becomes coarser or larger, the elongation, fatigue strength, impact transition temperature, etc. decrease. This is correlated to dislocation cracks that result from the coalescence of dislocations increasing with the grain size. Coarse-grained steels are also inferior when it comes to bending and fatigue testing. Moreover, coarse-grained materials are more prone to distortion and are more prone to crack during quenching or grinding. In a normalized condition, the coarse-grained steel is preferred during machining, but when finishing the part like grinding and polishing, fine grain is preferred. In this process, an equiaxed grain with the least possible grain size is desired as further heat treatment would be necessary to change the microstructure depending on the application.

Tough and wear-resistant characteristics would be required for part repair; the part should be case hardened with high dimensional stability and wear resistance for die making.

9.6.2 Selection of factors and levels

A total of seven factors were chosen as control factors with three levels except for the inner gas pressure, which was set at two levels as shown in Table 9.26. To decide upon the actual values of the levels, a series of experiments were conducted, and the range of each factor was determined. A feasible range of a factor would always give a satisfactory deposition that would bond with the substrate and not delaminate. A set of extreme factor combinations (e.g., lowest laser power, highest traverse speed, highest gas pressure, highest powder flow rate, lowest overlap factor, and lowest spot size) were chosen and tested for a satisfactory deposition with good bonding onto the substrate and no delamination.

Table 9.26 Selection of factors and levels

	Control factors	Levels 1	2	3
1	Laser power (P) (W)	500	750	1,000
2	Spot size (D) (mm)	0.71	0.74	0.81
3	Inner gas pressure (IG) (psi)	3	4	–
4	Outer gas pressure (OG) (psi)	8	10	12
5	Feed rate (F) (ipm)	20	25	30
6	Powder flow rate (PF) (g/min)	8	10	12
7	Percentage overlap (O) (%)	25	35	45

If not, the suitable factor range was reduced and the process repeated until a good result was obtained.

9.6.3 *Orthogonal array*

The number of factors and their levels determines the choice of an OA. For this setup, the L18 OA as shown in Table 9.27 was chosen with the remaining column allocated for the noise factors and the interactions. By doing this, a good estimate of the contribution of the noise and interactions can be done and the process compensated to behave desirably in their presence, that is, their effect minimized on the process. The numbers in the error column can be treated as the levels of the error control factors.

Table 9.27 Orthogonal array

Exp. no.	Laser power (W)	Spot size (mm)	Feed rate (ipm)	Powder flow rate (g/min)	Overlap factor (%)	Inner gas pressure (psi)	Outer gas pressure (psi)	Error
1	500	0.21	20	8	25	4	8	1
2	500	0.25	25	10	35	4	10	2
3	500	0.3	30	12	45	4	12	3
4	750	0.25	20	8	45	4	10	3
5	750	0.3	25	10	25	4	12	1
6	750	0.21	30	12	35	4	8	2
7	1,000	0.3	20	10	35	4	8	3
8	1,000	0.21	25	12	45	4	10	1
9	1,000	0.25	30	8	25	4	12	2
10	500	0.25	20	12	35	5	12	1
11	500	0.3	25	8	45	5	8	2
12	500	0.21	30	10	25	5	10	3
13	750	0.21	20	10	45	5	12	2
14	750	0.25	25	12	25	5	8	3
15	750	0.3	30	8	35	5	10	1
16	1,000	0.3	20	12	25	5	10	2
17	1,000	0.21	25	8	35	5	12	3
18	1,000	0.25	30	10	45	5	8	1

9.6.4 Sample preparation

The sample depositions were 0.5″ × 0.5″, and the deposition was continued until at least a 0.2″ height was obtained. Three samples were made for each experiment to ensure repeatability. The build rate was calculated for all the samples, and they were sectioned using an abrasive water-jet for cross-sectional evaluation as well as evaluation on the top surface. Then they were hot-mounted in bakelite. The mounts were then ground and polished to 0.5 mm surface finish.

There are around 248 etching methods that have been used until now. For the samples, preferential etching was conducted. Preferential etching, also known as anisotropic etching, is done by using an etchant that attacks different crystallographic planes at different rates and produces an image controlled by those planes. Initially, 2% Nital (2 ml HNO_3 + 98 mL ethanol) was used to etch the samples. Later 4% picric acid (96% water as a base) was tried out, and the microstructural images were found to be slightly better than with Nital.

9.6.5 Responses

Many responses, also known as criteria, were used in this study. They are the build rate in mm^3/s, microhardness (Knoop hardness), grain sizes, SDAS, cracks, and porosity. For the build rate, the dimensions of the sample (build volume) were measured using an optical microscope, and the time duration of the deposition was measured using a LabVIEW program that measured the total movement time of the CNC machine axes very accurately. The SDAS as shown in Figures 9.20 and 9.21 and other microstructural images were obtained using an optical microscope. The SDAS reduces with an increase in cooling rate of the deposit. The material properties improve when the SDAS gets finer. Long secondary arms would result in inter-dendritic shrinkage or shrinkage porosity. Smaller primary arms would help avoid this condition. Fe3C forming in steel increases the strength of the material. The Knoop hardness was measured on the top surface as well as the cross sections where the hardness was measured at various levels to detect any trend in the

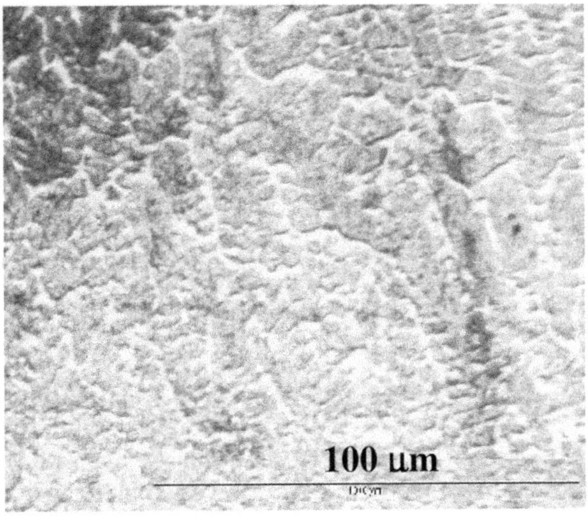

Figure 9.20 Large SDAS. 500X Picral etch (Image 1).

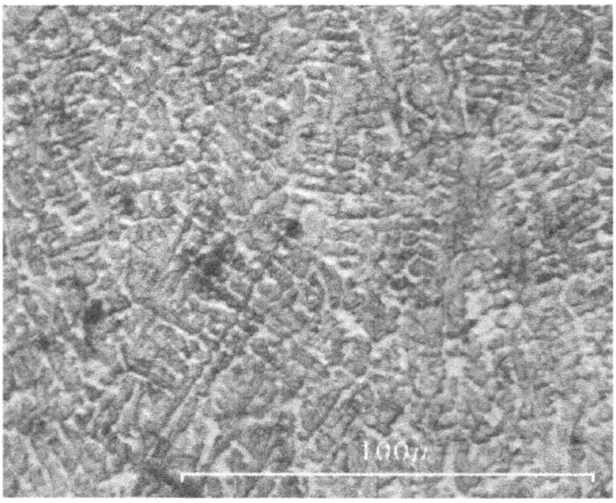

Figure 9.21 Large SDAS. 500X Picral etch (Image 2).

process. The cracks were measured from the photomicrograph of the sample at 100x magnification using image processing. As the cracks were darker than the deposit, the area covered by the cracks could be calculated. The samples were ranked on a scale of 1–5 based on the percentage of cracks (ranging from 0% to around 15%) with 5 being the best and 1 being the poorest. The ASTM grain size was calculated using the Heyn intercept counting [Dehoff68, Voort84]. This method is faster than other methods because only the grains in the perimeter are counted. The number of grain boundary intercepts per unit length N_L is given by

$$N_L = \frac{M\Sigma P_i}{L_t} \tag{9.43}$$

where

ΣP_i = total number of grain boundary intercepts
M = magnification
L_t = length of the reference line used for counting the intercepts

The ASTM grain size is given by

$$G = (6.6353 \log N_L) - 12.6 \tag{9.44}$$

All these responses were combined into one single entity called the OEC.

9.6.6 Formulation of the OEC

As discussed before, in order to form the OEC, a certain weight percentage must be allocated to each QC using the weight distribution method discussed in Chapter 2. Assume that the build rate, hardness, SDAS, grain size, and the cracks or porosity were each given the same weight percentage. The individual sample readings were not listed in the experiment table (Table 9.28). There were three readings taken for each QC on each sample. Only their mean values are shown in Table 9.28. The OEC is formulated as a larger-the-better

Table 9.28 Design of experiments using L18 array

									Average values						Responses			
															Equal OEC		OEC 10:10:20:20:20:20	
Exp no.	IG (psi)	P (W)	F (ipm)	PF (g/min)	OG (psi)	d (mm)	O (%)	Err	Build rate (mm³/s)	Hardness Knoop (HK)	Porosity rank	Dendrite size (μ)	SDAS (μ)	Grain size (ASTM)	Mean	S/N (db)	Mean	S/N (db)
1	4	500	20	8	8	0.71	25	1	0.37	562.48	2.57	22.19	1.45	2.5	18.64	30.18	21.57	31.45
2	4	500	25	10	10	0.74	35	2	0.85	580.83	1.50	68.09	1.60	5.0	37.06	36.15	40.47	36.91
3	4	500	30	12	12	0.81	45	3	0.81	591.12	4.25	44.59	2.39	6.5	54.44	39.49	60.09	40.35
4	4	750	20	8	10	0.74	45	3	1.19	610.50	2.50	26.53	2.41	9.5	51.26	38.97	53.49	39.34
5	4	750	25	10	12	0.81	25	1	1.27	576.58	3.17	60.26	1.68	7.0	49.35	38.64	54.59	39.51
6	4	750	30	12	8	0.71	35	2	3.36	596.87	3.50	58.56	3.12	8.5	74.12	42.17	76.70	42.47
7	4	1,000	20	10	8	0.81	35	3	3.40	612.17	3.50	59.15	2.89	5.5	69.69	41.63	69.64	41.63
8	4	1,000	25	12	10	0.71	45	1	4.17	633.56	4.33	61.35	2.96	9.0	89.45	43.80	89.08	43.77
9	4	1,000	30	8	12	0.74	25	2	3.41	580.88	4.67	52.03	2.11	6.0	65.21	41.06	67.58	41.37
10	5	500	20	12	12	0.74	35	1	2.18	587.17	1.50	43.26	2.15	10.5	51.54	39.01	53.70	39.37
11	5	500	25	8	8	0.81	45	2	0.82	649.33	1.66	22.61	2.32	4.5	40.12	36.84	36.97	36.13
12	5	500	30	10	10	0.71	25	3	2.98	569.25	3.50	49.77	1.05	3.5	43.71	37.58	44.14	37.67
13	5	750	20	10	12	0.71	45	2	1.04	580.97	3.66	43.73	2.62	5.0	48.02	38.40	53.11	39.27
14	5	750	25	12	8	0.74	25	3	4.20	577.61	2.50	10.45	0.33	5.0	33.63	35.31	27.96	33.70
15	5	750	30	8	10	0.81	35	1	1.59	562.71	3.50	26.75	1.82	6.0	39.42	36.69	43.30	37.50
16	5	1,000	20	12	10	0.81	25	2	2.52	555.00	3.83	7.85	2.26	5.0	38.88	36.57	41.05	37.04
17	5	1,000	25	8	12	0.71	35	3	1.67	608.29	4.33	37.81	0.00	8.5	50.77	38.88	51.88	39.07
18	5	1,000	30	10	8	0.74	45	1	2.24	628.94	4.33	54.98	2.87	2.5	64.48	40.96	64.65	40.93
														Total	919.78	692.32	949.98	697.53
														Average	51.10	38.46	52.78	38.75

case for the hardness (H) and build rate (B), and a smaller-the-better case for the remaining QCs, such as porosity (P), dendrite (D), etc. From Equations 9.41, one can obtain

$$\text{OEC} = \left\{ \frac{QC_H - QC_{\min(H)}}{QC_{\max(H)} - QC_{\min(H)}} \times W_H\% \right\} + \left\{ \frac{QC_B - QC_{\min(B)}}{QC_{\max(B)} - QC_{\min(B)}} \times W_B\% \right\}$$
$$\times \left\{ \left(1 - \frac{QC_{\max(P)} - QC_P}{QC_{\max(P)} - QC_{\min(P)}} \right) \times W_P\% \right\} + \left\{ \left(1 - \frac{QC_{\max(D)} - QC_D}{QC_{\max(D)} - QC_{\min(D)}} \right) \times W_D\% \right\} \ldots \quad (9.45)$$

9.6.7 *Experiment*

The experiment was conducted in random order. The order was randomly generated by Minitab software, and each experiment was repeated three times. The results of the experiment are shown in Table 9.28. The weightage of the criteria in the OEC was changed to different combinations, and the result studied for the sake of comparison. Only two OECs are shown in the experiment in Table 9.28.

9.6.8 *Analysis of the means*

The ANOM is used to do a two-step optimization by reducing the variation in the process first (using the S/N ratio and maximizing the slope) and then shifting the mean or target performance to get an optimized result or response. Reducing the variation in a process is often more difficult than shifting the mean. The ANOM was conducted next, as shown in Tables 9.29 and 9.30, and the factor level plots of the means were constructed as shown in

Table 9.29 Analysis of the means

Mean	Inner gas pressure (psi)	Laser power (W)	Feed rate (ipm)	Powder flow rate (g/min)	Outer gas pressure (psi)	Spot diameter (mm)	Overlap factor (%)
1	**56.58**	40.92	46.34	44.24	49.41	**54.12**	41.57
2	60.23	49.30	50.06	52.05	49.96	50.53	53.76
3	–	**63.08**	**56.90**	**57.01**	**53.22**	48.65	**57.96**

Table 9.30 Analysis of the means for the S/N ratio

S/N	Inner gas pressure (psi)	Laser power (W)	Feed rate (ipm)	Powder flow rate (g/min)	Outer gas pressure (psi)	Spot diameter (mm)	Overlap factor (%)
1	39.12	36.54	37.46	37.10	37.85	38.50	36.56
2	**40.10**	38.36	38.27	38.89	38.29	**38.58**	39.09
3	–	**40.48**	**39.66**	**39.39**	**39.25**	38.31	**39.74**

Figures 9.22 and 9.23. The levels of the control factors with the highest S/N ratio were used in corroboration of the predicted values for the confirmation experiment. The optimal levels for an equal OEC are laser power III, feed rate III, powder flow rate III, inner gas II, outer gas III, spot size II, and overlap factor III.

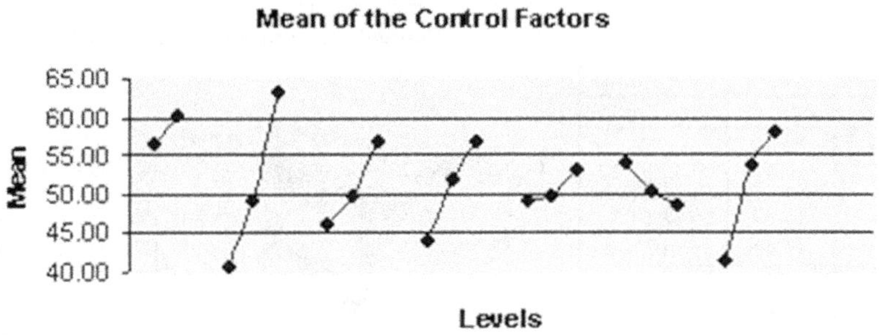

Figure 9.22 Factor level plots for the mean.

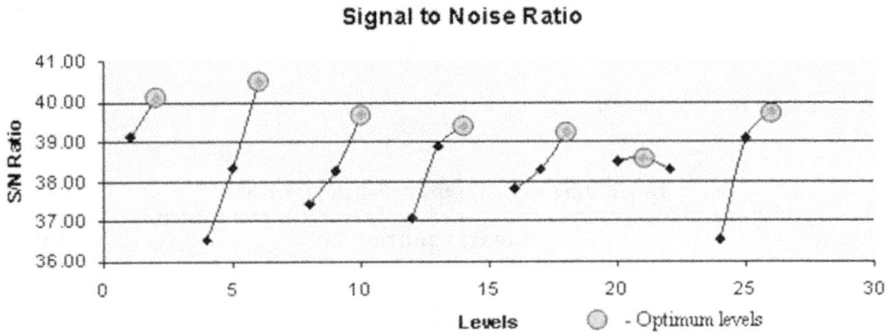

Figure 9.23 Factor level plots for the S/N ratio.

9.6.9 *Analysis of the variance*

The SS for this experiment is shown in Table 9.31. The percentage contributions of the factors are shown in Figure 9.24. It can be seen that the laser power, overlap factor, and the inner gas contribute 68% of the total process influence. This implies that they have a fair deal of control over this process when compared to the others. The degree of freedom (or df) of a factor effect is one less than the number of levels for that factor. For this experiment, the total df is 39 for three repetitions of each experiment. The L18 array has an empty column with 3 levels (df = 2). This column can be allocated to the interactions and error to study their effects on this process.

Table 9.31 Sum of squares

	$(dB)^2$
Grand total sum of squares	26789.45
Total sum of squares	161.20
Sum of squares due to the mean	26628.25

The laser power, overlap factor, inner gas, feed rate, and powder flow rate have been chosen as significant control factors, and the remaining factors have been pooled together as the error. Table 9.32 shows the mean squares of the various factors and their F-ratios. It can be seen that the inner gas, laser power, and overlap factor are very strong when compared to the experimental error, and the powder flow rate and feed rate are reasonably significant.

SS for Factors		Percentage Contribution
SS_{IG}	28.10	17.43
SS_{P}	46.71	28.98
SS_{F}	14.83	9.20
SS_{PF}	17.40	10.79
SS_{OG}	6.13	3.80
SS_{D}	0.23	0.14
SS_{O}	34.02	21.10
SS_{INT}	0.60	0.37
SS_{Error}	13.19	8.19

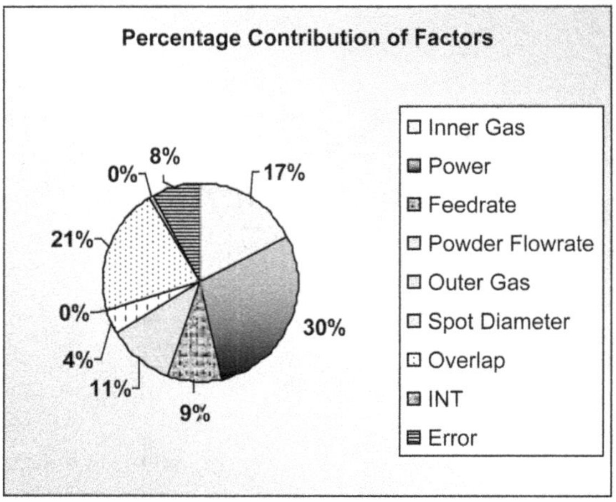

Figure 9.24 Percentage contribution.

Table 9.32 Mean squares and F-ratios

Factor	Mean square	F-ratio
Se^2	3.36	–
MS_P	23.35	6.95
MS_O	17.01	5.06
MS_{IG}	28.10	8.37
MS_{PF}	8.70	2.59
MS_F	7.41	2.21

The confidence interval (CI) is given by [Ross17]:

$$CI_1 = \sqrt{\frac{F_{\alpha;v_1;v_2} V_{ep}}{n}} \quad (9.46)$$

where $F_{\alpha;v_1;v_2}$ is the F-ratio required for α risk.

The confidence is 1-risk

$v_1 = 1$

v_2 = degrees of freedom for the pooled error which is 6

V_{ep} = pooled error variance which is 3.36

n = number of tests in that condition which is 3

The optimized mean lies in a range given by $\pm CI_1$. The confidence interval estimate around the mean has a range of ±2.05 with 90% confidence, ±2.59 with 95% confidence, and ±3.91 with 99% confidence. The confirmation experiment gave values of 77.19 for the mean and 42.52 for the S/N ratio showing that the experiment was successful and the additivity of the process is substantial enough for its control even in the presence of errors and interactions.

The predicted value for the mean was 78.64, and the S/N ratio was 43.87. Another set of experiments was conducted for unequal distribution of weights to form the OEC function, and the results were listed in Table 9.28.

The laser deposition process was thus optimized for build rate, microhardness, porosity, SDAS, and ASTM grain size, and the optimal parameter combinations were found for different weightage of OECs. The optimal values were predicted using the predictive model, and the confidence intervals were calculated. The confirmation experiments verified the model and fell within the confidence interval limits estimated. The system was made robust and controllable in the presence of interactions, noises, and experimental errors, and the microstructure was also studied. For further improvement, more iterative experimentation and honing in on the best parameter combination is needed.

References

[DeHoff68]. DeHoff, R.T. and F.N. Rhines, *Quantitative Microscopy*, McGraw-Hill, New York, 1968.

[Fowlkes98]. Fowlkes, W.Y. and C.M. Creveling, *Engineering Methods for Robust Product Design Using Taguchi Methods in Technology and Product Development*, Addison Wesley Publishing Company, Boston, MA, 1998.

[Pignatiello91]. Pignatiello, J.J., Jr. and J.S. Ramberg, "The top-ten triumphs and tragedies of Genichi Taguchi," *Quality Engineering*, Vol. 4, No. 2, pp. 221–225, 1991–1992.

[Prakash04]. Prakash, S., Optimization of the laser aided manufacturing process (LAMP) using the design of experiments approach—Taguchi way, Thesis T, University of Missouri-Rolla, 8431, 2004.

[Ranjit01]. Ranjit, K.R., *Design of Experiments using the Taguchi Approach—16 Steps to Product and Process Improvement*, John Wiley & Sons, Inc., Hoboken, NJ, 2001.

[Ross17]. Ross, R.J., *Taguchi Techniques for Quality Engineering*, McGraw-Hill, New York, 1989.

[StatSoft07]. StatSoft, ANOVA/MANOVA, Retrieved on September 2, 2017 from www.statsoft.com/textbook/stanman.html#basic.

[Voort84]. Voort, G.F.V., *Metallography: Principles and Practice*, McGraw-Hill, New York, 1984.

chapter ten

Prototype optimization

> The greatest challenge to any thinker is stating the problem in a way that will allow a solution.
>
> **Bertrand Russell**

It was previously discussed that initial product prototypes need to be tested to identify how closely engineered product variables match consumer needs and perceptions to find out what are the possible issues with the design. The prototype testing, evaluations, and consumer input should be used to fix these issues. Anyone of those wrong attributes can overcome consumer appreciation for any number of positive quality attributes. There will be trade-offs when all product quality variables are considered, and this process is called engineering optimization.

Optimization is, according to *Merriam-Webster*, an act, process, or methodology of making something (as a design, system, or decision) as perfect, functional or effective as possible. Different people may have different perspectives on optimization. For some, it is the process of doing iterations on a product until it achieves the set conditions of the product definition, and for others, it is to find values of the variables that minimize or maximize the objective function while satisfying the constraints. In product or prototype design, optimization is the design of a product that is, say, the most economical and efficient design possible, while also fulfilling all of the customer requirements. By using product models either by hand or through several different software programs, optimization of a product can be done.

The following examples are the situations in which one wants to find the best way to do something. The design variables are shown in Table 10.1.

Table 10.1 Optimization models and design variables

Model	Design variables
Maximize company's prototype performance	Design attributes
Minimize company's product shipping	Amount produced at each plant each month that is shipped to each customer
Maximize product profit	Sales, prices, time, capitals, etc.
Maximize company profit	Projects selected, money, and skilled engineers
Maximize production profit	Amount of each product produced during the month
Minimize production cost	Labor, materials, equipment, utilities, etc.

- How does a design company improve a prototype design?
- If a company produces products at three locations, how can they minimize the cost of meeting demand for the product?
- What price for the products will maximize profit from product sales?

- A manufacturing company would like to undertake 20 strategic initiatives that will tie up money and skilled engineers for the next 5 years. They do not have enough resources to undertake all 20 projects. Which projects should they undertake?
- How can a manufacturing company determine the monthly product mix at their plant that maximizes corporate profitability?
- How should a company minimize the production costs?

Optimization can help us determine the "best" solution without actually testing all possible cases. Before the 1940s, optimization techniques on the systems of equations were essentially heuristics, simply educated guesses. Not until the simplex method was first proposed in 1947 by Dantzig, did optimization become a defined process resulting in determinate answers. Optimization problems became linear programs, and systems of equations that must be satisfied yet are underdetermined, thus allowing for an infinite number of solutions.

Optimization is being used in various fields such as engineering, medicine, and transportation.

Examples of some optimization problems are as follows:

i. How should the transistors and other devices be laid out in a computer chip so that it occupies the minimum area?
ii. How should the telephone line be routed between two cities so that the maximum number of simultaneous calls is possible?

Figure 10.1 shows a typical optimization process. The input to the system is the values of the design parameters, and the output is the functional performance of the system. The functional performance is then compared to the desired functional requirements, and with the optimization algorithm, one decides the values of the new design parameters. This process repeats until the functional performance can meet the desired functional requirements. To evaluate the functional performance of the design variables, one can use a set of equations, an analytical model, such as a finite element model, a set of experiments, or even prototype evaluation. This chapter uses the equation approach as an example. Different optimization algorithms have different strengths and are more efficient in a specific domain. It is very important to use the appropriate optimization algorithm.

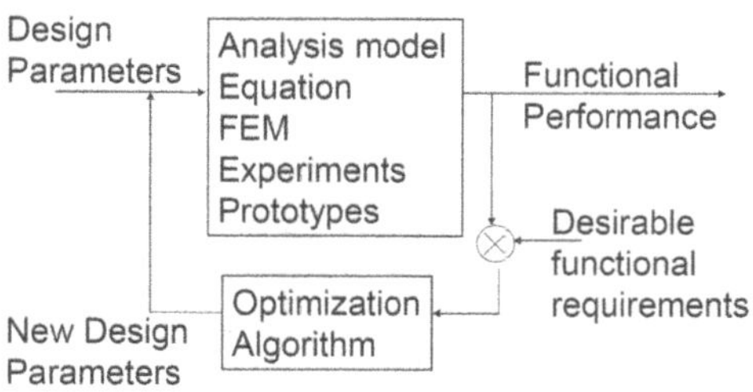

Figure 10.1 A typical engineering design process.

A lot of application software has the capability of engineering optimization. Modern optimization methods perform shape optimizations on components generated from a computer-aided design (CAD) tool. Ideally, there is seamless data exchange via direct memory transfer between the CAD and finite element analysis (FEA) applications without the need for file translation. Furthermore, if associability between the CAD and FEA software exists, any changes made in the CAD geometry are immediately reflected in the FEA model. In the approach, the design optimization process begins before the FEA model is generated. The user simply selects which dimension in the CAD model needs to be optimized and the design criterion, which may include maximum stresses, temperatures, or frequencies.

The analysis process appropriate for the design criteria is then performed. The results of the analysis are compared with the design criterion, and, if necessary without any human intervention, the CAD geometry is updated. Care is taken such that the FEA model is also updated using the principle of associability, which implies that constraints and loads are preserved from the prior analysis. The new FEA model, including a new high-quality solid mesh, is now analyzed, and the results are again compared with the design criterion. This process is repeated until the design criterion is satisfied. The optimization method also allows for global constraints to enforce weight and volume criteria. These global calculations require little additional effort because of the tight integration among the applications, including the weight, center of gravity, and mass moment of inertia processor. Even though global constraints can be used to optimize material usage, engineering expertise is generally required to optimize costs.

When one carries out a series of tests on a design prototype, the main reasons for doing them are to gain a firsthand understanding of the behavior of the system, assess how far the prototype is from the design objectives, and then to adjust unconstrained design parameters in order to achieve those objectives. It is usually very painstaking work, requiring the handling of large amounts of empirical data, and not necessarily accurate. The purpose of this chapter is to provide an overview of engineering optimization so that even without the commercial software, engineering optimization can be properly conducted.

10.1 Formulation of engineering problems for optimization

This section helps answer the following questions:

- Are there any special techniques that can be used for optimization?
- What kind of approach can one use when a design is based on several different parameters? Apparently, this makes the optimization process more complicated, but how can one come up with a system to include all of these parameters in a calculation?
- How can one tell if a product definition is unreachable, or in other words, an optimal solution cannot be found, so one would not waste time trying to achieve the impossible?
- What is the best way to optimize the development process? Are techniques such as the house of quality considered an optimization process?

10.1.1 Definitions

Finding optimal solutions in product prototyping requires one to first define the problem and then to define the required results. From this first step, one must identify all the variables of the problem and determine which of the variables must be fixed and at what value to fix them, in order to meet customer requirements. Once the fixed parameters have been established, determine equations or parameters for the remaining variables and also equate the relationship between the variables, so as one parameter varies, the results show

how it affects the other variables when defining the final configuration. One also needs to define the constraints, limits, and scope of the problem. The third step is to find the solution to the problem. Once the solution is found, one needs to assess the solution by answering the following questions:

- Is the design optimal?
- Is the design feasible?
- Is the design reasonable?
- How can further modification be done for the optimization model?

As shown in Figure 10.2, there is only a global optimal (maximum or minimum) solution for a function, and there may be several local optima. Therefore, a global maximum (minimum) is also a local maximum (minimum) solution, but a local maximum (minimum) may not necessarily be a global maximum (minimum) solution. Although in most cases, a global optimum is preferred, in some cases, it is not feasible to prove that a locally optimal solution exists. A point is Pareto optimal if the only way to improve any of its components is by worsening other components. It is actually the definition of a local optimal and is a solution to a modeled design problem. In case a Pareto optimal is found, but it is not possible to validate that it is the global optimal, then a decision needs to be made to see if more solutions need to be found. The previous assessment criteria, as well as the available resources to find additional solutions, will be used to make a decision. For engineering applications, sometimes a Pareto optimal that is feasible and reasonable is a sufficient solution. Sometimes, several Pareto optimal solutions are needed to find a solution, which is closer to the global optimal.

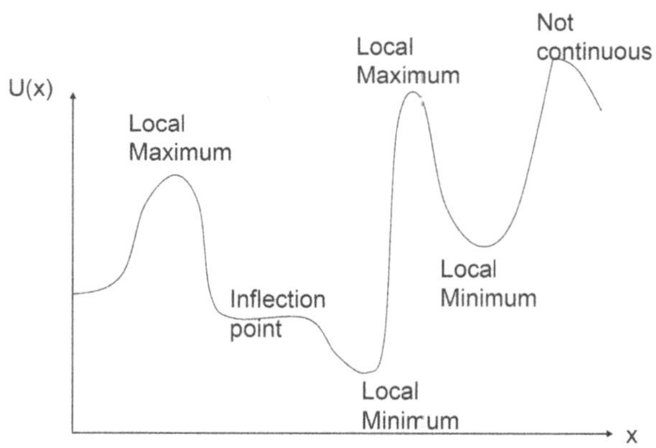

Figure 10.2 Various optimal solution points.

How can one find the optimal solutions in product prototyping? The first step is to define the problem properly. Typically, the second step is to put it into a mathematical form if possible. In this case, the system of equations would define the system and the constraints. The third step is to solve the problem using the optimization techniques.

10.1.2 Problem formulation

Problem formulation is very critical to the success of the design. One needs to know the problem well. If one does not have a detailed understanding of the problem at hand, one cannot expect to find the optimal solution. It is critical to fully understand the needs of one's customers. If customers' requirements are not known, it is not possible to fully satisfy them. The informal models discussed in Chapter 9 are very useful in defining the problem.

In most textbooks, minimization of an objective function instead of maximization is conducted for optimization. However, there are some references which use the maximization of an objective function. Either maximization or minimization should end up with the same result. However, one has to formulate the constraints accordingly when using classical optimization methods. The formulation of engineering problems is conducted by basically specifying the design variables, the problem definition parameters (PDP), the objective function, and the constraints. It is a good practice to formulate the problem into the following standard:

$$\begin{aligned}
&\text{Minimize } f(x) && \text{Objective function}\\
&\text{Subject to: } g_j(x) \leq 0,\ j = 1,2,3,\ldots,m && \text{Inequality constraints}\\
&h_k(x) = 0,\ k = 1,2,3,\ldots,l && \text{Equality constraints}\\
&\text{where } x = (x_1, x_2, x_3, \ldots, x_n)^T && \text{Design variables}\\
&x_i^{\,l} < x_i \leq x_i^{\,u} && \text{Side constraints}
\end{aligned} \tag{10.1}$$

Note that a constraint, say, $2 < L < 10$ can be broken into $2 - L < 0$ and $L - 10 < 0$.

The steps to come up with the above formulation include the following:

1. Specify the design variables: Design variables are the attributes of a design that require values, and one needs to define the name, symbols, units, and range of allowed values, and whether they are continuous or discrete.
2. Specify PDP: PDP is a parameter value imposed by a specific condition that the component will encounter in use. One needs to define its name, symbols, units, and limits on feasible assigned values.
3. Specify objective function f(x): An objective function is a single function, written in terms of the design variables, to express the overall quality or goodness of a trial design. One can use weighting factors to obtain a single criterion function from multiple criteria.
4. Specify constraints: Constraints are the limits on the ranges of values of a design variable, and the limits of a required relationship among the design variables and PDPs.

Example 10.1

Design a 1 L container shaped like a right circular cylinder, as shown in Figure 10.3, to hold a fixed volume. Design the container to use the least amount of material. Give the dimensions that use the least amount of material. Formulate this problem in the standard format.

Solution:

To formulate this problem, from the problem statement, one can find

$$\pi r^2 H = 1\ \text{L}$$

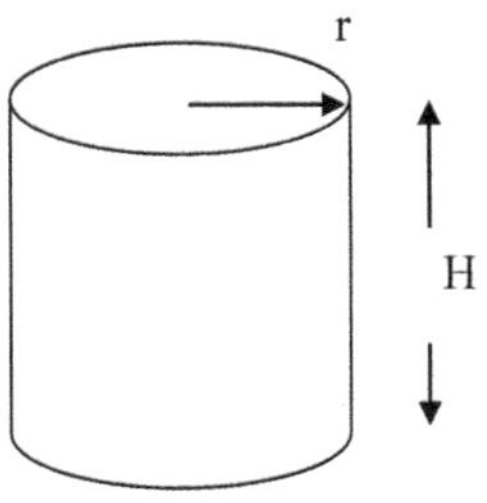

Figure 10.3 Design a container to hold a fixed volume of material.

and to minimize the amount of material, or surface area:

$$\text{Surface area} = \text{Area of outer surface} + \text{area of top} + \text{area of bottom}$$
$$= 2\pi rH + \pi r^2 + \pi r^2 = 2\,\pi r(H + r)$$

$$\text{Therefore, surface area} = 2\pi r^2 + 2\pi rH$$

To put the problem into standard format, the following equations can be obtained:

$$\text{Minimize } f(r, H) = 2\pi r^2 + 2\pi rH$$
$$\text{Subject to: } h(r, H) = \pi r^2 H - 1 = 0$$
where
$$r > 0$$
$$H > 0$$

Example 10.2

Given a cylinder with a top and bottom, maximize the volume inside it so that the total surface area is $A_0 = 24\pi$. Formulate this problem in the standard format.

Solution:

To formulate this problem, one needs to calculate the formula for total surface area:

$$\text{Surface area} = \text{Area of outer surface} + \text{area of top} + \text{area of bottom}$$
$$= 2\pi rH + \pi r^2 + \pi r^2 = 2\pi r(H + r) = 24\pi$$

To put the problem into the standard format, the following equations can be obtained:

$$\text{Minimize } f(r, H) = -\pi r^2 H$$
$$\text{Subject to: } h(r, H) = 2\pi r(H+r) - 24\pi = 0$$
where
$$r > 0$$
$$H > 0$$

Example 10.3

A company is designing an advertisement poster for a new product. The logo must have a printed area of $200\,cm^2$, a 0.5 cm margin at the two sides of the text, and 3 cm margins each along the top and the bottom. If it is desired to use the minimum material due to large quantity, find the best choice for the poster dimensions width (W) and height (H). Formulate this problem in the standard format.

Solution:

The objective function of this problem is

$$\text{The poster area} = WH,$$

$$\text{and the constraint } (W-1)(H-6) = 200$$

To put the problem into standard format, the following equations can be obtained:

$$\text{Minimize } f(W,H) = WH$$

$$\text{Subject to: } h(W,H) = (W-1)(H-6) - 200 = 0$$

where

$$W > 1$$

$$H > 6$$

Example 10.4

According to the regulation of an airline company, the maximum size of each carry-on case is 62 in./157 cm total for length + width + height. The length/width ratio for this design is 1.5. Determine the dimensions which will maximize the volume of the bag. Formulate this problem in the standard format.

Solution:

The objective function is to maximize the volume of the bag = abc,
where

a represents length
b represents width
c represents height

To put the problem into standard format, the following equations can be obtained:

$$\text{Minimize } f(a,b,c) = -abc$$

$$\text{Subject to: } g_1(a,b,c) = a+b+c \le 157$$

$$\text{and } h_1(a,b,c) = a/b - 1.5 = 0$$

$$a > 0, b > 0, \text{and } c > 0$$

Example 10.5

Build a rectangular enclosure as shown in Figure 10.4 with three parallel partitions using a total of 500 ft of fencing. What are the dimensions that will maximize the total area of the fence? Formulate this problem in a standard format.

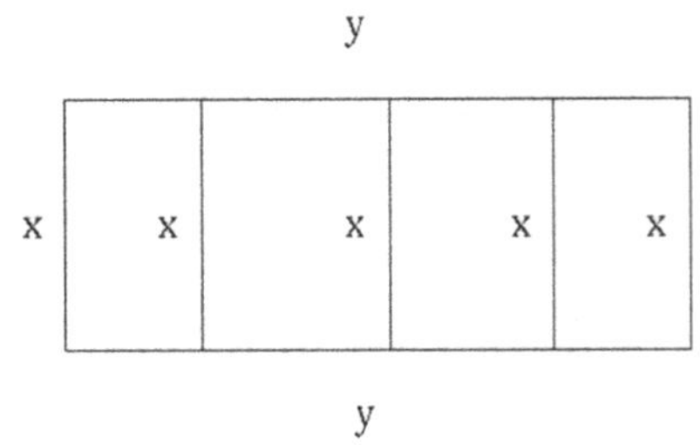

Figure 10.4 An example to design a rectangular enclosure with three parallel partitions using a total of 500 ft of fencing.

Solution:

To formulate this, let variable x be the width of the enclosure and variable y the length of the enclosure. The total amount of fencing is given to be

$$500 = 5(\text{width}) + 2(\text{length}) = 5x + 2y$$

To maximize the total area of the enclosure

$$\text{Area} = (\text{width})(\text{length}) = xy$$

To put the problem into standard format, the following equations can be obtained:

$$\text{Minimize } f(x, y) = -xy$$

$$\text{Subject to: } h(x, y) = 5x + 2y - 500 = 0$$

$$\text{and } x > 0, y > 0$$

Example 10.6

As shown in Figure 10.5, find a hollow round tube with the least mass that will support a 50,000 lbs load cantilevered 36 in. from the support with maximum bending stress of 30,000 psi. Formulate this problem in the standard format.

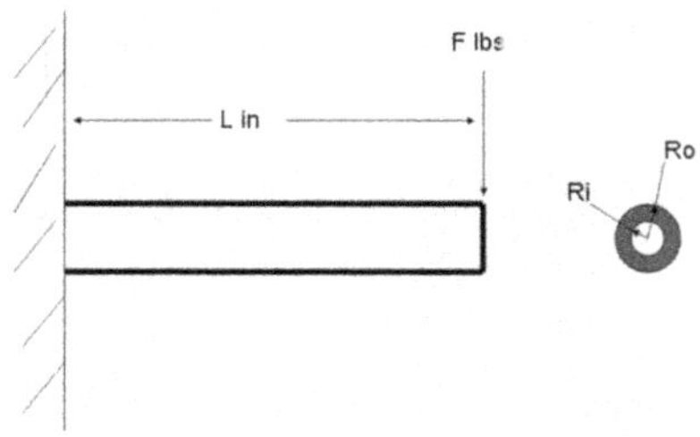

Figure 10.5 An example to design a hollow round tube with the least mass.

Solution:

Let:

K = volume
σ = stress
m = moment

R_o = outer radius
R_i = inner radius
I = area moment of inertia
L = length of bar
F = force
s = yield stress

$$c = \frac{R_o}{2}$$

$$\sigma = \frac{mc}{I}$$

$$m = FL$$

$$I = \frac{\pi\left(R_o^4 - R_i^4\right)}{4}$$

$$\text{Therefore, } \sigma = \frac{2LFR_o}{\pi\left(R_o^4 - R_i^4\right)}$$

To put the problem into the standard format, the following equations can be obtained:

$$\text{Min volume } K = \pi\left(R_o^2 - R_i^2\right)L$$

$$\text{Subject to: } g_1 = \frac{2LFR_o}{\pi\left(R_o^4 - R_i^4\right)} - s \le 0$$

$$R_o = \text{outer radius} > 0\ (R_o > R_i)$$

$$R_i = \text{inner radius} > 0$$

Review Problems

1. What is engineering optimization?
2. What is Pareto optimality?
3. What are the steps in the formulation of engineering problems?
4. Fill in the blank space in the diagram as shown in Figure 10.6.
5. Formulate the following problem: Find the dimensions of a box with the largest volume if the total surface area is less than 150 cm^2.
6. Design a cylindrical container with a fixed surface area. Maximize the volume inside the container such that the total surface is $A_0 = 56\pi$. Formulate this problem into the standard format for a container with a top and bottom, so that it is enclosed.

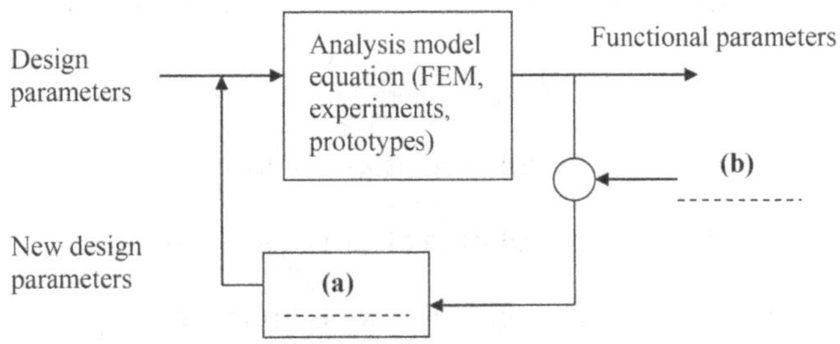

Figure 10.6 A general representation of an optimization problem.

10.2 *Optimization using differential calculus*

I don't have any solution, but I certainly admire the problem.

Ashleigh Brilliant

This section helps answer the following questions:

- How to perform a hand calculation for a simple optimization problem?
- How to find an optimal solution?
- When does one know that there is an absolute maximum or minimum solution?

Once the problem is in the form of a mathematical equation, the next step is to find the solution. One solution method is to set up a spreadsheet and iteratively search the design space defined by the range of design variables. However, for 10 design variables, and to consider 10 possible valid values per variable, how many combinations must be calculated? The answer is 10^{10}! This task is difficult, if not impossible. The optimization process reduces the trial and error spreadsheet operation. During problem formulation, one needs to consider typical values for every noise variable and simplify noise space to a single point to minimize (or maximize) the performance metrics.

There are many types of optimization problems such as continuous, discrete, constrained, unconstrained, and sometimes even a binary (Y/N) problems, which can also be used for optimization. Differential calculus is used for simpler kinds of problems like first-order and second-order equations. The derivative is taken, and then, it is decided whether one should go for maximum or minimum. The second derivative determines whether the value attains a maximum or minimum depending upon whether the value of the derivative is positive or negative. Differential calculus is also called classical optimization, to find the optimal solution to a set of equations. For a smooth and continuous function

$$y = U(x) \tag{10.2}$$

An extreme point on the function, either maximum or minimum, exists if the following conditions are met, for both first-order and second-order equations:

1. First-order condition: The function may contain an extreme point, that is, maximum or minimum, if the slope of the function is zero at that point.
2. Second-order condition: The value attains a maximum or minimum depending upon whether the value of the second derivative is negative or positive, respectively.

$$\text{In mathematical form, let objective function} = U(x) \tag{10.3}$$

$$dU/dx = 0 \text{ at } x = x_0 \text{ and}$$

$$\text{If } d^2U/dx^2 < 0 \text{ at } x = x_0 \text{ the solution at } x_0 \text{ is a local maximum}$$

$$\text{If } d^2U/dx^2 > 0 \text{ at } x = x_0 \text{ the solution at } x_0 \text{ is a local minimum} \tag{10.4}$$

$$\text{If } d^2U/dx^2 = 0 \text{ at } x = x_0 \text{ the solution at } x_0 \text{ is an inflection point which is neither local maximum or local minimum}$$

Example 10.7

Using the same conditions as Example 10.1, apply differential calculus to solve the problem.

Solution:

First, the volume of the can must be calculated. Knowing the height, h, and radius, r, the volume is computed using the equation:

$$\pi r^2 H = 1\text{ L}$$

It is known that 1 L = 1,000 cm^3, therefore, $\pi r^2 H$ = 1,000 cm^3. Next, the surface area must be calculated.

$$\text{Surface area} = 2\pi r^2 + 2\pi r H$$

$$H = \frac{1{,}000}{\pi r^2} \Rightarrow \text{surface area,} \quad A = 2\pi r^2 + 2\pi r\left(\frac{1{,}000}{\pi r^2}\right) = 2\pi r^2 + \frac{2{,}000}{r}$$

Differentiating and setting equal to zero gives

$$\frac{dA}{dr} = 4\pi r - \frac{2{,}000}{r^2} = 0 \Rightarrow 4\pi r^3 = 2{,}000$$

Solving for r:

$$r = \sqrt[3]{\frac{500}{\pi}} = 5.42$$

Therefore, the optimum radius is 5.42 cm.

Similarly, solving for H, yields

$$H = \frac{1{,}000}{\pi r^2} = \frac{1{,}000}{\pi\left(\sqrt[3]{\frac{500}{\pi}}\right)^2} = 10.84\text{ cm}$$

Therefore, the optimum height is 10.84 cm, or H = 2r.

Example 10.8

Using the same conditions as Example 10.2, apply differential calculus to solve the problem.

Solution:

To calculate the formula for the total surface area:

$$2\pi r(H + r) = 24\pi$$

Solving the above equation, $H = (12 - r^2)/r$

As the formula for the volume of the cylinder is given by $V = \pi r^2 h$

$$V = \pi r^2\left(\frac{12 - r^2}{r}\right) \Rightarrow V = \pi r(12 - r^2) \Rightarrow \frac{dV}{dr} = \pi(12 - 3r^2)$$

In order to maximize or minimize the given problem, one has to equate dV/dr to 0.

Hence, dV/dr = 0 gives $\pi\,(12 - r^2)$ or r = 2

Since d^2V/dr^2 = negative, the resultant value will be the local or in this case the global maxima.

By substitution, the value of r in the equation of H., that is, substituting r = 2 in the previous equation gives a result of H = 4. Hence, the volume of the cylinder will be maximum when r = 2 and H = 4.

Hence, the maximum volume of the cylinder with a surface area of 24π is

$$V = \pi r^2 H$$

$$V = \pi(2)^2(4)$$

$$V = 16\pi$$

Example 10.9

Using the same conditions as Example 10.3, apply differential calculus to solve the problem.

Solution:

$$WH - H - 6W + 6 = 200$$

$$H = \frac{194 + 6W}{W - 1}$$

$$U = WH = W\left(\frac{194 + 6W}{W - 1}\right)$$

$$W > 1$$

$$\frac{\partial U}{\partial W} = \frac{(194 + 12W)(W - 1) - (194W + 6W^2)}{(W - 1)^2}$$

$$6W^2 - 12W - 194 = 0$$

$$W = 6.8\,\text{cm}$$

$$H = 40.5\,\text{cm}$$

$$\frac{\partial U^2}{\partial W^2}(@W = 6.8) > 0$$

Therefore, the solution is OK.

Example 10.10

Using the same conditions as Example 10.4, apply differential calculus to solve the problem.

Solution:

$$U = abc$$

where

a represents length
b represents width
c represents height

and set $a + b + c = 157$ and $a/b = 1.5$

$$\text{so } 2.5b + c = 157$$

$$c = 157 - 2.5b$$

$$U = abc = 1.5b^2(157 - 2.5\,b) = 235.5\,b^2 - 3.75b^3$$

$$\frac{dU}{db} = 0,\ 471b - 11.25b^2 = 0$$

$b=0$ or $b=41.87$

$$\frac{d^2U}{db^2} = 471 - 22.5b$$

$$\text{when } b=0,\ \frac{d^2U}{db^2} = 471 - 22.5b = 471 > 0$$

$$\text{when } b = 41.87,\ \frac{d^2U}{db^2} = 471 - 22.5b = -471.075 < 0$$

Therefore, $b = 41.87$ cm, $a = 62.81$ cm, $h = 52.32$ cm.

Example 10.11

Using the same conditions as Example 10.5, apply differential calculus to solve the problem.

Solution:

The total amount of fencing is given to be

$$500 = 5(\text{width}) + 2(\text{length}) = 5x + 2y,$$

so that

$$2y = 500 - 5x$$

Or

$$y = 250 - (5/2)x$$

To maximize the total area of the enclosure,

$$A = (\text{width})(\text{length}) = xy$$

However, before differentiating the right-hand side, one will write it as a function of x only.

Substitute for y getting

$$A = xy = x\left[250 - (5/2)x\right] = 250x - (5/2)x^2$$

Now differentiate this equation, getting

$$A' = 250 - (5/2)2x = 250 - 5x = 5(50 - x) = 0$$

$$x = 50.$$

If $x = 50$ ft and $y = 125$ ft, then $A = 6{,}250\ \text{ft}^2$ is the largest possible area of the enclosure.

Example 10.12

Find the combination of L and K that maximizes output subject to the constraint that the cost of resources used is C; that is, maximize Q with respect to L and K subject to the constraint that wL + rK = C. Q = output, L = labor input, and K = capital input, where $Q = L^{2/3} K^{1/3}$. The cost of resources used is C = wL + rK, where w is the wage rate and r is the rental rate for capital.

Solution:

From the expression of Q, ln(Q) = (2/3) ln(L) + (1/3) ln(K), a more convenient expression, is the same as maximizing Q.

Therefore, the objective function for the optimization problem is

$$\ln(Q) = (2/3)\ \ln(L) + (1/3)\ \ln(K).$$

$$\text{Set } G = \ln(Q) - \lambda(wL + rK - C)$$

$$G = (2/3)\ln(L) + (1/3)\ln(K) - \lambda(wL + rK - C)$$

$$\partial G/\partial L = (2/3)(1/L) - \lambda w = 0$$

$$\partial G/\partial K = (1/3)(1/K) - \lambda r = 0$$

$$\text{If } (2/3)(1/L) = \lambda w, L = (2/3)/(\lambda w)$$

$$\text{If } (1/3)(1/K) = \lambda r, K = (1/3)/(\lambda r)$$

$$wL + rK = (2/3)(1/\lambda) + (1/3)(1/\lambda) = 1/\lambda = C, \text{so } \lambda = 1/C.$$

$$L = (2/3)(C/w)$$

and

$$K = (1/3)(C/r).$$

$$\text{Therefore, } Q = (2/3)(C/w)^{2/3} \times (1/3)(C/r)^{1/3}$$

Review Problems

1. What conditions of the derivatives have to be met to find a local maximum and local minimum?
2. Find two nonnegative numbers whose sum is 9, so that the product of one number and the square of the other number is a maximum. Use differential calculus to solve this problem.
3. Using the optimization methods discussed in class, find the point P = (x, y) on a line y = kx + b that is closest to a given point $Q = (x_0, y_0)$. The point Q lies above the line. Use differential calculus to solve this problem.
4. A civil engineer's intention is to construct an office building at the lowest possible cost per floor. As the building gets taller, the cost of the material decreases because of bulk rates, but the cost per floor increases because of increased labor and insurance. The following expressions may be taken as valid measures of these various costs.

$$C_1 = 100,000/(f/2)$$

$$C_2 = 1,000f$$

where

C_1 = cost of materials (dollars)
C_2 = cost of labor and insurance (dollars)
f = number of floors

Assuming that the total cost is the sum of materials, labor, and insurance, find the optimum number of floors up to which the building should be built.

5. As shown in Figure 10.7, a box with a square base with no top has a surface area of 108 ft^2. Find the dimensions that will maximize the volume.
6. Using the analytical method and given a cube with a width, height, and depth of X, determine the value of X that obtains the maximum volume of the cube AND the minimum total surface area of the cube.

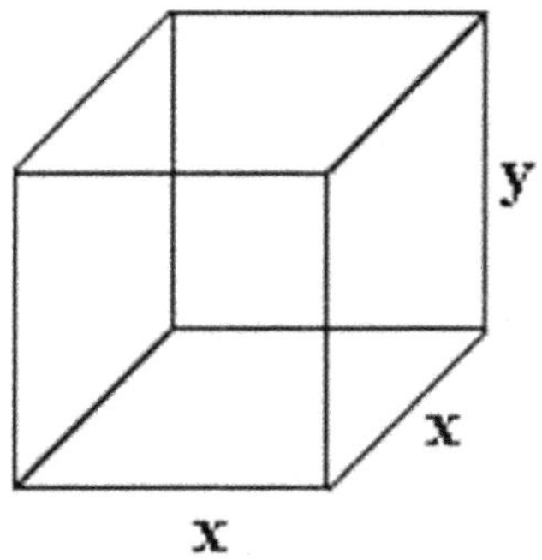

Figure 10.7 Find the dimensions of a box with a square base with no top.

10.3 *Lagrange's multiplier method*

Every great and deep difficulty bears in itself its own solution.

It forces us to change our thinking in order to find it.

Niels Bohr

This section helps answer the following questions:

- How to perform hand calculations for a constrained optimization problem with inequality constraints?
- How to perform hand calculations for a constrained optimization problem with equality constraints?
- When could one know that there is a redundant constraint?

One common problem in calculus is that of finding the maximum or minimum of a function, but it is often difficult to find a closed form for the function being optimized. Such difficulties often arise when one wishes to maximize or minimize a function subject to fixed outside conditions or constraints. The method of Lagrange multipliers is a powerful tool to solve this class of problems without the need to explicitly solve the conditions and use them to eliminate extra variables [Jensen17]. This method converts a constrained problem to an unconstrained problem. First, the Lagrangian form is obtained by combining the Lagrange multiplier λ, the objective function, the inequality constraint g, and the equality constraint h. For each of the constraints, a variable λ is introduced, and the constraint is rearranged so that the expression is equal to zero. The objective function is modified and multiplies each constraint with the Lagrange multiplier. The second step involves the derivation of the first-order conditions. Then, the equations are solved. The steps to solve the problem using the Lagrange multiplier method are summarized below:

1. Format the equations into the standard format, as in Equation 10.1.
2. Form the Lagrangian function: For each of the constraints in the problem, one can introduce a variable, say λ_i. Each constraint is rearranged so that the expression is equal to 0. The original objective function is modified by adding each constraint function multiplied by a corresponding Lagrange multiplier.

$$L = f(x) + \lambda_1 g_1 + \lambda_2 g_2 + \cdots \lambda_m g_m + \lambda_{m+1} h_1 + \lambda_{m+2} h_2 + \cdots \lambda_{m+k} h_k \tag{10.5}$$

3. Derive the first-order conditions.

$$\frac{\partial L}{\partial x_i} = 0 \quad \text{and} \quad x_i = 1, \ldots, n \tag{10.6}$$

4. Use the three equations to solve for x^*, y^*, and λ^*.
5. λ_1–λ_m need to be nonnegative. If λ_j is negative, one needs to ignore g_j, as it is a redundant constraint and resolve the problem again.
6. There is no need to check $\lambda_{m+1} \ldots \lambda_{m+k}$ for equality constraints.

The following examples are used to illustrate this method.

Example 10.13

With the same conditions as Example 10.3, apply Lagrange's multiplier method to solve the problem.

Solution:

$$\text{Minimize area } U = WH$$

$$\text{Subject to: } (W-1)(H-6) = 200$$

$$\text{Let equality constraint equation } h = (W-1)(H-6) - 200$$

Using Lagrange's multiplier method, one can define

$$L = (U + \lambda h)$$

or the problem is equivalent to

$$\text{Minimize } L = (U + \lambda h) = WH + \lambda\left[(W-1)(H-6) - 200\right]$$

Take the derivatives with respect to W and H, the following can be obtained:

$$LW = H + \lambda(-H-6) = 0$$

$$LH = W + \lambda(-W-1) = 0$$

Removing "λ" from the above two equations, one can obtain

$$H = 6W$$

Taking the derivatives of L with respect to λ, we have,

$$L_\lambda = [(W-1)(H-6)-200] = 0, \text{and thus } W = 6.8, H = 40.8$$

As there is only one equality constraint, we do not need to check the sign of λ, that is, the above results are the solution to this problem.

Example 10.14

Using the same conditions as Example 10.2, apply Lagrange's multiplier method to solve the problem.

Solution:

$$\text{Minimize } f = -\pi R^2 H$$

$$\text{Subject to: } g = 2\pi R^2 + 2\pi RH - 24\pi \leq 0$$

$$\text{Let le} = f + \lambda g = -\pi R^2 H + \lambda(2\pi R^2 + 2\pi RH - 24\pi)$$

$$\frac{\partial le}{\partial R} = 0 \Rightarrow -\pi 2RH + \lambda(4\pi R + 2\pi H) = 0 \text{ and thus } \lambda = RH/(2R+H)$$

$$\frac{\partial le}{\partial H} = 0 \Rightarrow -\pi R^2 + 2\pi R\lambda = 0 \text{ and thus } \lambda = \tfrac{1}{2}R \text{ as } R \geq 0. \text{ Therefore, g is active.}$$

$$\lambda = \tfrac{1}{2}R = RH/(2R+H), \text{ and thus } H = 2R$$

$$\frac{\partial le}{\partial \lambda} = 0 \Rightarrow 2\pi R^2 + 2\pi RH - 24\pi = 0 \text{ and thus } R^2 + R(2R) - 12 = 0$$

$$R = 2$$

$$H = 2R = 4$$

Example 10.15

Using the same conditions as Example 10.4, apply Lagrange's multiplier method to solve the problem.

Solution:

$$\text{Minimize } f = -abc$$

$$\text{Subject to: } g_1 = a + b + c - 157 < 0$$

$$h_1 = a - 1.5b = 0$$

$$\text{Let } L = f + \lambda_1 g_1 + \lambda_2 h_1 = -abc + \lambda_1(a + b + c - 157) + \lambda_2(a - 1.5b)$$

$$\frac{\partial L}{\partial a} = 0, -bc + \lambda_1 + \lambda_2 = 0$$

$$\frac{\partial L}{\partial b} = 0, -ah + \lambda_1 - 1.5\lambda_2 = 0$$

$$\frac{\partial L}{\partial h} = 0, -ab + \lambda_1 = 0$$

$$\frac{\partial L}{\partial \lambda_1} = 0, \ a + b + c - 157 = 0$$

$$\frac{\partial L}{\partial \lambda_2} = 0, \ a - 1.5b = 0$$

Therefore, $\lambda_1 = 2629.85 > 0$; $\lambda_2 = -437.58 > 0$ (ok as h is an equality constraint); b = 41.87 cm; a = 62.81 cm; c = 52.32 cm

Example 10.16

In an engineering design, one needs to find the largest rectangular area that can be inscribed in the ellipse $x^2 + 2y^2 = 1$, as shown in Figure 10.8. Use the Lagrange multiplier method to find it.

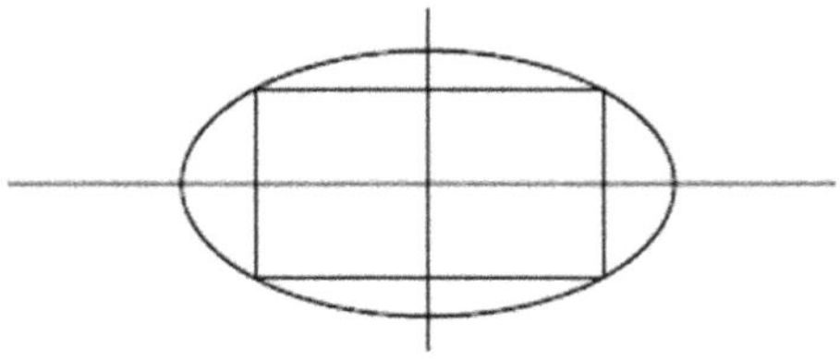

Figure 10.8 Find the largest square area that can be inscribed in the ellipse.

Solution:

Clearly, the largest rectangle will also be centered at the origin. Let the corner of the rectangle in the first quadrant be (x, y). The area of the rectangle is 4xy.

$$\text{Minimize} f = -4xy$$

$$\text{Subject to}: h = x^2 + 2y^2 = 1$$

Define the Lagrange equation as $L = -4xy + \lambda(x^2 + 2y^2 - 1)$.

$$\frac{\partial L}{\partial x} = 0 \Rightarrow -4y + 2x\lambda = 0 \Rightarrow \lambda = 2y/x$$

$$\frac{\partial L}{\partial y} = 0 \Rightarrow -4x + 4y\lambda = 0 \Rightarrow \lambda = x/y = 2y/x \Rightarrow x^2 = 2y^2$$

$$\frac{\partial L}{\partial \lambda} = 0 \Rightarrow x^2 + 2y^2 - 1 = 0 \Rightarrow 4y^2 = 1 \Rightarrow y = 1/2 \Rightarrow x = \frac{1}{\sqrt{2}}$$

So when $x = 1/\sqrt{2}, y = 1/2$, the maximum area of the rectangle is $\sqrt{2}$

Example 10.17

Using the same conditions as Example 10.6, can you use Lagrange's multiplier method to solve the problem?

Solution:

Recall that

K = volume
σ = stress
m = moment
R_o = outer radius
R_i = inner radius
I = area moment of inertia
L = length of bar
F = force
s = yield stress

$$\sigma = \frac{mc}{I}$$

$$m = FL$$

$$c = \frac{R_o}{2}$$

$$I = \frac{\pi\left(R_o^{\,4} - R_i^{\,4}\right)}{4}$$

$$\text{Therefore, } \sigma = \frac{2LFR_o}{\pi\left(R_o^{\,4} - R_i^{\,4}\right)}$$

Let us assume that we formulate the problem as follows:

$$\text{Minimize volume } K = \pi\left(R_o^{\,2} - R_i^{\,2}\right)L$$

$$\text{Subject to } g_1 = \frac{2LFR_o}{\pi\left(R_o^{\,4} - R_i^{\,4}\right)} - \sigma < 0$$

$$le = K + \lambda_1 g_1$$

$$le = \pi\left(R_o^{\,2} - R_i^{\,2}\right)L + \frac{\lambda_1 2LFR_o}{\pi\left(R_o^{\,4} - R_i^{\,4}\right)} - \lambda_1\sigma$$

Partial derivative done by using the quotient rule

$$\frac{\partial le}{\partial R_o} = 2\pi R_o L + \left[\frac{\left(R_o^{\,4} - R_i^{\,4}\right) - R_o\left(4R_o^{\,3}\right)}{\left(R_o^{\,4} - R_i^{\,4}\right)^2}\right]\frac{\lambda_1 2LF}{\pi} = 0 \tag{10.7}$$

$$\frac{\partial le}{\partial R_i} = -2\pi R_i L + \left[\frac{4R_i^{\,3}}{\left(R_o^{\,4} - R_i^{\,4}\right)^2}\right]\frac{\lambda_1 2LFR_o}{\pi} = 0 \tag{10.8}$$

(Solve Equation 10.7 for λ_1)

$$\lambda_1 = \frac{-\pi^2 R_o}{F}\left[\frac{\left(R_o^{\,4} - R_i^{\,4}\right)^2}{\left(R_o^{\,4} - R_i^{\,4}\right) - 4R_o^{\,4}}\right] \tag{10.9}$$

(Solve Equation 10.8 for λ_1)

$$\lambda_1 = \frac{\pi^2 R_i}{F}\left[\frac{\left(R_o^{\,4} - R_i^{\,4}\right)^2}{R_o 4R_i^{\,3}}\right] \tag{10.10}$$

Set Equations 10.9 and 10.10 equal to each other

$$\frac{-\pi^2 R_o}{F}\left[\frac{\left(R_o^{\,4} - R_i^{\,4}\right)^2}{\left(R_o^{\,4} - R_i^{\,4}\right) - 4R_o^{\,4}}\right] = \frac{\pi^2 R_i}{F}\left[\frac{\left(R_o^{\,4} - R_i^{\,4}\right)^2}{R_o 4R_i^{\,3}}\right]$$

Simplify

$$\left(\frac{R_o}{3R_o^{\,4} + R_i^{\,4}}\right) = \left(\frac{1}{R_o 4R_i^{\,2}}\right)$$

$R_o = R_i$ (This will maximize the objective function instead of minimizing and thus is not the right answer), or $3\,R_o^{\,2} - R_i^{\,2}$ (not reasonable as $R_o > R_i$)

When examining the original equation, one can find that one important side constraint was missed:

$$\text{Minimize volume } K = \pi\left(R_o^{\,2} - R_i^{\,2}\right)L$$

$$\text{Subject to } g_1 = \frac{2LFR_o}{\pi\left(R_o^{\,4} - R_i^{\,4}\right)} - \sigma < 0$$

$$g_2 = R_i - R_o < 0$$

$$R_i > 0 \text{ and}$$

$$R_o > 0$$

At this point, one should go back and substitute the above equations back into the original equation to find the λ's, R_o in terms of R_i, and then solve σ to get a value. However, as the problem becomes more complex, we decide to get help from the computer. This problem will be solved in the next section using Excel.

Review Problems

1. Find the stationary points for the following constrained problem using the method of Lagrange multipliers.

$$\text{Minimize}: y(x) = x_1 x_2$$

$$\text{Subject to}: f(x) = x_1^{\,2} + x_2^{\,2} - 1 = 0$$

2. Using the optimization methods discussed in class, find the point P = (x, y) on a line y = kx + b that is closest to a given point Q = (x_0, y_0). The point Q lies above the line. Use the Lagrange multiplier method to solve this problem.
3. Find the combination of L and K that maximizes output subject to the constraint that the cost of resources used is C; that is, maximize Q with respect to L and K subject to

the constraint that $wL + rK = C$, where w is the wage rate, and r is the rental rate for capital. Q = output, L = labor input, and K = capital input and $Q = L^{2/3} K^{1/3}$.
4. The capacity of the tank is to be 4,000 L. The objective is to minimize the surface area and to minimize the material required to build the tank. Use the Lagrange multiplier method for the analytical solution. For the purpose of this example, do not need to consider the effect of the height of the tank on the pressure applied to the material at the bottom, and the impact on the thickness requirement of the material.
   R = radius of the tank
   H = height of the tank

10.4 Optimization using Microsoft Excel

> Life is not a problem - Life is the closest God has yet come to a solution.
>
> **Ashleigh Brilliant**

This section helps answer the following questions:

- Are there software programs that can be used to optimize engineering solutions?
- How can one use a computer to solve an optimization problem?
- How does one specify an objective function, constraints, a constant, and a design variable?
- What is the algorithm behind Microsoft Excel Solver?

Microsoft Excel Solver is a Microsoft Excel Add-in, one of the features that allows the creation of engineering and financial models in a spreadsheet [Barreto17, Solver17, Wittwer17]. Microsoft Excel Solver uses the generalized reduced gradient (GRG2) algorithm for optimizing nonlinear problems. This algorithm was initially developed by Leon Lasdon, the University of Texas at Austin, and Allan Waren, Cleveland State University. Linear and integer problems use the simplex method, with bounds on the variables and the branch and bound method, implemented by John Watson and Dan Fylstra, of Frontline Systems, Inc. Solver may be used to solve problems with up to 200 decision variables, 100 explicit constraints, and 400 simple constraints.

On the Excel Tools menu, click Solver to solve an optimization problem. A window will appear as shown in Figure 10.9. The optimization model has three parts: the target cell, changing cells, and constraints. If Solver is not found, one may use the Add-ins command on the Tools menu to load the Add-in into Excel. Solver uses iterative numerical methods that involve plugging in trial values for the changing cells and observing the results calculated by the constraint cells and the optimum cell. Each trial is called an iteration. Because a pure trial and error approach would be extremely time-consuming, Solver performs extensive analyses on the observed outputs and their rates of change as the inputs are varied, to guide the selection of new trial values. In a typical problem, the constraints and the target cell are functions of the changing cells. The first derivative of a function measures its rate of change as the input is varied. When several values are entered, the function has several partial derivatives measuring its rate of change with respect to each of the input values together; the partial derivatives form a vector called the function gradient.

In the dialog box as shown in Figure 10.9, **Set Target Cell** is where one indicates the objective function to be optimized (maximized, minimized, or targeted to a certain value). The option **Equal To** gives one the option of treating the target cell in three alternative

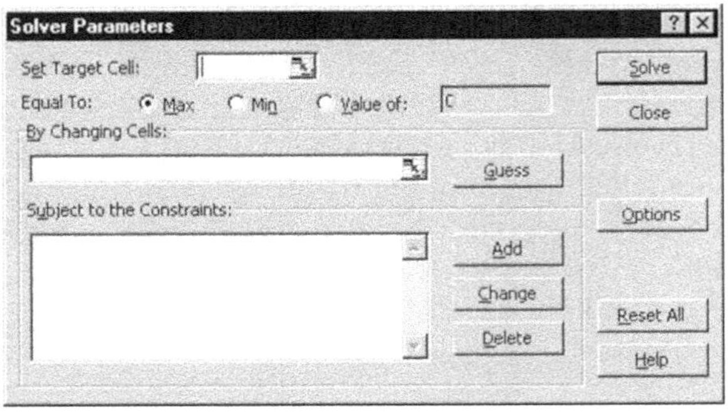

Figure 10.9 Solver parameters dialog box.

ways. **Max** tells Solver to maximize the target cell and **Min**, to minimize it, whereas **Value** is used if one wants to reach a certain particular value of the target cell by choosing a particular value of the endogenous variable. If one chooses **Value**, one must enter the particular value one wants to attain in the box to the immediate right unless one wants the value to be zero (default). This cell must contain a formula that depends on one or more of the other cells. In general, it should include at least one changing cell. Changing cells are the spreadsheet cells that we can change or adjust to optimize the target cell. **By Changing Cells** permits one to indicate which cells are the changing cells (i.e., design variables). Each noncontiguous choice variable is separated by a comma. If one uses the mouse technique (clicking on the cells), the comma separation is automatic. **Guess** controls the initial position of the changing cells. Excel uses the current values of the cells as the default. Solver is sensitive to the initial values. If a solution cannot be found, try different starting values. Because Excel is sensitive to starting values, many should be tried to distinguish between a local and global optimum. Using multiple new starting values will search more of the solution space. Granted, a global optimum most likely will never be found if the function is complex. However, using multiple points will give a better solution because the GRG2 method is in the hill climbing class of algorithms, which means it will stop when the gradient gets to zero or near zero.

In the example of a design company that wants to improve a prototype design, as shown in Table 10.1, the design engineering manager may want to maximize the prototype performance by changing the design attributes. The cell that measures performance would be the target cell. The target cells and changing cells for each situation described in Table 10.1 are listed in Table 10.2.

Derivatives, or gradients, play a crucial role in iterative methods in Solver. They provide clues as to how the changing cells should be varied. For example, if the target cell is being maximized and its partial derivative with respect to one changing cell is a large positive number, whereas another partial derivative is near zero, Solver will probably increase the first changing cell's value on the next iteration. A negative partial derivative suggests that the related changing cell's value should be varied in the opposite direction. Because the first derivative of the target cell measures its rate of change with respect to the changing cells, when all of the partial derivatives of the target cell are zero, the first-order conditions for optimality have been satisfied, the highest (or lowest) possible value for the target cell having been found.

Some problems have many locally optimum points, where the partial derivatives of the optimum cell are zero. A graph of the optimum cell function in such cases would show many hills and valleys of varying heights and depths as shown in Figure 10.2. When

Table 10.2 Target cells and changing cells for situations

Target cell	Maximize or minimize	Changing cell
Company's prototype performance	Maximize	Design attributes
Company's product shipping	Minimize	Amount produced at each plant each month that is shipped to each customer
Product profit	Maximize	Sales, prices, time, capitals, etc.
Company profit	Maximize	Projects selected, money, and skilled engineers
Product profit	Maximize	Amount of each product produced during the month
Production cost	Minimize	Labor, materials, equipment, utilities, etc.

started at a given set of changing cell values, the methods used by Solver will tend to converge on a single hilltop or valley floor close to the starting point. However, Solver has no sure way of knowing whether a global optimum has been reached. One way to find the global optimum is to apply external knowledge to the problem, either through common sense reasoning about the problem or through experimentation. Alternatively, one can use Monte Carlo optimization methods to come closer to a global optimum using Excel.

Constraints are restrictions to place on the changing cells. In the company's prototype performance example, some design attributes have certain limitations. For example, the maximum dimensions of the product are limited by the actual use of the product, and the minimum size of the product is limited by the size of the motors within. Table 10.3 lists the constraints for the problems presented in Table 10.1.

Table 10.3 Constraints for situations in Table 10.1

Model	Constraints
Maximize company's prototype performance	Certain design attributes have limitations
Minimize company's product shipping	Do not ship more units each month from a plant than plant capacity Make sure that each customer receives the number of units they need
Maximize product profit	Prices cannot be too far out of line from competitors' prices
Maximize company profit	Projects selected cannot use more money or skilled workers than are available
Maximize product profit	Product mix uses no more resources than are available Do not produce more of a product than can be sold
Minimize production cost	Obtain an expected return of at least 10% on investments

Any specification of the changing cells that satisfies the model's constraints is known as a feasible solution. Essentially, Solver searches over all feasible solutions and finds the feasible solution that has the optimal target cell value. The best way to understand how to use Solver is by looking at detailed examples. The constraints must be specified in the **Subject to the Constraints** box by clicking on **Add**. **Change** allows the user to modify a constraint already entered, and **Delete** allows the user to delete a previously entered constraint. **Reset All** clears the current problem and resets all parameters to their default values. **Options** invokes, the Solver options dialog box. When the **Add** button is clicked, the **Add Constraint** dialog box appears, as shown in Figure 10.10.

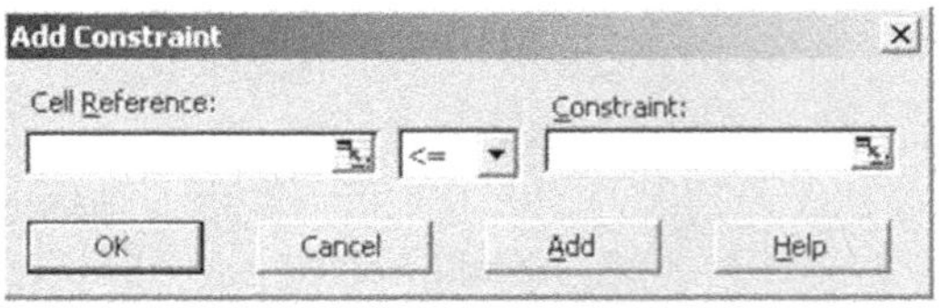

Figure 10.10 Solver add constraint dialog box.

Clicking on the **Cell Reference** box allows one to specify a cell location. The constraint type may be set by selecting the down arrow (< =, > =, =, int, where int refers to an integer, or bin, where bin refers to binary). The **Constraint** box may contain a formula of cells, a simple cell reference, or a numerical value. The **Add** button adds the currently specified constraint to the existing model and returns to the Add Constraint dialog box. The **OK** button adds the current constraint to the model and returns one to the Solver dialog box.

If the **Options** button is selected from the Solver parameters dialog box, the dialog box appears as shown in Figure 10.11. **Max Time** allows one to set the number of seconds before Solver will stop. The icon **Iterations** allows one to specify the maximum number of iterations before stopping. **Precision** is the degree of accuracy of the solver algorithm. **Tolerance** is used for integer programs. It specifies a percentage within which the solution is guaranteed to be optimal. If one seeks the optimal solution, this must be set to zero. If run-time becomes too long, one may wish to set this to a higher value. If a model is a linear program or a linear integer program, one should check **Assume Linear Model**. This tells Solver to use the simplex algorithm rather than the GRG2 method. **Assume Non-Negative**

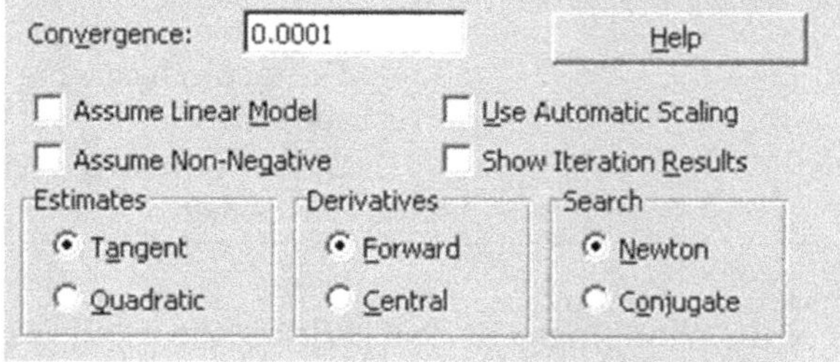

Figure 10.11 Solver option dialog box.

should be checked if all the changing cell values are greater than zero. Check **Show Iterations Results** if one wants to see information iteration by iteration. Use **Automatic Scaling** is useful if one's model is poorly scaled or if the inputs are of drastically different orders of magnitude. The bottom section of the dialog box concerns options for the nonlinear algorithm, namely, how it estimates nonlinearities, how rates of change are estimated and the type of search technique employed.

The best way to explain Solver is to go through an example. We have the following equation, and we want to find the value of x and H that minimize f.

$$\text{Minimize } f = x^2H$$

$$\text{Subject to } 2x + H \leq 96$$

$$x > 0, \text{ and } H > 0$$

First, key in the following in Excel as shown in Figure 10.12: A1 = x, A2 = H, A3 = volume. Initially set A1(x) = 1; A2(H) = 1; and constraint A3 = A1 * A1 * A2. Make sure that A3 = 1 after calculation. Then, key in the dimension constraint: A4 = 2 * A1 + A2. Make sure that A4 = 3 after calculation.

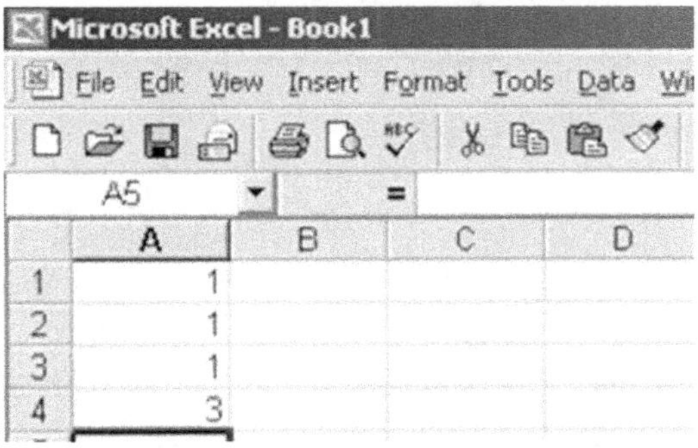

Figure 10.12 A screenshot of the example problem.

By clicking on **Tools**, then **Solver** a small "solver parameters" window should pop up, as shown in Figure 10.13. In the **Set Target Cell** box (objective function), enter a cell reference or name for the target cell (i.e., A3). The target cell (i.e., A3) must contain a formula.

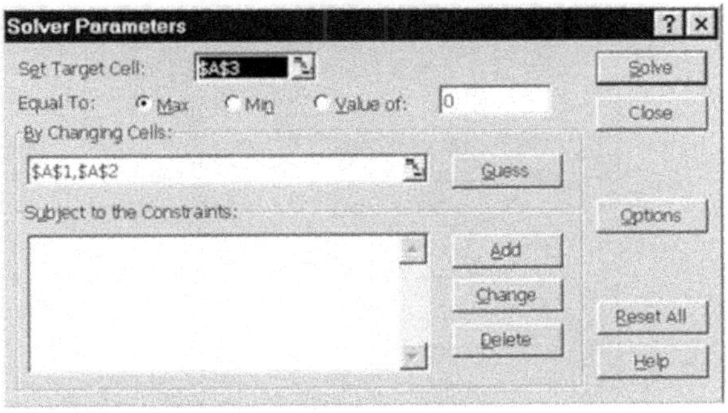

Figure 10.13 A screenshot of the Solver parameters of the example problem.

To have the value of the target cell be as large as possible, click **Max** in the following line. For values as small as possible, click **Min**. To have the target cell be a certain value, click **Value of**, and then type in the value. In the **By Changing Cell**, key in "A1, A2," separating references with commas. The adjustable cells must be related directly or indirectly to the target cell. One can specify up to 200 adjustable cells. To have Solver automatically propose the adjustable cells based on the target cell, click **Guess**. In the **Subject to the Constraints** box, click **Add** and enter any constraints one wants to apply, as shown in Figure 10.14. The resulting screen is shown in Figure 10.15.

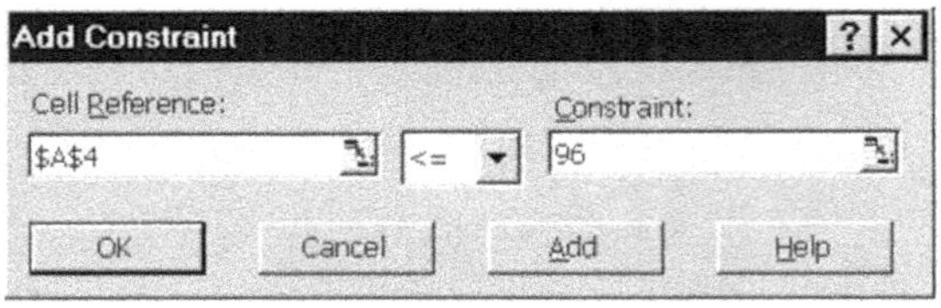

Figure 10.14 A screenshot of the Solver add constraint of the example problem.

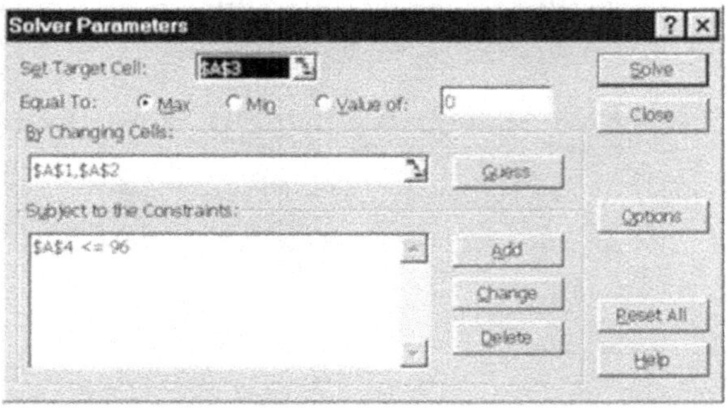

Figure 10.15 A screenshot of the Solver parameter of the example problem.

i.e., Add(constraint), and follow by typing "A4 < = 96"

Click Solve on the right-hand corner to solve the problem. To keep the solution values on the worksheet, click **Keep Solver Solution Results** dialog box. To restore the original data, click **Restore Original Values.** Now A1 and A2 should be equal to 32. A3 = 32,768, and A4 = 96, as shown in Figure 10.16.

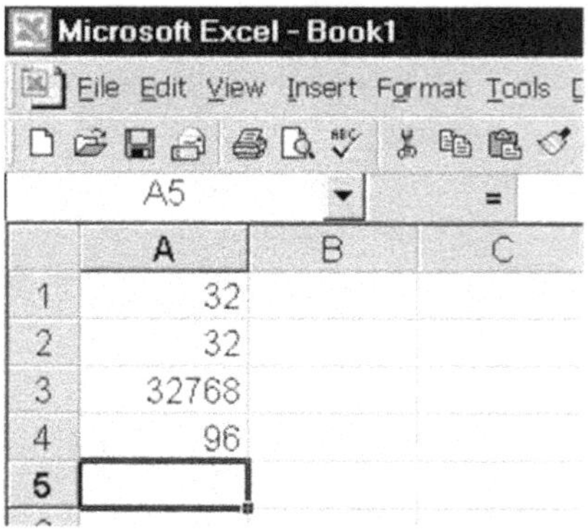

Figure 10.16 A screenshot of the result of the example problem.

Example 10.18

With the same conditions as Example 10.4, use Microsoft Excel to solve the problem.

Solution:

Use Excel Solver as shown in Table 10.4.

Table 10.4 Data input for Example 10.18 in Excel Solver

	A	B
1	A	1.5
2	B	1
3	C	1
4	a/b	1.5
5	a + b + c	3.5
6	Max V	1.5

The parameters are set as below: Set Target Cell: B5
By Changing Cells: B1, B2, B3
Subject to the constraints: B4 = 1.5 and B5 < = 157 Then, we can obtain the results in Table 10.5.

Table 10.5 Data output for Example 10.18 in Excel Solver

	A	B
1	A	62.8
2	B	41.86667
3	C	52.33333
4	a/b	1.5
5	a + b + c	157
6	Max V	137596.2

Example 10.19

With the same conditions as Example 10.5, use Microsoft Excel to solve the problem.

Solution:

Applying optimization method through Excel Solver,

The variables x and y are given initial values of 1, and the objective function is to maximize the area xy subject the constraint of fencing to $5x + 2y = 500$.

Hence, the input data in Excel Solver is shown in Table 10.6.

Table 10.6 Data input for Example 10.19 in Excel Solver

Width (X)	1
Length (Y)	1
Total area (A)	1
Amount of fencing	7

When the Solver is used to maximize the objective function, the solution obtained, as shown in Figure 10.14, which is consistent with the previous method.

We can obtain the final result as shown in Table 10.7.

Table 10.7 Data output for Example 10.19 in Excel Solver

Width (X)	50
Length (Y)	125
Total area (A)	6,250
Amount of fencing	500

Example 10.20

Use Excel to find the largest rectangle that can be inscribed in the ellipse $x^2 + 2y^2 = 1$, as shown in Figure 10.17.

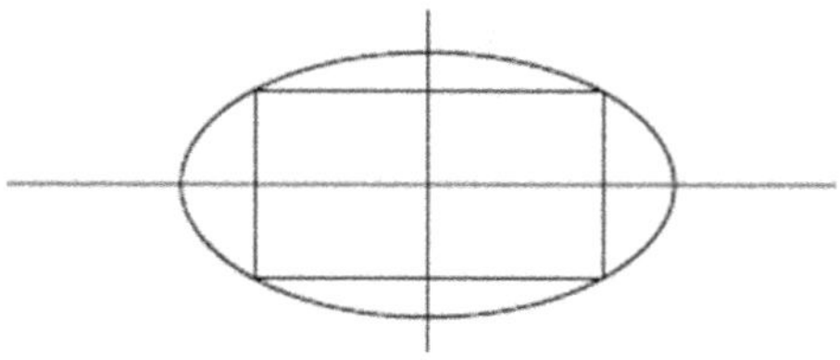

Figure 10.17 Example 10.20 to find the largest square that can be inscribed in the ellipse $x^2 + 2y^2 = 1$.

Solution:

Clearly the largest rectangle will also be centered at the origin. Let the corner of the rectangle in the first quadrant be (x, y). The area of the rectangle is 4xy.

Use Excel, and the same results are obtained as shown in Figure 10.18.

Example 10.21

With the same conditions as Examples 10.6 and 10.16, use Microsoft Excel to solve the problem. Also, assume the outer radius R_o is no more than 5 in.

Solution:

K = volume
σ = stress
m = moment
R_o = outer radius
R_i = inner radius
I = area moment of inertia
L = length of bar
F = force
s = yield stress

$$\sigma = \frac{mc}{I}$$

$$m = FL$$

$$c = \frac{R_o}{2}$$

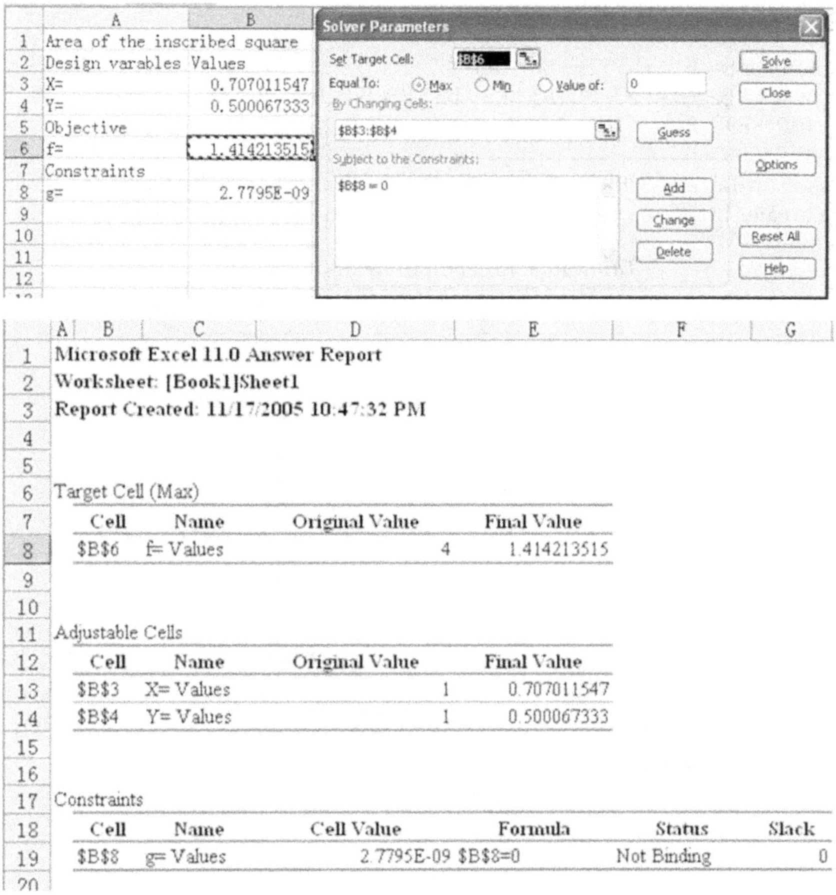

	A	B
1	Area of the inscribed square	
2	Design varables	Values
3	X=	0.707011547
4	Y=	0.500067333
5	Objective	
6	f=	1.414213515
7	Constraints	
8	g=	2.7795E-09

Microsoft Excel 11.0 Answer Report
Worksheet: [Book1]Sheet1
Report Created: 11/17/2005 10:47:32 PM

Target Cell (Max)

Cell	Name	Original Value	Final Value
$B$6	f= Values	4	1.414213515

Adjustable Cells

Cell	Name	Original Value	Final Value
$B$3	X= Values	1	0.707011547
$B$4	Y= Values	1	0.500067333

Constraints

Cell	Name	Cell Value	Formula	Status	Slack
$B$8	g= Values	2.7795E-09	$B$8=0	Not Binding	0

Figure 10.18 The Excel results for Example 10.20.

$$I = \frac{\pi\left(R_o^4 - R_i^4\right)}{4}$$

$$\text{Therefore, } \sigma = \frac{2LFR_o}{\pi\left(R_o^4 - R_i^4\right)}$$

The standard format is as follows:

$$\text{Minimize volume } K = \pi\left(R_o{}^2 - R_i{}^2\right)L$$

$$\text{Subject to } g_1 = \frac{2LFR_o}{\pi\left(R_o^4 - R_i^4\right)} - s \le 0$$

R_o = outer radius > 0, $R_o > R_i$
R_i = inner radius > 0

Using Excel, the results can be obtained.

Define:

Volume: B8 = B1 * (B5 * B5–B6 * B6) * B2

σ: B9 = 2 * B2 * B3 * B5/B1/(B5 * B5 * B5 * B5–B6 * B6 * B6 * B6)

g_1: B10 = B9 – B4

g_2: B10 = B6 – B5

Based on Table 10.8, the initial values: $R_o = 0.125$ and $R_i = 0.025$. Assuming R_o should not be greater than 5 in., the Solver input is shown in Table 10.8 and Figure 10.19.

Table 10.8 Excel input for Example 10.21

	A	B
1	π	3.1416
2	L	36
3	F	50,000
4	s	30,000
5	R_o	0.125
6	R_i	0.025
7		
8	K	1.696464
9	σ	587,647,646.5
10	g_1	587,617,646.5
11	g_2	–0.1

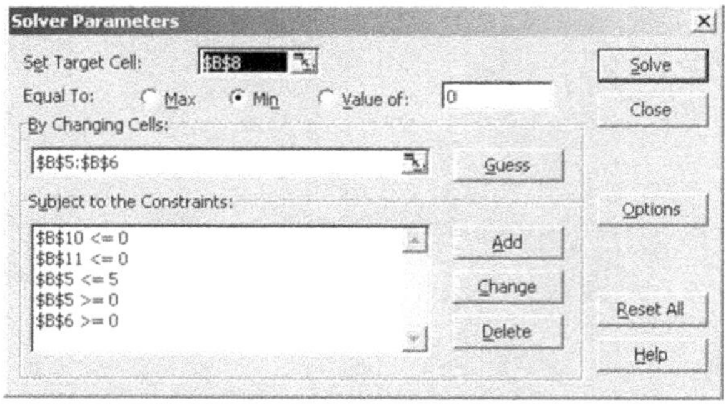

Figure 10.19 Defining Solver parameters in Excel for Example 10.21.

We can obtain the final result as shown in Table 10.9.

Therefore, $R_o = 5$ in., $R_i = 4.56$ in., and $K = 471.3$ $in.^3$

Review Problems

1. How is a constraint entered in an Excel sheet while solving an optimization problem using Excel Solver? Use the example of the constraint $h = 2 + (4 - r^2)^{1/2}$, where h and r are design variables.
2. With the conditions as Example 10.1, use Microsoft Excel to solve the problem.
3. Find two nonnegative numbers whose sum is 9, so that the product of one number and the square of the other number is a maximum. Use Microsoft Excel to solve the problem.
4. Find the maximum values of $f(x, y) = 4x^2 + 10y^2$, on the disk $x^2 + y^2 < 4$.

Table 10.9 Excel output for Example 10.21

	A	B
1	Π	3.1416
2	L	36
3	F	50,000
4	S	30,000
5	R_o	5
6	R_i	4.564319776
7		
8	K	471.2760003
9	σ	30,000
10	g_1	–1.18598E–08
11	g_2	–0.435680224

5. Find the minimum values of $f(x, y) = 4x^2 + 10y^2$, on the disk $x^2 + y^2 < 4$.
6. Find the maximum values of $f(x, y, z) = xyz$ subject to the constraint $x + y + z = 1$.
7. Microsoft Excel Solver primarily uses what three solution search heuristics?
8. Use Excel to solve the following problem: find the dimensions of a cylindrical tin (with top and bottom) made of sheet metal to maximize its volume such that the total surface area is equal to $A = 24\pi$.
9. A manufacturing company has a budget of \$10,000 to hire the necessary labor and purchase the raw materials for producing end products. The price of raw material is \$15 per unit and the cost of the labor is \$20 per unit. If K units of raw materials and L units of labor are used, the company can produce KL units of the end product. The question is to formulate a mathematical model for the optimization of the end product that can be manufactured by the company and how the company can solve this problem by using Excel software.
10. A civil engineer's intention is to construct an office building at the lowest possible cost per floor. As the building gets taller, the cost of the material decreases because of bulk rates, but the cost per floor increases because of increased labor and insurance. Assuming the maximum floor number is 15 due to regulation. The following expressions may be taken as valid measurements of these various costs.

$$C_1 = 100{,}000/(f/2)$$

$$C_2 = 1{,}000f$$

where
C_1 = cost of materials (dollars)
C_2 = cost of labor and insurance (dollars)
f = number of floors

Assuming that the total cost is the sum of materials, labor, and insurance, find the optimum number of floors up to which the building should be built.

11. A company has two grades of inspectors, 1 and 2, who are to be assigned to a quality control inspection. It is required that at least 1,800 pieces be inspected 8h per day. Grade 1 inspectors can check pieces at the rate of 25 per hour, with an accuracy of 98%. Grade 2 inspectors can check at the rate of 15 pieces per hour, with an accuracy

of 95%. The wage rate of a grade 1 inspector is \$4.00 per hour, while that of a grade 2 inspector is \$3.00 per hour. Each time an error is made by an inspector, the cost to the company is \$2.00. The company wants to determine the optimal assignment of inspectors that will minimize the total cost of the inspection.

x_1: number of grade 1 inspectors assigned to inspection.

x_2: number of grade 2 inspectors assigned to inspection.

12. A farmer has 2,400 linear feet of fencing that he can put up. He wants to fence off a rectangular field that borders a rock bluff. No fencing is required along the bluff. What are the dimensions of the field that contains the largest area?
13. Use Excel optimization for the following problem: you want to find the dimensions of a cylindrical container that would hold 500 gal of liquid. The container should use the least amount of material possible to keep the material costs down. You first have to convert 500 gal to 66.8 ft^3. The radius is labeled "R," and the length of the container is labeled "L."

$$\text{Volume} = \pi R^2 L$$

$$\text{Surface area} = 2\pi R^2 + 2\pi RL$$

14. A welded tank is to be constructed with a cylindrical body and spherical ends. The material cost is \$0.025/$in.^2$, and the welding costs \$0.10/linear in. The height of the tank must not be more than 96 in. The volume of the tank should be 50 ft^3 (86,400 $in.^3$). Solve to optimize cost.
15. Find out the minimum value of y from the equation $y = 4x^2 + 6x + 1$.
16. You are designing an I-beam as shown in Figure 10.20 to act as a support for a floor. The I-beam will span 16 ft. The maximum weight of the I-beam should be no more than 500 lbs. Also, due to the amount of room available for head clearance, the beam can be no taller than 10 in. and should have an aspect ratio (total height/base width) of no more than 1.5. The wall thickness of the I-beam is set at 3/8 in. The I-beam will be constructed of steel which has a density of 0.30 $lbs/in.^3$. Design the beam to maximum load carrying ability (maximize the moment of inertia). Use the Excel problem solver method.

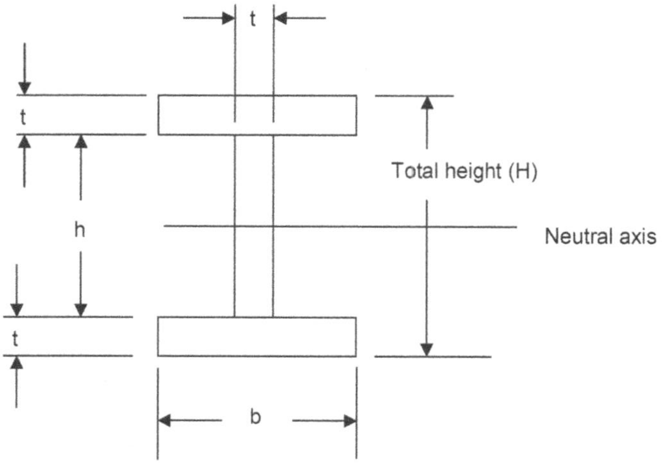

Figure 10.20 Design an I-beam to act as a support for a floor.

10.5 Case study: application of optimization in fixture design

Optimization techniques can be used not only to solve simple problems such as those in the previous sections, but also to solve complex problems. To provide more insight into this capability, this section summarizes a research case study using linear programming optimization for fixture planning [Meyer97]. Linear programming is an optimization problem in which the objective function and the constraints are all linear. The algorithm can be implemented by writing computer code or using tools such as Excel, as discussed in the previous section.

In general, to use the optimization technique to solve the engineering problem, the proper formulation of the problem as discussed in Section 10.1 is critical. To properly formulate the problem, the modeling techniques discussed in Chapter 3 may be useful. Once the problem is formulated, it can be solved using an appropriate optimization method. This case study uses linear programming to find the locating and clamping points for a modular fixture. It takes into account dynamic machining conditions which occur when the machining forces and moments travel or change with respect to time.

10.5.1 Development of a fixture generation methodology

A modular fixturing system mainly consists of locators and clamps. The function of locators is to properly locate a work part to provide repeatable manufacturing operations. A work party, with the same geometry and placed in the same orientation, will be uniquely positioned when it remains in contact with all the locators. The clamps are used to secure the work part from being deviated away from the manufacturing position by applying external forces on the work part. For simplicity, in this example, each locator and clamp is treated as a point force normal to the contact surface of the work part. Friction between the work part and the locators and clamps is not considered in this example.

In general, the sequential steps in the creation of a fixture are creation of a mesh over the surface of the work part as shown in Figure 10.21; determination of the positions of the clamps and locators; checking deterministic positioning as shown in Figure 10.22; and checking for clamping stability, a positive clamping sequence, and that the fixture reaction forces are nonnegative for all time, as shown in Figure 10.23 [Chou89]. Deterministic positioning, as shown in Figure 10.22, means that in order for the work part to be uniquely positioned, the work part needs to remain in contact with all the locators; reaction forces at the locators must not be zero. In other words, positive, or nonnegative check on the reaction forces at the locators can ensure deterministic positioning. Once the work part is in position, clamps are added in sequence from all possible clamping locations, as shown in Figure 10.23. However, each clamping activity should make sure the work part remains in contact with all the locators to achieve clamping stability. Clamps should be designed to ensure the overall rigidity of the work part during manufacturing in which manufacturing loads can vary from position to position. This process requires extensive search and optimization, and thus is a good candidate to use optimization technology.

The primary synthesis and analysis of the above stated fixturing sequence and configuration can be implemented using the solution of a set of equations using linear programming methods. Linear programming allows the most optimal solutions to indeterminant systems of equations to be easily found. However, all values for the variables that are solved for must be nonnegative. If a variable has a negative solution value, then the solution is infeasible. In this case study, the most optimal solution is the one that minimizes

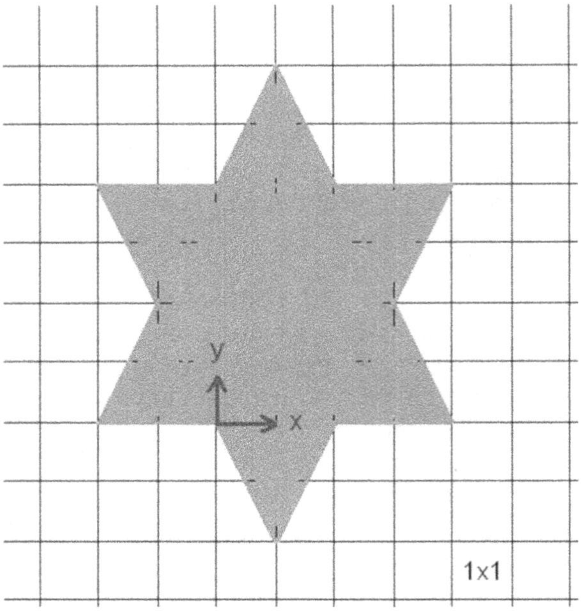

Figure 10.21 The first step in modeling a fixture is setting up the grid. For simplicity, a star-shaped part in 2D is used as an example. The grid can represent the locating holes in a modular fixturing system.

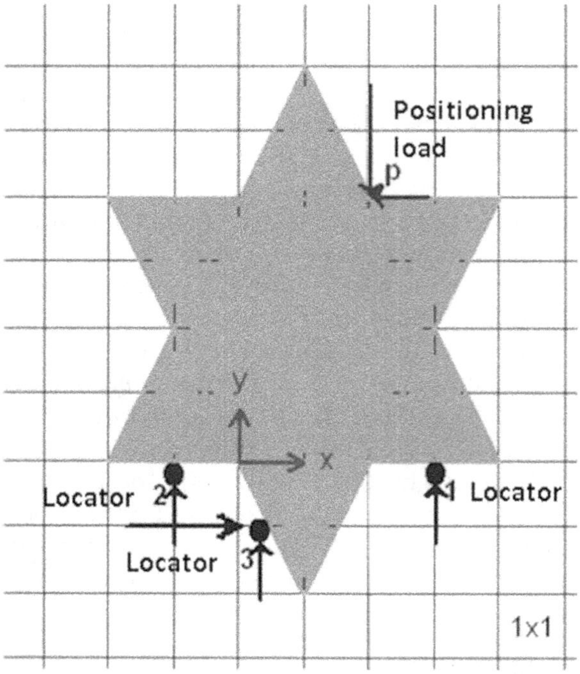

Figure 10.22 Three locators restrict four degrees of freedom. In order for the work part to be uniquely positioned, the work part needs to remain in contact with all the locators. The position load can be applied by an operator to make sure that the work part will be in contact with all the locators.

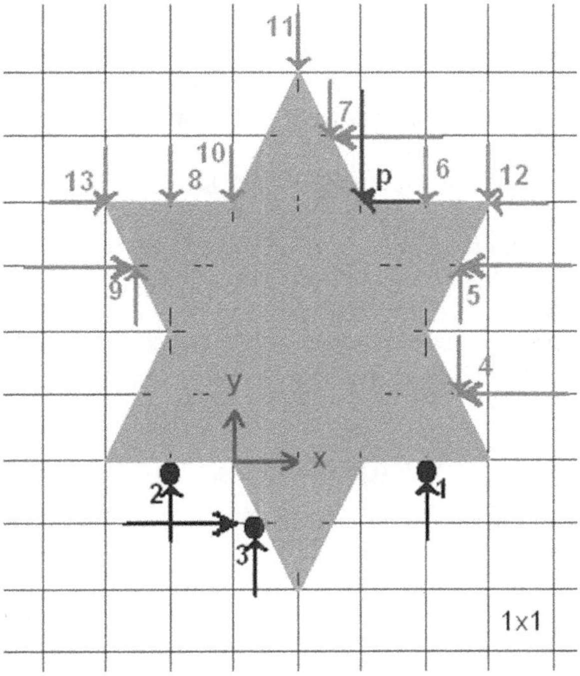

Figure 10.23 Selection of clamping locations (points 4–13 as shown). Each clamping activity should make sure the work part to remain in contact with all the locators to achieve clamping stability.

the magnitudes of all of the forces acting on the work part during the machining process. This function can be expressed as follows:

$$\text{Minimize}: \sum_{i=1}^{n} \lambda_i \tag{10.11}$$

where

λ_i = ith force acting on the work part (clamping or locating)
n = total number of unknown clamping and locating forces

Equation 10.11 is the objective function of a system of equations. The system of equations is composed of the summation of forces and moments in static equilibrium, and in some cases, variables are set to a minimum clamping force to minimize workpiece deflection or deformation. Typically, equilibrium of forces in the fixture-work part system, located in Cartesian coordinate space, can be expressed as follows:

$$\sum_{j=1}^{m} F_{x,j} = 0 \quad \sum_{j=1}^{m} F_{y,j} = 0 \quad \sum_{j=1}^{m} F_{z,j} = 0 \tag{10.12}$$

$$\sum_{j=1}^{p} M_{x,j} = 0 \quad \sum_{j=1}^{p} M_{y,j} = 0 \quad \sum_{j=1}^{p} M_{z,j} = 0 \tag{10.13}$$

where

$F_{x,j}, F_{y,j}, F_{z,j}$ = Cartesian components of a force acting on the work part
m = total number of forces in the work part–fixture system
$M_{x,j}, M_{y,j}, M_{z,j}$ = components of a moment acting on the work part about the Cartesian axes
p = total number of moments in the work part–fixture system

These equations are reformed for the linear programming method by placing the known forces on the right-hand side of the equations and introducing the fact that the unknown force magnitudes are dependent upon the point in the machining process at which they are solved. Every unknown force magnitude in the equations is multiplied by given force and moment coefficients which correspond to the unit wrench at the force's point of action.

$$\sum_{k=1}^{q} F_k(t) n_{x,k} = -F_{x,b} - M_{Fx}(t)$$

$$\sum_{k=1}^{q} F_k(t) n_{y,k} = -F_{y,b} - M_{Fy}(t)$$

$$\sum_{k=1}^{q} F_k(t) n_{z,k} = -F_{z,b} - M_{Fz}(t)$$

$$\sum_{k=1}^{q} F_k(t)\left(n_{z,k} l_{y,k} - n_{y,k} l_{z,k}\right) = -\left(F_{z,b} l_{y,b} - F_{y,b} l_{z,b}\right) - M_{Mx}(t) \tag{10.14}$$

$$\sum_{k=1}^{q} F_k(t)\left(n_{x,k} l_{z,k} - n_{z,k} l_{x,k}\right) = -\left(F_{x,b} l_{z,b} - F_{z,b} l_{x,b}\right) - M_{My}(t)$$

$$\sum_{k=1}^{q} F_k(t)\left(n_{y,k} l_{x,k} - n_{x,k} l_{y,k}\right) = -\left(F_{y,b} l_{x,b} - F_{x,b} l_{y,b}\right) - M_{Mz}(t)$$

where

$F_k(t)$ = kth unknown magnitude of a force acting on the work part at some time, t
$n_{x,k}, n_{y,k}, n_{z,k}$ = axial components of a unit vector at the position of kth force application, normal to the work part
$l_{x,k}, l_{y,k}, l_{z,k}$ = respective moment arms of the unknown force
$F_{x,b}, F_{y,b}, F_{z,b}$ = components of the body force of the work part
$l_{x,b}, l_{y,b}, l_{z,b}$ = respective moment arms of the body force's components
$M_{Fx}(t), M_{Fy}(t), M_{Fz}(t), M_{Mx}(t), M_{My}(t), M_{Mz}(t)$ = force and moment components due to the machining process at some time, t
q = total number of unknown clamping and locating forces acting on the work part.

Dynamic terms, such as acceleration, are omitted from Equation 10.14, since the work part must be in static equilibrium at all times for the fixture to be successful. However, the machining forces and moments need to be further defined. The machining forces and moments at any time, t, in Equation 10.14, can be represented as follows:

$$
\begin{aligned}
&M_{Fx}(t) = F_{x,m}(t) \\
&M_{Fy}(t) = F_{y,m}(t) \\
&M_{Fz}(t) = F_{z,m}(t) \\
&M_{Mx}(t) = \left[F_{z,m}(t)l_{y,m}(t) - F_{y,m}(t)l_{z,m}(t)\right] + M_{x,m}(t) \\
&M_{My}(t) = \left[F_{x,m}(t)l_{z,m}(t) - F_{z,m}(t)l_{x,m}(t)\right] + M_{y,m}(t) \\
&M_{Mz}(t) = \left[F_{y,m}(t)l_{x,m}(t) - F_{x,m}(t)l_{y,m}(t)\right] + M_{z,m}(t)
\end{aligned}
\tag{10.15}
$$

where

$F_{x,m}(t)$, $F_{y,m}(t)$, $F_{z,m}(t)$ = axial components of the machining forces at time, t
$l_{x,m}(t)$, $l_{y,m}(t)$, $l_{z,m}(t)$ = moment arms of the machining force components at solution time, t
$M_{x,m}(t)$, $M_{y,m}(t)$, $M_{z,m}(t)$ = components of a machining moment at solution time, t

The first step in synthesizing a fixture is to generate a mesh over the external surface planes of the work part on which machining is to take place. The nodes on this mesh represent all of the possible points that fixture elements, and their forces can act on the work part. However, if any of these points are within a specified minimum distance from the machining path(s), then they are disregarded as possible fixture element locations. The mesh of points is spaced with reference to the plane that they lie on. Also, unit vectors normal to the work part at the nodal points on its surface are found so that the only unknowns in the system of equations are the magnitudes of the forces at the nodes. The planes that the mesh nodal points lie on are sorted into clamping or locating planes according to whether or not machining forces are directed into them. Since the machining forces are time-dependent, the choice is made with the maximum machining forces that occur in each axial direction. The maximum machining forces and moments are the maximum values, based on magnitude, on the right-hand side of each equation in Equation 10.14, which occur during the entire machining process. For example, the maximum value of the machining force in the x-direction may be found at a time equal to 10 s; however, for the y-direction it may be found at 20 s. If a plane of the work part has an average normal vector within or at 458 of at least one maximum machining force Cartesian component vector, then it is chosen as a clamping plane. However, if there is no maximum machining force in a direction, meaning it is zero, then the maximum moments are used to find the clamping planes in that axial direction. The choice is conducted so that the maximum machining force directions are all mutually orthogonal. Planes that do not meet the clamping criteria are defaulted as locating planes. This is done to ensure that the machining forces are directed into the most rigid fixture elements: the locators.

Now that the clamping and locating planes are known, as well as the maximum forces and moments due to machining, it is possible to begin the algorithm that generates a viable fixture for work parts experiencing dynamic machining conditions. Clamping positions are chosen iteratively on the clamping planes with no duplication of their axial restraint force contribution. This means that if three clamps are necessary, only one is considered to be applying restraint in the z-direction, one in the y-direction, and the last in the x-direction. The clamps that can contribute the most axial directions of force are applied first to attempt to minimize the number of clamps used. Subsequent clamps are applied in decreasing amount of axial force restraint contribution. The number of clamps can range between one and three, depending on the geometry of the work part. Furthermore, because the application of the clamping positions is an iterative procedure, it is desired to reach a valid fixture configuration in a reasonable amount of time. Therefore, the selection of clamping positions on the planes is incremented according to the integer of the overall maximum number of nodal points on any clamping plane divided by 10. If no valid fixture configuration is found with this increment, it is halved until it reaches one, and then, every possible clamping point is iteratively chosen. This allows every possible valid combination of clamping positions to be considered and a large amount of widely different ones to be selected relatively quickly.

The next stage in the generation of a valid fixture is to solve Equation 10.14 using all of the possible locator positions on the locating planes, the current clamping positions, and the maximum machining forces and moments. Using only the maximum forces and moments of machining at this stage of the fixture generation process allows the removal of the model's time dependence. If this is not done, a different possible fixture configuration would be produced at every point in time which the model is solved. To generate a possible fixture configuration, the model needs to be solved twice using linear programming before fixture validation procedures can be carried out. The first time it is solved, the value of each of the applied clamping forces is set. If the force for a clamp from this first solution is found to be between the specified minimum and maximum clamping forces, then it is allowed to remain at this value. However, if it is found to be below the minimum force considered to be safe, the clamping force is raised to this minimum magnitude. If the clamping force is above the given maximum, then it is decreased to it. By specifying a region that the clamping force is to remain within, clamping forces that are too little or large can be avoided. Also, the restriction of clamping forces to a maximum helps reduce the deformation of the work part by disallowing excessive clamping forces. Since machining forces and moments cannot usually be reduced to prevent large work part deformation, the only other set of applied forces that can be kept to a minimum are the clamping forces. Setting clamps to their applied force values involve the addition of the following equation for each clamp to the linear programming model:

$$F_{ci} = Q_i \tag{10.16}$$

where

F_{ci} = magnitude of the ith clamp
Q_i = z-clamping value to which the ith clamp is set

The second solution then involves Equations 10.14 and 10.16 with the same parameters used in the first solution. The result of this solution is the production of the possible fixture configuration. Since the positions of the clamps are known, the only unknown is the position of the locators. From all of the possible locator positions on the work part, the ones with positive forces are chosen as the locators. The number of locators found by this method for a valid fixture is six. The known force and moment coefficients of the chosen locators from Equation 10.14 are formed into an active locator matrix, which serves the same purpose as the left-hand side of Equation 10.14, to make further manipulation easier. An advantage of linear programming is that there can be no negative values for the forces at locators or clamps, if there is a feasible solution to the equations. A feasible solution with linear programming automatically means that the work part is restrained. This is very important in fixturing, where the presence of a negative clamping or reaction force indicates that the work part is actually detaching from the fixture. The positions and forces of the clamps and locators are checked against valid fixture criteria that include deterministic positioning, accessibility, the stability of the work part in the fixture, positive clamping sequence, and ability to retain positive reaction forces in the fixture during all machining. If any of these criteria are not met, then the next clamping positions are chosen, and the generation process continues until there is a fixture configuration that meets all of them. This means that the system stops once the first valid fixture configuration is reached. If all possible clamping positions are chosen, and no valid fixture configuration has been reached, the minimum clamping force is raised by an additional one-fifth of the difference between the maximum and original minimum clamping force values. If the current minimum clamping force reaches the maximum and all clamping positions are chosen without a valid configuration, several steps can be taken to obtain a valid fixture. These include increasing the number of possible fixture element points by decreasing the planar mesh spacing, raising the minimum clamping force, or decreasing the number of machining operations that the fixture is expected to be valid for. However, if a fixture meets all of the validity criteria, then force planes are created that are based on the forces at the locators over time. The critical algorithms that determine the characteristics of a fixture are summarized in the following section.

10.5.2 Modeling deterministic positioning using linear programming

A work part is deterministically positioned when the Jacobian matrix derived from the functions that represent the positions of the locators and the final position of the work part has full rank equal to six. The approach presented in this example does not rely on the actual Jacobian matrix, whose creation can be both time-consuming and tedious, to determine whether or not a work part is deterministically positioned. Instead, it has been observed that a matrix formed from the force and moment coefficients in Equation 10.14 of the terms that represent the locators of the fixture to be checked can be used in place of the previously defined Jacobian matrix. This makes it much easier to determine the deterministic positioning and accessibility characteristics of a work part with complex geometry since its piecewise differentiable representation does not have to be found and only a matrix based on the active locator matrix needs to be created. Also, this means that the nature of deterministic positioning and accessibility can be determined for work parts without a differentiable representation. The matrix formed from the coefficients of the locator terms in the active locator matrix will be known as the equivalent Jacobian, J0. Its rows are made up of the following which can be thought of as the transpose of the vector that represents the chosen locators in Equation 10.14 if they were put into matrix form:

$$H_i' = \left[x_i, y_i, z_i\ m_{x_i}, m_{y_i}, m_{z_i}\right] \tag{10.17}$$

where

H_i' is the equivalent gradient vector

x_i, y_i, z_i = components of the unit vector that the ith locator force acts through

m_{x_i}, m_{y_i}, m_{z_i} = moment coefficients of the ith locator force about the axes

Therefore, the equivalent Jacobian matrix is formed from the gradient vector in the following manner:

$$J' = \left[H_1' \ldots H_n'\right] \tag{10.18}$$

where

J' = equivalent Jacobian matrix

n = number of locators

The creation of the equivalent Jacobian results in an $n \times 6$ matrix. As long as n is greater than or equal to the number of degrees of freedom of the unfixtured work part, there are no rows or columns of zero, every row is unique, and no row is the negative of another, then the work part is deterministically located. Physically, this means that locators contribute resisting forces and moments along and about each axial direction to keep the work part positioned by restricting at least one direction of every degree of freedom. Therefore, the work part cannot move out its deterministic location without losing contact with at least one locator.

10.5.3 Modeling accessibility of a fixture determined with linear programming

A work party is accessible in a fixture if the following relation holds true for some small displacement of the work part, δq, and the equivalent Jacobian:

$$J'\delta q \geq 0 \tag{10.19}$$

Satisfaction of Equation 10.19 results in weak accessibility which means that during unloading, the work part will detach from at least one locator while remaining in contact with the others. Equation 10.19 can be modified to determine if the fixture has strong accessibility. This is given by

$$J'\delta q > 0 \tag{10.20}$$

A fixture that fulfills Equation 10.20 allows the work part to be detached from all of the locators at once. These relations are solved using linear programming techniques, but they first must be modified. To determine weak accessibility, an additional equation must be included along with Equation 10.19. This extra equation simply states that the sum of the displacements must be greater than zero so that the displacements are all not set to zero in the effort to minimize them. If they were all set to zero, as they would be if only Equation 10.19 was solved, this would give no information about the accessibility characteristics of the fixture.

10.5.4 Modeling clamping stability of the work part in the fixture

The stability of the work part in the fixture when it is not subject to external forces, such as clamping or machining, is necessary so that the work part does not require extra support until the clamps are applied for it to remain in the fixture, against the locators. Checking for stability is accomplished by setting the active locator matrix equal to the body forces and solving for the force magnitudes at the locators. This is the same as setting all clamps to zero, all forces due to machining to zero, and disallowing any forces at locators that are not being tested for validity and solving Equation 10.14. If the solution is feasible, then the work part is stable in the fixture.

10.5.5 Modeling positive clamping sequence using linear programming

The positive clamping sequence is the order in which clamps should be applied to the work part so that it does not detach from the locators. To determine this order, the work part is only under the influence of its body forces, and the current clamps are applied one by one with their previously found clamping force. The locators in the formulation are the active ones in the fixture configuration currently being tested.

The order of clamping depends on the infeasibility of the solution given by the application of a clamp. The first clamp in the sequence will be the first that yields a feasible solution. Once the first clamp has been found, its applied clamping force value stays with the linear programming formulation from then on, meaning that it is never again set to zero. The other possible clamps are tested one by one in conjunction with the first clamp, and the next feasible solution yields the second clamp in the sequence. This process continues until all of the clamps are accounted for in the sequence, or until all possible clamping application configurations have been tested and there is no order in which they can be applied that will not result in the work part detaching from the fixture. If there is no viable clamping order, then the fixture configuration being tested is failed.

10.5.6 Modeling positive fixture reaction to all machining forces

Since the machining forces and moments travel along the surface of or through the work part, it is necessary to ensure that the fixture forces are positive for all time. Any negative reaction force experienced by the fixture indicates that the work part is free from restraint. This includes any number of machining passes made on the work part for which the fixture is supposed to be valid. Each pass is characterized by a changing position of the machine tool, and there may be changing machining forces, moments, and feed rates. To decrease the infinite number of possible machining positions, forces, and moments that can be used in the solution of Equations 10.14–10.16 for each pass, a discrete number of machining force and moment sets are used. Between the start and end of each machine tool direction, and change of machining forces, moments, and feed rates, an incremental number of machining force and moment solutions is found and tested in the linear programming model. The solutions are taken at the start and stop points of the change and a chosen number of incremental points. The transition between changes is modeled as being linear and the incremental solutions reflect this. Solutions that exist between the discrete solution points can be found by linear interpolation since the system's equations are linearly based. Therefore, for a fixture to be considered valid for all time by this methodology, it must have positive reaction forces at the beginning and end of each machining change and for each incremental solution. The feed rate and distance the tool has traveled since the beginning of a pass are used to calculate the times at which machining force and moment solutions are taken. The beginning of each machining pass starts at time zero.

10.5.6.1 Numerical example

The case study is shown in Figures 10.24 and 10.25 with machining passes one and two, respectively. The first machining operation is drilling through the work part, from (11.5, 5, 8 cm) to (11.5, 5, 0 cm), with force $[0, 0, -50\text{ N}]^T$ and moment $[0, 0, 75\text{ N cm}]^T$. The second machining pass is across a small section of the top of the work part to clean up the area around the just drilled hole. It travels between (11.5, 7, 8 cm) and (11.5, 3, 8 cm) and has machining forces of $[100, -100, -50\text{ N}]^T$. The feed rate for the first pass is 0.5 cm/s and for the second 0.75 cm/s. The machining time for the first pass is 16 s, whereas for the second, it is 5.33 s. The time between machining force and moment solutions is one-fifth the total per pass. Fixture elements cannot be placed within 1.5 cm of the machining paths. The minimum and maximum clamping forces are 40 and 80 N. The body force for a steel work part is $[0, 0, -67.237\text{ N}]^T$ at (12.266, 5, 4.688 cm), its centroid.

Figure 10.26 shows the fixture designed (acting positions of locators (L1–L6) and clamps (C1–C3)) for the present work part geometry and dynamic machining conditions. The forces at the locators are graphed in Figure 10.27 for pass one and in Figure 10.28 for

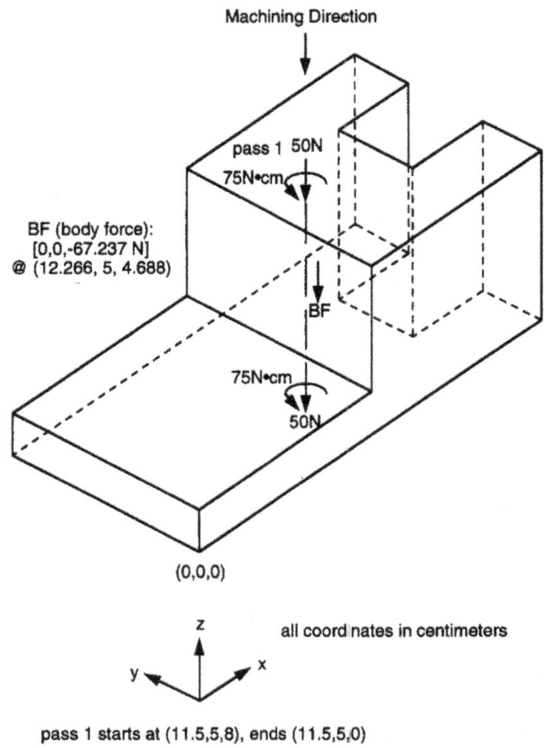

Figure 10.24 Work part and machining operation 1.

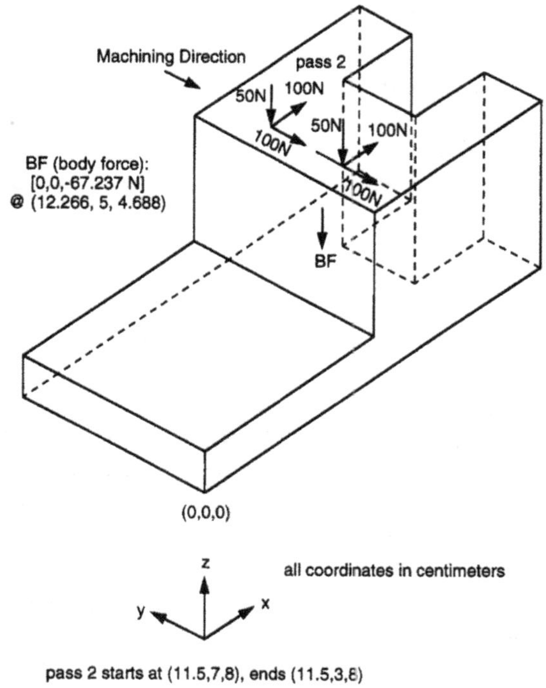

Figure 10.25 Work part and machining operation 2.

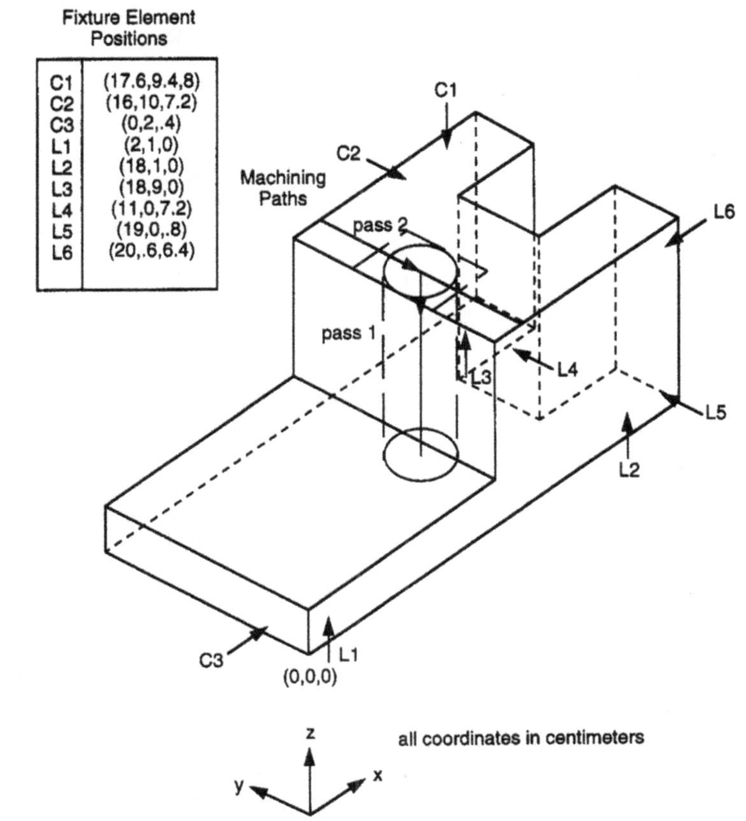

Figure 10.26 Fixture configuration: locators (L1–L6) and clamps (C1–C3).

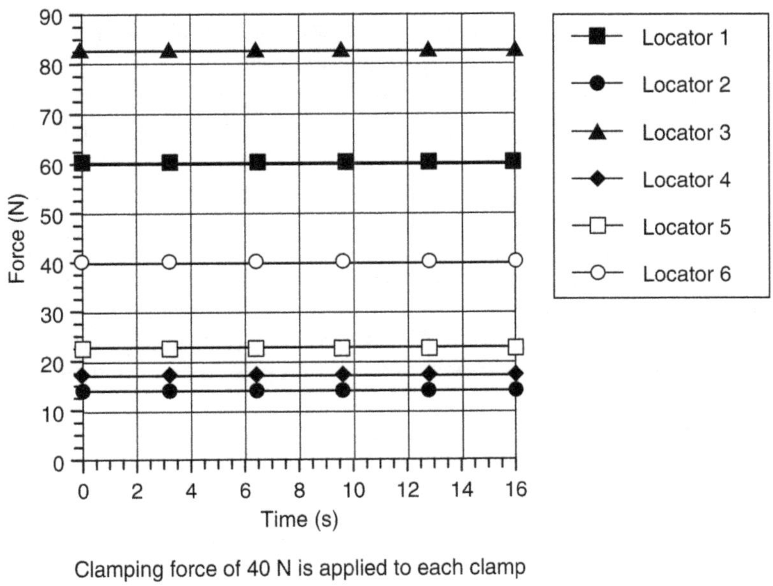

Figure 10.27 Fixture forces overall time for pass 1.

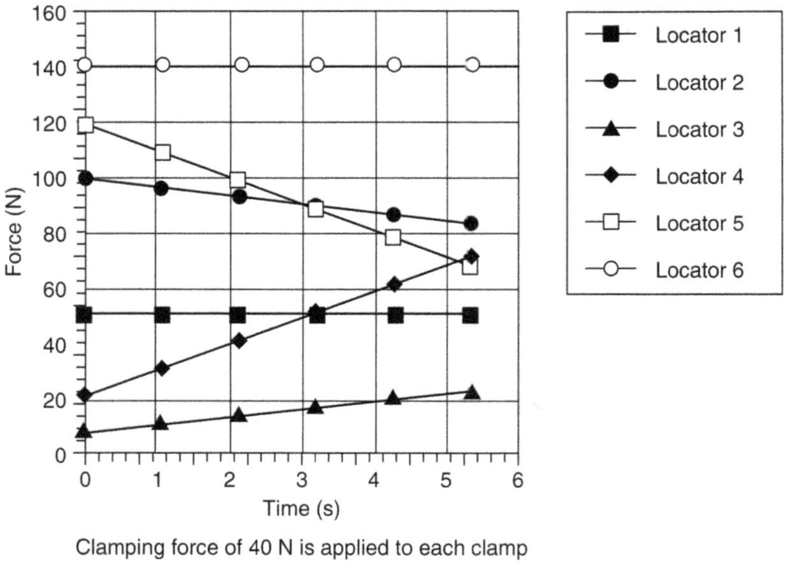

Figure 10.28 Fixture forces overall time for pass 2.

pass two. The minimum clamping force of 40 N is utilized by the system, thus reducing the possibility of excessive work part deformation. The overall positions of the locators follow the 3-2-1 locating principle and the clamping order is sequential. The fixture configuration deterministically positions the work part, is strongly accessible, allows the work part to be at rest in the fixture while under no external forces, has a positive clamping sequence, and provides for nonnegative reactions at the locators for all time.

This case study demonstrates the use of linear programming optimization to solve a complex engineering problem by first formulating the work part and fixturing system into the linear program format. Then, linear programming techniques were used to determine the most optimal positions of locators, given a certain clamping configuration, the clamping forces, the forces at the locators, deterministic positioning, level of accessibility, whether or not the work part is stable in the fixture with no external forces, and the presence of a positive clamping sequence. This allowed valid fixture configurations to be created.

References

[Barreto17]. Barreto, H. and F. Howland, The excel solver add-in, Retrieved on September 4, 2017 from www3.wabash.edu/econometrics/EconometricsBook/Basic%20Tools/ExcelAddIns/DummyDepVar/Solver.doc.

[Chou89]. Chou, Y-C., V. Chandru, and M.M. Barash, "A mathematical approach to automatic configuration of machining fixtures: Analysis and synthesis," *ASME Journal of Engineering for Industry*, Vol. 111, pp. 299–306, 1989.

[Jensen17]. Jensen, S., An introduction to Lagrange multipliers, Retrieved on September 4, 2017 from www.slimy.com/~steuard/teaching/tutorials/Lagrange.html.

[Meyer97]. Meyer, R.T. and F.W. Liou, "Fixture analysis under dynamic machining," *International Journal of Production Research*, Vol. 35, No. 5, pp. 1471–1489, 1997.

[Solver17]. Solver tutorial for optimization users, Retrieved on September 4, 2017 from www.Solver.com.

[Wittwer17]. Wittwer, J.W., Excel solver examples from Vertex42.com, Retrieved on September 4, 2017 from http://vertex42.com/ExcelArticles/excel-solver-examples.html.

Appendix A-1

Percentage points of the F-distribution ($\alpha = 0.01$)									
V_1/V_2	1	2	3	4	5	6	7	8	9
1	39.86	49.50	53.59	55.83	57.24	58.20	58.91	59.44	59.86
2	8.53	9.00	9.16	9.24	9.29	9.33	9.35	9.37	9.38
3	5.54	5.46	5.39	5.34	5.31	5.28	5.27	5.25	5.24
4	4.54	4.32	4.19	4.11	4.05	4.01	3.98	3.95	3.94
5	4.06	3.78	3.62	3.52	3.45	3.40	3.37	3.34	3.32
6	3.78	3.46	3.29	3.18	3.11	3.05	3.01	2.98	2.96
7	3.59	3.26	3.07	2.96	2.88	2.83	2.78	2.75	2.72
8	3.46	3.11	2.92	2.81	2.73	2.67	2.62	2.59	2.56
9	3.36	3.01	2.81	2.69	2.61	2.55	2.51	2.47	2.44
10	3.29	2.92	2.73	2.61	2.52	2.46	2.41	2.38	2.35
11	3.23	2.86	2.66	2.54	2.45	2.39	2.34	2.30	2.27
12	3.18	2.81	2.61	2.48	2.39	2.33	2.28	2.24	2.21
13	3.14	2.76	2.56	2.43	2.35	2.28	2.23	2.20	2.16
14	3.10	2.73	2.52	2.39	2.31	2.24	2.19	2.15	2.12
15	3.07	2.70	2.49	2.36	2.27	2.21	2.16	2.12	2.09
16	3.05	2.67	2.46	2.33	2.24	2.18	2.13	2.09	2.06
17	3.03	2.64	2.44	2.31	2.22	2.15	2.10	2.06	2.03
18	3.01	2.62	2.42	2.29	2.20	2.13	2.08	2.04	2.00
19	2.99	2.61	2.40	2.27	2.18	2.11	2.06	2.02	1.98
20	2.97	2.59	2.38	2.25	2.16	2.09	2.04	2.00	1.96
21	2.96	2.57	2.36	2.23	2.14	2.08	2.02	1.98	1.95
22	2.95	2.56	2.35	2.22	2.13	2.06	2.01	1.97	1.93
23	2.94	2.55	2.34	2.21	2.11	2.05	1.99	1.95	1.92
24	2.93	2.54	2.33	2.19	2.10	2.04	1.98	1.94	1.91
25	2.92	2.53	2.32	2.18	2.09	2.02	1.97	1.93	1.89
26	2.91	2.52	2.31	2.17	2.08	2.01	1.96	1.92	1.88
27	2.90	2.51	2.30	2.17	2.07	2.00	1.95	1.91	1.87
28	2.89	2.50	2.29	2.16	2.06	2.00	1.94	1.90	1.87
29	2.89	2.50	2.28	2.15	2.06	1.99	1.93	1.89	1.86
30	2.88	2.49	2.28	2.14	2.05	1.98	1.93	1.88	1.85
31	2.87	2.48	2.27	2.14	2.04	1.97	1.92	1.88	1.84

(Continued)

Percentage points of the F-distribution ($\alpha = 0.01$)									
V_1/V_2	1	2	3	4	5	6	7	8	9
32	2.87	2.48	2.26	2.13	2.04	1.97	1.91	1.87	1.83
33	2.86	2.47	2.26	2.12	2.03	1.96	1.91	1.86	1.83
34	2.86	2.47	2.25	2.12	2.02	1.96	1.90	1.86	1.82
35	2.85	2.46	2.25	2.11	2.02	1.95	1.90	1.85	1.82
36	2.85	2.46	2.24	2.11	2.01	1.94	1.89	1.85	1.81
37	2.85	2.45	2.24	2.10	2.01	1.94	1.89	1.84	1.81
38	2.84	2.45	2.23	2.10	2.01	1.94	1.88	1.84	1.80
39	2.84	2.44	2.23	2.09	2.00	1.93	1.88	1.83	1.80
40	2.84	2.44	2.23	2.09	2.00	1.93	1.87	1.83	1.79

Notes: V_1, degrees of freedom for the numerator; V_2, degrees of freedom for the denominator.

Appendix A-2

Percentage points of the F-distribution ($\alpha = 0.05$)									
V_1/V_2	1	2	3	4	5	6	7	8	9
1	161.45	199.50	215.71	224.58	230.16	233.99	236.77	238.88	240.54
2	18.51	19.00	19.16	19.25	19.30	19.33	19.35	19.37	19.38
3	10.13	9.55	9.28	9.12	9.01	8.94	8.89	8.85	8.81
4	7.71	6.94	6.59	6.39	6.26	6.16	6.09	6.04	6.00
5	6.61	5.79	5.41	5.19	5.05	4.95	4.88	4.82	4.77
6	5.99	5.14	4.76	4.53	4.39	4.28	4.21	4.15	4.10
7	5.59	4.74	4.35	4.12	3.97	3.87	3.79	3.73	3.68
8	5.32	4.46	4.07	3.84	3.69	3.58	3.50	3.44	3.39
9	5.12	4.26	3.86	3.63	3.48	3.37	3.29	3.23	3.18
10	4.96	4.10	3.71	3.48	3.33	3.22	3.14	3.07	3.02
11	4.84	3.98	3.59	3.36	3.20	3.09	3.01	2.95	2.90
12	4.75	3.89	3.49	3.26	3.11	3.00	2.91	2.85	2.80
13	4.67	3.81	3.41	3.18	3.03	2.92	2.83	2.77	2.71
14	4.60	3.74	3.34	3.11	2.96	2.85	2.76	2.70	2.65
15	4.54	3.68	3.29	3.06	2.90	2.79	2.71	2.64	2.59
16	4.49	3.63	3.24	3.01	2.85	2.74	2.66	2.59	2.54
17	4.45	3.59	3.20	2.96	2.81	2.70	2.61	2.55	2.49
18	4.41	3.55	3.16	2.93	2.77	2.66	2.58	2.51	2.46
19	4.38	3.52	3.13	2.90	2.74	2.63	2.54	2.48	2.42
20	4.35	3.49	3.10	2.87	2.71	2.60	2.51	2.45	2.39
21	4.32	3.47	3.07	2.84	2.68	2.57	2.49	2.42	2.37
22	4.30	3.44	3.05	2.82	2.66	2.55	2.46	2.40	2.34
23	4.28	3.42	3.03	2.80	2.64	2.53	2.44	2.37	2.32
24	4.26	3.40	3.01	2.78	2.62	2.51	2.42	2.36	2.30
25	4.24	3.39	2.99	2.76	2.60	2.49	2.40	2.34	2.28
26	4.23	3.37	2.98	2.74	2.59	2.47	2.39	2.32	2.27
27	4.21	3.35	2.96	2.73	2.57	2.46	2.37	2.31	2.25
28	4.20	3.34	2.95	2.71	2.56	2.45	2.36	2.29	2.24
29	4.18	3.33	2.93	2.70	2.55	2.43	2.35	2.28	2.22
30	4.17	3.32	2.92	2.69	2.53	2.42	2.33	2.27	2.21
31	4.16	3.30	2.91	2.68	2.52	2.41	2.32	2.25	2.20

(Continued)

Percentage points of the F-distribution ($\alpha = 0.05$)

V_1/V_2	1	2	3	4	5	6	7	8	9
32	4.15	3.29	2.90	2.67	2.51	2.40	2.31	2.24	2.19
33	4.14	3.28	2.89	2.66	2.50	2.39	2.30	2.23	2.18
34	4.13	3.28	2.88	2.65	2.49	2.38	2.29	2.23	2.17
35	4.12	3.27	2.87	2.64	2.49	2.37	2.29	2.22	2.16
36	4.11	3.26	2.87	2.63	2.48	2.36	2.28	2.21	2.15
37	4.11	3.25	2.86	2.63	2.47	2.36	2.27	2.20	2.14
38	4.10	3.24	2.85	2.62	2.46	2.35	2.26	2.19	2.14
39	4.09	3.24	2.85	2.61	2.46	2.34	2.26	2.19	2.13
40	4.08	3.23	2.84	2.61	2.45	2.34	2.25	2.18	2.12

Notes: V_1, degrees of freedom for the numerator; V_2, degrees of freedom for the denominator.

Appendix A-3

Percentage points of the F-distribution ($\alpha = 0.01$)									
V_1/V_2	1	2	3	4	5	6	7	8	9
1	4052.18	4999.34	5403.53	5624.26	5763.96	5858.95	5928.33	5980.95	6022.40
2	98.50	99.00	99.16	99.25	99.30	99.33	99.36	99.38	99.39
3	34.12	30.82	29.46	28.71	28.24	27.91	27.67	27.49	27.34
4	21.20	18.00	16.69	15.98	15.52	15.21	14.98	14.80	14.66
5	16.26	13.27	12.06	11.39	10.97	10.67	10.46	10.29	10.16
6	13.75	10.92	9.78	9.15	8.75	8.47	8.26	8.10	7.98
7	12.25	9.55	8.45	7.85	7.46	7.19	6.99	6.84	6.72
8	11.26	8.65	7.59	7.01	6.63	6.37	6.18	6.03	5.91
9	10.56	8.02	6.99	6.42	6.06	5.80	5.61	5.47	5.35
10	10.04	7.56	6.55	5.99	5.64	5.39	5.20	5.06	4.94
11	9.65	7.21	6.22	5.67	5.32	5.07	4.89	4.74	4.63
12	9.33	6.93	5.95	5.41	5.06	4.82	4.64	4.50	4.39
13	9.07	6.70	5.74	5.21	4.86	4.62	4.44	4.30	4.19
14	8.86	6.51	5.56	5.04	4.69	4.46	4.28	4.14	4.03
15	8.68	6.36	5.42	4.89	4.56	4.32	4.14	4.00	3.89
16	8.53	6.23	5.29	4.77	4.44	4.20	4.03	3.89	3.78
17	8.40	6.11	5.19	4.67	4.34	4.10	3.93	3.79	3.68
18	8.29	6.01	5.09	4.58	4.25	4.01	3.84	3.71	3.60
19	8.18	5.93	5.01	4.50	4.17	3.94	3.77	3.63	3.52
20	8.10	5.85	4.94	4.43	4.10	3.87	3.70	3.56	3.46
21	8.02	5.78	4.87	4.37	4.04	3.81	3.64	3.51	3.40
22	7.95	5.72	4.82	4.31	3.99	3.76	3.59	3.45	3.35
23	7.88	5.66	4.76	4.26	3.94	3.71	3.54	3.41	3.30
24	7.82	5.61	4.72	4.22	3.90	3.67	3.50	3.36	3.26
25	7.77	5.57	4.68	4.18	3.85	3.63	3.46	3.32	3.22
26	7.72	5.53	4.64	4.14	3.82	3.59	3.42	3.29	3.18
27	7.68	5.49	4.60	4.11	3.78	3.56	3.39	3.26	3.15
28	7.64	5.45	4.57	4.07	3.75	3.53	3.36	3.23	3.12
29	7.60	5.42	4.54	4.04	3.73	3.50	3.33	3.20	3.09
30	7.56	5.39	4.51	4.02	3.70	3.47	3.30	3.17	3.07
31	7.53	5.36	4.48	3.99	3.67	3.45	3.28	3.15	3.04

(Continued)

Percentage points of the F-distribution ($\alpha = 0.01$)

V_1/V_2	1	2	3	4	5	6	7	8	9
32	7.50	5.34	4.46	3.97	3.65	3.43	3.26	3.13	3.02
33	7.47	5.31	4.44	3.95	3.63	3.41	3.24	3.11	3.00
34	7.44	5.29	4.42	3.93	3.61	3.39	3.22	3.09	2.98
35	7.42	5.27	4.40	3.91	3.59	3.37	3.20	3.07	2.96
36	7.40	5.25	4.38	3.89	3.57	3.35	3.18	3.05	2.95
37	7.37	5.23	4.36	3.87	3.56	3.33	3.17	3.04	2.93
38	7.35	5.21	4.34	3.86	3.54	3.32	3.15	3.02	2.92
39	7.33	5.19	4.33	3.84	3.53	3.30	3.14	3.01	2.90
40	7.31	5.18	4.31	3.83	3.51	3.29	3.12	2.99	2.89

Notes: V_1, degrees of freedom for the numerator; V_2, degrees of freedom for the denominator.

Short Answers to Selected Review Problems

Section 1.2
3. 155%; **4.** 31%; **5.** 116%; **8.** $ROI_{Scenario\ 1} = 300\%$, $ROI_{Scenario\ 2} = 218\%$.

Section 2.1
2. A, D; **11.** (1) When the prototype satisfies all the final system requirements, (2) when the purpose of the prototype has been served, (3) when the developers and users agree to move to the next stage, or (4) when run out of time/money.

Section 2.2
1. Specific, measurable, achievable, relevant, time dimensioned; **9.** The best is to build a physical— Selective Laser Sintering (SLS) model, followed by a virtual—CAD model, then a Physical—wood mock-up, and finally a virtual—hand sketch.

Section 2.3
3. $y = 134.29 + 43.57(9) = \$526.42$; **4.** $Y = 186.2 + 99.1X$.

Section 2.4
6. Absolute importance for battery is the highest (351). The rests are housing (327), screen (324), and interface (294).

Section 2.5
1. Cost (1/21), accuracy (6/21), material selection (2/21), strength (2/21), product size (3/21), lead time (5/21), and heat resistance (2/21); **4.** Specialty (3), college ranking (2) points, expense (1), and location (0); **5.** College C highest (2.89), and then B (2.74), D (2.63), and A (1.56).

Section 2.6
1. 1-c, 2-b, 3-a, 4-e, 5-d.

Section 2.7
4. 1-b, 2-c, 3-d, 4-a; **5.** 1-d, 2-b, 3-c, 4-a.

Section. 3.1
1. Maximum bending stress = My/I = (600,000 in.-lb)*(3 in.)/(36 in.^{4}) = 50,000 lb/in.^{2}; **5.** 32.03 m; **7.** 8 mg min^{-1}; **8.** 35 min; **9.** (a) 385 ft-lbs, (b) 6.015 ft; **10.** $\sigma = 250$ psi, $\varepsilon = 0.00025$ in./in.; **11.** 18 bolts.

Section 3.2

1. 8,000 psi; **6.** $\omega_{full} = 1.8\,\text{Hz}$, $\omega_{empty} = 18.5\,\text{Hz}$.

Section 3.3

1. 1-b & d, 2-a & c; **2.** a, b, c, d.

Section 3.4

5. (1) Better visualization with a greater detail; (2) engineering analysis could be performed; (3) computer simulation can be done; (4) transfer of data files between different software, and it also makes it easier for Computer-Aided Manufacturing (CAM) to simulate the manufacturing of the part.

Section 4.1

5. 1-d, 2-b, 3-e, 4-a, 5-c; **16.** 1-(b, e, g); 2-(b, e, g); 3-(f, h); 4-(c, e, g); 5-(a); 6-(b, c, e, g); 7-(d, f, h); 8-(f, h).

Section 4.2

1. High-carbon spring steel (327); **7.** 1,000,000 times.

Section 4.3

1. To sustain shear stress 382 Mpa steel is a choice; **3.** 1-e, 2-c, 3-a, 4-d, 5-b, 6-j, 7-i, 8-f, 9-g, 10-h; **10.** 1117 steel or similar low to medium carbon steel.

Section 5.1

1. $p(t) := \frac{1}{2} \cdot \begin{pmatrix} t^2 & t & 1 \end{pmatrix} \cdot \begin{pmatrix} 1 & -2 & 1 \\ -2 & 2 & 0 \\ 1 & 1 & 0 \end{pmatrix} \cdot \begin{pmatrix} p_0 \\ p_1 \\ p_2 \end{pmatrix}$, error (t = 0.5) = 2.56%; **2.** error (t = 0.5) = 0.515%; **3.** error (t = 0.5) = 25%; **4.** Pc(0), Error = 1.0725%, Pc(1/3), Error = 1.12% Pc(2/3), Error = 1.11%, Pc(1), Error = 1.095%; **5.** t = 0.5, error = 6.07%.

Section 5.2

1. a-d-c-b; **5.** E; **7.** A point cloud is a set of three-dimensional points describing the outlines or surface features of an object.

Section 5.2

1. a-d-c-b; **5.** E; **7.** A point cloud is a set of three-dimensional points describing the outlines or surface features of an object.

Section 5.3

4. A, C, D; **5.** B-E-C-D-A.

Section 5.4

1. A feature is a physical constituent of a part; **2.** A, B, C, D, E; **4.** B, C, D; **6.** Slot, hole, pocket, and C-slot; **7.** Recognition can be made a lot easier if the user can tell the program what to look for and also pick one or more entities of the feature to get the process started on the right track. This is easy to do when a graphical image of the part is available to the designer.

Section 6.1
7. A, B, C, D; **8.** Tooling (time, cost, life), tolerance, materials.

Section 6.2
5. C-E-B-G-A-F-D.

Section 6.3
5. d; **6.** Internal cavities.

Section 6.4
3. C; **5.** Extrusion-based process: a, c, d; contour-cutting process: b, c, g, h, e.

Section 6.5
8. B, D; **9.** (a) Laser sintering; (b) 3D inject printing; (c) DLD; (d) Ultrasonic Consolidation (UC); (e) contour cutting; (f) stereolithography.

Section 7.1
2. Before you begin to purchase parts the entire process should be planned out; Ensure that the main components of the project are identified; Create a flow chart to lay out the process; Ensure that the proper operating system is identified; Verify that the proper components have been selected to work with your system.

Section 7.2
1. A, B, C, and D; **2.** D-C-E-A-B-F; **3.** A, B, D.

Section 7.3
1. A; **2.** A, C, D.

Section 7.4
1. Maximum clearance between shaft and hole = 0.0034″, minimum clearance between shaft and hole = 0.0014; **2.** Hole: 2.000″–2.00075″, shaft: 2.0005″–2.00125″; **3.** Hole: 3.5″–3.50091″, shaft: 3.49939″–3.5″; **4.** Between 2.2496″ and 2.2491″.

Section 7.5
1. $X_{max} = 7.9$, $X_{min} = 5.8$; **2.** Bushing max = 23.4; Bushing min = 21.1; **3.** [9.65, 10.35]; **4.** [14.9, 15.04]; **5.** [14.95, 15.01]; **6.** [1.6, 2.6]; **7.** [8.6, 9.1]; **8.** [14.6, 15.4]; **9.** [8.6, 9.4]; **10.** [2.4, 3.35].

Section 7.6
1. $X_{max} = 4.2$; $X_{min} = 3.7$; **3.** S = [1.003″, 1.005″]; **4.** P = [1.000″, 1.001″], S = [1.250″, 1.251″], B = [0.245″, 0.247″]; **5.** 9.18.

Section 7.7
1. $\mu = 7.4$; **3.** $\sigma = 2.72$; **4.** UCL = 15.56, LCL = −0.76 => 0; **5.** Yes.

Section 7.8
1. (a) 68.26%; (b) 95.44%; (c) 99.74%; (d) 99.994%; (e) 99.99994%; (f) 99.9999998%; **3.** $(UCL - LCL)/\sigma = 0.010/0.005 = 2$ times the process deviation (total) = $\pm\sigma$ leads to 68.26% of nonmissed parts; **4.** If set at $\pm 3\sigma$, all assemblies fall within ±0.0045″; If set at $\pm 4\sigma$,

all assemblies fall within ±0.006″; **5.** ±0.0052″ for each part; **7.** It is skewed to the left meaning more parts are made smaller than larger.

Section 8.1

1. A, D; **6.** Low cost, lightweight, small size, no oil leakage, power supply (compressed air) often readily available; **10.** d; **11.** 1-e, 2-f, 3-a, 4-b, 5-d, 6-c; **12.** Servo motors.

Section 8.2

1. 1-d, 2-f, 3-a, 4-c, 5-b, 6-e; **5.** 1-c, 2-b, 3-d, 4-a; **6.** 1-d, 2-a, 3-b, 4-c.

Section 8.3

1. D; **3.** 1-d, 2-b, 3-a, 4-c; **5.** A-4, B-3, C-2, D-1; **6.** 1000(D) = 3E8(H); B35F(H) = 45919(D) **7.** 1803.78515625; **8.** 1-b, 2-c, 3-a, 4-e, 5-d; **9.** A; **11.** Parallel ports: A, B, D, E; Serial ports: A, C, F, G; **12.** D; **14.** C; **16.** C; **19.** C.

Section 8.4

1. 1-b, 2-d, 3-c, 4-a; **3.** (a) index time = 1.125 s, (b) dwell time = 1.875 s, (c) ideal production rate = 1,200 pieces per hour; **4.** Index time = 1.25 s, dwell time = 3.75 s, ideal production rate = 5,600 pieces per 8 hour shift; **5.** 640 lb; **6.** Diameter = 720 mm; velocity ratio = 4; **7.** Driver speed = 171.4 rpm; **8.** (1) 2 dof; (2) 1 dof; (3) 3 dof; **9.** F = 1 lb; **11.** Cycle time = 0.056 s.

Section 9.1

4. a, b, c, & d; **5.** a, b, c, and d; **7.** Titanium $\sigma = 29.4$; H13 $\sigma = 31.2$; steel $\sigma = 25.3$; 8. $\sigma = 119.9$; **9.** ±1.13°; **10.** $450; **11.** 3.8 months.

Section 9.2

1. Factor C, fuel pressure has the highest effect, and factors E and F have the least effect; **2.** Factor A, oven temperature has the highest effect, and factors C and D have the least effects equally (factor G represents errors and thus is not counted here); **3.** Factor A has the greatest influence, and thus the tolerance should be tightened. As level 2 can yield more scores, we should pick A2 (level 2) as the optimal factor; **4.** L_9 array.

Section 9.3

1. Total SS = 251.3, F-ratio = 0.6; **2.** $SS_T = 536.67$, $SS_V = 106.17$, $SS_P = 156$, $SS_e = 274.5$, and $F_V = 1.16$, and $F_P = 1.14$; **3.** $SS_T = 2$, $SS_A = 1$, $SS_B = 1$, $SS_C = 0$, $SS_e = 0$, and $F_A = 1/0$, $F_B = 1/0$; **4.** $PI_A = 38.1\%$, $PI_B = 61.3\%$, $F_A = 200$, $F_B = 321$; **5.** $S_N = -14.57$.

Section 9.4

5. 0.0063; **6.** 0.5260.

Section 9.5

1. Y = 6; **2.** Y = 94.5;

3. $$\text{OEC} = \left\{\frac{QC_B - 0.37}{4.2 - 0.37} \times 16.7\%\right\} + \left\{\frac{QC_H - 555}{649.33 - 555} \times 16.7\%\right\} + \left\{\left(1 - \frac{4.67 - QC_P}{4.67 - 1.50}\right) \times 16.7\%\right\} +$$

$$\left\{\left(1 - \frac{68.09 - QC_D}{68.09 - 7.85}\right) \times 16.7\%\right\} + \left\{\left(1 - \frac{3.12 - QC_S}{3.12 - 0}\right) \times 16.7\%\right\} + \left\{\left(1 - \frac{10.5 - QC_G}{10.5 - 2.50}\right) \times 16.7\%\right\}.$$

Section 10.1

5. Maximize V = xyz; Subject to 2xy + 2xz + 2yz < 150; x > 0, y > 0, z > 0; **6.** Minimize f (r, H) = – 2 * π * r * H; Subject to: h (r, H) = 2rπ * (H + r) – 56π = 0; r > 0, and H > 0.

Section 10.2

1. x = 3 and y = 6, P = 108; **4.** floor # = 14, cost = $28,284; **5.** x = 6, y = 3, V = 108; **6.** X = 4.

Section 10.3

1. $x_1 = (½)^{½}$, $x_2 = -(½)^{½}$, λ = ½ > 0; $x_1 = -(½)^{½}$, $x_2 = (½)^{½}$, λ = ½ > 0; **3.** Q = $(2/3)(C/w)^{2/3}$ X $(1/3)(C/r)^{1/3}$; **4.** R = 86 cm; H = 172 cm.

Section 10.4

4. x = 0, y = 2, f = 40; **5.** x = 0, y = 0, f = 0; **6.** x = 0.3333, y = 0.3333, z = 0.3333, f = 0.0370; **8.** h = 4, r = 2, V_{max} = 16π; **11.** x_1 = 8, x_2 = 1.67; **13.** Radius = 2.20 ft, length = 4.40 ft, volume = 66.80 ft^3, surface area = 91.14 ft^2; **14.** Radius = 26.2 in., height of rolled cylinder = 5.2 in., height of tank = 57.6 in.; **15.** x = –0.75, y_{min} = –1.25; **16.** h = 9.24, b = 6.802, t = 0.38.

Index

D

P

www.ingramcontent.com/pod-product-compliance
Lightning Source LLC
Chambersburg PA
CBHW080239190225
22177CB00026B/30
* 9 7 8 1 4 9 8 7 9 8 9 2 1 *